The Midbrain Periaqueductal Gray Matter

Functional, Anatomical, and Neurochemical Organization

NATO ASI Series

Advanced Science Institutes Series

A series presenting the results of activities sponsored by the NATO Science Committee, which aims at the dissemination of advanced scientific and technological knowledge, with a view to strengthening links between scientific communities.

The series is published by an international board of publishers in conjunction with the NATO Scientific Affairs Division

A	**Life Sciences**	Plenum Publishing Corporation
B	**Physics**	New York and London
C	**Mathematical and Physical Sciences**	Kluwer Academic Publishers
D	**Behavioral and Social Sciences**	Dordrecht, Boston, and London
E	**Applied Sciences**	
F	**Computer and Systems Sciences**	Springer-Verlag
G	**Ecological Sciences**	Berlin, Heidelberg, New York, London,
H	**Cell Biology**	Paris, Tokyo, Hong Kong, and Barcelona
I	**Global Environmental Change**	

Recent Volumes in this Series

Volume 209—Molecular Basis of Human Cancer
edited by Claudio Nicolini

Volume 210—Woody Plant Biotechnology
edited by M. R. Ahuja

Volume 211—Biophysics of Photoreceptors and Photomovements in Microorganisms
edited by F. Lenci, F. Ghetti, G. Colombetti, D.-P. Häder, and Pill-Soon Song

Volume 212—Plant Molecular Biology 2
edited by R. G. Herrmann and B. Larkins

Volume 213—The Midbrain Periaqueductal Gray Matter: Functional, Anatomical, and Neurochemical Organization
edited by Antoine Depaulis and Richard Bandler

Volume 214—DNA Polymorphisms as Disease Markers
edited by D. J. Galton and G. Assmann

Volume 215—Vaccines: Recent Trends and Progress
edited by Gregory Gregoriadis, Anthony C. Allison, and George Poste

Series A: Life Sciences

The Midbrain Periaqueductal Gray Matter

Functional, Anatomical, and
Neurochemical Organization

Edited by

Antoine Depaulis

Centre de Neurochimie du CNRS
Strasbourg, France

and

Richard Bandler

The University of Sydney
Sydney, Australia

Plenum Press
New York and London
Published in cooperation with NATO Scientific Affairs Division

Based on the proceedings of a NATO Advanced Research Workshop
on The Midbrain Periaqueductal Gray Matter,
held July 10–15, 1990,
at Château de Bonas, France

Library of Congress Cataloging-in-Publication Data

NATO Advanced Research Workshop on the Midbrain Periaqueductal Gray
   Matter (1990 : Château de Bonas)
    The midbrain periaqueductal gray matter : functional, anatomical,
 and neurochemical organization / edited by Antoine Depaulis and
 Richard Bandler.
       p.   cm. -- (NATO ASI series. Series A, Life science ; vol.
 213)
    "Based on the proceedings of a NATO Advanced Research workshop on
 the Midbrain Periaqueductal Gray Matter, held July 10-15, 1990, at
 Château de Bonas, France"--T.p. verso.
    "Published in cooperation with NATO Scientific Affairs Division."
    Includes bibliographical references and index.
    ISBN 0-306-44033-4
    1. Periaqueductal gray matter--Congresses.   I. Depaulis, Antoine.
 II. Bandler, Richard.   III. North Atlantic Treaty Organization
 Scientific Affairs Division.   IV. Title.   V. Series: NATO ASI
 series.   Series A, Life sciences ; v. 213.
    [DNLM: 1. Periaqueductal Gray--anatomy & histology--congresses.
 2. Periaqueductal Gray--physiology--congresses.   WL 310 N279m 1990]
 QP378.4.N38   1990
 599'.0188--dc20
 DNLM/DLC
 for Library of Congress   91-28948
                                                                    CIP

ISBN 0-306-44033-4

© 1991 Plenum Press, New York
A Division of Plenum Publishing Corporation
233 Spring Street, New York, N.Y. 10013

Printed in the United States of America

Preface

This book constitutes the proceedings of a NATO Advanced Research Workshop held at Château de Bonas (France) from 10-15 July 1990 on the Midbrain Periaqueductal Gray Matter (PAG). The aim of this meeting was to review and integrate our knowledge about the functional, anatomical and neurochemical organization of the PAG. The PAG has been the subject of many investigations during the last decade usually on different topics (e.g., pain modulation, defensive and sexual behavior) and generally there has been little interchange between the different research areas. The main purpose of this meeting was to bring together, for the first time, scientists who have worked on the PAG from different perspectives.

This book does not pretend to present an exhaustive review of the data collected during the last 20 years of research on the PAG. The contributors to this book have been selected because their data provide key elements in the search to understand both the organization of the PAG and the role of this structure in the integration of behavior. We believe that this book will provide clues that will assist in unraveling the organization of the PAG in the coming years.

This book is composed of summaries of the main lectures given at the meeting, as well as commentaries addressing important issues or points of controversy raised during the meeting. The contributions have been organized in two parts. In the first part functional aspects of the PAG are considered. The second part looks at the anatomical and neurochemical organization of the PAG. However, this organization does not preclude multidisciplinary approaches and a combination of physiological, anatomical and neurochemical methodologies have been used in many of the chapters. An introductory chapter summarizes, from our viewpoint, certain of the conclusions that were reached during the discussions and roundtables of the workshop.

Although it had somewhat delayed its publication, all of the contributions to this book have been sent for outside review. We believe that this process has helped to improve the clarity of the individual chapters and to integrate better the contents of the book.

The workshop was sponsored by the Scientific Affairs Division of the North Atlantic Treaty Organization (NATO) and we are grateful to NATO for their financial support. We wish also to thank The Harry Frank Guggenheim

v

Foundation, the Institut National de la Santé et de la Recherche Médicale (INSERM), Servier, the International Brain Research Organization (IBRO), Hoffmann-La Roche, Merrell Dow and the Université Louis Pasteur of Strasbourg for their generous financial assistance. A special note of thanks to Christine Depaulis and Pascal Carrive for their assistance in the preparation of the workshop and this book.

Antoine Depaulis
Richard Bandler
March 1991

Contents

Introduction: Emerging Principles of Organization of the Midbrain
Periaqueductal Gray Matter
Richard Bandler, Pascal Carrive and Antoine Depaulis 1

PART I: FUNCTIONAL ASPECTS

Neurochemical Study of PAG Control of Vocal Behavior
Uwe Jürgens 11

Activity of PAG Neurons During Conditioned Vocalization in the Macaque
Monkey
Charles R. Larson 23

Discharge Relationships of Periaqueductal Gray Neurons to Cardiac
and Respiratory Patterning During Sleep and Waking States
Ronald M. Harper, Huifang Ni and Jingxi Zhang..41

What is the Role of the Midbrain Periaqueductal Gray in Respiration
and Vocalization? (commentary)
Pamela J. Davis and Shi Ping Zhang 57

Functional Organization of PAG Neurons Controlling Regional Vascular Beds
Pascal Carrive...67

Interactions Between Descending Pathways from the Dorsal and Ventrolateral
Periaqueductal Gray Matter in the Rat
Thelma A. Lovick 101

Analgesia Produced by Stimulation of the Periaqueductal Gray Matter:
True Antinoceptive Effects Versus Stress Effects
Jean-Marie Besson, Véronique Fardin and Jean-Louis Olivéras...........................121

Differences in Antinociception Evoked from Dorsal and Ventral Regions
of the Caudal Periaqueductal Gray Matter (commentary)
Michael M. Morgan 139

The Midbrain Periaqueductal Gray as a Coordinator of Action in Response
to Fear and Anxiety
Michael S. Fanselow 151

Midbrain Periaqueductal Gray Control of Defensive Behavior in the Cat
and the Rat
Richard Bandler and Antoine Depaulis 175

Does the PAG Learn about Emergencies from the Superior Colliculus ?
(commentary)
Peter Redgrave and Paul Dean..199

Midbrain PAG Control of Female Reproductive Behavior: *In Vitro* Electrophysiological Characterization of Actions of Lordosis-Relevant Substances
Sonoko Ogawa, Lee-Ming Kow, Margaret M. McCarthy, Donald W. Pfaff,
and Susan Schwartz-Giblin 211

PART II: ANATOMICAL AND NEUROCHEMICAL ASPECTS

Descending Pathways from the Periaqueductal Gray and Adjacent Areas
Gert Holstege 239

Induction of the Proto-Oncogene *c-fos* as a Cellular Marker of Brainstem
Neurons Activated from the PAG
Jürgen Sandkühler..267

The Nociceptive Modulatory Effects of Periaqueductal Gray Activation are
Mediated by Two Neuronal Classes in the Rostral Ventromedial Medulla
Peggy Mason..287

Localization of Putative Amino Acid Transmitters in the PAG and their
Relationship to the PAG-Raphe Magnus Pathway
Alvin J. Beitz and Frank G. Williams..305

GABAergic Neuronal Circuitry in the Periaqueductal Gray Matter
David B. Reichling..329

Organization of Spinal and Trigeminal Input to the PAG
Anders Blomqvist and A.D. Craig..345

Somatosensory Input to the Periaqueductal Gray: A Spinal Relay
to a Descending Control Center
Robert P. Yezierski 365

Hypothalamic Projections to the PAG in the Rat: Topographical,
Immuno-Electronmicroscopical and Functional Aspects
J. Veening, P. Buma, G.J. Ter Horst, T.A.P. Roeling, P.G.M. Luiten
and R. Nieuwenhuys 387

**Topographical Specificity of Forebrain Inputs to the Midbrain Periaqueductal
Gray: Evidence for Discrete Longitudinally Organized Input Columns**
Michael T. Shipley, Matthew Ennis, Tilat A. Rizvi
and Michael M. Behbehani..417

**Regional Subdivisions in the Midbrain Periaqueductal Gray of the Cat
Revealed by *In Vitro* Receptor Autoradiography**
Andrew L. Gundlach 449

Participants...465

Index..471

Introduction
Emerging Principles of Organization of the Midbrain Periaqueductal Gray Matter

Richard Bandler*, Pascal Carrive* and Antoine Depaulis**

*Department of Anatomy, The University of Sydney
Sydney, Australia
**L.N.B.C., Centre de Neurochimie du CNRS
Strasbourg, France

Throughout this book the term periaqueductal gray (PAG) has been used in preference to the term central gray, since PAG refers specifically to the portion of the ventricular gray matter which surrounds the midbrain aqueduct. Rostrally, the PAG is continuous with the periventricular gray matter surrounding the third ventricle in the hypothalamus and thalamus. Caudally, it is continuous with the periventricular gray matter which in the dorsal pons forms the ventral and ventrolateral border of the fourth ventricle. Although the oculomotor related group of nuclei (i.e., oculomotor and trochlear nuclei, the Edinger-Westphal nucleus, the nucleus of Darkschewitsch and the interstitial nucleus of Cajal) and the dorsal raphe nucleus, constitute the major part of the gray matter ventral to the midbrain aqueduct, they are usually considered, on functional and anatomical grounds, separable from the PAG. The midbrain tegmentum laterally adjacent to the PAG has also usually been considered to be a separate entity. However, as discussed by Holstege (this volume), the PAG and the laterally adjacent tegmentum together, likely form a common neuronal pool which is divided by the fiber stream formed by the tectobulbospinal fibers and the fibers of the mesencephalic trigeminal tract.

Although it is generally recognized that the PAG is not a homogeneous structure, there has been little agreement about how best to subdivide the PAG. Some researchers have suggested subdivisions based largely on cytoarchitectonic criteria (e.g., Beitz, 1985; Hamilton, 1973; Paxinos and Watson, 1986); but, the validity and functional significance of such subdivisions have been questioned frequently by others (e.g., Bandler and Carrive, 1988; Gioia et al., 1984; 1985; Mantyh, 1982). The lack of agreement on the validity of any subdivisional scheme for the PAG is a continuing problem. For example, the "dorsal" PAG has been defined as the region of the PAG found dorsal to the aqueduct, dorsal to the midpoint of the aqueduct, or dorsal to the base of the aqueduct. Similarly, the ventral PAG may include or exclude the dorsal raphe nucleus.

This meeting, the first ever on the topic of the PAG, has brought together scientists who study the PAG, from a variety of perspectives. One of the goals of the meeting was to determine an acceptable way to approach the question of subdivisions within the PAG. Out of the diversity of interests and experimental approaches a number of principles of organization of the PAG emerged. At the risk of gross oversimplification, we would suggest that much of the organization of the PAG can be subsumed under a single fundamental principle. Namely, that *anatomical and functional specificity is expressed in the form of longitudinal neuronal columns extending for varying distances along the rostrocaudal axis of the PAG.*

Overview of Longitudinal Columnar Organization within the PAG

The idea that the PAG is organized primarily along a longitudinal dimension is very much in its infancy (see also Bandler et al., 1991). In this section, as a way of introduction, we will discuss briefly the evidence for the existence of, at least, four longitudinal columns within the PAG. Figures 1 to 3 illustrate schematically, in a series of coronal sections, along the rostrocaudal extent of a "typical mammalian PAG", the approximate position of each of these proposed longitudinal neuronal columns

DORSOLATERAL NEURONAL COLUMN: A dorsolateral neuronal column, which appears "wedge-shaped" in coronal section, has a significant presence throughout the rostral and intermediate thirds of the PAG, but gradually diminishes in the caudal third of the PAG (Fig. 1). The dorsolateral neuronal column possesses a special set of afferent and efferent connections. Thus, the neurons of the dorsolateral column do not project extensively to the brain stem (Carrive; Holstege, this volume), apart from strong projections to the cuneiform nucleus and the periabducens region (Redgrave and Dean, this volume; Redgrave et al., 1988). Brain stem afferents to the dorsolateral column include projections from the nucleus prepositus hypoglossi, cuneiform nucleus and deep layers of the superior colliculus (Dean and Redgrave; Holstege, this volume). The dorsolateral neuronal column also contains large number of GABA immunoreactive cells (Reichling, this volume) and has higher densities of muscarinic, kainate and $GABA_A$/benzodiazepine binding sites (Gundlach, this volume) than other parts of the rostral and intermediate PAG. Nothing is known at present about the function of the dorsolateral neuronal column.

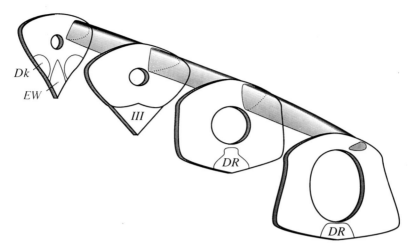

Figure 1. The shaded area iilustrates schematically the extent of the dorsolateral neuronal column within (from left to right) the rostral PAG, the rostral intermediate PAG, the caudal intermediate PAG and the caudal PAG. Abbreviations: EW, Edinger Westphal nucleus; DK, nucleus of Darkschewitsch; DR, dorsal raphe nucleus; III, oculomotor nucleus.

DORSOMEDIAL NEURONAL COLUMN: The presence of the dorsolateral PAG defines the lateral limit of a dorsomedial neuronal column situated immediately above the aqueduct. This column is present along the entire rostrocaudal extent of the PAG (Fig. 2). In contrast to the dorsolateral neuronal column, the neurons of the dorsomedial column project extensively to the caudal brainstem, in particular to the rostral ventromedial and rostral ventrolateral medulla (Carrive; Holstege, this volume). The functional significance of these projections is not presently known. There exist highly organized cortical and subcortical forebrain inputs to both the dorsomedial and the dorsolateral columns (Shipley et al.; Veening et al., this volume). However, neither the dorsomedial nor the dorsolateral columns receives significant direct spinal dorsal horn or laminar spinal trigeminal input (Blomqvist and Craig, this volume).

LATERAL NEURONAL COLUMN: The presence of the dorsolateral neuronal column in the rostral and intermediate thirds of the PAG helps to separate the "dorsal PAG" (i.e., dorsomedial and dorsolateral columns) from the rest of the PAG. Within the rostral third of the PAG, the number, extent and position of longitudinal columns, in the region ventral to the dorsal PAG, has yet to be determined (although see Shipley et al., this volume).

Within the intermediate third of the PAG, both functional and anatomical data, support the existence of a neuronal column in the region immediately lateral to the aqueduct (Fig. 3). This lateral neuronal column projects extensively to the ventromedial, ventrolateral and the dorsal medulla i.e., nucleus of the solita-

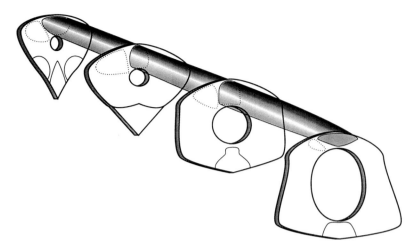

Figure 2. The shaded area illustrates schematically the extent of the dorsomedial neuronal column within (from left to right) the rostral PAG, the rostral intermediate PAG, the caudal intermediate PAG and the caudal PAG.

ry tract (Bandler and Törk, 1987; Carrive; Holstege; Lovick, this volume). The lateral neuronal column of the rostral intermediate PAG receives a significant afferent input from the laminar spinal trigeminal nucleus, and the lateral neuronal column of the caudal intermediate PAG receives a significant afferent input from the cervical enlargement of the spinal cord (Blomqvist and Craig, this volume). Additional afferents to the lateral column of the intermediate PAG include the anterior hypothalamic/medial preoptic region, the central nucleus of the amygdala and the anterior cingulate cortex (Shipley et al.; Veening et al., this volume). Microinjections of excitatory amino acid (EAA) within the lateral neuronal column of the intermediate PAG evoke increased autonomic (sympathetic) and somatomotor activity associated with specific forms of defensive behavior (Bandler and Depaulis; Carrive; Lovick, this volume).

As the midbrain aqueduct begins to expand within the caudal third of the PAG, the dorsoventral extent of the lateral neuronal column also increases. Anatomical studies indicate that within the caudal PAG, the lateral neuronal column continues to project extensively to the medulla and to receive substantial afferent inputs from the same forebrain cortical and subcortical fields as do the neurons of the lateral column in the intermediate third of the PAG. However, in the caudal PAG, a unique somatosensory input to the lateral column arises from lumbar enlargement of the spinal cord (Blomqvist and Craig, this volume). Within the caudal third of the PAG, EAA microinjections made into lateral

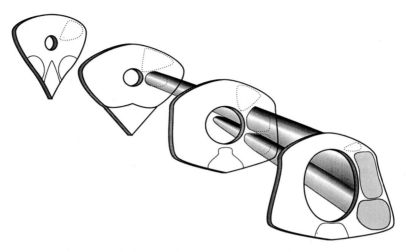

Figure 3. The shaded area illustrates schematically the extent of the lateral and ventrolateral neuronal columns within (from left to right) the rostral PAG, the rostral intermediate PAG, the caudal intermediate PAG and the caudal PAG.

column evoke also increased somatomotor and autonomic (sympathetic) activity associated with defensive behavior, although the patterns of defensive behavior, and the associated somatic and autonomic adjustments, are distinct from those evoked from the lateral neuronal column of the intermediate PAG (Bandler and Depaulis; Carrive, this volume).

VENTROLATERAL NEURONAL COLUMN: Functional studies support the existence, in the caudal half of the PAG, of a ventrolateral neuronal column (Fig. 3). The anatomical connections of the ventrolateral column are seemingly identical to those of the dorsally adjacent lateral neuronal column. Thus, ventrolateral PAG neurons project extensively to both the rostral and caudal medulla; and receive significant afferent input from the same forebrain cortical and subcortical areas, as well as from the lumbar and cervical enlargements of the spinal cord (Blomqvist and Craig; Carrive; Holstege; Lovick; Shipley et al.; Veening et al., this volume). However, injections of EAA within the ventrolateral neuronal column evoke effects which are the exact opposite of those evoked from the lateral column, namely decreased autonomic (sympathetic) and somatomotor activity (Bandler and Depaulis; Carrive; Lovick, this volume). As well, considerable data indicate that analgesia evoked from the ventrolateral neuronal column is opioid-dependent, whereas analgesia evoked from the lateral neuronal column is not (Fanselow; Lovick; Mason; Morgan, this volume).

To summarize, longitudinal columnar organization within the PAG is manifest in a number of different ways:

I. COLUMNAR ORGANIZATION

Efferents from the PAG originate from neurons which form longitudinal columns that extend for varying distances along its rostrocaudal axis.

Forebrain, brain stem and spinal cord afferents to the PAG terminate as longitudinally organized terminal fields.

Chemical stimulation studies support the existence, within the PAG, of functional longitudinal columns of varying rostrocaudal extent.

Afferent input columns co-distribute with certain of the functionally and anatomically defined longitudinal output columns.

II. TOPOGRAPHICAL ORGANIZATION OF LONGITUDINAL COLUMNS

Some classes of afferents terminate along the entire rostrocaudal extent of a longitudinally organized output column. Such afferents likely modulate the neural activity of an entire longitudinal column.

Other classes of afferents terminate in a restricted rostrocaudal zone within a longitudinally organized column. Such afferents likely modulate the activity of a subset of neurons within a longitudinal column.

Both functional and anatomical studies indicate some degree of somatotopic and viscerotopic organization within specific longitudinal columns.

III. FUNCTIONAL INTEGRATION AND COLUMNAR OVERLAP

Studies of a number of different functions (e.g., arterial pressure, regional vasoconstrictor tone, respiration/vocalization, analgesia, defensive behavior) suggest that functionally distinct longitudinal columns have partially overlapping distributions.

Chemical stimulation, within the regions of overlap of functionally distinct longitudinal columns, evoke co-ordinated patterns of behavior which are important for the survival of the organism.

Integrative Functions of the PAG

A perusal of the table of contents of this volume immediately reveals the two largely independent streams which have dominated research on the PAG during the last 10-15 years. Namely, the role of the PAG: (i) in mediating nociceptive inhibition (Besson et al.; Beitz and Williams; Lovick; Mason; Morgan; Reichling; Yezierski, this volume) and (ii) in integrating behavioral responses to potentially threatening or stressful stimuli, i.e., defensive reactions; lordosis (Bandler and Depaulis; Fanselow; Ogawa et al., this volume). Although, no agreement was reached at this workshop concerning either the physiological role(s) of the different types of analgesia evoked from the PAG; or the conditions under which specific analgesic systems become activated; a general consensus emerged that the analgesic and behavioral reactions evoked from the PAG are, in fact, best conceptualized as components of co-ordinated responses necessary for the animal's survival. A more integrative approach to analgesic and behavioral responses, which have often been studied as isolated functions of the PAG, is well illustrated by the contributions of Fanselow, Lovick, Morgan and Yezierski, to this volume (although for an alternative viewpoint see Besson et al., this volume).

The integral role played by the PAG in co-ordinating both cardiovascular and respiratory adjustments emerged as a new and significant aspect of PAG function. Attention was drawn (Carrive; Davis and Zhang; Holstege; Lovick, this volume) to the extensive projections of PAG neurons to "cardiorespiratory" regions of the rostral ventrolateral, caudal ventrolateral, dorsal and ventromedial medulla. Together these PAG-afferent recipient zones of the medulla provide major sources of excitatory or inhibitory drive on sympathetic vasomotor tone, and contain also the dorsal and ventral respiratory groups of neurons. Secondly, EAA microinjections within the lateral or ventrolateral columns of PAG neurons were shown to evoke respectively: (i) pressor or depressor responses associated with selective changes in regional vasomotor tone (Carrive; Lovick, this volume); and (ii) profound changes in respiratory rate and inspiratory depth, as well as the initiation of vocalization (Davis and Zhang; Jürgens, this volume). Finally, single cell studies undertaken in the intact, awake cat and monkey revealed that large numbers of PAG neurons show specific discharge relationships with respiratory, laryngeal and cardiac activity (Harper et al.; Larson, this volume). Interestingly, it was during rapid eye movement (REM) sleep that discharge relationships to cardiac and respiratory activity were seen in the greatest proportion of PAG cells (Harper et al., this volume), suggesting for the first time that the PAG plays a prominent role in cardiorespiratory regulation, not only during intense emotional states, but also during REM sleep.

Future Directions

The integral role played by the PAG in the generation of an animal's responses to threatening or stressful stimuli has become well recognized (e.g., Bandler, 1988; Bandler et al., 1991; Jordan, 1990). These responses included rapid, co-ordinated skeletal and autonomic adjustments associated with significant alterations in "pain" thresholds. We have suggested, in summarizing the presentations of this workshop, that underlying the expression of the co-ordinated responses evoked from the PAG lies an anatomical specificity organized along the longitudinal dimension of the structure.

The number of longitudinal columns that exist within the PAG; the different types and properties (morphological, physiological, immunocytochemical) of neurons within specific longitudinal neuronal columns; the extent to which the processes of neurons spread across columns; the extent to which longitudinal columns overlap; the distribution and function of extrinsic inputs to, and intrinsic anatomical linkages between, longitudinal neuronal columns; and the significance of a probable, complimentary radial (i.e., aqueduct to periphery) organization within the PAG; are just a few of the important unanswered questions which need to be studied. For too long now, studies have considered only the dorsal-ventral dimension of the PAG and usually within a limited rostrocaudal extent. Fundamental to future research, we would suggest, is the need to focus on functional, anatomical and neurochemical representations along the longitudinal extent of the PAG.

Acknowledgements

We wish to thank Dr K. Keay and Dr S.P. Zhang for their critical comments on earlier versions of the manuscript.

References

Bandler, R., Brain mechanisms of aggression as revealed by electrical and chemical stimulation: Suggestion of a central role for the midbrain periaqueductal grey region, In: Progress in Psychobiology and Physiological Psychology, Vol. 13, Epstein A. and Morrison A. (Eds.) Academic Press, New York, 1988, pp. 67-154.

Bandler, R. and Carrive, P., Integrated defence reaction elicited by excitatory amino acid microinjection in the midbrain periaqueductal grey region of the unrestrained cat, Brain Res., 439 (1988) 95-106.

Bandler, R., Carrive, P. and Zhang, S.P., Integration of somatic and automatic reactions within the midbrain periaqueductal grey: Viscerotopic, somatotopic and functional organization, Prog. Brain Res., 87 (1991) 269-305.

Bandler, R. and Törk, I., Midbrain periaqueductal grey region in the cat has afferent and efferent connections with solitary tract nuclei, Neurosci. Lett., 74 (1987) 1-6.

Beitz, A.J., The midbrain periaqueductal gray in the rat. I. Nuclear volume, cell number, density, orientation, and regional subdivisions, J. Comp. Neurol., 237 (1985) 445-459.

Gioia, M., Bianchi, R. and Tredici, G., Cytoarchitecture of the periaqueductal gray matter in the cat: A quantitative Nissl study, Acta Anat., 119 (1984) 113-117.

Gioia, M., Tredici, G. and Bianchi, R., A Golgi study of the periaqueductal gray matter in the cat. Neuronal types and their distribution, Exp. Brain Res., 58 (1985) 318-332.

Hamilton, B. L., Cytoarchitectural subdivisions of the periaqueductal gray matter in the cat, J. Comp. Neurol., 152 (1980) 45-58.

Jordan, D., Autonomic changes in affective behavior, In: Central regulation of autonomic functions, Loewy A.D. and Spyer K.M. (Eds.), Oxford Univ. Press, New York, 1990, pp. 349-366.

Mantyh, P.W., The midbrain periaqueductal gray in the rat, cat and monkey: A Nissl, Weil and Golgi analysis, J. Comp. Neurol., 204 (1982) 349-363.

Redgrave, P., Dean, P., Mitchell, I.J., Odekunle, A. and Clark, A., The projection from superior colliculus to cuneiform area in the rat. I. Anatomical studies, Exp. Brain Res., 72 (1988) 611-625.

Paxinos, G. and Watson, C., The rat brain in stereotaxic coordinates, Academic Press, New York, 1986.

PART I
FUNCTIONAL ASPECTS

Neurochemical Study of PAG Control of Vocal Behavior

Uwe Jürgens

Max-Planck-Institute of Psychiatry
Munich, Germany

Electrical stimulation experiments

The relationship between periaqueductal gray (PAG) and vocal behavior has a rather long history. It goes back to a report made by T.G.Brown in 1915. Brown stimulated the cut surface of the rostral brainstem in a chimpanzee. When the electrode was positioned near the aqueduct, a vocalization was obtained Brown referred to as laughter. It then took more than 20 years until the first, more systematic study on this topic appeared. It was carried out by Magoun and co-workers (1937) using the, at that time newly developed, stereotaxic technique. Magoun and his colleagues explored the whole midbrain, pons and upper medulla with movable electrodes for sites yielding vocalization when electrically stimulated. They found numerous vocalization-eliciting sites in the PAG of both cats and rhesus monkeys. The vocalization-eliciting sites, however, were not limited to the PAG. They invaded also the laterally adjacent tegmentum underneath the superior and inferior colliculus and could be followed along the medial edge of the lateral lemniscus down into the ventrolateral pons. The call types elicitable were soft cries, screeches and barks in the case of the rhesus monkey, and spitting, screams and howls in the case of the cat. Since then, numerous authors have confirmed the electrical elicitability of vocalization in the PAG. Apart from the chimpanzee, rhesus monkey and cat, vocalization also was obtained in the rat (Waldbillig, 1975), guinea pig (Martin, 1976), bat (Suga et al., 1973), squirrel monkey (Jürgens and Ploog, 1970) and gibbon (Apfelbach, 1972).

The Midbrain Periaqueductal Gray Matter, Edited by A. Depaulis and
R. Bandler, Plenum Press, New York, 1991

Lesioning Experiments

It was mentioned already that the PAG is not the only brain area yielding vocalization when electrically stimulated. A number of other structures, such as the anterior cingulate cortex, amygdala, hypothalamus and midline thalamus also produce vocalization (for review see Sutton and Jürgens, 1988). From this the question arises of the specific role the PAG plays in vocal control in relation to other vocalization-eliciting areas. In a combined stimulation/lesioning study we carried out several years ago (Jürgens and Pratt, 1979a), it was found that destruction of the PAG blocks all vocalizations elicitable from other brain structures. This holds for structures rostral as well as caudal to the PAG. The only exception is the parvocellular reticular formation in the lower brainstem. Vocalizations elicitable from this area still can be produced after PAG lesions. Vocalizations in this area, however, differ from those elicitable from other vocalization-eliciting areas, including the PAG, in that they do not sound like natural calls but have a clearly abnormal acoustic structure. We interpret this in the way that the parvocellular reticular formation represents part of the motor coordination mechanism for vocalization. The artificial acoustic structure, in this view, would be due to the fragmentation of the output to the expiratory, laryngeal and articulatory motoneuron pools. As the PAG-evoked vocalizations resemble natural calls, we assume that the PAG is not the site of the phonatory motor coordination mechanism, but represents a higher integration level.

The fact that destruction of the PAG blocks all naturally sounding calls elicitable from other brain structures suggests that the PAG represents something like a bottle-neck of the vocalization system through which all other vocalization-eliciting structures must pass their activity in order to induce vocalization. Our neuroanatomical studies have shown that, in fact, almost all vocalization-eliciting brain areas project directly into the PAG (Jürgens and Müller-Preuss, 1977; Jürgens and Pratt, 1979a). In our combined stimulation/lesioning experiments, we were able to confirm the functional relevance of some of these projections. For instance, alarm calls elicitable from the amygdala could be blocked by lesions all along a pathway leading from the amygdala via the ventral amygdalofugal fiber bundle into the hypothalamus, and from here along the periventricular fiber system into the PAG (Jürgens, 1982). Another example is the cingular vocalization pathway. Vocalizations elicitable from the anterior cingulate cortex can be blocked along a pathway running through the corona radiata and descending within the internal capsule down to the posterior diencephalon. Here, the pathway leaves the internal capsule, traverses the substantia nigra and rostral midbrain tegmentum to reach again the PAG (Jürgens and Pratt, 1979b).

Neurochemical Approaches

The fact that lesions in the PAG can abolish vocalizations elicitable from other structures does not necessarily mean that the PAG is a relay station of the vocalization system, in the sense that here converging information from different areas is integrated. It could be as well that crucial fiber bundles of the vocalization system pass through the PAG without being relayed there. It was one of the editors of this book, Richard Bandler, who could show that injection of glutamate into the PAG also yields vocalization (Bandler, 1982). Glutamate, in contrast to electrical stimulation, however, does not activate fibers-en-passage, but only cell bodies, dendrites and terminals, that is, regions containing synapses. The elicitability of vocalization by glutamate, therefore, suggests that the PAG is a relay station of the vocalization system, presumably integrating vocalization-relevant information from different brain areas.

Table I

NEUROTRANSMITTER	ANTAGONIST(NMOLS)	EFFECT	N
GLUTAMATE	KYNURENIC ACID (3-53;7)	i	9
	APV(0.3-5;0.5)	i	6
	CNQX(0.4-43;4)	i	3
	GAMS(4-42;21)	i	5
	APB(55)	-	3
GABA	BICUCULLINE (0.2-0.4)	f	5
	PICROTOXIN (0.08-0.3)	f	3
	PHACLOFEN(40)	-	1
GLYCINE	STRYCHNINE(14-27)	f,-	6
ACETYLCHOLINE	ATROPINE(7-15)	-	7
	SCOPOLAMINE(29)	-	3
	BENZTROPINE(24)	-	1
	QNB(15-30)	-	2
	MECAMYLAMINE(49)	-	5
DOPAMINE	HALOPERIDOL(13-27)	-	6
NORADRENALINE	BENEXTRAMINE(14)	-	1
	PHENOXYBENZAMINE(29)	-	3
	YOHIMBINE(10-26)	-	3
	TOLAZOLINE(51)	-	3
	PROPRANOLOL(17-34)	-	5
SEROTONIN	CYPROHEPTADINE(12-31)	-	4
	METHYSERGIDE(28)	-	1
HISTAMINE	DIPHENHYDRAMINE(34)	-	4
	CHLORPHENIRAMINE(26)	-	2
	PROMETHAZINE(16-31)	-	3
OPIOIDS	NALOXONE(14-28)	f,-	7

i inhibition; f facilitation; - no effect; N number of sites tested. In parentheses, the range of doses is given in nmols; the lowest effective dose is underlined.

Our current work is concerned with the question of which neurotransmitters are involved in the PAG's control of vocalization. In this study, we inject antagonists of transmitters into the PAG and look for the effects on the electrical elicitability of vocalization from other structures. Table I gives a summary of the present status of our experiments. Up-to-now, 27 antagonists of 9 transmitters have been tested. All 9 transmitters have been reported to be likely transmitters in the PAG. Evidence for this partly stems from immunohistochemical identification of these transmitters in terminals of the PAG (Moore and Bloom, 1979; Yamano et al., 1986; Barbaresi and Manfrini, 1988; Inagaki et al., 1988; Reichling et al., 1988; Airaksinen et al., 1989; Ibuki et al., 1989), partly from autoradiographic or immunocytochemical receptor-binding studies (McLennan, 1983; Greenamyre et al., 1984; Monaghan and Cotman, 1985; Pazos and Palacios, 1985; Mash and Potter, 1986; Spencer et al., 1986; Stuart et al., 1986; Bouthenet et al., 1987; Araki et al., 1988; Bouthenet et al., 1988; Mansour et al., 1988; Reichling et al., 1988; Camps et al., 1990; Chu et al., 1990), partly from transmitter-specific retrograde transport studies (Beart et al., 1990), and partly from immunocytochemical identification of nerve cell bodies retrogradely labeled from the PAG with a conventional retrograde tracer (Beitz, 1989). The vocalization-eliciting electrode positions were distributed over several forebrain structures that are listed in Table II. The exact injection site within the PAG was chosen according to the following criteria: 1). Electrical stimulation of the site had to yield vocalization. This was to insure that the site was within the vocalization-relevant part of the PAG. 2) Injection of 2% lidocaine, a local anaesthetic, into the site had to block vocalization elicitable from the vocalization electrode in the forebrain. This was to insure that the site lay within the projection system of the vocalization-eliciting position. 3) Injection of the glutamate analogue homocysteic acid had to induce vocalization at that site. In this way, we made sure that the antagonists were delivered near synapses of the vocalization system and not into by-passing fiber bundles of this system. All injections were made ipsilateral to the vocalization-eliciting forebrain site, as earlier studies (Jürgens and Pratt, 1979a) have shown that ipsilateral PAG lesions are sufficient to block forebrain-elicited vocalizations.

Altogether 13 injection sites were tested in 9 squirrel monkeys. The distribution of the injection sites is shown in Fig. 1. Twelve of the injection sites were tested with one vocalization-eliciting forebrain electrode only; the remaining injection site was tested with two. Table II shows which injection site was paired with which vocalization-eliciting position. It also lists the types of elicited calls (for a sonographic representation and functional interpretation of the calls the reader is referred to Jürgens, 1979). It can be seen that dependent upon the stimulation site very different call types were obtained. This is in contrast to the injection site in the PAG the electrical stimulation of which always yielded cackling calls. However, for 5 of the 13 injection sites, cackling was not the sole call type elicited but was accompanied by one or two additional calls. A systematic relationship between such additional calls and the stimulation site within the PAG existed only in as far as purring and growling calls (which express comfort and dominance, respectively) could be obtained only from the caudal ventrolateral PAG (injection sites 6 and 10). Other accompanying calls, such as groaning, shrie-

Table II

INJECTION SITE	ANIMAL	SITE OF ELECTRICAL STIMULATION	VOCALIZATION TYPE ELICITED
1	16	NUCL.CENTRALIS AMYGDALAE	GROANING
2	16	NUCL.CENTRALIS AMYGDALAE	GROANING
3	32	NUCL.CENTRALIS AMYGDALAE	GROANING
4	30	NUCL.STRIAE TERMINALIS	PURRING
5	19	PEDUNCULUS INFERIOR THALAMI	CLUCKING
6	28	GENU OF INTERNAL CAPSULE	YELLING
7	22	GENU OF INTERNAL CAPSULE	CACKLING
7	22	VENTROMEDIAL HYPOTHALAMUS	SHRIEKING
8	29	ANTERIOR HYPOTHALAMUS	SQUEALING
9	29	ANTERIOR HYPOTHALAMUS	SQUEALING
10	26	ANTERIOR HYPOTHALAMUS	PURRING
11	23	POSTERIOR HYPOTHALAMUS	GROWLING
12	28	PERIVENTRICULAR GRAY	CACKLING
13	28	PERIVENTRICULAR GRAY	CACKLING

king and peeping (expressing uneasiness, defense and alertness, respectively) are not restricted to a specific part of the PAG. The same holds for the vocalizations elicited by injection of homocysteic acid into the PAG. Thus, purring and growling seem to be the only call types in the squirrel monkey the elicitation of which is restricted to a small (caudal ventrolateral) part of the PAG. This part may correspond to an area described by Bandler and Depaulis (this volume) to cause a decrease in spontaneous movements when chemically stimulated with homocysteic acid or kainic acid - an effect contrary to that obtained when more dorsal parts are stimulated. Furthermore, Lovick (this volume) reports that in the rat, stimulation of this area produces attenuation of the defense response elicited from more dorsal parts of the PAG. As purring and growling is uttered by the squirrel monkey only in the absence of flight motivation, it is suggested from this obser-

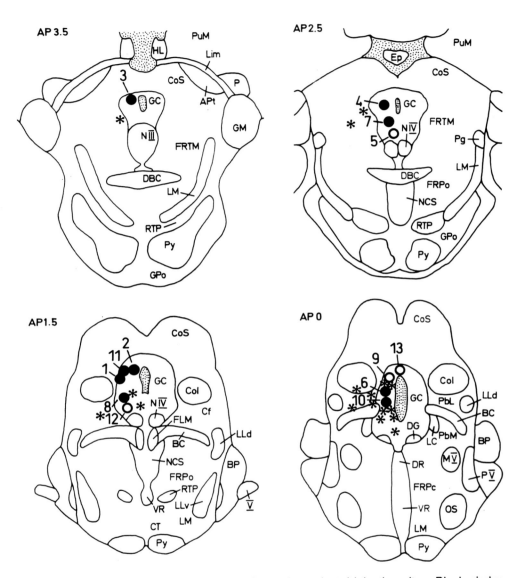

Figure 1. Frontal brain sections showing the periaqueductal injection sites. Black circles represent sites at which vocalizations electrically elicitable from the forebrain could be blocked with lidocaine as well as kynurenic acid. Open circles indicate sites without effect on forebrain vocalization. Black as well as open circle sites yielded vocalization when stimulated electrically and chemically with homocysteic acid. Stars indicate sites yielding vocalization with electrical but not chemical stimulation. The numbers relate to Table 2 and serve to identify the forebrain vocalization sites tested with each injection site.

vation together with those of Lovick and of Bandler and Depaulis that the caudal ventrolateral PAG may be involved in the induction of relaxed emotional states.

Injections in the present study were made by the aid of a chronically implanted cannula and a microliter syringe. All antagonist injections had a volume of 500 nl. The maximal doses were 10 µg, except in the cases of bicuculline and picrotoxin in which it was 200 ng. Injections at each site started with the maximal dose. If there was an effect, the dose was lowered stepwise to determine the threshold. The range of tested doses (in nmols) is given in Table I. Some of the sites injected with antagonists were also tested with the corresponding transmitter agonists. In this case, the volume was 200 nl; the maximal doses were 20 µg for glutamate (corresponding to 118 nmol), aspartate (129 nmol), glycine (266 nmol), dopamine (105 nmol), noradrenaline (97 nmol), serotonin (52 nmol), histamine (108 nmol) and morphine (30 nmol); 5 µg for homocysteic acid (27 nmol), N-methyl-D-aspartate (34 nmol) and carbachol (27 nmol); 1 µg for baclofen (5 nmol) and 200 ng for muscimol (2 nmol).

Table I shows that the only antagonists capable of blocking vocalization are glutamate antagonists, namely, the non-selective glutamate antagonist kynurenic acid as well as the NMDA receptor-specific antagonist 2-amino-5-phosphonovaleric acid (APV) and the quisqualate and kainate receptor-specific antagonists 6-cyano-7-nitroquinoxaline-2,3-dione (CNQX) and glutamylaminomethylsulfonic acid (GAMS). No effect is obtained with 2-amino-4-phosphonobutyric acid (APB), the antagonist of a recently proposed fourth type of glutamate receptor that seems to be located presynaptically (Cotman et al., 1986). These findings suggest that glutamate, or an analogue of it, plays a crucial role in the neural transmission processes underlying the PAG's control of vocalization. In this processing, NMDA and quisqualate/kainate receptors seem to be involved. This suggestion is supported by the fact that at all sites at which glutamate antagonists block vocalization, injection of glutamate agonists - NMDA as well as non-NMDA ones - produce vocalization.

In contrast to glutamate antagonists, bicuculline and picrotoxin, two GABA antagonists, have facilitatory effects. That is, both lower the threshold for elicitation of vocalization from the forebrain. With higher doses, vocalization is even obtained without electrical stimulation. On the other hand, injection of the GABA agonist muscimol blocks electrically elicited vocalization completely. This means that there is a continuous inhibitory GABAergic input on the periaqueductal vocalization-controlling neurons which, when removed by GABA antagonists, increase the readiness of the animal to vocalize. As neither baclofen nor phaclofen, that is, $GABA_B$ receptor agonist and antagonist, show any effect, we conclude that the GABAergic control of vocalization in the PAG is exerted via $GABA_A$ receptors.

The effects of glycine and its antagonist strychnine are similar to those of GABA and its antagonists bicuculline and picrotoxin, except that the effects are less consistent. More specifically, not all sites facilitating vocalization with GABA antagonists also facilitate vocalization with strychnine, whereas all strychnine-effective sites are also effective with GABA antagonists. At the moment, we are not yet able to characterize what distinguishes the strychnine-effective from the strychnine-ineffectice sites.

As concerns acetylcholine, neither muscarinic nor nicotinic antagonists have any effect. This is somewhat unexpected as injection of carbachol, an acetylcholine agonist, into the PAG can induce vocalization. We cannot exclude, therefore, that acetylcholine may have a modulatory effect on PAG vocalization-controlling neurons under specific circumstances. Our experiments make clear, however, that the activation of the vocalization-controlling neurons does not depend on acetylcholine; nor does acetylcholine exert a tonic facilitatory or inhibitory influence on the periaqueductal vocalization mechanism. The results are similar for the monoamines. Neither dopamine nor noradrenaline, serotonin or histamine antagonists cause a change of vocalization threshold. On the other hand, injection of noradrenaline, serotonin and dopamine, at some sites, increase vocalization threshold. Whether these effects are of functional relevance in normal vocal control, or simply reflect the fact that many PAG neurons possess monoaminergic receptors serving other than phonatory purposes, remains unclear.

Finally with respect to the endogenous opioids, we found a facilitatory effect of the non-selective opioid antagonist naloxone in three out of seven cases. When we compared the naloxone-effective with the naloxone-ineffective cases, we found that naloxone was only effective when the vocalization elicited from the forebrain expressed an aversive emotional state. Thus, shrieking calls, squealing and groaning, expressing defense, frustration and uneasiness in the squirrel monkey, showed a lowered threshold under naloxone. Calls representative of non-aversive situations, such as feeding calls or calls normally uttered during huddling of group mates, remained unchanged. Injection of morphine into the PAG caused an increase in threshold in two of the three naloxone-sensitive sites and had no effect in all naloxone-insensitive sites.

Conclusions

In summary, we may conclude that the PAG represents a crucial relay station in the vocalization control system. It gets its input partly directly from sensory systems. Visual input reaches the PAG from the deeper layers of the superior colliculus (Grofová et al., 1978; Beitz, 1982; Mantyh, 1982; Meller and Dennis, 1986), auditory input from the pericentral part of the inferior colliculus (Grofová et al., 1978; Beitz, 1982; Mantyh, 1982; Meller and Dennis, 1986), somatosensory input from the dorsal horn of the spinal cord and the spinal trigeminal nucleus (Harmann et al., 1988; Hayashi and Tabata, 1989; Wiberg et al., 1987; Blomqvist and Craig, this volume; Yezierski et al., 1987; Yezierski, this volume) and visceral input from the solitary tract nuclei (Bandler and Tork, 1987; Beitz, 1982; Mantyh, 1982; Meller and Dennis, 1986). These inputs may serve to trigger vocalization directly by external stimuli, for instance, to elicit pain-shrieking when a noxious stimulus is applied. In addition to this sensory input, there is an input from motivation-controlling limbic structures, such as the anterior cingulate cortex, amygdala, bed nucleus of stria terminalis, septum, hypothalamus and midline thalamus. This input probably serves to modulate the vocal reactivity to external stimuli according to prior experience and momentary emotional state. For instance the sight of food may induce a feeding call in food-deprived animals but leave the animals unresponsive when satiated. Or, an approaching group mate may elicit a contact call, while an approaching conspecific from a foreign group elicits an alarm call. The output of the periaqueductal vocalization center goes, among others, to the reticular formation of the lower brainstem, more specifically, the nucleus reticularis parvocellularis. We assume that this nucleus contains the vocal motor coordination mechanism, that is, the site from which the different motoneuron pools involved in phonation are integrated in their activity. The PAG, in this view, would represent an interface between sensory and motivational stimuli on the one hand and motor-coordinating structures on the other. In other words, the PAG probably serves to trigger vocalization selectively according to the incoming sensory and motivational information. In this process, glutamate (or an analogue of it) seems to be indispensable for the information transfer, while GABA, glycine and endogenous opioids play a modulatory role.

References

Airaksinen, M.S., Flügge, G., Fuchs, E. and Panula, P., Histaminergic system in the tree shrew brain, J. Comp. Neurol., 286 (1989) 289-310.

Apfelbach, R., Electrically elicited vocalizations in the gibbon Hylobates lar (Hylobatidae) and their behavioral significance, Z. Tierpsychol., 30 (1972) 420-430.

Araki, T., Yamano, M., Murakami, T., Wanake, A., Betz, H. and Tohyama, M., Localization of glycine receptors in the rat central nervous system: an immuno-cytochemical analysis using monoclonal antibody, Neuroscience, 25 (1988) 613-624.

Bandler, R., Induction of "rage" following microinjections of glutamate into midbrain but not hypothalamus of cats, Neurosci. Lett., 30 (1982) 183-188.

Bandler, R. and Tork, I., Midbrain periaqueductal grey region in the cat has afferent and efferent connections with solitary tract nuclei, Neurosci. Lett., 74 (1987) 1-6.

Barbaresi, P. and Manfrini, E., Glutamate decarboxylase-immunoreactive neurons and terminals in the periaqueductal gray of the rat, Neuroscience, 27 (1988) 183-191.

Beart, P.M., Summers, R.J., Stephenson, J.A., Cook, C.J. and Christie, M.J., Excitatory amino acid projections to the periaqueductal gray in the rat: a retrograde transport study utilizing D[³H]aspartate and [³H]GABA, Neuroscience, 34 (1990) 163-176.

Beitz, A.J., The organization of afferent projections to the midbrain periaqueductal gray of the rat, Neuroscience, 7 (1982) 133-159.

Beitz, A.J., Possible origin of glutamatergic projections to the midbrain periaqueductal gray and deep layer of the superior colliculus of the rat, Brain Res. Bull., 23 (1989) 25-35.

Bouthenet, M.-L., Martres, M.-P., Sales, N. and Schwartz, J.C., A detailed mapping of dopamine D-2 receptors in rat central nervous system by autoradiography with [¹²⁵I] Iodosulpride, Neuroscience, 20 (1987) 117-155.

Bouthenet, M.-L., Ruat, M., Sales, N., Garbarg, M. and Schwartz, J.C., A detailed mapping of histamine H1-receptors in guinea pig central nervous system established by autoradiography with [¹²⁵I] Iodobolpyramine, Neuroscience, 26 (1988) 553-600.

Brown, T.G., Note on physiology of basal ganglia and midbrain of anthropoid ape especially in reference to act of laughter, J. Physiol. (Lond.), 49 (1915) 195-207.

Camps, M., Kelly, P.H. and Palacios, J.M., Autoradiographic localization of dopamine D1 and D2 receptors in the brain of several mammalian species, J. Neural Trans., 80 (1990) 105-127.

Chu, D.C., Albin, R.L., Young, A.B. and Penney, J.B., Distribution and kinetics of GABA-B binding sites in rat central nervous system: a quantitative autoradiographic study, Neuroscience, 34 (1990) 341-357.

Cotman, C.W., Flatman, J.A., Ganong, A.H. and Perkins, M.N., Effects of excitatory amino acid antagonists on evoked and spontaneous excitatory potentials in guinea-pig hippocampus, J. Physiol (Lond.), 378 (1986) 403-415.

Greenamyre, J.T., Young, A.B. and Penney, J.B., Quantitative autoradiographic distribution of L-[³H]glutamate-binding sites in rat central nervous system, J. Neurosci., 4 (1984) 2133-2144.

Grofová, J., Ottersen, O.P. and Rinvik, E., Mesencephalic and diencephalic afferents to the superior colliculus and periaqueductal gray substance demonstrated by retrograde axonal transport of horseradish peroxidase in the cat, Brain Res., 146 (1978) 205-220.

Harmann, P.A., Carlton, S.M. and Willis, W.D., Collaterals of spinothalamic tract cells to the periaqueductal gray: a fluorescent double-labeling study in the rat, Brain Res., 441 (1988) 87-97.

Hayashi, H. and Tabata, T., Physiological properties of sensory trigeminal neurons projecting to mesencephalic parabrachial area in the cat, J. Neurophysiol., 61 (1989) 1153-1160.

Ibuki, T., Okamura, H., Miyazaki, M., Yanaihara, N., Zimmermann, E.A. and Ibata, Y., Comparative distribution of three opioid systems in the lower brain stem of the monkey (Macaca fuscata), J. Comp. Neurol., 279 (1989) 445-456.

Inagaki, N., Yamatodani, A., Ando-Yamamoto, M., Tohyama, M., Watanabe, T. and Wada, H., Organization of histaminergic fibers in the rat brain, J. Comp. Neurol., 273 (1988) 283-300.

Jürgens, U., Vocalization as an emotional indicator. A neuroethological study in the squirrel monkey, Behaviour, 69 (1979) 88-117.

Jürgens, U., Amygdalar vocalization pathways in the squirrel monkey, Brain Res., 241 (1982) 189-196.

Jürgens, U. and Müller-Preuss, P., Convergent projections of different limbic vocalization areas in the squirrel monkey, Exp. Brain Res., 29 (1977) 75-83.

Jürgens, U. and Ploog, D., Cerebral representation of vocalization in the squirrel monkey, Exp. Brain Res., 10 (1970) 532-554.

Jürgens, U. and Pratt, R., Role of the periaqueductal grey in vocal expression of emotion, Brain Res., 167 (1979a) 367-378.

Jürgens, U. and Pratt, R., The cingular vocalization pathway in the squirrel monkey, Exp. Brain Res., 34 (1979b) 499-510.

Magoun, H.W., Atlas, D., Ingersoll, E.H. and Ranson, S.W., Associated facial, vocal and respiratory components of emotional expression: an experimental study, J. Neurol. Psychopath., 17 (1937) 241-255.

Mansour, A., Khachaturian, H., Lewis, M.E., Akil, H. and Watson, S.J., Anatomy of CNS opioid receptors, Trends Neurosci., 11 (1988) 308-314.

Mantyh, P.W., The ascending input to the midbrain periaqueductal gray of the primate, J. Comp. Neurol., 211 (1982) 50-64.

Martin, J.R., Motivated behaviors elicited from hypothalamus, midbrain and pons of the guinea pig (Cavia porcellus), J. Comp. Neurol., 50 (1976) 1011-1034.

Mash, D.C. and Potter, L.T., Autoradiographic localization of M1 and M2 muscarine receptors in the rat brain, Neuroscience, 19 (1986) 551-564.

McLennan, H., Receptors for the excitatory amino acids in the mammalian central nervous system, Prog. Neurobiol., 20 (1983) 251-271.

Meller, S.T. and Dennis, B.J., Afferent projections to the periaqueductal gray in the rabbit, Neuroscience, 19 (1986) 927-964.

Monaghan, D.T. and Cotman, C.W., Distribution of N-methyl-D-aspartate-sensitive L-[³H] glutamate-binding sites in rat brain, J. Neurosci., 5 (1985) 2909-2919.

Moore, R.Y. and Bloom, F.E., Central catecholamine neuron systems: anatomy and physiology of the norepinephrine and epinephrine systems, Ann. Rev. Neurosci., 2 (1979) 113-168.

Pazos, A. and Palacios, J.M., Quantitative autoradiographic mapping of serotonin receptors in the rat brain, Brain Res., 346 (1985) 205-249.

Reichling, D.B., Kwiat, G.C. and Basbaum, A.I., Anatomy, physiology and pharmacology of the periaqueductal gray contribution to antinociceptive controls, Prog. Brain Res., 77 (1988) 31-46.

Spencer, D.G. Jr., Horvath, E. and Traber, J., Direct autoradiographic determination of M1 and M2 muscarinic acetylcholine receptor distribution in the rat brain: relation to cholinergic nuclei and projections, Brain Res., 380 (1986) 59-68.

Stuart, A.M., Mitchell, I.J., Slater, P., Unwin, H.L.P. and Crossman, A.R., A semi-quantitative atlas of 5-hydroxytryptamine-1 receptors in the primate brain, Neuroscience, 18 (1986) 619-639.

Suga, N., Schlegel, P., Shimozawa, T. and Simmons, J., Orientation sounds evoked from echo-locating bats by electrical stimulation of the brain, J. Acoust. Soc. Am., 54 (1973) 793-797.

Sutton, D. and Jürgens, U., Neural control of vocalization, In: Comparative primate biology, Vol. 4: Neurosciences, Steklis H.D. and Erwin J. (Eds.), Alan R. Liss, New York, 1988, pp. 625-647.

Waldbillig, R.J., Attack, eating, drinking and gnawing elicited by electrical stimulation of rat mesencephalon and pons, J. Comp. Physiol. Psychol., 89 (1975) 200-212.

Wiberg, M., Westman, J. and Blomqvist, A., Somatosensory projection to the mesencephalon: an anatomical study in the monkey, J. Comp. Neurol., 264 (1987) 92-117.

Yamano, M., Inagaki, S., Kito, S., Matsuzaki, T., Shinohara, Y. and Tohyama, M., Enkephalinergic projection from the ventromedial hypothalamic nucleus to the midbrain central gray matter in the rat: an immunocytochemical analysis, Brain Res., 398 (1986) 337-346.

Yezierski, R.P., Sorkin, L.S. and Willis, W.D., Response properties of spinal neurons projecting to midbrain or midbrain-thalamus in the monkey, Brain Res., 437 (1987) 165-170.

Activity of PAG Neurons During Conditioned Vocalization in the Macaque Monkey

Charles R. Larson

Depts. of Communication Sciences and Disorders
and Neurobiology and Physiology
Northwestern University
Evanston, IL, USA

Introduction

Considering the wide variety of behaviors in which the midbrain PAG is involved and the lack of clearly defined cytoarchitectural organization (Beitz, 1985), it is not surprising to note the lack of a clear understanding of its unified function, if such exists. Various functions ascribed to the PAG include defense reactions (Bandler and Carrive, 1988; Bandler and Depaulis, this volume), antinociception (Amit and Galina, 1986; Hosobuchi, 1981; Besson et al., this volume), reproductive behaviors (Ogawa et al., this volume) and vocalization (Jürgens and Pratt, 1979; Larson and Kistler, 1984; 1986; Jürgens, this volume). It is also possible that the PAG may be involved in the integration of various components of complex behaviors into functional entities.

From the standpoint of understanding the function of the PAG, the study of vocalization is quite interesting. Vocalization is often elicited as part of defense reactions (Bandler and Carrive, 1988), but vocalization may also occur in natural settings (Grimm, 1967) irrespective of any obvious defense reaction. In macaque monkeys, vocalization may be conditioned by pairing with a food reward (Sutton et al., 1973). Moreover, since the PAG is the recipient of many higher limbic system projections, many of which are related to vocalization (Jürgens and Ploog, 1970; Jürgens, 1982; Lamandella, 1977; Müller-Preuss and Jürgens, 1976; Robinson, 1967; Sutton et al., 1974), there is the very real possibility that the PAG's role in vocalization may extend beyond that of integrating vocalization into behaviors such as defense reactions. In other words, the PAG may be involved in vocal production for a wide variety of social reasons and if so, it is possible that the PAG itself is involved in a wider variety of behaviors than heretofore suspected. Moreover, by the study of the PAG's involvement in vocalization, we may not only learn more about neural mechanisms underlying vocalization, but about general functions of the PAG. In the present study, chronic single unit recordings were made in awake, trained monkeys with the aim of learning how the PAG functions during vocalization.

Methods

Two *Macaca fascicularis* and four *Macaca nemestrina* monkeys were first trained in an operant paradigm to vocalize. The animals were given a normal food ration on a 24 hour schedule. About two hours before feeding, when they were hungry, the animals were taken to a sound-treated room, placed in a restraining chair and reinforced with apple juice for vocalization. The animal's voice was transduced with an Electro-Voice microphone and amplified conventionally. Following four to six months of training, electromyographic (EMG) electrodes were surgically implanted in muscles and the lead wires attached to a plastic connector fastened to the skull with miniature screws and dental acrylic. From the first three animals, EMGs were recorded from right and left cricothyroid (CT) and right thyroarytenoid (TA) muscles. From the last three animals, EMGs from the CT, TA, posterior cricoarytenoid (PCA), intercostal (IC), diaphragm (D) and rectus abdominus (RA) muscles were recorded. In the first three animals, as we were developing techniques, we only sampled muscles from the laryngeal system that were easy to access. In the later animals, we also chose the PCA, because it is the only vocal fold abductor muscle and is generally thought to be active on inhalation. We sampled a group of respiratory muscles to represent activity on inhalation and exhalation. The IC and D were chosen because of their accessibility and known relation to inhalation. The RA was chosen because of its representation of the abdominal musculature group and because of its involvement with exhalation.

Following 10 days recovery, a second procedure was performed to fuse the upper four cervical vertebrae with the skull and reduce movement of the brainstem during recording. In the third surgery, which followed the second by 10 days, a chronic single unit recording chamber was fastened to the top of the skull. Surgeries were performed using aseptic techniques and with the animals fully anesthetized with halothane or Surital. Following surgery, analgesics were administered, and antibiotic ointment was applied to the wound margins. After recovery from the final surgery, the animals were again brought to the testing chamber.

With the animals held in the restraining chair and vocalizing, a tungsten microelectrode, insulated with epoxylite, except for the tip, was advanced towards the PAG at a 30° angle from the sagittal plane. Unit potentials were amplified with Grass P511 amplifiers, displayed on an oscilloscope and an audio monitor. As the electrode was advanced, potentials related to eye movements (superior colliculus) and jaw opening (mesencephalic nucleus of the trigeminal nerve) were identified and used to locate the position of the tip and guide electrode movement. When cells were found that changed their discharge rate before vocalization, voice, unit and EMG potentials were recorded on magnetic tape (Hewlett Packard 3968A). Each day, the electrode was moved to a new position by way of an x-y stage on the hydraulic microdrive, and another penetration was made.

After a few months of recording, it usually became difficult to record from cells because of tissue damage. At that time, the animals were anesthetized, an electrode positioned in the middle of the area from where most cells were recorded, and DC current (30 μA, 30 sec) passed through the electrode to mark the recording site. The animals were then sacrificed with an overdose of Nembutal and perfused transcardially with normal saline and 10% formalin. The brains were then removed for histological verification of the recording site and the muscles dissected to locate the position of the EMG recording electrodes.

For data analysis, the recorded signals were digitized onto magnetic tape and analyzed with a PDP 11/73 computer at 200 or 500 Hz sampling frequency. Ensemble averages of vocalization, rectified EMGs and unit firing frequency were created by triggering an averaging program on onset of vocalization. Spike triggered averaging was done by triggering the program on discharge of the unit and averaging the rectified EMGs. Parametric correlations were done by displaying all signals on a computer terminal and marking the beginning and end of EMG episodes, vocalization, and unit activity with a cursor. A program then made measurements of level of activity, durations and latencies. Cross correlation functions were also performed on the digitized signals.

Results

The animals produced mainly three different types of calls, barks, shrieks and coos. During the training sessions, it was found that some animals produced one type of call more than another, e.g. barks or coos. The animals were reinforced for either type of call they made. The result was that two animals emitted coos, another gave a mixture of coos and barks and the rest made barks almost exclusively. Shrieks were only made by some of the animals who were more excitable and in response to the approach of an investigator or displaying the capture gloves to the animals.

Although the detailed muscle activity patterns were unique for each animal, the general properties of the muscles were constant across the animals. Figure 1 illustrates ensemble averages of rectified vocalization, EMGs, unit firing frequency and raster displays of unit activity for four different units. The cricothyroid (CT) and thyroarytenoid (TA) became active slightly before vocal onset and remained active throughout the vocal episode. The posterior cricoarytenoid (PCA) became active earlier than the CT and TA, at about the same time as the intercostal (IC) and diaphragm (D) muscles. In some animals, the PCA and D showed a decrease in activity just before vocalization and then increased activity during the vocalization. The IC was active throughout vocalization in some ani-

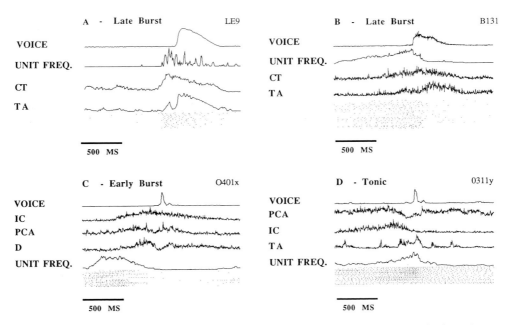

Figure 1. Ensemble averages of voice, EMGs and unit activity. Raster display of unit discharge at bottom of each panel. A. Late burst unit active during vocalization. B. Late burst unit active before vocalization. C. Early burst unit. D. Tonic cell increases rate before and during vocalization. The EMGs in A appear different than in B, C and D because of a lower digitizing sampling frequency in A. Abbreviations: CT, cricothyroid; TA, thyroaryte-noid; IC, intercostal; PCA, posterior cricoarytenoid; D, diaphragm.

mals, but in others decreased activity with vocal onset. The rectus abdominus (RA) muscle (not shown) began its activity shortly before vocal onset, about the same time as the CT, and then ceased discharging near vocal onset. A complicating factor describing the muscle activity is that muscle patterns differed depending on the type of vocalization emitted. An important point is that both the laryngeal and respiratory muscles may be divided into those that begin activity early, with inhalation, and those that begin late, with exhalation and glottal closure. A detailed description of laryngeal and respiratory muscle activity during monkey vocalization is in preparation (West and Larson, submitted).

In each of the animals, single PAG cells related to vocalization were isolated in the dorsolateral region of the PAG. In some animals a very small area was identified wherein there was a generalized increase in background "noise" preceding and during vocalization. Within this area there was a high concentration of vocalization-related cells. In other animals, such an area was not as evident, but a region was found from which a high density of vocalization-related cells could be found. Most vocalization-related cells were located between AP-0 and AP +2 from ear-bar zero (Fig. 2). The size of this area was 1-2 mm³. Outside of this area and extending for 1 -2 mm in each direction, a few vocalization-related

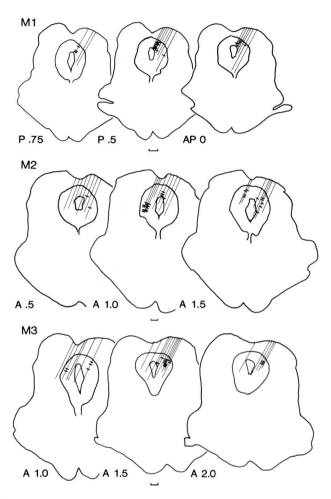

Figure 2. Coronal sections through the midbrain region of a *Macaca fascicularis* (M1) and two *Macaca nemestrina* monkeys (M2 and M3). Diagonal lines indicate representative electrode tracks. Cross hatches indicate location of vocalization-related cells. Locations of cells were determined in part from marker lesions made at the conclusion of the experiments as well as notes of electrode depth and the position of the mesencephalic nucleus of the V nerve (Mes V) made at the time of recording. Scale = 1 mm. Reprinted with permission from Larson and Kistler (1986) Experimental Brain Research.

cells could be found, but they were scattered among cells having very different discharge properties. While searching for the area containing the high concentration of vocalization-related cells, electrode penetrations were made throughout a region extending 2 mm medial and lateral as well as anterior and posterior of the high concentration area. Thus, a fairly wide area of the PAG was explored. In one animal, difficulty was experienced in locating cells in the dorsolateral PAG and the electrode penetrations were extended deeper. In this animal, another group of cells was located in the lateral PAG, ventral to the dorsolateral area. In no animal were penetrations made into the most ventral aspects of the PAG or at the extreme rostral and caudal limits of the PAG.

From the six monkeys, a total of 380 cells was recorded and identified as being related to vocalization. Of this group, 30 were located predominently in the lateral and the remaining in the dorsolateral PAG. There was no difference in the discharge properties between cells of the lateral and the dorsolateral PAG. The most common type of cell found was that which was normally quiet, began discharging shortly before vocal onset, and then ceased discharging just after vocalization onset (late burst, N = 168). Among this group there was considerable variability. Some such as those depicted in Fig. 1B, were active for several hundred ms before vocalization, increased in frequency towards vocal onset and then ceased discharge with vocalization. Others were only active briefly near the onset of vocalization, and some such as that depicted in Fig. 1A were active throughout the vocalization. Also, while some cells were completely silent in the absence of vocalization, others were sporadically active.

The second most common type of cell was the tonic (N= 80). Tonic cells had a very regular discharge rate that either increased or decreased before or during vocalization. The example in Fig. 1D illustrates a cell that increased its rate about 500 ms before vocalization and then decreased back to a normal rate following vocalization.

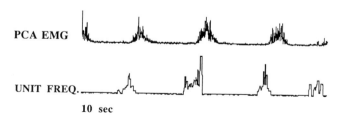

PCA EMG

UNIT FREQ.

10 sec

Figure 3. Illustration of the relationship between an early burst unit (discharge frequency, bottom trace) and posterior cricoarytenoid muscle (rectified EMG, top trace) activity in a monkey during quiet respiration. The unit becomes active shortly before the PCA on inhalation. Total trace duration 10 sec.

The third type of cell was that which became active well *before* vocalization and ceased discharging before vocalization (early burst, N = 55). Some, such as that shown in Fig. 1C, became active and then ceased discharging about 200 ms before vocalization. As with the late burst group, there was considerable variability in the pattern of activity and in the resting firing rate of the early burst cells. Some were active and then ceased discharging hundreds of ms before vocalization, while others stopped just at vocal onset. Also, some were completely silent in the absence of vocalization, while others were not. Upon closer examination, it was found that many (42%) of the early burst cells discharged on inhalation (Fig. 3). When a monkey vocalizes, it normally inhales deeper than during quiet respiration, and these cells showed a greater modulation of their firing rate before vocalization than during quiet respiration. In the last monkey studied, when our techniques became perfected, it was found that cells of the early burst type were located just lateral to late burst cells in the dorsolateral PAG. Figure 4 illustrates through boxplots, differences in latencies between onset of unit activity, or change in discharge rate, and vocal onset for early, late burst and tonic cells. It is clear that the late burst cells were active later than cells of the other two groups. In group comparisons, late burst cells were significantly different from early burst cells (t=5.9, df 198, p<.05).

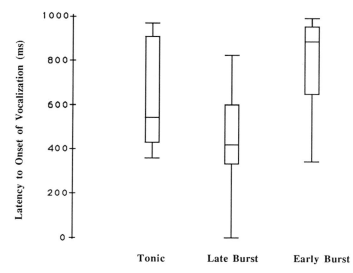

Figure 4. Boxplot display of latency from onset or change in discharge frequency of cells to onset of vocalization. Limits of vertical rectangles represent 75th and 25th percentile points. Horizontal line in rectangle is median latency. Cross bars above and below rectangles represent range of latencies. Tonic cells, N = 51; Late burst cells, N = 150; Early burst cells, N = 50.

Finally, 20 cells were not easily categorized because they showed complex patterns of discharge involving increases and decreases in rate before vocalization. Also, 57 cells could not be categorized because insufficient data were collected from them before the cells were lost.

Of considerable interest in three monkeys was the finding that some cells discharged for one type of vocalization but not another. Figure 5 is from an oscillograph record illustrating a cell discharging with bark vocalizations (left) but not with shrieks (right). There were only a few instances where such dramatic differences in cell discharge occurred with different vocalizations. In most instances the monkeys did not make calls of a different nature, and hence comparisons routinely could not be made between cell activity with different vocalizations.

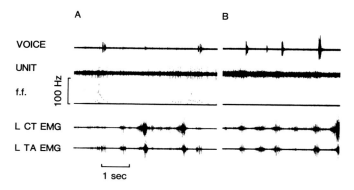

Figure 5. Oscillograph traces of cell activity for two different types of vocalization. On the left, A, cell is active with bark vocalizations, while in B, cell was not active with shriek vocalizations. Abbreviations: LCT, right cricothryoid; LTA, left thyroarytenoid. Reprinted with permission from Larson (1985) Journal of Speech and Hearing Research.

With some cells, there was a remarkable resemblance in the discharge pattern with temporal features of one of the muscles studied. Such similarities suggest the units may be functionally related to one or more muscles. Through the techniques of spike triggered averaging (STA) and parametric correlations, this suggestion was supported. Positive STA results were obtained from 113 cells. In 57/113 cases, EMG changes in single muscles were revealed by STA (Fig. 6A), while in 56/113 cases, changes were noted in two or three muscles (Fig. 6B and C). These examples are of interest because they suggest ways in which PAG cells function for the coordination of muscles. The example in Figure 6B illustrates facilitation in the IC and suppression in the TA muscle from STA. This reciprocal relationship with a respiratory and laryngeal muscle is similar to the different EMG patterns accompanying vocalization (e.g., Fig. 1D). The example in Figure 6C shows facilitation in the IC and D muscles, reflecting the finding that these two muscles normally are active synchronously on inhalation.

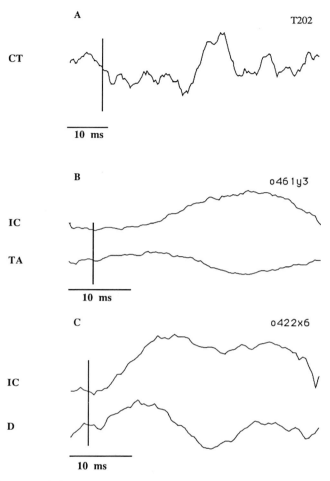

Figure 6. Spike triggered averages (STA) of rectified EMGs for three different units. In the STA technique, a single unit discharge triggers a computer, which then averages rectified EMGs following the unit potential. This technique has been used to demonstrate functional connections between corticomotoneuronal cells and EMGs in specific muscles of the forelimb (Fetz and Cheney, 1980). In each plot, the vertical line represents the time of unit discharge, or the time when the computer was triggered. EMG amplitudes are scaled in arbitrary units. Abbreviations as in Fig. 1. Number of averages for each unit are A - 151, B- 72, C -140.

The results of parametric correlation analysis were complimentary to those of STA. A total of 100 cells was found through parametric correlations to be related to one or more muscles. Of this group, 48 were related to a single muscle, the remainder to two or more muscles. The parametric correlation analysis also illustrated in an indirect fashion how PAG cells are related to the final

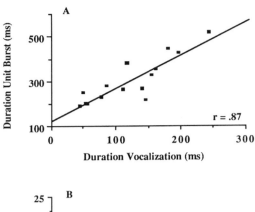

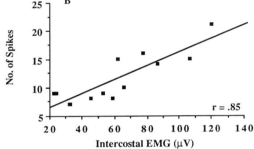

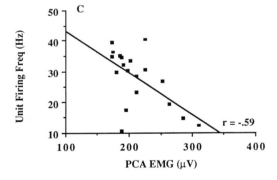

Figure 7. Scatterplots of unit activity and parametric measures of vocal activity. Lines of linear regression and correlation coefficients determined by a computer program. A, duration of burst of unit activity correlated with duration of vocalization. B, number of spikes in a burst of activity correlated with mean IC EMG before vocalization. C, mean unit firing frequency negatively correlated with mean PCA EMG before and during vocalization.

output of the process, i.e., vocalization. Most cells related to vocalization were correlated with vocal intensity, however, some were correlated with vocal duration (Fig. 7A). In those monkeys from which laryngeal and respiratory muscles were recorded, approximately equal numbers of cells were correlated with laryngeal and respiratory muscles (46 laryngeal and 44 respiratory). Figure 7B shows a scatterplot of a cell positively correlated with IC EMG and Figure 7C a cell negatively correlated with the PCA muscle. Negative correlations were rare (13/100) relative to positive correlations.

An important observation of both the STA and parametric correlation analyses was that the early burst cells were more frequently correlated with muscles active on inhalation, e.g., IC, D and PCA, while late burst cells were more frequently correlated with muscles active during exhalation, e.g., TA, CT and RA. This relationship is illustrated in Figure 8, which is a bar graph display of percent of cells from which positive results were obtained with both STA and parametric correlations, as a function of the muscles with which the cells were correlated.

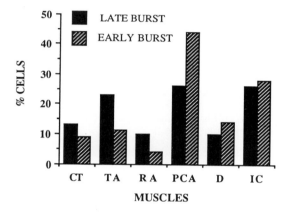

Figure 8. Bar graph representation of percent of cells of early and late burst groups correlated through parametric analysis or STA with six muscles. For this figure, the percent of cells correlated with a muscle, through parametric analysis and STA were combined, and the percentage from both techniques was plotted for each muscle. Thus, each bar represents the combined percentage from parametric and STA analysis. The CT, TA and RA muscles were more frequently correlated with late burst cells while the PCA, D and IC muscles were more frequently correlated with early burst cells. For a cell to be considered correlated using parametric analyses, the correlation coefficient, r, had to reach significance at the .02 level. Significance of STAs was determined by comparing data before and after the trigger point and testing the value at the .025 level.

Microstimulation was attempted several times after recording from a cell. Figure 9 illustrates an example of averaged excitation in the PCA and TA muscles following 20 trains of biphasic, constant current pulses of 100 μA. In only a few instances was it possible to detect changes with currents as low as 20 μA; in most cases, currents of 60 or 100 μA were necessary. Considering the wide spread of current with these levels, this technique probably does not reveal useful information about the effects of a single cell's activity on other neurons. This technique does show, however, that the shortest latency from beginning of a train of pulses to onset of EMG activity was 12 ms for the CT muscle and 15-20 ms for other laryngeal and respiratory muscles.

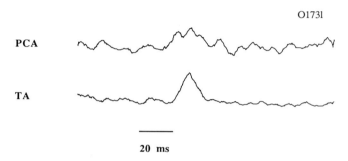

Figure 9. Results from microstimulation in PAG adjacent to a vocalization-related cell. Stimulus pulses delivered at rate of 200 Hz (0.1 ms biphasic square wave, 100 μA) during time bar. Average of 20 stimulus trains.

Discussion

The present studies (Larson and Kistler, 1984; 1986; Larson, submitted) have demonstrated the nature of PAG cellular activity associated with vocalization in the awake monkey. The anatomical location of most cells involved in vocalization appear to lie in the dorsolateral PAG. Within this area, it appears that there is a smaller region containing a fairly dense population of cells involved primarily in vocalization. Outside of this area, vocalization-related cells were sometimes recorded, but they were few in number and sparsely scattered among cells with quite different discharge patterns.

Observations from two animals suggest further work is needed to more precisely determine the anatomical extent of vocalization-related cells. In one animal in which it was difficult to locate cells in the dorsolateral PAG, the electrode penetrations were extended further ventrally into the lateral PAG, and here another group of vocalization-related cells was located. Also, in the final animal of the series, it was found that within the dorsolateral PAG, late burst cells were found to lie just slightly medial to early burst cells. The reason why this differentiation was found in the last animal and not in previous subjects probably is because our techniques of electrode penetration control had improved, and we became aware of the early burst cells and their different discharge properties. We then began searching for cells with these properties and identifying these neurons in greater numbers. Had we been aware of that type of cell earlier, we may have looked for and found them in earlier animals. Nevertheless, the data indicate the dorsolateral PAG may be the most important location for vocalization-related cells, but there is the suggestion that the lateral PAG may also be important, and even within the dorsolateral PAG, there may be a further differentiation of unit types.

The classification of unit types based upon discharge properties in the PAG, as elsewhere in the CNS, is always a difficult process. This process may presuppose independent knowledge of how the CNS functions, and without other knowledge, such presuppositions are inherently faulty. The decision to categorize units as we have done is in part based upon categorizations done for other areas in the CNS (Luschei et al., 1971; Hoffman and Luschei, 1980), and upon an understanding of activity of the muscles studied. An inherent hypothesis underlying these studies is that PAG neurons are involved in motor activity related to vocalization. According to this hypothesis, unit activity should ultimately influence the activity of muscles involved in vocalization, and therefore, there may be some relationship between discharge properties of the neurons and muscle activity. Our observations on muscle activity patterns indicated that some were active well in advance of vocalization, some at or near onset of vocalization and some throughout vocalization. Moreover, in one animal it was found that late burst neurons were located slightly medial to the early burst type. Tonic cells were categorized separately primarily because this has been done in other studies. Functionally, the tonic cells may be no different than phasic cells, perhaps only possessing a tonic activity level.

It should be stressed that, following recording in six animals and analyzing the data of some 380 neurons, the results presented here most likely define the extent of neuronal activity patterns one would expect to find in this population of cells in this behavioral paradigm. Recording in other species and in other behavioral contexts may lead to the discovery of neurons with different discharge characteristics. Therefore, as a summary, most units related to vocalization are quiet, or nearly so, and become active shortly before vocalization. Units are also related to respiratory activity exclusive of vocalization, but are necessarily active with the respiratory events that accompany vocalization.

Previous studies have shown that dorsolateral PAG neurons receive afferents from medial and lateral cortical fields (Hardy and Leichnetz, 1981; Shipley et al., this volume), hypothalamus, cuneiform nucleus, deep layers of superior colliculus, zona incerta (Beitz, 1982; Mantyh, 1982; Dean and Redgrave, this volume) and nucleus prepositus hypoglossi (Holstege et al., 1990). The dorsolateral PAG has rather limited efferent projections, primarily to the cuneiform nucleus (Redgrave and Dean, this volume). On the other hand, it has been shown that the lateral PAG projects to lower brainstem areas including the nucleus ambiguus (Yoshida et al., 1985) and nucleus retroambiguus (Holstege, 1989). Nucleus retroambiguus neurons project to the vicinity of laryngeal moto-neurons in nucleus ambiguus and respiratory motoneurons in the spinal cord (Holstege, 1989). Considering that most PAG cells reported here were recorded in the dorsolateral PAG, an area which evidently does not project to the medulla, it is possible that dorsolateral PAG cells make connections with those in the late-ral region (Tredici et al., 1983), which then project to the medulla and change the activity levels of medullary cells involved in vocalization. The critical question in this description is the precise function of the PAG in vocalization. Does it function as a nuclear relay, or does it play some other role in vocalization?

The answer to this question may be found in the results of this and other studies. Electrical and chemical stimulation of the PAG leads to "natural soun-ding" vocalization (Magoun et al., 1937; Jürgens and Richter, 1986). By increa-sing intensity of electrical stimulation, vocalizations become louder and are accompanied by more strenuous muscular contractions (Sapir et al., 1981). Increased electrical stimulation probably activates more cells, which may in part explain increased vocal loudness. The fact that such vocalizations sound natural suggests that the PAG may actually be coordinating the various muscle groups involved in vocalization. Although the cells described here may be involved in coordinating muscle groups, it is possible that PAG cell discharge activates other sets of neurons that handle this coordination process. Lesions of the PAG lead to total or partial mutism (Adametz and O'Leary, 1959), which supports either the suggestion that the PAG links upper limbic system structures and the lower brainstem or that the PAG helps coordinate muscles involved in vocalization. At the very least these data illustrate the extreme importance of the PAG to vocali-zation.

In the studies described here and elsewhere (Larson and Kistler, 1984; 1986), it was shown that PAG neurons are correlated with vocalization and laryngeal and respiratory EMGs. Specifically, early burst cells were preferential-ly correlated with activity in the PCA, IC and D EMGs. Usually, number of spikes preceding the vocalization or mean discharge rate of the cells was correla-ted with mean amplitude of the rectified EMG burst preceding vocalization. The same types of measures for late burst cells led to significant correlations with the CT, TA and RA muscles, which are active closer to onset of vocalization. Less fre-quently, duration of the unit burst was correlated with duration of vocalization or EMGs. Based on these data, it would thus appear that PAG cells are involved in determining the amount of contraction of vocalization-related muscles. Since strength of contraction relates to measures of subglottal pressure, vocal loudness and vocal fundamental frequency (Hirano et al., 1970; Hixon, 1973), it is sugges-ted that these cells help in determining loudness, fundamental frequency and duration of vocalization. It is possible that the PAG also determines more quali-tative aspects of vocalization such as harmonics to noise ratio, harshness, or tem-poral varying parameters of species-specific vocalizations.

The results of STA analysis support the above conclusions. Specifically, through STA, PAG cell discharge was related to facilitation or suppression in single muscles, facilitation in pairs of synergistic muscles or facilitation coupled with suppression in antagonistic muscles. The observation of patterned STA responses in two or more muscles that are normally either active simultaneously or reciprocally active during vocalization, leads to the suggestion that PAG cells are involved in some aspects of motor coordination for vocalization. PAG cells thus do more than relay information from higher limbic system structures to the medulla, but in addition affect the level of activity of medullary cells and thus the degree of muscle contraction.

There are still important unresolved questions about the PAG's function in the above described processes. The specific pathway and number of intervening synapses occurring between PAG cells and motoneurons is not resolved. The microstimulation data, indicating a minimal 12 ms latency from PAG stimulation to CT activation, suggests a fairly direct pathway. The latency from nucleus ambiguus excitation to laryngeal muscle activity is 5 ms (Zealear and Larson, 1988), which leaves 7 ms for PAG to nucleus ambiguus transit. Considering many PAG axons are nonmyelinated (Gioia et al., 1984), a slow conduction speed from the PAG to nucleus ambiguus is reasonable. Estimating a distance of 10 cm from the PAG to nucleus ambiguus, leads to a conduction speed of 1 - 2 m/s, without additional synapses. If the pathway involves more synapses, such as in the PAG, nucleus retroambiguus (Holstege, 1989) or reticular formation (Jürgens, 1986; Thoms and Jürgens, 1987), then the conduction speed over the axons would be higher.

The analysis of the PAG's function in vocalization may provide clues to its function in other activities as well. Bandler and Carrive (1988) injected excitatory amino acids into the PAG and noted a difference in the vocalization types elicited from rostral vs. caudal injections. In the monkey it has also been demonstrated that acoustically different vocalizations may be elicited from different rostrocaudal locations of the PAG (DeRosier et al., 1987). Such calls are known to accompany different social situations (Grimm, 1967), and the stimulation may reflect the artificial activation of groups of neurons normally active in different motivational situations. In the present paper, it was also reported that individual cells may be active for one vocalization type but not another, which further strengthens the suggestion that various groups of PAG cells are involved in different behaviors. Thus, the data from unit recording and those from electrical and chemical stimulation support the notion that various groups of PAG cells, and perhaps different regions within the PAG, are involved in disparate affective states, with vocalization being just one component of a larger behavioral response.

From the anatomical standpoint, the present results provide important information to be considered in the context of other behaviors as well. It was found in the present study that most vocalization-related cells are found concentrated in the dorsolateral PAG. A smaller group was also found in the lateral PAG, but had this area been explored in more animals, additional cells may have been located there. Cytoarchitecturally, cells of the dorsolateral and lateral PAG are larger than those found in other areas (Beitz, 1985; Beitz and Shepard, 1985).

This size difference plus the variations in cell morphological structure that are found throughout the PAG (Gioia et al., 1984) may underlie different functions of cells in the dorsolateral vs. other areas of the PAG. The fact that functional differences in cell activity were found and that there have been relatively few chronic, single unit recording studies from the PAG (Ni et al. 1990), suggests that functional concentrations of cells might be found for other behaviors as well. The anatomical connections between closely adjacent cells observed by Tredici et al. (1983) may form the basis for the intercellular communication underlying the PAG's role in coordination of responses.

As a final comment, it is apparent that some PAG cells involved in vocalization are also related to respiration. Since some such cells are active with respiration regardless of vocalization, these cells should properly be termed respiratory cells. Their involvement with vocalization may only be related to the fact that respiration is an integral component of vocalization. The observations that respiratory related cells are also located in the PAG (Harper, this volume), further reinforces the earlier comments on the function of the PAG. Namely, the PAG appears to be involved in coordination of laryngeal and respiratory muscular groups participating in vocalization.

Acknowlegments

I would like to thank the following individuals for assistance in data collection and analysis: Michael Kistler, Sungmin Park, David Niemann, Rob West, Elisabeth DeRosier and Pamela Ko. This research was supported by a grant from the NIH, DC00207-08.

References

Adametz, J. and O'Leary, J.L., Experimental mutism resulting from periaqueductal lesions in cats, Neurol., 9 (1959) 636-642.

Amit, Z. and Galina, Z. H., Stress-induced analgesia: Adaptive pain suppression, Physiol.Rev., 66 (1986) 1091-1120.

Bandler, R. and Carrive, P., Integrated defence reaction elicited by excitatory amino acid microinjection in the midbrain periaqueductal grey region of the unrestrained cat, Brain Res., 439 (1988) 95-106.

Beitz, A. J., The Organization of Afferent Projections to the Midbrain Periaqueductal Gray of the Rat, Neuroscience, 7 (1982) 133-159.

Beitz, A. J., The Midbrain Periaqueductal Gray in the Rat. I. Nuclear Volume, Cell Number, Density, Orientation, and Regional Subdivisions, J. Comp. Neurol., 237 (1985) 445-459.

Beitz, A.J. and Shepard, R.D., The midbrain periaqueductal gray in the rat. II. A Golgi analysis, J. Comp. Neurol., 237 (1985) 460-475.

DeRosier, E.A., Ortega, J.D., Park, S. and Larson, C.R., Effects of PAG stimulation on laryngeal EMG and vocalization in the awake monkey, Neurosci Abst., 17 (1987) 855.

Fetz, E.E. and Cheney, P.D., Postspike facilitation of forelimb muscle activity by primate corticomotoneuronal cells, J. Neurophysiol., 44 (1980) 751-772.

Gioia, M., Bianchi R. and Tredici, G., Cytoarchitecture of the periaqueductal gray matter in the cat: A quantitative nissl study, Acta anat., 119 (1984) 113-117.

Grimm, R. J., Catalogue of Sounds of the Pigtailed Macaque *(Macaca nemestrina)*, J. Zool. Lond., 152 (1967) 361-373.

Hardy, S.G.P. and Leichnetz, G.R., Cortical projections to the periaqueductal gray in the monkey: A retrograde and orthograde horseradish peroxidase study, Neurosci. Lett., 22 (1981) 97-101.

Hirano, M., Vennard, W. and Ohala, J., Regulation of register, pitch and intensity of voice, Folia phoniat., 22 (1970) 1-20.

Hixon, T. J., Respiratory function in speech, In: Normal Aspects of Speech, Hearing and Language, Minifie F. D., Hixon T. J. and Williams F. (Eds.), Prentice-Hall, Englewood Cliffs, 1973, pp. 73-125.

Hoffman, D. S. and Luschei, E. S., Responses of monkey precentral cortical cells during a controlled jaw bite task, J. Neurophysiol., 44 (1980) 333-348.

Holstege, G., An anatomical study on the final common pathway for vocalization in the cat, J. Comp. Neurol., 284 (1989) 242-252.

Holstege, G., Cowie, R.J. and Gerrits, P.O., Nucleus prepositus hypoglossi projects to the dorsolateral periaqueductal gray (PAG): A link between visuomotor and limbic systems, Neurosci. Abst., 16 (1990) 729.

Hosobuchi, Y., Periaqueductal gray stimulation in humans produces analgesia accompanied by elevation of β-endorphin and ACTH in ventricular CSF, Mod. Probl. Pharmacopsychiat., 17 (1981) 109-122.

Jürgens, U. and Ploog, D., Cerebral representation of vocalization in the Squirrel monkey, Exp. Brain Res., 10 (1970) 532-554.

Jürgens, U., Amygdalar Vocalization Pathways in the Squirrel Monkey, Brain Res., 241 (1982) 189-196.

Jürgens, U., The Squirrel monkey as an experimental model in the study of cerebral organization of emotional vocal utterances, Eur. Arch. Psychiatr. Neurol. Sci., 236 (1986) 40-43.

Jürgens, U. and Pratt, R., Role of the periaqueductal grey in vocal expression of emotion, Brain Res., 167 (1979) 367-378.

Jürgens, U. and Richter, K., Glutamate-Induced vocalization in the Squirrel monkey, Brain Res., 373 (1986) 349-358.

Lamandella, J. T., The limbic system in human communication, In: Studies in Neurolinguistics, Whitaker J. and Whitaker H. A. (Eds.), Academic Press, New York, 1977, pp. 157-222.

Larson, C. R. and Kistler M. K, Periaqueductal gray neuronal activity associated with laryngeal EMG and vocalization in the awake monkey, Neurosci. Lett., 46 (1984) 261-266.

Larson, C. R. and Kistler, M. K., The relationship of periaqueductal gray neurons to vocalization and laryngeal EMG in the behaving monkey, Exp. Brain Res., 63 (1986) 596-606.

Larson, C.R., On the relation of PAG neurons to laryngeal and respiratory muscles during vocalization in the monkey (Submitted).

Luschei, E. S., Garthwaite, C. R. and Armstrong, M. E., Relationship of firing patterns of units in face area of monkey precentral cortex to conditioned jaw movements, J. Neurophysiol., 34 (1971) 552-561.

Magoun, H. W., Atlas, D., Ingersoll, E. H. and Ranson, S. W., Associated facial, vocal and respiratory components of emotional expression: An experimental study, J. Neurol. Psychopath., 17 (1937) 241-255.

Mantyh, P. W., Forebrain Projections to the Periaqueductal Gray in the Monkey, with Observations in the Cat and Rat, J. Comp. Neurol., 206 (1982) 146-158.

Müller-Preuss, P. and Jürgens, U., Projections from the cingular vocalization area in the squirrel monkey, Brain Res., 103 (1976) 29-34.

Ni., H., Zhang, J. and Harper, R.M., Respiratory-related discharge of periaqueductal gray neurons during sleep-waking states, Brain Res., 511 (1990) 319-325.

Robinson, B. W., Vocalization evoked from forebrain in *Macaca mulatta*, Psychol. Behav., 2 (1967) 345-354.

Sapir, S., Campbell, C. and Larson, C., Effect of geniohyoid, cricothyroid and sternothyroid muscle stimulation on voice fundamental frequency of electrically elicited phonation in rhesus macaque, Laryng., 91 (1981) 457-468.

Sutton, D., Larson, C., Taylor, E. M. and Lindeman, R. C., Vocalization in Rhesus monkeys: Conditionability, Brain Res., 52 (1973) 225-231.

Sutton, D., Larson, C. and Lindeman, R. C., Neocortical and limbic lesion effects on primate phonation, Brain Res., 71 (1974) 61-75.

Thoms, G. and Jürgens, U., Common input of the cranial motor nuclei involved in phonation in Squirrel monkey, Exp. Neurol., 95 (1987) 85-99.

Tredici, G., Bianchi, R. and Gioia, M., Short Intrinsic Circuit in the Periaqueductal Gray Matter of the Cat, Neurosci. Lett., 39 (1983) 131-136.

West, R. and Larson, C.R., Measurements of the variability in EMGs of laryngeal and respiratory muscles during vocalization in the monkey. (Submitted).

Yoshida, Y., Mitsumasu, T., Hirano, M. and Kanaseki, T., Afferent connections to the nucleus ambiguus in the brain stem of the cat: an HRP study, Presented at the 4th Int. Conf. on Vocal Fold Physiology, Yale Univ., New Haven, CT (1985).

Zealear, D. L. and Larson, C. R., A microelectrode study of laryngeal motoneurons in the nucleus ambiguus of the awake vocalizing monkey, In: Vocal Physiology: Voice Production, Mechanisms and Functions, Fujimura O. (Ed.), Raven Press, New York, 1988, pp. 29-37.

Discharge Relationships of Periaqueductal Gray Neurons to Cardiac and Respiratory Patterning During Sleep and Waking States

Ronald M. Harper, Huifang Ni and Jingxi Zhang

Department of Anatomy and Cell Biology
and the Brain Research Institute
University of California at Los Angeles
Los Angeles, California, USA

Introduction

Respiratory muscle activity assists many bodily functions for purposes other than mere tissue oxygenation. Respiratory musculature, for example, are used in vocalization, for providing thoracic and abdominal pressure for defecation and urination, and for maintaining thoracic pressure and body position in somatic movements. Respiratory musculature are used heavily in some species for temperature regulation; tachypnea and upper airway dilation, for example, are primary mechanisms used for cooling in selected animals. Use of the respiratory musculature for these actions involves precise coordination of a number of brain structures, and those brain structures incorporate a number of rostral brain regions in addition to the classical brain stem "dorsal" (e.g., nucleus of solitary tract) and "ventral" (e.g., nucleus ambiguus) respiratory groups typically outlined in the respiratory control literature.

Neurons in the PAG provide considerable potential to modify respiratory patterns. PAG neurons discharge to vocalization efforts (Larson and Kistler, 1986) which must recruit respiratory musculature for these functions. The PAG has major reciprocal projections to rostral brain regions, including particular cortical, preoptic and amygdala structures which have, by stimulation, cooling and recording evidence, demonstrated the potential to modify respiratory and cardiac patterning. The cingulate gyrus and medial temporal cortex (Shipley et al., this volume), for example, project to the PAG; the cingulate cortex contains neurons which discharge both tonically and on a breath-by-breath basis with the respiratory cycle, and on a beat-by-beat basis with the cardiac cycle (Frysinger and Harper, 1986). Massive reciprocal projections exist between the central nucleus of the amygdala and the PAG; the central nucleus contains neurons which discharge with both the respiratory and cardiac cycles (Zhang et al., 1986a; Frysinger et al., 1988), and, on single-pulse electrical stimulation, will pace respiration (Harper et al., 1984), an effect which is abolished during quiet sleep (QS). Cooling of the central nucleus of the amygdala will abolish an aversive conditioned respiratory and blood pressure response (Zhang et al., 1986b). The medial preoptic region projects strongly to the PAG; this projection is reciprocal (Shipley et al., this volume). The medial preoptic region contains temperature-sensitive neurons which have the potential to greatly modify respiratory patterning. Thus, PAG-rostral brain interconnections provide the possibility of integrating respiratory patterning as required by particular physiological needs. The PAG has reciprocal projections to the nucleus of the solitary tract (Bandler and Tork, 1987), and projects heavily to the nucleus retroambiguus (Bandler, 1988; Bandler et al., 1991; Holstege, 1989; Holstege, this volume) and thus is positioned strategically for integrating rostral and caudal brain regions which regulate aspects of cardiac and respiratory control. The PAG may integrate respiratory responses as part of physiological patterns for which it is principally responsible. Electrical or chemical stimulation of the PAG, for example, can elicit defensive reactions which are accompanied by major changes (usually tachypnea) in respiratory patterns.

Sleep states exert profound effects on temperature regulation, skeletal motor control, and respiratory patterning, among a number of other physiological activities. Among state modifications of motor functioning, respiratory patterning is perhaps the most dramatic. For example, QS prolongs inspiratory efforts and total respiratory cycle durations (Orem et al., 1977). It also greatly reduces breath-to-breath variation, i.e., the state markedly regularizes the respiratory rhythm. Rapid eye movement (REM) sleep greatly increases respiratory variability and respiratory rate over QS states, resulting in respiratory patterns similar to those found in waking (AW) states. REM sleep abolishes tone in a number of upper airway and thoracic wall musculature, except for phasic activation in the former muscles (Sauerland and Harper, 1976).

The increase in variability and rates of respiration observed during REM sleep is far in excess of the metabolic demands that would be expected from the muscle atonia that characterizes that state or the phasic muscular twitches which occur with episodes of rapid eye movements. Substantial influences are exerted on respiratory musculature during that state, and these influences are not obviously related to mechanical or metabolic demands associated with that stage of sleep. The neural attributes underlying the variability and timing changes found in REM sleep are not known. The mechanisms for at least a portion of the respiratory changes occurring during REM sleep lie caudal to collicular structures, since considerable respiratory variability remains after transection at that level (Jouvet, 1965), although a complete description of potential differences in respiratory patterning between intact and transected preparations has yet to be made.

The REM sleep state often has been associated with periods of extreme affect. A REM sleep disorder has been described in humans which is characterized by outbursts of negative emotional and combative behaviors (Schenk et al., 1987). A "REM without atonia" state has been described for cats, which has been characterized as a REM state with partial or near total restoration of skeletal tone during that state following bilateral sub-locus coeruleus lesions; this state is often associated with excessive affective behavior of a "predatory nature" (Morrison, 1988). Animals with these lesions also often exhibit particular types of excessive aggressive behavior toward other cats or to their handlers during AW; the behavior toward other cats often includes species-specific posturing and vocalization. The normal REM state exhibits epochs of "autonomic storms" with extreme alterations of sympathetic activity, and marked alterations in blood pressure and heart rate (Mancia and Zanchetti, 1980). We use states such as REM to elicit conditions which can demonstrate the role which particular brain structures exert during different behaviors.

Sleep states thus provide an appropriate "experiment of nature" to study brain structures which have the potential to exert differential influences on physiological mechanisms. In the case under consideration here, respiratory and cardiovascular patterning are grossly affected by state, and both measures are affected in such a fashion as to suggest that the PAG plays a prominent role in these pattern modifications.

Thus, the PAG may contribute to a portion of the respiratory patterning observed during different sleep states. The mechanisms integrated by the PAG in mediating affective defense reactions may underlie some of the respiratory pattern changes observed during REM sleep. Furthermore, the medial preoptic region described earlier as having heavy reciprocal projections to the PAG (Shipley et al., this volume) underlies a portion of the control of QS; destruction of the preoptic region results in at least a partial loss of QS, while both high and low frequency electrical stimulation results in QS induction. A subset of preoptic neurons discharges selectively during onset of this state (Sterman and Clemente, 1974). The anatomical relationship between the PAG and preoptic areas may mediate a portion of the respiratory patterning changes observed during this QS state as well.

The control of cardiovascular activity during sleep is of great interest because of the close interaction between cardiovascular and respiratory systems. Particular conditions related to circulation have profound effects on breathing during sleep; delayed circulation times, for example, lead to periodic breathing, a dramatic sequence of gradually increasing inspiratory efforts followed by a long pause; on the other hand, obstructive sleep apnea, characterized by continued diaphragmatic efforts with a closed upper airway, results in exaggerated blood pressure variation with extreme transient hypertension and tachycardia-bradycardia sequences. An understanding of the neural mechanisms underlying control of cardiovascular and respiratory patterns during sleep is thus of major interest, and an examination of the potential PAG contributions is of particular concern.

Methods

We examined the potential contributions of the PAG to respiratory patterning by studying the discharge properties of a subset of PAG neurons during spontaneous sleep-waking states. We examined that portion of the PAG (the caudal two thirds) known to receive heavy projections from the central nucleus of the amygdala (Shipley et al., this volume), because the central nucleus can exert pronounced effects on cardiovascular and respiratory patterning, and these effects may be mediated through the PAG. Because anesthesia greatly affects neuronal discharge in a number of structures with rostral brain projections, the studies were carried out in drug-free, freely moving preparations using chronic single neuron monitoring procedures. Detailed procedures for animal surgery, electrode and microdrive construction, electrode implantation, and recording have been described elsewhere (Harper and McGinty, 1973; Sieck and Harper, 1980). Three female and two male cats weighing 2.7 to 3.7 kg were prepared with electrodes for recording sleep parameters, including electrical activity of cortical, hippocampal, and lateral geniculate regions; eye movement; and nuchal muscle activity. Bundles of nine microwire electrodes (62.5 μm, Formvar-insulated Nichrome) were inserted into miniature microdrives (Zhang and Harper, 1984) stereotaxically aimed bilaterally at the PAG (A 1.5, L 1.2, V 1.0; Snider and Niemer, 1970). Respiratory effort was recorded as the root-mean-square (RMS, time constant 55 ms) amplitude of the electromyogram (EMG) from stranded, fine-wire electrodes sewn into the costal portion of the diaphragm. Electrocardiographic activity was recorded between a diaphragm electrode and a skull screw reference. All sleep parameter electrodes and diaphragmatic electrodes were soldered to a 20-pin Winchester female connector and secured to the skull with dental acrylic.

Animals recovered from surgery for 7 to 10 days before recordings were initiated. The animals were adapted to a sound-attenuated chamber, with food and water, for 2 to 3 h before recordings were initiated. All recordings occurred within a period of daylight (1100 h to 1600 h) and encompassed at least two complete sleep cycles. During recording sessions, a flexible recording cable, attached to the head connector through a commutator, permitted relatively free movement of the animal. The filtered and amplified physiologic signals were simultaneously written onto polygraph paper, digitized using an LSI-11/73 computer, and recorded with an FM tape recorder. The digitization program acquired analog data at variable rates, with sampling rates at least twice the highest relevant frequency present in the signal (Harper et al., 1974). The RMS diaphragmatic EMG was lowpass filtered (20 Hz) and digitized at 32 samples/s. Peaks and troughs of the RMS diaphragmatic EMG were detected off-line using software that determined the inspiratory on and off times based on derivative changes exceeding defined amplitude and time criteria. The onset of inspiration was stored by the computer as time of occurrence and used to correlate respiratory timing with unit discharge. The PAG was probed for single neurons using a miniature microdrive to adjust the microelectrode bundle position. Neuronal spikes were fed to a voltage-time discriminator, and the output pulses from this device triggered circuits to indicate times of occurrence of neuronal discharges on the computer. A similar discrimination procedure was used to detect the occurrence of cardiac R waves. A time code was used to synchronize the polygraph, digital, and FM tape records. The animals behavior was continuously monitored during recording.

Data were analyzed with programs written for the LSI-11/73 computer. The paper records were used to sleep-score the recording sessions. Artifact-free periods 2 to 10 min in length were selected from each digital record for quiet AW, QS, and active or REM sleep (including both phasic and tonic REM periods), using state classification criteria based on electroencephalographic activity, eye movement, and nuchal muscle tone.

Cross-correlation histograms (Perkel et al., 1967) were used to assess the timing dependency of neuronal discharge and the inspiratory and cardiac triggers. Discharge timing relationships between neuronal discharge and the onset of inspiration were examined using bin widths of 25 ms; with cardiac timing relationships, a 5 ms bin width was used. This bin width produced histograms spanning several respiratory or cardiac cycles, permitting any phase-locked periodicity in cell discharge to be readily observed. The histograms were digitally smoothed by multiplying the count in each successive bin by a factor derived from a sliding Gaussian bell to reduce high-frequency components in the histogram (Abeles, 1982) and to allow easier visualization of periodicity in the correlation patterns, and then plotted. To verify a correlation between PAG neuronal activity and the respiratory or cardiac cycles, the neuronal spike train data were shuffled to randomize the times of occurrence of the neuronal events. Cross-correlation functions were then recalculated between the shuffled train and the respiratory or cardiac reference.

Results

General Findings

We recorded sixty-eight neurons from histologically confirmed sites in the PAG of five cats. Figure 1 illustrates these loci. PAG neurons fired slowly; the discharge rates ranged from nearly 0 to 25 Hz, and a majority of cells fired at rates of less than 5/s. Discharge rates varied with sleep-waking states, with highest rates during REM sleep, lowest rates during QS, and with AW rates intermediate (7.2/s, 3.9/s, and 4.4/s respectively). Changes in rate and pattern as the animal made a transition from one state to another occurred gradually over the time of state transitions, which often required 20-60 s. The difficulty in classifying the precise onset and termination of a state, together with the slow spontaneous discharge of the neurons, precluded an assessment of exact timing relationships of cell discharge during state transition periods; thus, only well-defined state periods were used for analysis.

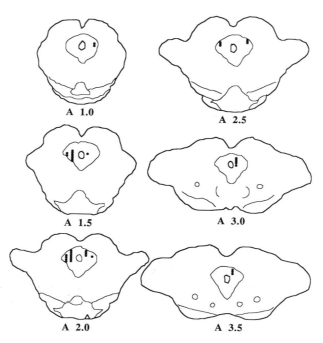

Figure 1. Histological location of PAG neurons recorded in this study. Sixty-eight recordings were made from 41 loci; on occasion, two or more neurons were recorded from the loci indicated.

Respiratory Relationships

A substantial portion of recorded PAG cells, nearly a third (28%), discharged on a breath-by-breath basis with the respiratory cycle; slightly over a third (34%) discharged on a correlated-rate basis with respiratory rate (Table I). All of the tonic rate relationships were state-related, i.e., the correlations occurred during particular states and disappeared or were greatly reduced in other states. A substantial portion of the breath-by-breath relationships (74%) were also state related.

Table I: Distribution of Respiratory-Related Neurons

	Resp.-Related	State Dependent/Respiratory	AW	QS	REM
Timing	19/68	14/19	9/19	9/19	7/19
Tonic	23/68	23/23	5/23	3/23	15/23

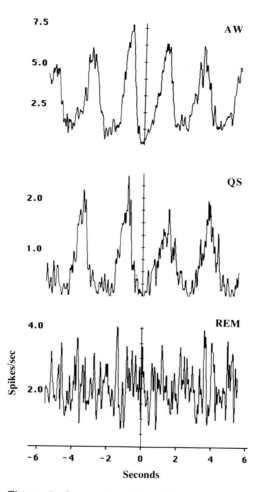

Figure 2. Cross correlation histogram showing a breath-by-breath relationship between discharge of a PAG neuron and respiratory onset. These histograms illustrate
strong correlations during waking (AW) and
quiet sleep (QS), and no dependency during
rapid eye movement (REM) stages.
Reprinted with permission from Ni et al.
(1990) Brain Research.

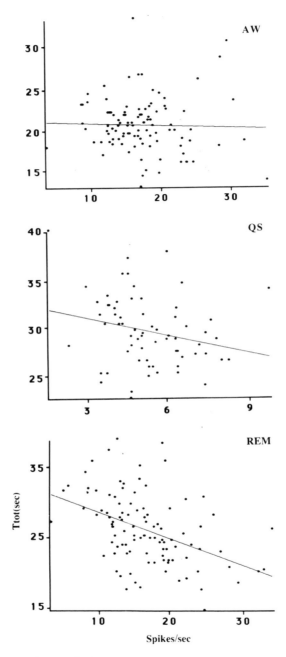

Figure 3. Linear regression analysis between respiratory period and PAG rate discharge; these histograms illustrate patterns for a cell in which no relationship appeared during AW and QS but a negative correlation developed during REM. Pearson's r values: AW=-0.05, df=103; QS=-0.25, df=67; REM=-0.44, df=105. Reprinted with permission from Ni et al. (1990) Brain Research.

Phasic discharge timing relationships were nearly equally distributed in AW (9/19), QS (9/19), and REM sleep (7/19) ($\chi^2 = 0.4$, n.s.). Histograms for one such neuron are presented in Figure 2. A rhythmicity at the respiratory frequency occurred for this neuron during AW and QS, and was absent in REM sleep.

Tonic discharge relationships with the respiratory cycle were observed most frequently in REM sleep (15/23), less often in AW (5/23), and even less often in QS (3/23). The differences between REM over AW and QS states were significant ($\chi^2 = 12.1$, p<0.01). An example of a tonic rate correlation with Ttot is shown in Figure 3. This neuron showed a significant tonic rate correlation with Ttot in REM sleep, but not in AW or QS.

Cardiac Relationships

The discharge patterning relationships to the cardiac cycle produced a most remarkable set of findings; seventy-four percent of the neurons showed a discharge timing relationship and/or a tonic correlation with cardiac activity, and nearly all of these relationships were state dependent (Table II). Over a third (35%) of recorded cells discharged on a cycle-by-cycle relationship in at least one sleep or AW state, and nearly all (22 of 24) of these relationships were state-dependent. Fourteen of 24 cycle-by-cycle relationships occurred during QS (14/24), 12/24 in AW and 9/24 in REM sleep; these differences in counts were not significant ($\chi^2 = 1.3$, p<0.05). Histograms for one such neuron are presented in Figure 4. Neuronal discharge of the cell in Figure 4 was heavily influenced by timing of the cardiac cycle during AW, somewhat influenced in QS, and not at all influenced in the REM sleep state.

Over half of the cells showed a tonic rate relationship with cardiac activity. All of these correlations were positive, and all were state-dependent, i.e., none showed a significant correlation across all three states.

Table II: Distribution of Cardiac-Related Neurons

	Cardiac-Related	State Dependent/Cardiac	AW	QS	REM
Timing	24/68	22/24	12/24	14/24	9/24
Tonic	41/68	41/41	10/41	8/41	30/41

An example of a relationship between unit discharge rate and heart rate is shown in Figure 5. This neuron showed a significant tonic rate correlation with heart rate in REM sleep, but not in AW or QS. This figure demonstrates the most common finding, i.e., tonic relationships were found more frequently in REM sleep (30/41) over AW (10/41) or QS (8/41) ($\chi^2 = 24.2$, p<0.001); the number of cells showing tonic relationships did not differ between AW and QS ($\chi^2 = 0.3$, n.s.). Pearson's r correlations ranged from 0.15 (df=205) to 0.58 (df=41), with a mean of 0.35 for all cells showing significant correlations.

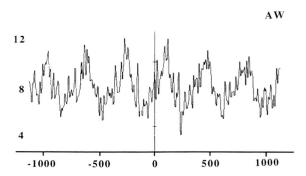

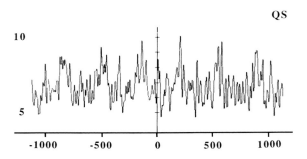

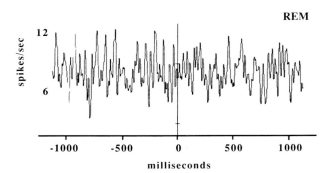

Figure 4. Cross correlation histograms between
discharge of PAG neurons and the cardiac cycle in
three states. This neuron showed a dependency in AW
and QS but not in REM. Reprinted with permission from
Ni et al. (1991) Brain Research.

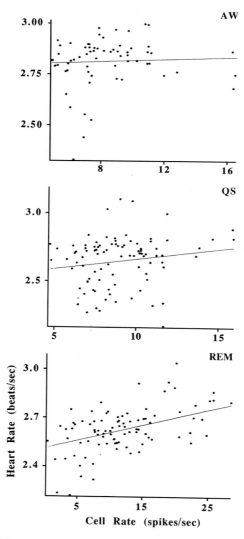

Figure 5. Linear regression plots between cell discharge rate and cardiac rate in three states; note the higher positive correlation in REM sleep. Pearson's r values: AW=0.11, df=64; QS=0.15, df=93; REM=0.41, df=90. Reprinted with permission from Ni et al. (1991) Brain Research.

Of the 24 neurons showing a timing relationship and the 41 showing a rate correlation, 15 displayed both a timing and a rate relationship with the cardiac cycle. Six of these occurred during the same state: one in AW, one in QS, and four in REM sleep.

Nearly half (29/68) of recorded cells showed either timing or tonic dependencies to both the cardiac and respiratory cycles. If the total number of respiratory cells showing either timing or tonic changes is considered (35), the percentage of cells showing dependency to both the cardiac and respiratory cells is substantial (29/35, 83%).

Discussion

The PAG contains neurons which exhibit state-bound cardiac and respiratory patterning relationships on both breath-by-breath and cycle-by-cycle dependencies as well as tonic interactions with the cardiac and respiratory cycles. A substantial proportion of cells exhibiting respiratory relationships also exhibits a discharge dependency to cardiac activity.

A most significant aspect of PAG neurons was the exceptionally high proportion (74%) of cells which showed a timing or rate dependency in discharge to the cardiac cycle. The high proportion (35%) of cells showing a timing dependency approximates that found for the parabrachial pons (33%; Sieck and Harper, 1980), and is higher than that described for the central nucleus of the cat amygdala (12%; Frysinger et al., 1988), and human amygdala and hippocampus (19%; Frysinger and Harper, 1990), and much higher than the cat anterior cingulate (7%; Frysinger and Harper, 1986). A simplistic interpretation would place a rostral-caudal differentiation of cycle-by-cycle dependencies, with more rostral structures exhibiting fewer timing relationships; however, a more plausible interpretation is that the PAG and parabrachial pons are more closely involved in cardiovascular control.

Most of the PAG cardiac tonic relationships were found in REM sleep, suggesting a unique state relationship between neural discharge in the PAG and cardiovascular activity during that state.

A similar finding of tonic discharge relationships was found for the respiratory cycle during REM sleep; nearly 65% of tonically active cells showed a dependency in the REM state. This state relationship suggests a significant role for the PAG in mediating some of the respiratory effects, such as the enhanced respiratory rates in that state.

The state specificity of both tonically and phasically active neurons suggests a multiplicity of functions served by the different cells. It is of interest that a proportion of neurons discharged both phasically and tonically with the respiratory and cardiac cycles during QS. The cells discharging phasically, i.e., on a breath-by-breath basis to the respiratory cycle during that state, were not the same cells that were phasically active during REM sleep. It may be that the phasically-active cells observed during QS may be targets of neurons projecting from the preoptic area; we did not, however, identify these neurons by stimulation or other procedures. The high proportion of neurons which show dependencies to both cardiac and respiratory activity suggests that at least a subset of these cells receives simultaneous projections from "traditional" respiratory and cardiac areas. Furthermore, the cardiac and respiratory action suggests an integrative role for these neurons, a speculation that must be confirmed.

The potential role that the PAG can play in mediating sudden death from limbic seizure discharge should not be overlooked. Some of the potential interactions between thresholds for defensive reactions and amygdala kindled seizures have been discussed (Hiyoshi et al., 1990); it is apparently the case that a portion of amygdala seizure discharge effects can be mediated through projections to the PAG, and then from PAG structures to other brainstem cardiorespiratory regions. Kainic acid administration into the dorsal hippocampus of the cat results in the development of seizure activity which begins in the hippocampus and amygdala (Griffith et al., 1987); the development of such a seizure focus elicits a variety of ictal and interictal emotional behavioral disturbances, including explosive rage reactions. The role of the PAG becomes particularly important in understanding sudden death in infants who succumb to rapid desaturation following extreme affective responses; such responses apparently are associated with very rapid increases in pulmonary arterial pressure which result in a shunting of blood and extraordinarily rapid desaturation; loss of action of the respiratory musculature follows this sequence, i.e., the rapid desaturation appears to result from extreme constriction of the pulmonary vasculature, mediated by excessive sympathetic outflow to the vasculature, followed by motor outflow alterations to the respiratory musculature (Southall, personal communication). This sequence is most likely mediated by seizures or "seizure-like" extreme activation of brain structures integrating aversive affect; the PAG would be a prime candidate for mediating these actions.

In summary, a subset of neurons in the PAG discharge in patterns which suggest a dependency to both respiratory and cardiac activity. The discharge dependencies occur during particular states for particular neurons, suggesting a multiplicity of specific cardiac or respiratory relationships, each function perhaps unique for a state. We speculate that the contributions from the PAG lie in aspects of affective integration to respiratory and cardiovascular control, and possibly in integration of aspects of temperature control during sleep.

References

Abeles, M., Quantification, smoothing, and confidence limits for single-units' histograms, J. Neurosci. Methods, 5 (1982) 317-325.

Bandler, R., Brain mechanisms of aggression as revealed by electrical and chemical stimulation: suggestion of a central role for the midbrain periaqueductal grey region, In: Progress in Psychobiology and Physiological Psychology (Vol. 13), Epstein A. and Morrison A. (Eds.), Academic Press, New York, 1988, pp. 67-154.

Bandler, R., Carrive, P., and Zhang, S.P., Integration of somatic and autonomic reactions within the midbrain periaqueductal grey: Viscerotopic, somatotopic and functional organization, Prog. Brain Res., 57 (1991) 269-305.

Bandler, R. and Tork, I., Midbrain periaqueductal grey region in the cat has afferent and efferent connections with solitary tract nuclei, Neurosci. Lett., 74 (1987) 1-6.

Frysinger, R.C. and Harper, R.M., Cardiac and respiratory relationships with neural discharge in the anterior cingulate cortex during sleep-waking states, Exp. Neurol., 94 (1986) 247-263.

Frysinger, R.C. and Harper, R.M., Cardiac and respiratory correlations with unit discharge in epileptic human temporal lobe, Epilepsia, 31 (1990) 162-171.

Frysinger, R.C., Zhang, J. and Harper, R.M., Cardiovascular and respiratory relationships with neuronal discharge in the central nucleus of the amygdala during sleep-waking states, Sleep, 11 (1988) 317-332.

Griffith, N., Engel, J. and Bandler, R., Ictal and enduring interictal disturbances in emotional behaviour in an animal model of temporal lobe epilepsy, Brain Res., 400 (1987) 360-364.

Harper, R.M., Frysinger, R.C., Trelease, R.B. and Marks, J.D., State-dependent alteration of respiratory cycle timing by stimulation of the central nucleus of the amygdala, Brain Res., 306 (1984) 1-8.

Harper, R.M. and McGinty, D.J., A technique for recording single neurons from unrestrained animals, In: Brain unit activity during behavior, Phillips M.I. (Ed.), Charles C. Thomas, Springfield, Illinois, 1973, pp. 80-104.

Harper, R.M., Sclabassi, R.J. and Estrin, T., Time series analysis and sleep research, IEEE Trans. Automatic Control, 19 (1974) 932-943.

Holstege, G., Anatomical study on the final common pathway for vocalization in the cat, J. Comp. Neurol., 284 (1989) 242-252.

Hiyoshi, T., Matsuda, M., and Wada, J.A., Centrally induced feline behavior and limbic kindling, Epilepsia, 31 (1990) 359-369.

Jouvet, M., Paradoxical sleep - a study of its nature and mechanisms, Prog. Brain Res., 18 (1965) 20-62.

Larson, C.R. and Kistler, M.K., The relationship of periaqueductal gray neurons to vocalization and the laryngeal EMG in the behaving monkey, Exp. Brain Res., 63 (1986) 596-606.

Mancia, G. and Zanchetti, A., Cardiovascular regulation during sleep, In: Physiology in sleep, Orem J. and Barnes C.D. (Eds.), Academic Press, New York, 1980, pp. 2-55.

Morrison, A.R., Paradoxical sleep without atonia, Arch. Ital. Biol., 126 (1988) 275-289.

Ni, H., Zhang, J.X. and Harper, R.M., Respiratory-related discharge of periaqueductal gray neurons during sleep-waking states, Brain Res., 511 (1990) 319-325.

Ni, H., Zhang, J.X. and Harper, R.M., Cardiovascular-related discharge of periaqueductal gray neurons during sleep-waking states, Brain Res., 532 (1991) 242-248.

Orem, J., Netick, A. and Dement, W.C., Breathing during sleep and wakefulness in the cat, Respir. Physiol., 30 (1977) 265-289.

Perkel, D.H., Gerstein, G.L. and Moore, G.P., Neuronal spike trains and stochastic point processes. II. Simultaneous spike trains, Biophys. J., 7 (1967) 419-440.

Sauerland, E.K. and Harper, R.M., The human tongue during sleep: electromyographic activity of the genioglossus muscle, Exp. Neurol., 51 (1976) 160-170.

Schenk, C.H., Bundle, S.R., Patterson, A.L. and Mahowald, M.W., Rapid eye movement sleep behavior disorder. A treatable parasomnia affecting older adults, JAMA, 257 (1987) 1786-1789.

Sieck, G.C. and Harper, R.M., Pneumotaxic area neuronal discharge during sleep-waking states in the cat, Exp. Neurol., 67 (1980) 79-102.

Snider, R.S. and Niemer, W.T., A Stereotaxic atlas of the cat brain, University of Chicago Press, Chicago, 1970.

Sterman, M.B. and Clemente, C.D., Forebrain mechanisms for the onset of sleep, In: Basic sleep mechanisms, Petre-Quadens O. and Schlag J.D. (Eds.), Academic Press, New York, 1974, pp. 83-97.

Zhang, J.X. and Harper, R.M., A new microdrive for extracellular recording of single neurons using fine wires, Electroencephalogr. Clin. Neurophysiol., 57 (1984) 392-394.

Zhang, J.X., Harper, R.M. and Frysinger, R.C., Respiratory modulation of neuronal discharge in the central nucleus of the amygdala during sleep and waking states, Exp. Neurol., 91 (1986a) 193-207.

Zhang, J.X., Harper, R.M. and Ni, H., Cryogenic blockade of the central nucleus of the amygdala attenuates aversively conditioned blood pressure and respiratory responses, Brain Res., 386 (1986b) 136-145.

What is the Role of the Midbrain Periaqueductal Gray in Respiration and Vocalization?

P.J. Davis* and S.P. Zhang**

*Cumberland College of Health Sciences
and **Department of Anatomy
The University of Sydney, Australia

Introduction

It has been established that the midbrain periaqueductal gray (PAG) and adjacent tegmentum is a critical area, in all vertebrates studied, for the control of vocalization. In previous chapters, Jürgens and Larson have reviewed much of the experimental data including the findings that vocalization can be evoked by electrical or chemical stimulation of the PAG (for review see also Bandler, 1988) and the evidence that destruction of the PAG blocks the ability to elicit *acoustically normal* vocalization from all other central sites.

There are many questions which have not been resolved about the role of PAG in vocalization and respiration.

1. Does the PAG integrate the activity of the respiratory, laryngeal, pharyngeal, palatal and perioral muscle groups associated with vocalization? Or, alternatively, does stimulation of the PAG activate a vocalization integration center located elsewhere in the brainstem?

2. Is the PAG involved with voluntary as well as involuntary (emotional) vocalization?

3. Does the PAG generate patterns of activity for respiratory states other than vocalization?

Does the PAG Integrate Patterns of Muscle Activity for Vocalization?

Microinjections of excitatory amino acids (EAA) into the PAG of freely moving or unanesthetized, decerebrate cats (Bandler, 1982; Bandler and Carrive, 1988; Carrive et al., 1987; Zhang et al., 1990) evoke a number of discrete patterns

The Midbrain Periaqueductal Gray Matter, Edited by A. Depaulis and
R. Bandler, Plenum Press, New York, 1991

of vocalization including howling, hissing, mewing and growling. Vocalizations which can be elicited from the caudal 2/3 of the lateral PAG have a natural acoustic structure and are indistinguishable from spontaneous vocalizations. As well, vocalization is usually evoked as a component of coordinated defense reactions (Bandler and Carrive, 1988).

Jürgens (this volume) considered the specific role of the PAG in vocal control, in relation to other vocalization-eliciting areas in the central nervous system. On the basis of current and previous studies, he proposed that the PAG integrates relevant information from forebrain structures including the anterior cingulate cortex, the amygdala and bed nucleus of stria terminalis, and the hypothalamus. Jürgens also provided new data about the neurochemistry of vocalization pathways from limbic forebrain structures which: (i) highlighted the possible importance of glutamate agonists in the PAG for the production of vocalization; and (ii) suggested that there is a tonic inhibitory GABAergic input onto PAG neurons mediating vocalization.

Jürgens (this volume) included the PAG and the diffuse parvocellular pontine and medullary reticular formation in the caudal brainstem as "final" components of the vocalization pattern generator on the basis that vocalization, albeit unnatural acoustically, can be evoked: (i) by chemical stimulation of the pontine or medullary reticular formation (Jürgens and Richter, 1986); (ii) by electrical stimulation of the parvocellular reticular formation after lesions made in the PAG (Jürgens & Pratt, 1979). A more specific pathway was suggested by Holstege (1989; this volume). His work provided anatomical evidence that the descending pathway for vocalization in the cat includes a projection from the caudal two-thirds of the lateral PAG, to premotor neuronal pools in the nucleus retroambiguus (NRA), which in turn, project to abdominal, intercostal, pharyngeal, palatal, lingual, facial and masticatory motoneurons. A projection from NRA to the nucleus ambiguus was also demonstrated by Holstege (1989), but it is not known whether this includes the laryngeal motoneurons as the later are dispersed in the ventrolateral tegmentum (Davis and Nail, 1984). It is probable that such a projection exists, at least to those laryngeal motoneurons which discharge in the expiratory phase. Future work could clarify this point. It will also be interesting in future studies to compare the muscle activity patterns evoked by excitation of PAG and NRA neurons, as this may reveal the output organizations of the two areas.

Previous studies have shown that distinct types of vocalization are evoked from different parts of the PAG. Thus, in the freely moving or decerebrate cat, hissing and howling were evoked from the intermediate third of the lateral PAG, whereas growling and mewing were evoked from a larger region extending from the intermediate to the caudal part of the lateral PAG (Bandler and Carrive, 1988; Bandler and Depaulis, this volume; Bandler et al., 1991; Zhang et al., 1990). In the rat, sonic vocalization was evoked from the intermediate lateral PAG (Keay et al, 1990); whereas ultrasonic vocalization was evoked predominantly from the caudal lateral PAG (Bandler and Depaulis, this volume). Larson (1988) has also observed a rostro-caudal difference in the types of vocalization evoked

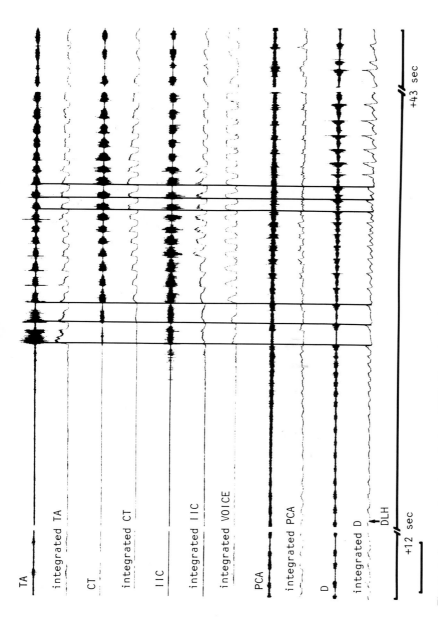

Figure 1. Laryngeal and respiratory EMG (filtered 0.1-5KHz) during part of a sequence of 38 howls evoked by DLH microinjection into the lateral PAG of the unanesthetized, decerebrate cat. To facilitate comparisons of the records, vertical lines marking several voice onsets have been added. Raw and integrated (0.1 s time constant) data from the posterior crico-arytenoid (PCA); thyroarytenoid (TA); cricothyroid (CT); internal intercostal (IIC); and diaphragm (D) muscles, were plotted by a laboratory computer after digital acquisition (2 KHz per channel). The acoustic microphone signal of the voice has been smoothed, and 12 and 43 seconds of the record have been omitted as indicated on the figure. Time calibration: 10 seconds.

by electrical stimulation of the PAG in the awake monkey with rostral sites evoking a greater percentage of clear calls (coos) and stimulation of the caudal PAG eliciting more rough sounding calls (barks).

We have recently observed (Zhang et al., in preparation) that the various types of vocalization such as howling, hissing, mewing and growling which can be evoked in the cat from different parts of the PAG, are each associated with a specific pattern of laryngeal and respiratory electromyographic (EMG) activity. Microinjections of D,L-homocysteic acid (DLH 40nmol in 200nl) were made into sites within the PAG in the unanesthetized, decerebrate cat. EMG activity was recorded from laryngeal and respiratory muscles using either hooked or sewn wire techniques. A typical EMG pattern, resulting from a microinjection within the lateral PAG (A1.6) which evoked a sequence of 38 episodes of howling, each lasting 1-2 seconds, is shown in Figure 1. It can be seen that the first response to DLH injection was often an increase in both the rate and depth of respiration, for several breaths prior the onset of vocalization. This effect is indicated by an increase in both the amplitude and frequency of posterior crico-arytenoid (PCA) and diaphragm (D) EMG. Howling was accompanied by a clear and significant increase in the activity of the thyroarytenoid (TA), cricothyroid (CT), and internal intercostal (IIC) muscles. As well, D and PCA EMG activity continued to show an increased amplitude during the inspiratory phase and some lower level

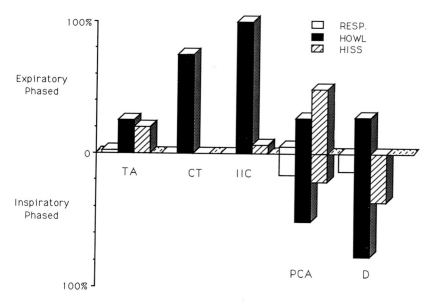

Figure 2. Histogram shows EMG activity for the muscles indicated in Figure 1 in one experimental animal during quiet respiration (resp.), and during howling and hissing evoked by DLH injection. The peak integrated EMG for each activity is expressed as a percentage of the maximum observed for each muscle electrode placement.

of expiratory phased activity. The activity of the expiratory muscles generally increased as the intensity and pitch of the howl increased to its maximum and then slowly decreased.

In the same cat, hissing was evoked by a microinjection in the more rostral part of the intermediate lateral PAG (A3.3). It was associated with a different pattern of muscle recruitment (Fig. 2). Note in particular the strong expiratory-phased PCA activity, the inactivity of CT and the greatly diminished pattern of inspiratory D and PCA activity.

These findings suggest that neurons in different parts of the PAG are involved in the control of distinct types of vocalization. It is important to note also that the effect evoked by PAG microinjection includes the integration of the appropriate depth of inspiratory effort for each specific type of vocalization.

The fundamental question, about how the specific coordinated EMG patterns characteristic of each specific type of vocalization are integrated, remains to be answered. As there exists little evidence of direct projections from the PAG to motoneurons (Holstege, this volume), it seems clear that such PAG integration must be achieved via projections to the relevant premotor neuronal pools. In its simplest form there are two possible ways this PAG-premotor neuronal organization could be structured. First, PAG neurons could have highly collateralized efferent projections which target diverse premotor neuronal pools controlling different groups of predominantly inspiratory- or expiratory-phased muscles. This is a "command neuron" form of organization (Fig. 3). Second, the relevant PAG region may contain largely coextensive populations of neurons with projections to only one specific premotor neuronal pool. This is the "segregated neuron" form of PAG organization (Fig. 3). There are no data currently available to suggest which form of organization predominates.

Does the PAG Mediate Voluntary as Well as Involuntary Vocalization?

The term involuntary vocalization usually includes: (i) sounds characteristic of various respiratory reflexes, e.g., coughing; (ii) emotional forms of human vocalization, e.g., groaning, laughing or crying; and (iii) the vocal component of emotional behaviors in animals, such as the howling and hissing sounds characteristic of the threat display of a cat. These types of vocalization are involuntary in the sense that they are often evoked without "conscious" control: and since they are present in newborns, they are not learned behaviors. At a more volitional level, vocalization is the carrier of human speech and song. Other species can communicate also with a limited degree of voluntary vocalization.

In his chapter, Holstege suggested that there is a clear anatomical dissociation of the voluntary (neocortical) and involuntary/emotional (limbic) innervation of vocalization motoneurons. Based largely on the clinical observations that emotional and voluntary vocalization may be separable, he suggests that emotional vocalization, for example, laughing and crying, may be mediated by the

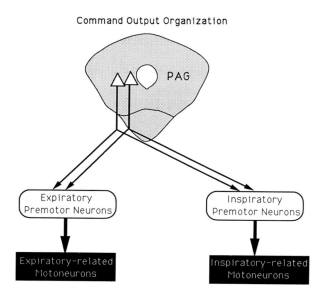

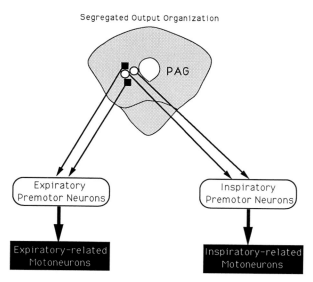

Figure 3. Diagrammatic representation of the two possible organizations which the PAG may use in the control of vocalization. For details see text.

PAG and its output pathway to the NRA (Holstege, 1989). As described by Holstege the impairment or loss of voluntary vocalization, a condition which occurs in pseudo-bulbar palsy as well as a number of other neurologic conditions (Haymaker and Kuhlenbeck, 1984), does not affect emotional laughing or crying. This combination of symptoms has also been described in the early stages of recovery from traumatic midbrain damage which may have initially produced a complete loss of all vocalization - reflexive, emotional and voluntary (Vogel and von Cramon, 1982; Cummings et al., 1983). The problem, however, is that although Holstege's dissociative hypothesis suggests that reflexive or emotional vocalization may be lost while normal voluntary vocalization is retained; clinical evidence suggests that this kind of symptomatology is never observed. That is, when emotional vocalization is impaired, voluntary vocalization at the minimum also lacks normal melodic or prosodic quality. This implies that the so-called "voluntary" and "emotional" vocalization pathways at least partly overlap. Indeed, other researchers have suggested that limbic structures facilitate neocortical (voluntary) vocalization and that lesions within the limbic lobe compromise both emotional and voluntary vocalization (Cummings et al., 1983; Sapir and Aronson, 1985).

Further, there is considerable evidence that the voluntary motor pathway for vocalization includes the anterior cingulate cortex (Jürgens and von Cramon, 1982), a limbic lobe structure which projects extensively to the PAG (Shipley et al., this volume). Other evidence suggests also that the voluntary and involuntary vocalization pathways likely converge, perhaps at the level of PAG. For example, anxiety states or "shock" may be associated with the sudden loss of voluntary vocalization. As well, at times of strong emotional upheaval, voice quality in speech is inevitably altered, usually in a manner which is very easily recognized by others. Finally, with practice the emotional tone of one's voice may be modified or disguised by conscious control, in a manner similar to the ease with which an actor conveys different emotions by facial expressions or body movement.

Does the PAG Generate Respiratory Patterns Specific for Certain States Other than Vocalization?

From the work of Harper and his colleagues (this volume) it is clear that the discharge of neurons in the lateral PAG of the cat is related to respiratory as well as cardiac activity and that these relationships are state-dependent. The data presented indicate, for example, that PAG neurons may be associated with respiratory patterning in some states and not in others (see Fig. 2 and 3, Harper et al., this volume). Harper and his colleagues also suggest that PAG neurons may integrate respiration as a component of more complex physiological patterns, for example the tachypneic responses associated with defense reactions. As well, respiratory patterning is profoundly modified by changes in core temperature. In the awake cat, increasing core temperature by 1 °C can result in a respiratory rate up to four times the basal rate (Harper, personal communication). The anterior hypothalamic/medial preoptic region plays an important role in detecting any change in temperature and the subsequent modification of the respiratory pattern (Jacobson and Squires, 1970; Kastella et al., 1974). In this regard, it is interesting to note the heavy projections from the anterior hypothalamic/medial preoptic region to the PAG (Shipley et al., this volume; Veening et al., this volume).

The respiratory related activity of PAG neurons, and their possible role in affecting respiratory patterning for different behavioral states are important issues which certainly warrant further investigation. It will be important to fully delineate the connections of functionally-defined groups of PAG neurons with respiratory structures, such as the ventral and dorsal respiratory groups of medullary neurons. Because PAG neurons do not project directly to respiratory "pumping" motoneurons in the spinal cord (Holstege, 1989), the role for the PAG as a respiratory pattern generator for certain behavioral states is likely brought about via the major medullary respiratory centers. The projection from PAG via the NRA to abdominal and intercostal motor neurons is one such possible pathway (Holstege, 1989; this volume).

Larson (this volume) reported that neurons in the dorsolateral PAG of the awake monkey show discharge relationships with the respiratory cycle and/or laryngeal muscle activity. He considered PAG neuronal activity to be "early or late stage" with respect to the onset of vocalization and correlated the activity of these cells with inspiratory- and expiratory-phased muscles of the respiratory system and larynx. A possible mediolateral topographical organization of "late-" and "early-stage" neurons in the dorsolateral PAG was also suggested. These neurons, some of which exhibited a greater modulation of their discharge during the initial deep inhalations which precede vocalization, may well be discharging primarily in relation to the ongoing respiratory activity associated with vocalization. This would be similar to the respiratory-related neurons recorded by Harper and his colleagues.

Clearly, vocalization, or the production of sound by the respiratory system and larynx, represents a modified form of respiration. During speech, respiratory patterning is altered, particularly the timing of inspiratory and expiratory durations, as well as the responsiveness to chemoreceptor input (for review see von Euler, 1986; Phillipson et al., 1978). The close association between the neural control of vocalization and respiration is highlighted also by observations that loss or recovery of voluntary vocalization, after traumatic midbrain damage, occurs in parallel with impairment of the ability to modify the timing or depth of the respiratory cycle (Vogel and von Cramon, 1982). Although not studied directly, Vogel and von Cramon reported that those head-injured patients were conscious and they were not artificially ventilated. Presumably then, their control of respiration was adequate to maintain tidal ventilation although they were not able to control voluntary breathing or speech. It has often been assumed that if the respiratory system is capable of providing ventilation, it is capable of generating vocalization. However, the clinical observation mentioned above indicates that is not always the case. That is, some patients with head injuries, although they are able to breath normally, have great difficulty in the voluntary control of speech and breathing. This raises the possibility then, that different neural mechanisms are responsible on the one hand for the maintenance of normal respiratory rhythm, and on the other hand, for the co-ordination of respiratory (and laryngeal) muscle activity in vocalization. In support of this view is the recent suggestion, based on comparisons of the spectral characteristics of the chest wall EMG pattern during speech and breathing tasks, that the primary respiratory pattern generator is not the main source of drive to the respiratory motoneuron pools during speech (Smith and Denny, 1990).

Since the PAG is the only brain stem region known to be essential for generating respiratory patterns specific for vocalization, it is possible that it represents an absolutely critical region for the coordination of breathing during both involuntary (emotional) and voluntary vocalization.

References

Bandler, R., Induction of rage following microinjections of glutamate into midbrain but not hypothalamus of cats, Neurosci. Lett., 30 (1982) 183-188.

Bandler, R., Brain mechanisms of aggression as revealed by electrical and chemical stimulation: Suggestion of a central role for the midbrain periaqueductal grey region, In: Progress in Psychobiology and Physiological Psychology (Vol.13), Epstein A. and Morrison A. (Eds.), Academic Press, New York, 1988, pp. 67-153.

Bandler, R. and Carrive, P., Integrated defence reaction elicited by excitatory amino acid microinjection in the midbrain periaqueductal grey region of the unrestrained cat, Brain Res., 439 (1988) 95-106.

Bandler, R., Carrive, P. and Zhang, S.P., Integration of somatic and autonomic reactions within the midbrain periaqueductal grey: viscerotopic, somatotopic and functional organization, Prog. Brain Res., 87 (1991) 269-305.

Berman, A.L., The brain stem of the cat: A cytoarchitectonic atlas with stereotaxic coordinates, The University of Wisconsin Press, Madison, 1968.

Carrive, P., Dampney, R.A.L. and Bandler, R., Excitation of neurons in a restricted portion of the midbrain periaqueductal grey elicits both behavioral and cardiovascular components of the defence reaction in the unanesthetized decerebrate cat, Neurosci. Lett., 81 (1987) 273-278.

Cummings, J.L., Benson, D.F., Houlihan, J.P. and Gosenfeld, L.F., Mutism: Loss of neocortical and limbic vocalization, J. Nerv. Ment. Dis., 171 (1983) 255-259.

Davis, P. and Nail, B., On the location and size of laryngeal motoneurons in the cat and rabbit, J. Comp. Neurol., 230 (1984) 13-32.

von Euler, C., Brain stem mechanisms for generation and control of breathing pattern, In: Handbook of physiology, Sect. 3, Vol. II, Fishman A. P. (Ed.), Am. J. Soc., Bethesda, 1986, pp. 1-66.

Haymaker, W. and Kuhlenbeck, H., Disorders of the brainstem and its cranial nerves, In: Clinical Neurology, Vol. 3, Baker A.B. and Baker L.H. (Eds.), Harper & Row, Philadelphia, 1984, pp. 1-82.

Holstege, G., An anatomical study on the final common pathway for vocalization in the cat, J. Comp. Neurol., 284 (1989) 242-252.

Jacobson, F.H. and Squires, R.D., Thermoregulatory responses of the cat to preoptic and environmental temperatures, Am. J. Physiol., 218 (1970) 1575-1582.

Jürgens, U. and von Cramon, D., On the role of the anterior cingulate cortex in phonation: A case report, Brain Lang., 15 (1982) 234-248.

Jürgens, U. and Pratt, R., Role of the periaqueductal grey in vocal expression of emotion, Brain Res., 167 (1979) 367-378.

Jürgens, U. and Richter, K., Glutamate-induced vocalization in the squirrel monkey, Brain Res., 373 (1986) 349-358.

Kastella, K.G., Spurgeon, H.A. and Weiss, G.K., Respiratory-related neurons in anterior hypothalamus of the cat, Am. J. Physiol., 227 (1974) 710-713.

Keay, K.A., Depaulis, A., Breakspear, M. and Bandler, R., Pre- and subtentorial periaqueductal grey of the rat mediates different defense responses associated with hypertension, Soc. Neurosci. Abst., 16 (1990) 247.6.

Larson, C.R., Brain mechanisms involved in the control of vocalization, J. Voice, 2 (1988) 301-311.

Phillipson, E. A., McClean, P.A., Sullivan, C. E. and Zamel, N., Interaction of metabolic and behavioral respiratory control during hypercapnia and speech, Am. Rev. Respir. Dis., 117 (1978), 903-909.

Sapir, S. and Aronson, A.E., Aphonia after closed head injury: Aetiologic considerations, Br. J. Dis. Comm., 20 (1985) 289-296.

Smith, A. and Denny, M., High-frequency oscillations as indicators of neural control mechanisms in human respiration, mastication, and speech, J. Neurophysiol., 63 (1990) 745-758.

Vogel, M. and von Cramon, D., Dysphonia after traumatic midbrain damage: A follow-up study, Folia Phoniatr. (Basel), 34 (1982) 150-159.

Zhang, S.P., Bandler, R. and Carrive, P., Flight and immobility evoked by excitatory amino acid microinjection within distinct parts of the subtentorial midbrain periaqueductal grey of the cat, Brain Res., 520 (1990) 73-82.

Zhang, S. P., Davis, P., Carrive, P. and Bandler, R., Integration of respiration and vocalization: I The midbrain periaqueductal grey, in preparation.

Functional Organization of PAG Neurons Controlling Regional Vascular Beds

Pascal Carrive

Department of Anatomy
The University of Sydney, Australia

Introduction

There are recent physiological and anatomical findings which indicate the existence of regional differences within the PAG. These regional differences reflect some degree of functional organization within the PAG. The aim of this chapter is to present certain of these new findings which are related to cardiovascular control and to propose a model of functional organization for PAG control of the circulation. It is hoped that this model will help to understand the organization of other functions represented in the PAG.

The history of cardiovascular studies in the PAG can be briefly summarized as follows. Kabat et al. (1935) were probably the first to report, in the cat, that electrical stimulation of the midbrain central gray (and adjacent tegmentum) evoked rises in blood pressure. They also noted that the rises in blood pressure were associated with other autonomic changes (pupillary dilation, emptying of the bladder). In the 1950's, Swedish cardiovascular physiologists (Eliasson et al., 1954; Lindgren, 1955; Lindgren et al., 1956) also explored this area, and for the first time measured regional blood flows. They found that the electrically-evoked rises in blood pressure were associated with a very specific pattern of blood flow changes: vasodilation in the muscles of the hindlimb accompanied by vasoconstriction in skin and viscera. The significance of these cardiovascular changes was not understood at the time.

Kabat et al. (1935) also noted that the rise in blood pressure was accompanied by clear signs of behavioral arousal in their lightly anesthetized preparation. But it was Abrahams et al. in 1960 who first demonstrated that the same sites at which electrical stimulation produced a rise in blood pressure and hindlimb vasodilation also produced defence reactions in the freely moving cat. Their

The Midbrain Periaqueductal Gray Matter, Edited by A. Depaulis and
R. Bandler, Plenum Press, New York, 1991

study suggested that the cardiovascular changes evoked from the PAG region were well adapted to the defence reaction: redistribution of blood flow to skeletal muscles, at the expense of the viscera, would seem ideal to meet the metabolic demands of the strenuous exercise of a fight or flight. Because only this single pattern of blood flow changes could be identified in the anesthetized preparation, it was thought to be characteristic of all stages of the defence reaction, from the early alerting stage and onwards (Abrahams et al., 1964).

In the work to be presented here, the role of PAG in control of blood pressure and regional vasculature is re-examined in more detail. There were two major methodological innovations in this study. First, in order to selectively excite PAG neurons but not axons of passage, electrical stimulation was replaced by microinjection of excitatory amino acids (Goodchild et al., 1982; Lipski et al., 1988). Second, to circumvent the problem of anesthesia and to allow a simultaneous observation of cardiovascular and somatomotor changes, experiments were carried out in acute, unanesthetized, precollicular decerebrate preparations instead of intact anesthetized animals. It had been shown previously that reflex activation of the blood flow pattern specific for defence was not affected by high decerebration (Abrahams et al., 1960; Schramm and Bignall, 1971).

Methods

Physiological experiments
DECEREBRATE PREPARATION. Experiments were carried out in a total of 62 decerebrated cats (either sex, 2.5-4.5 kg). Precollicular decerebration was performed under deep halothane anesthesia using a suction-diathermy technique (see Carrive and Bandler, 1991a). The hypothalamus was usually left intact as described in Abrahams et al. (1960). We have observed, however, that removal of the hypothalamus does not modify PAG evoked responses. After decerebration, the bony tentorium which covers the caudal third of the midbrain was removed and the anesthesia was discontinued. Within 30 to 60 minutes the effect of halothane dissipated. Respiration returned to normal and mean blood pressure stabilized (usually around 120 mm Hg).

PHYSIOLOGICAL PARAMETERS. Pulsatile blood pressure was measured via a catheter in the right femoral artery. Mean blood pressure and heart rate were derived from this signal. In most experiments, an electromagnetic flow probe, placed on the left external iliac artery recorded blood flow to the hindlimb. In some experiments, an additional electromagnetic flow probe was placed around the left renal artery or around the left common carotid artery. Conductances were calculated either manually or with a custom built divider (mean conductance = mean flow / mean pressure). In 40 experiments, the decerebrate preparation was kept unparalyzed to observe the somatomotor effects accompanying the PAG-evoked cardiovascular changes. In the remaining 22 experiments, the preparation was paralyzed and artificially ventilated.

MICROINJECTION PROCEDURE. A stainless steel cannula (o.d. 200 μm, insulated except for 0.3 mm at the tip) was inserted vertically into the midbrain and microinjections of sodium D,L-homocysteic acid (DLH, 40 nmoles in 200 nl) were made into sites within and surrounding the PAG. Injections (5-15 sec in duration) made along a vertical track were separated by a period of at least 10 minutes. Injections were made at 1 to 6 sites in each track, and 2 to 6 tracks were usually explored in each experiment.

HISTOLOGY. At the completion of each experiment the cat was perfused intracardially with physiological saline followed by a solution of 4% formaldehyde. The midbrain was removed and 50 μm sections were cut. The tracks were reconstructed from the sections (counterstained with Cresyl Violet) and the coordinates of each site of injection were determined with reference to Berman's atlas of the cat brainstem (Berman, 1968).

Anatomical Experiments

Experiments were carried out to trace the anatomical connections of the PAG with a vasopressor center located in the rostral ventrolateral medulla.

In a first series of experiments the efferent projections of PAG neurons were studied with the anterograde tracer wheatgerm agglutinin-horseradish peroxidase (WGA-HRP) (Carrive et al., 1988). Briefly, WGA-HRP (50-150 nl, 1.5% solution) was injected into the PAG of 3 intact animals via a chronically implanted cannula. Two days later, the animal was anesthetized, perfused and sections of the lower brainstem examined for anterograde labeling.

An additional series of experiments was carried out to identify more precisely the location of PAG neurons projecting to this vasopressor center (Carrive et al., 1988; 1989b). In 14 intact anesthetized (α-chloralose 70 mg/kg) cats, the surface of the medulla was exposed using a ventral approach. Each animal was instrumented for the recording of blood pressure. Iliac and renal flows were also recorded in 5 experiments. The vasopressor center of the rostral ventrolateral medulla was localized physiologically with microinjections of sodium glutamate (GLU, 5-25 nmoles in 10-50 nl made through a glass micropipette), and microinjections of the retrograde tracers WGA-HRP (10-50 nl, 1.3-1.5%) or fluorescent rhodamine latex beads (30-100 nl, undiluted solution) were then made at these physiologically identified sites. The animal was kept anesthetized for a further 24 hours and then perfused. Sections of the midbrain (30-50 μm) were examined for retrograde labeling within the midbrain. Labeled cells were plotted on standard sections taken from Berman's atlas of the cat brainstem (Berman, 1968).

PAG-Evoked Changes in Arterial Blood Pressure

DLH microinjection in the PAG evoked clear changes in blood pressure (Fig. 1). The effect usually started within a few seconds of the end of the injection and lasted for aproximately 1 to 2 minutes. Dependent on the site of injection, either an increase (Fig. 1 A, B) (Carrive et al., 1987; 1989a) or a decrease in blood pressure (Fig. 1 C, D) (Carrive and Bandler, 1991a) was observed.

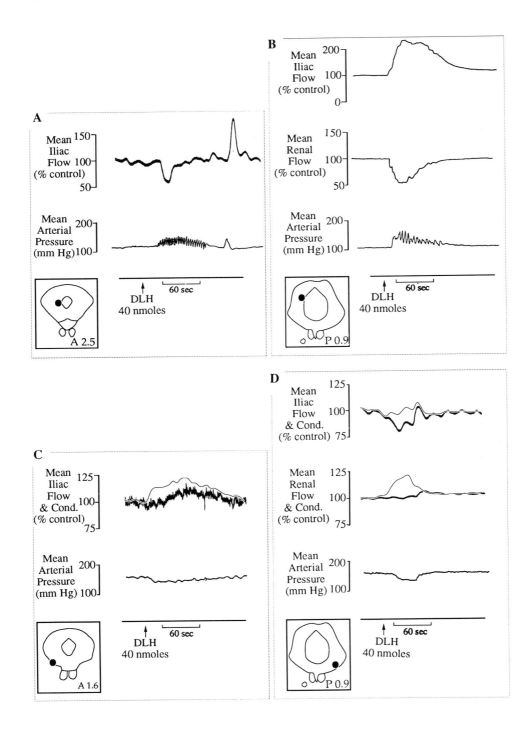

The two types of effects were evoked from distinct regions of the PAG. As can be seen on Figure 2, sites at which DLH evoked an increase in blood pressure (hypertensive sites) were located in a restricted area of the PAG, just lateral to the aqueduct, whereas sites at which DLH evoked a decrease in blood pressure (hypotensive sites) were located in an area ventral and lateral to the aqueduct. This dorsoventral gradient was very clear with vertical track explorations: hypotensive effects were always evoked at sites ventral to hypertensive sites.

Figure 2 shows also that hypertensive and hypotensive sites were distributed along a longitudinal axis. Hypertensive sites formed a longitudinal column lying in the caudal 2/3 of the PAG (A3.3 - P1.5) just lateral to the aqueduct. Hypotensive sites formed another longitudinal column lying in the caudal half of the PAG (A2.5 - P1.5), in the region ventrolateral to the aqueduct. On the basis of these observations one can postulate the existence of two functional columns: a hypertensive column and a hypotensive column (Fig. 3).

Unlike DLH, electrical stimulation in the PAG almost always evoked rises in blood pressure. The effects of both types of stimulation were tested at nearly all sites represented on Figure 2. Electrical stimulation evoked rises in blood pressure from the rostral third of the PAG (A5.2-A3.3) and from the dorsolateral region of the intermediate PAG. DLH had no effect in these two regions. Rises in blood pressure were evoked also from the hypotensive area of the ventrolateral PAG (i.e., at sites at which DLH evoked hypotensive effects). These observations indicate that the effects of excitatory amino acids were more specific than those evoked by electrical stimulation. The functional heterogeneity revealed by excitatory amino acid could not have been detected with electrical stimulation.

Figure 1 (opposite page). Typical cardiovascular effects evoked by DLH microinjections in the PAG of the unanesthetized, decerebrate cat. Panels A and B: hypertensive effects evoked from the lateral part of the PAG. Panels C and D: hypotensive effects evoked from the ventrolateral part of the PAG. Different patterns of blood flow changes were evoked from the lateral and ventrolateral PAG. Panel A: iliac vasoconstriction characterized the hypertensive effect evoked from intermediate, lateral PAG (pretentorial half of the hypertensive column). Panel B: renal vasoconstriction (associated with iliac vasodilation) characterized the hypertensive effect evoked from caudal, lateral PAG (subtentorial half of the hypertensive column). Panel C: moderate increase in iliac conductance characterized the hypotensive effect evoked from intermediate, ventrolateral PAG (pretentorial half of the hypotensive column). Panel D: moderate increase in renal conductances (but not iliac conductance) characterized the hypotensive pattern evoked from caudal, lateral PAG (subtentorial half of the hypotensive column). In panels C and D, the thin lines plotted over the mean flow traces represent the calculated mean conductances (mean flow/mean pressure). The locations of the centers of each microinjections are indicated on coronal sections of the PAG. The arrow indicates the start of a DLH microinjection (40 nmoles).

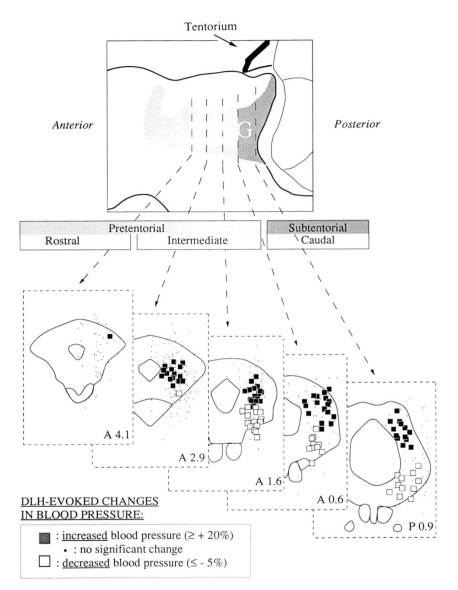

Figure 2. Top: schematic parasagittal view of the midbrain showing the rostro-caudal subdivisions of the PAG. The pretentorial and subtentorial portions are defined with respect to the bony tentorium. The pretentorial portion corresponds to the rostral and intermediate third of the PAG. The subtentorial portion corresponds to the caudal third of the PAG. Bottom: a series of 5 representative coronal sections drawn with reference to the atlas of Berman (1968) showing the location of the centers of the sites at which arterial blood pressure changes were recorded following microinjection of DLH, in the decerebrate cat. Sites at which hypertensive, hypotensive or no significant effects were evoked are represented with different symbols.

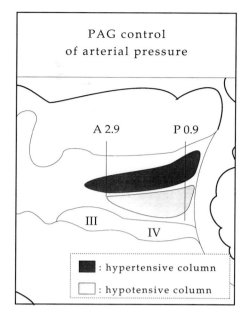

Figure 3. Parasagittal schematic representation of the PAG hypertensive and hypotensive columns. III, oculomotor nucleus; IV: trochlear nucleus.

PAG-evoked Changes in Regional Vasoconstrictor Tone

Arterial blood pressure depends largely on two parameters: cardiac output and peripheral vascular resistance. Thus, increases in regional vasoconstrictor tone contribute to rises in blood pressure and decreases in regional vasoconstrictor tone contribute to falls in blood pressure. The next step in our investigation was to determine if significant changes in regional vasoconstrictor tone were associated with the PAG-evoked hypertensive and hypotensive effects. Two vascular beds were studied: the hindlimb (external iliac artery) and the kidney (renal artery).

Hypertensive effects are associated with different patterns of increased regional vasoconstrictor tone

The most significant finding was that hypertensive effects were associated with very different patterns of changes in hindlimb blood flow. At some hypertensive sites, there was a marked reduction in iliac flow (up to 50% decrease in conductance) (Fig. 1 A) (Carrive et al., 1987), while at other hypertensive sites, there was instead a marked increase in iliac flow (up to 280% increase in conductance) (Fig. 1 B) (Carrive et al., 1989a). Experiments in which renal blood flow was measured then revealed that the hypertensive effect characterized by iliac vasodilation was associated with a marked decrease in renal flow (up to 60% decrease in conductance; mean, 39% ± 4%; n=12) (Fig. 1 B) (Carrive et al., 1989a), whereas the hypertensive effect characterized by iliac vasoconstriction was associated with only a small to moderate decrease in renal flow and conductance (mean, 22% ± 3.5%; n=13). The two groups were significantly different (p< 0.001).

Thus, at least two patterns of hypertensive effects could be evoked from the lateral PAG: one characterized by a marked increase in hindlimb vasoconstrictor tone, the other characterized by a marked increase in renal vasoconstrictor tone. As well, the two types of hypertensive effects were evoked from different portions of the hypertensive column. As can be seen on Figure 4, iliac vasoconstrictor sites were found in the rostral or pretentorial half of the hypertensive column whereas renal vasoconstrictor sites were located in the caudal or subtentorial half of the hypertensive column (except for one site).

The results demonstrate that PAG-evoked hypertensive effects are associated with a selective rather than a diffuse increase in regional vasoconstrictor tone. Furthermore, there is evidence of a crude viscerotopic organization within the PAG hypertensive column, in the sense that increased vasoconstrictor tone in the hindlimb or in the renal vasculature were evoked from pools of lateral PAG neurons located at different rostrocaudal levels.

Hypotensive effects are associated with different patterns of decreased regional vasoconstrictor tone

As with the hypertensive effects, it was found that hypotensive effects were associated with different patterns of changes in hindlimb blood flow (Carrive and Bandler, 1991a). Thus, at some hypotensive sites, there was a small increase in iliac blood flow (Fig. 1 C), while at other hypotensive sites there was a small decrease in iliac blood flow (Fig. 1 D). In the first case, the flow change reflected a real increase in iliac conductance (between 10 and 30%) (Fig. 1 C, the thin line plotted over the mean flow trace represents mean conductance), but in the second case there was no change in conductance (the decrease in iliac blood flow followed the decrease in arterial blood pressure) (Fig. 1 D). Experiments in which renal blood flow was measured then revealed that the hypotensive effect characterized by no significant change in iliac conductance was in fact associated with a small to moderate increase in renal conductance (between 10 and 30%; mean, 12% ± 2%; n=11) (Fig. 1 D) (Carrive and Bandler, 1991a). In contrast, the hypotensive effect characterized by increased iliac conductance was not associated with a significant change in renal conductance (mean, 2% ± 2%; n=8).

Thus, two types of hypotensive effects could be evoked from the ventrolateral PAG: one characterized by a decrease in hindlimb vasoconstrictor tone, the other characterized by a decrease in renal vasoconstrictor tone. Significantly, the two types of hypotensive effects were evoked from different portions of the

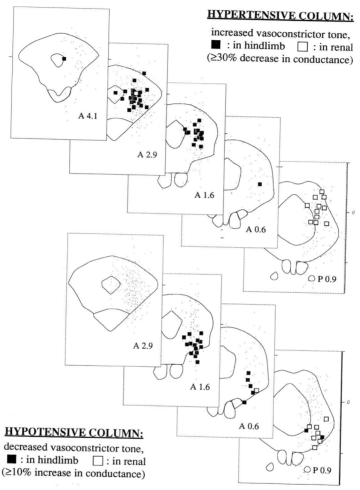

HYPERTENSIVE COLUMN:

increased vasoconstrictor tone,
■ : in hindlimb   □ : in renal
(≥30% decrease in conductance)

A 4.1

A 2.9

A 1.6

A 0.6

P 0.9

A 2.9

A 1.6

A 0.6

P 0.9

HYPOTENSIVE COLUMN:

decreased vasoconstrictor tone,
■ : in hindlimb   □ : in renal
(≥10% increase in conductance)

Figure 4. Series of representative coronal sections drawn with reference to the atlas of Berman (1968) showing the location of the sites at which the different patterns of hypertensive and hypotensive effects were evoked by microinjection of DLH in the decerebrate cat. Top: hypertensive sites characterized by increased vasoconstrictor tone in the hindlimb or by increased vasoconstrictor tone in the renal vasculature. Bottom: hypotensive sites characterized by decreased vasoconstrictor tone in the hindlimb or by decreased vasoconstrictor tone in the renal vasculature.

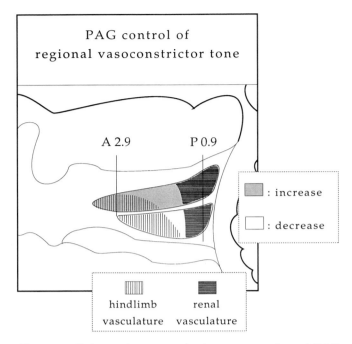

Figure 5. Schematic parasagittal representation of PAG control of regional vasoconstrictor tone. The hypertensive and hypotensive columns contain neurons mediating respectively increase and decrease in vasoconstrictor tone. The separate representations of hindlimb and renal vasculature in the 2 columns suggest the existence of a viscerotopic organization in the PAG.

hypotensive column. As can be seen in Figure 4, sites characterized by a decrease in hindlimb vasoconstrictor tone were located predominantly in the rostral or pretentorial half of the hypotensive column, whereas sites characterized by a decrease in renal vasoconstrictor tone were located predominantly in the caudal or subtentorial half of the hypotensive column. In other words, the hypotensive column possesses a viscerotopic organization which is seemingly identical to that of the hypertensive column.

In conclusion, a preliminary model of the PAG control of regional vasoconstrictor tone can be proposed (Fig. 5). According to this model, PAG neurons controlling regional vasoconstrictor tone are organized along two axis: a rostrocaudal axis representing the regional vasculature (hindlimb rostrally, renal caudally), and a dorsoventral axis representing the direction of change in vasoconstrictor tone (increase dorsally, decrease ventrally).

A Descending Pathway Mediating PAG-evoked Changes in Blood Pressure and Regional Vasoconstrictor Tone

Blood pressure and vasoconstrictor tone depend largely on the sympathetic outflow to the heart and blood vessels. It is well known also that the last central neurons controlling sympathetic outflow are the preganglionic sympathetic neurons located in the intermediolateral column (IML) of the thoracic and upper lumbar spinal cord. These neurons must therefore play a major role in the mediation of PAG-evoked changes in blood pressure and vasoconstrictor tone. There is no evidence, however, for a direct and significant projection from PAG to IML (see Holstege, this volume). This indicates that PAG output reaches the IML indirectly, presumably via synapse(s) in the pons and/or the medulla.

A direct projection from PAG to the rostral ventrolateral medulla pressor region

The PAG efferent connections to the pons and medulla were studied initially in a series of experiments in which large injections of wheat germ agglutinin-horseradish peroxidase (WGA-HRP) were made into the PAG (Bandler, 1988; Carrive et al., 1988). The major descending projection consisted of predominantly ipsilateral fibers which passed laterally in the dorsolateral midbrain tegmentum and then through the cuneiform region to the parabrachial region. In the pons, the fibers continued ventralward to the level of the superior olivary nucleus and then shifted to a more medial position, as they passed into the ventral medulla. The ipsilateral medial and paramedian part of the rostral ventral medulla were particularly well labeled (Fig. 6). The lateral part of the ventral medulla was also labeled: caudal to the facial nucleus, fibers continued to sweep laterally, ventral to the retrofacial nucleus first, then around the ambiguus and retroambiguus nuclei, all the way to the caudal medulla (Fig. 6). Some fibers shifted dorsomedially and terminated in the region of the solitary nuclear complex. Similar finding have been reported by others in studies using tritiated leucine (Holstege, this volume) and the lectin *Phaseolus vulgaris* leuco-agglutinin (Luiten et al., 1987; Van Bockstaele, personal communication).

Many of the regions innervated by the PAG are known to play a role in cardiovascular control (e.g., parabrachial region, rostral ventrolateral medulla, caudal ventrolateral medulla, nucleus of the tractus solitarius and perhaps ventromedial medulla), and each of them may be implicated in the coordination and mediation of the PAG-evoked cardiovascular patterns. One of these regions, however, could be of particular importance. The rostral ventrolateral medulla (RVLM) has received considerable attention in recent years because of its major role in the control and maintenance of resting arterial blood pressure (Ciriello et al., 1986; Calaresu and Yardley, 1988; Dampney et al., 1985; McAllen et al., 1987; Reis et al., 1988; Ross et al., 1984b). The RVLM contains bulbospinal neurons projecting to the IML (Ciriello et al., 1986; Dampney et al., 1987a; Ross et al., 1984a; Strack et al., 1989a; 1989b), and in this sense, it can be considered as a pre-sympathetic center regulating the spinal sympathetic outflow. The RVLM-IML path-

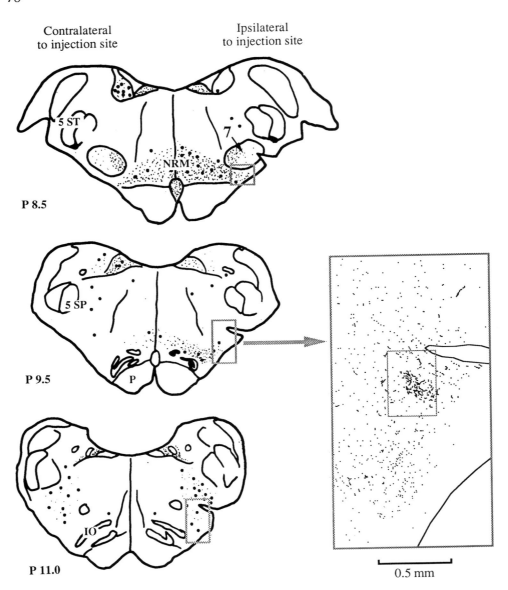

Figure 6. Anterograde labeling in the rostral medulla following a large injection of WGA-HRP (100 nl, 1.5% solution) in the PAG. The three coronal sections show the regions that contained the densest anterograde labeling (fine dots). It can be seen that the medial and paramedian parts of the rostral ventral medulla were particularly well labeled. Although not as dense, labeling was also present more laterally in the region that lies ventral and caudal to the facial nucleus. An enlarged view of this region shows some relatively dense labeling in and around a group of cell which has been identified as the subretrofacial nucleus. See next figure for an enlarged view of this area. IO: inferior olive; 5 SP: spinal trigeminal nucleus, 5 ST: spinal tract of the trigeminal nucleus; 7: facial nucleus. Adapted from Bandler, 1988 and Carrive et al., 1988.

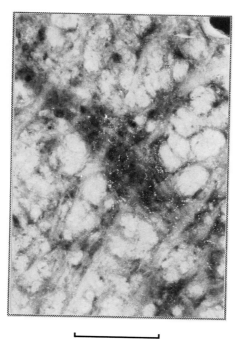

0.1 mm

Figure 7. Photomicrograph of the anterograde labeling in the ipsilateral subretrofacial nucleus. See previous Figure for
details. Reprinted with permission from
Carrive et al. (1988) Brain Research.

way is also thought to be a final common pathway from many brain regions to
the spinal sympathetic outflow, and the PAG may be one of them (Hilton et al.,
1983; Lovick, 1985).

Precise physiological and anatomical studies carried out in the cat, suggest
that the RVLM zone which specifically controls sympathetic outflow to the cardiovascular effectors (i.e., heart, blood vessels and adrenal medulla) is a 2 mm
long longitudinal column located below and behind the facial nucleus (Dampney
et al., 1987b; McAllen, 1986a; 1986b; McAllen et al., 1987). In the cat, this column
can be identified in Nissl stained material, and because of its location, it has been
called the subretrofacial (SRF) nucleus (Dampney et al., 1987b; McAllen, 1986c;
McAllen et al., 1987). Close examination of the PAG projections to the RVLM
revealed the existence of a quite specific projection to the SRF nucleus (Fig. 6 and
Fig. 7) (Carrive et al., 1988). It should be noted however, that the labeling in the
SRF was relatively moderate in comparison with labeling in other regions such as
the nucleus raphe magnus or the caudal ventrolateral medulla. The labeling in

the SRF nucleus was most marked ipsilaterally, although contralateral labeling was also found.

A series of retrograde tracing experiments was then carried out to identify precisely which regions of the PAG projected to the SRF pressor column (Carrive et al., 1988). In this series of experiments, WGA-HRP was injected at physiologically identified sites in the SRF region, i.e., at sites at which an injection of glutamate (5 nmoles, 10 nl) produced a rise in blood pressure (>30%). Figure 8 shows a typical experiment with a relatively large injection of WGA-HRP. Although the injection was centered in SRF, there was spread into the retrofacial nucleus. As can be seen, retrogradely labeled cells (Fig. 9) were found inside the PAG but also ventral to it, in the tegmentum rostrally (Fig. 8). Note that most of the labeling in this tegmental region (A5.2-A4.1) was contralateral to the site of injection and was in continuity with a group of supra-oculomotor cells also located on the contralateral side (A2.9). In contrast, most of the labeling in the PAG was found on the side ipsilateral to the site of injection. This is consistent with anterograde labeling studies which have shown the projection from PAG to SRF to be mainly ipsilateral (see above). The contralateral labeling in the tegmental and supra-oculomotor regions was probably due to spread outside the SRF, perhaps into the retrofacial area. In studies with more restricted injections into the SRF (fluorescent microspheres) there was significantly less labeling within these contralateral regions.

Within the PAG, two major columns of labeled neurons were identified: (i) a dorsomedial column extending along the whole length of the PAG (although particularly prominent in its middle third: A2.9-A0.6), and (ii) a large lateral/ventrolateral column within the middle and caudal thirds of the PAG (A2.9-P0.9). Note that this lateral/ventrolateral column expands dorsoventrally with the widening of the midbrain aqueduct.

There is a striking similarity between the lateral/ventrolateral column of neurons labeled from the SRF pressor region and the hypertensive and hypotensive columns defined physiologically in the unanesthetized decerebrate preparation (Fig. 2 and Fig. 8). It suggests that this direct projection from the PAG to the SRF pressor region could well play a significant role in the mediation of PAG-evoked changes in blood pressure. It is conceivable that the bulbospinal neurons of the SRF pressor region receive excitatory signals from the lateral PAG neurons and inhibitory signals from the ventrolateral PAG neurons (Carrive and Bandler, 1991a). Electrophysiological observations made by Lovick and collaborators support this hypothesis: activation of lateral PAG cells evokes a facilitation of SRF bulbospinal neurons while activation of ventrolateral PAG cells evokes an inhibition of these same neurons (Lovick, this volume).

Little is known about the function of neurons in the dorsomedial column. It has been reported recently that kainic acid injections in the dorsomedial PAG of the intact anesthetized rat evoked rises in blood pressure and heart rate (Jones et al., 1990). In the decerebrate cat, DLH microinjection in this area have this far failed to evoke consistent and significant changes in blood pressure, but very few sites have been studied (Fig. 2).

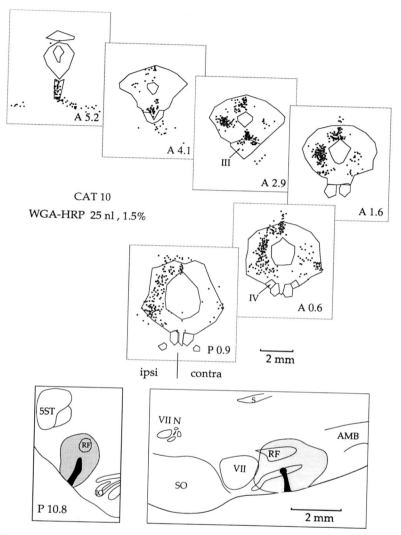

Figure 8. Series of representative coronal sections of the PAG showing the distribution of retrogradely labeled neurons following injection of WGA-HRP in the subretrofacial pressor region. Each representative section is a reconstruction from 10 consecutive 50 μm sections. The site of injection was identified as a site at which microinjection of sodium glutamate (5 nmoles, 10 nl) evoked a large rise in arterial blood pressure (+ 50 mm Hg). The drawing in the lower left indicates on a coronal section the center of injection and the mediolateral and dorsoventral extent of the apparent spread of WGA-HRP. The drawing in the lower right is a parasagittal reconstruction of the site of injection showing the rostrocaudal extent of the apparent spread of WGA-HRP. AMB: nucleus ambiguus; IO: inferior olive; RF: retrofacial nucleus; S: solitary tract; SO: superior olive; SRF: subretro-facial nucleus; VII: facial nucleus; VIIN: facial nerve; 5 ST: spinal tract of the trigeminal nucleus; III, oculomotor nucleus; IV, trochlear nucleus.

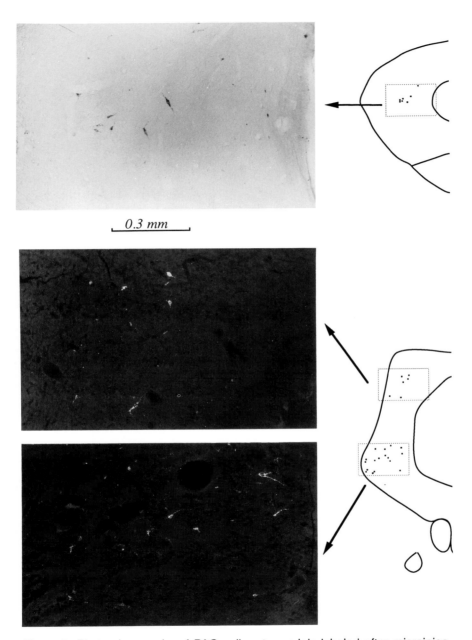

_0.3 mm_

Figure 9. Photomicrographs of PAG cells retrogradely labeled after microinjection of WGA-HRP (upper panel) or rhodamine latex beads (middle and lower panels) in the subretrofacial pressor region. Upper panel: labeling in the intermediate lateral PAG. Middle and lower panels, labeling in the lateral and ventrolateral caudal PAG.

The PAG-SRF projection is viscerotopically organized

There is good evidence, at least in the cat, that cardiovascular effectors are topographically represented within the SRF pressor region (Dampney and McAllen, 1988; Lovick, 1987; McAllen and Dampney, 1990; Carrive et al., 1989b). Lovick (1987) was the first to report the existence of a viscerotopic organization within the SRF pressor region: she found that microinjections of DLH in the caudal part of the SRF evoked a rise in blood pressure associated with vasoconstriction in the external iliac artery, while microinjections in the rostral part of the SRF evoked a rise in blood pressure associated with a vasoconstriction in the renal artery (Fig. 10). The renal vasoconstriction was often associated with a late-occurring iliac vasodilation that could be totally abolished by the intravenous injection of a β-adrenergic antagonist, suggesting that the adrenal medulla is also represented in the rostral part of the SRF pressor region (Fig. 10). Our observations are in total agreement with Lovick's report (Fig. 10).

An additional series of retrograde experiments were carried out (Carrive et al., 1989b). In these experiments, fluorescent latex microspheres were microinjected at physiologically identified sites in the iliac constrictor or renal constrictor parts of the SRF pressor region (Fig. 10) and the pattern of retrograde labeling within the PAG was examined. Fluorescent latex microspheres were chosen because of their very limited spread at the site of injection (see Fig. 10, lower panel).

It can be seen on the three-dimensional reconstruction of the brainstem in Figure 11, that retrograde labeling within the PAG was restricted to the three longitudinal columns previously described (dorsomedial, lateral, ventrolateral). The most significant finding of these experiments, however, was the rostrocaudal distribution, within the lateral and ventrolateral PAG, of cells projecting to the iliac and renal constrictor parts of the SRF. Specifically, cells projecting to the iliac constrictor part of the SRF (Fig. 12, green dots) were located rostrally (pretentorially) in the lateral and ventrolateral PAG whereas cells projecting to the renal constrictor part of the SRF (Fig. 12, red dots) were found caudally (subtentorially) in the lateral and ventrolateral PAG. Labeled cells in the dorsomedial PAG did not have any apparent viscerotopic distribution.

As summarized on Figure 13, there was a striking match between the distribution of lateral and ventrolateral PAG cells projecting to the iliac and renal constrictor parts of the SRF and the physiologically identified PAG regions from which increases and decreases in iliac or renal vasoconstrictor tone were evoked in the unanesthetized decerebrate preparation. This observation adds considerable weight to the physiological evidence that a crude viscerotopic organization exist within both the lateral and the ventrolateral PAG.

The organization of the PAG-SRF projections is schematically represented in Figure 14. These projections may well play an important role in the mediation of PAG-evoked changes in blood pressure and regional vasoconstrictor tone. This is not to say, however, that the PAG-SRF projection is the only pathway. Clearly, the PAG project to other parts of the pontomedullary cardiovascular network, i.e., parabrachial nucleus, caudal ventrolateral medulla, solitary complex

Parasagittal section of Rostral Ventrolateral medulla.

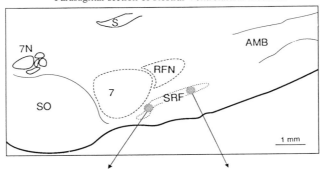

Cardiovascular effects evoked by Glutamate injection into:
the ROSTRAL SRF   the CAUDAL SRF

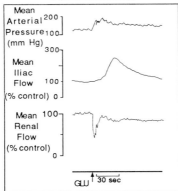   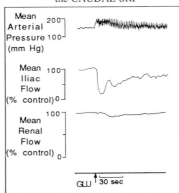

Retrograde fluorescent tracer injection sites in the rostral and caudal SRF.

region and ventromedial medulla (see also Holstege; Lovick, this volume). Nonetheless, of all these possible polysynaptic routes, the PAG-SRF-IML pathway is one of the shortest, and the only one known to be viscerotopically organized.

Finally and most importantly, these anatomical experiments show clearly that the organization of the PAG-SRF projection is remarkably consistent with the model of PAG control of vasoconstrictor tone that was drawn from physiological observations, i.e., that the lateral and ventrolateral PAG contain viscerotopically organized longitudinal columns.

Another Functional Column in the Lateral PAG Controls Regional Vasodilation

The model of PAG control of regional vasculature that has been sketched thus far, is based only on two vascular beds, the renal and the hindlimb. In an additional series of physiological experiments, blood flow was measured in another vascular bed, the extracranial vasculature of the head. This series of experiments were carried out in 7 decerebrated and paralyzed cats (Carrive and Bandler, 1991b). One flow probe was placed around the iliac artery, as in the previous series of experiments. A second flow probe was placed around the common carotid artery. Since most of the intracranial vasculature is removed in the decerebrate preparation, the common carotid artery supplies almost exclusively the extracranial vasculature of the head.

Figure 15 shows that different patterns of changes in extracranial flow were evoked from the different portions of the lateral PAG: at some site, DLH evoked a rapid extracranial vasodilation; at other sites DLH evoked an extracranial vasoconstriction. These changes were accompanied by the characteristic patterns of iliac conductance changes that were previously described for the lateral PAG, that is, iliac vasoconstriction in the intermediate portion and iliac vasodilation in the caudal portion.

Figure 16 shows the location of the sites at which extracranial vasodilation or vasoconstriction was evoked in the lateral PAG. First of all, it shows that the sites at which extracranial vasoconstriction was evoked are preferentially located

Figure 10 (opposite page). Upper panel: drawing of a parasagittal section (Lat 3.7) of the rostral ventrolateral medulla. Injection sites in the rostral and caudal parts of the subretrofacial (SRF) nucleus are indicated. Middle panel: patterns of cardiovascular changes evoked from the rostral and caudal SRF by microinjections of sodium glutamate (GLU, 11 nmoles, 22 nl). Lower panel: Photomicrograph of a parasagittal section of the rostral ventrolateral medulla showing injection sites of retrograde fluorescent tracers in the rostral (rhodamine-coated latex microspheres) and caudal (fluorescein-coated latex microspheres) parts of the SRF pressor region. At each site a glutamate microinjection had previously elicited the cardiovascular pattern characteristic of the rostral and caudal parts of the SRF. AMB: nucleus ambiguus; RFN: retrofacial nucleus; S: solitary tract; SO: superior olive; SRF: subretrofacial nucleus; 7: facial nucleus; 7N: facial nerve. Reprinted with permission from Carrive et al. (1989b) Brain Research.

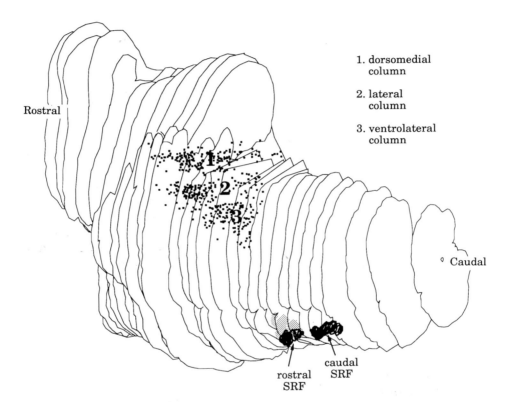

1. dorsomedial
   column

2. lateral
   column

3. ventrolateral
   column

Rostral

° Caudal

rostral   caudal
SRF   SRF

Figure 11. Computer assisted three-dimensional reconstruction of the distribution of labeled neurons in the PAG following injection of fluorescent
microspheres into the rostral and caudal parts of the SRF pressor nucleus
in the rostral ventrolateral medulla (reconstructed from two separate
experiments). The labeled midbrain PAG neurons form 3 longitudinal
columns. From dorsal to ventral: a dorsomedial column (1); a lateral
column (2) which corresponds to the PAG hypertensive area; and a ventrolateral column (3) which corresponds to the PAG hypotensive area. The
brainstem is viewed from below, behind and rotated 35° to the right.
Modified from Carrive and Bandler (1991) Brain Research.

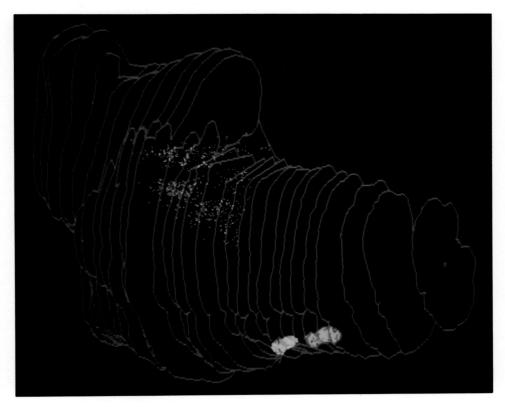

Figure 12. Color version of previous figure. It can be seen that cells located in the rostral (pretentorial) parts of the lateral and ventrolateral columns (green dots) are labeled after tracer injection into the caudal part of the SRF nucleus (green patch) and that cells located in the caudal (subtentorial) parts of the lateral and ventrolateral columns (red dots) are labeled after tracer injection into the rostral part of the SRF nucleus (red patch). Reprinted with permission from Bandler et al. (1991) Progress in Brain Research.

RETROGRADE LABELLING FROM SRF:

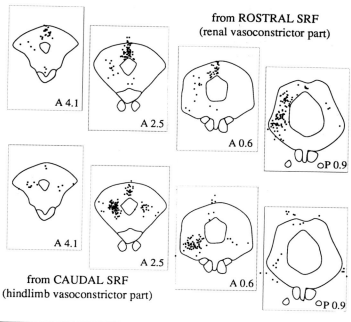

from ROSTRAL SRF
(renal vasoconstrictor part)

A 4.1

A 2.5

A 0.6

P 0.9

A 4.1

A 2.5

A 0.6

P 0.9

from CAUDAL SRF
(hindlimb vasoconstrictor part)

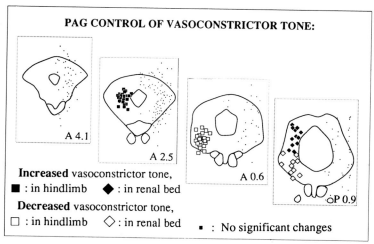

PAG CONTROL OF VASOCONSTRICTOR TONE:

A 4.1

A 2.5

A 0.6

P 0.9

Increased vasoconstrictor tone,
■ : in hindlimb ◆ : in renal bed
Decreased vasoconstrictor tone,
□ : in hindlimb ◇ : in renal bed

▪ : No significant changes

Figure 13. First and second row: retrograde labelling in the PAG following injection of rhodamine-labeled beads into the rostral or caudal SRF. Each section is a reconstruction from 15 consecutive 30 μm sections. Ipsilateral is to the left, contralateral is to the right. Third row: location of the centers of the sites at which significant changes in hindlimb and renal vasoconstrictor tone were evoked by DLH microinjection in the unanesthetized decerebrate cat. Note the similarity between the distribution of lateral and ventrolateral PAG cells projecting to the renal and iliac constrictor parts of the SRF and the physiologically identified PAG regions from which increases and decreases in renal or iliac vasoconstrictor tone were evoked.

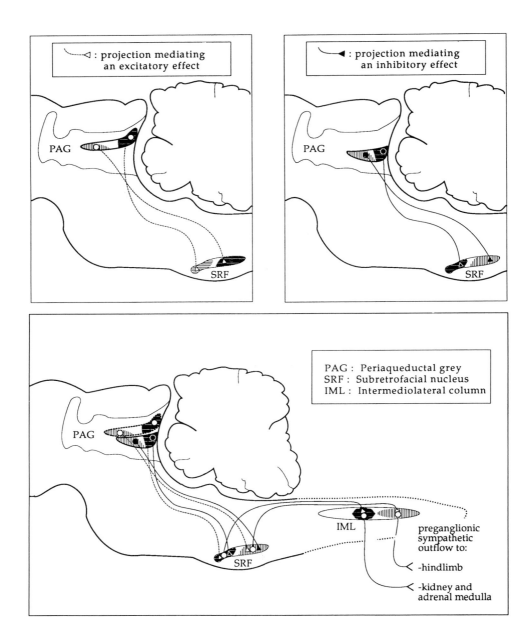

Figure 14. Schematic parasagittal diagram of the postulated pathways mediating the increase and decrease of vasoconstrictor tone in the renal and hindlimb vascular beds following excitation of neurons in the lateral and ventrolateral PAG. Modified from Carrive and Bandler (1991) Brain Research.

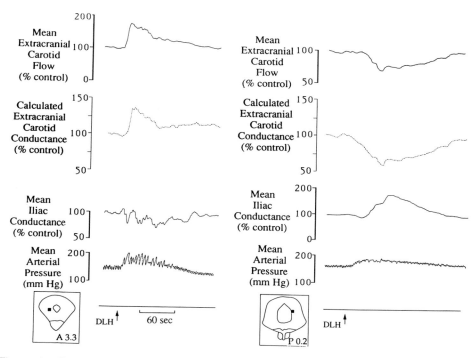

Figure 15. Patterns of blood flow changes in the extracranial and hindlimb vasculatures evoked by DLH microinjection in the lateral PAG of the paralyzed, unanesthetized decerebrate cat. Extracranial vascular conductance was calculated afterwards. The location of the centers of each microinjection are indicated on coronal sections of the PAG. Left panel: the site of injection is located in the intermediate lateral PAG. Right panel: the site of injection is located in the caudal lateral PAG. Arrows indicate the start of a DLH microinjection (40 nmoles). Reprinted with permission from Carrive and Bandler (1991b) Experimental Brain Research.

in the caudal part of the lateral PAG (P0.9). Extracranial vasoconstriction was rarely evoked at more rostral levels, unlike iliac vasoconstriction. These results further support the idea of a viscerotopic organization within the hypertensive column and indicate that the representation of the extracranial vasculature is relatively similar to that of the renal vasculature (Fig. 18 B). Figure 16 also shows that extracranial vasodilation was evoked exclusively at the most rostral levels, unlike iliac vasodilation which was preferentially evoked at the more caudal levels. These results suggest that regional vasodilation is also controlled by lateral PAG, and further, that different parts of the lateral PAG control vasodilation in different regional vascular beds. On the basis of these observations, one can postulate the existence of another functional column, a vasodilator column (Fig. 18 A). The vasodilator column lies in the lateral PAG and overlaps considerably

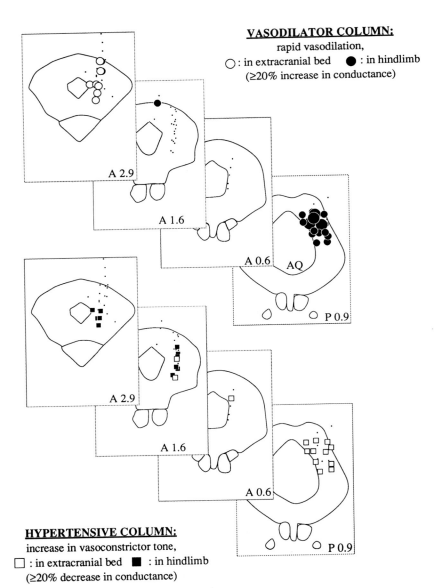

VASODILATOR COLUMN:
rapid vasodilation,
○ : in extracranial bed   ● : in hindlimb
(≥20% increase in conductance)

A 2.9

A 1.6

A 0.6   AQ

P 0.9

A 2.9

A 1.6

A 0.6

HYPERTENSIVE COLUMN:
increase in vasoconstrictor tone,
□ : in extracranial bed   ■ : in hindlimb
(≥20% decrease in conductance)

P 0.9

Figure 16. A series of 4 representative coronal sections showing the location of the sites at which extracranial and hindlimb blood flow were recorded following microinjection of DLH in the dorsal and lateral PAG of the paralyzed, unanesthetized decerebrate cat. Top row: the sites at which vasodilation in either extracranial or hindlimb vasculature was evoked define a vasodilator column. The large black circles indicate a ≥ 100% increase in hindlimb conductance. Bottom row: the sites at which vasoconstriction in either extracranial or hindlimb vasculature was evoked belong to the hypertensive column.

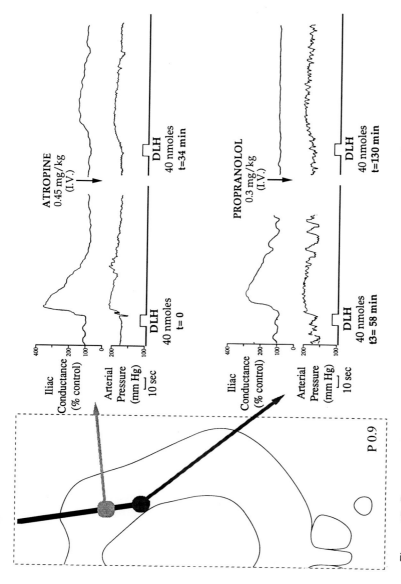

Figure 17. Demonstration of the two components mediating iliac vasodilation in the same experiment. Repeated microinjections of DLH were made at two sites along a track in the caudal lateral PAG of a paralyzed, unanesthetized decerebrate cat. Blood pressure and iliac conductance were recorded. At the first site, DLH evoked a short latency iliac vasodilation which was subsequently blocked with intravenously administered atropine. Still, a vasodilation of long latency and small amplitude remained. The cannula was then moved to a second site, 0.5 mm lower. At this site, DLH evoked also a long latency vasodilation, but this time of greater amplitude. This vasodilation was totally abolished by intravenously administered propranolol. The time at which each injection was made is indicated. The effect of a single intravenous injection of atropine (0.45 mg/kg) lasts for at least 6 hours.

with the hypertensive column, although its boundaries are not yet precisely defined. One definite feature about this column, however, is its viscerotopic organization: the head is represented rostrally and the hindlimb, caudally. Note that this viscerotopy is different from that of the hypertensive column (Fig. 18 A, B).

The output pathways of the vasodilator column have not yet been identified. Extracranial vasodilation is probably mediated by parasympathetic preganglionic neurons relaying in the pterygopalatine ganglion (Goadsby et al., 1984). A retrograde transneuronal viral study recently showed labeling in the PAG following pseudorabies virus injections in the pterygopalatine ganglion (Spencer et al., 1990). The mechanisms and the pathways mediating iliac vasodilation have been partly elucidated. Figure 17 shows that the iliac vasodilation evoked by microinjection of DLH in the caudal part of the lateral PAG has two components: one that is mediated by postganglionic cholinergic vasodilator fibers, and another that is mediated by circulating catecholamines (they are released from the adrenal medulla and act on the β-receptors present on the wall of skeletal muscle arterioles). In this experiment, intravenously administered atropine reduced the iliac vasodilation by blocking the fast, short latency iliac vasodilator effect (atropine blocks the cholinergic receptors). The remaining vasodilation, slower and characterized by a longer latency was totally abolished by intravenous propranolol (propranolol blocks the β-adrenoreceptors). Nothing is known about the pathway mediating the active, atropine sensitive component, except that it traverses the RVLM (Lindgren, 1955; Schramm and Bignall, 1971) without relaying there (Lovick and Hilton, 1985). There are good reasons, however, to believe that the slow, propranolol sensitive component relays there, and more precisely in the rostral SRF. It has been shown that the rostral SRF receives a direct projection from the caudal lateral PAG (see above), and mediates a long latency iliac vasodilation (Fig. 10) that is totally blocked by intravenous propranolol (Lovick, 1987).

Conclusion: Adaptive Values of PAG Vasomotor Control

Adaptive value of lateral PAG vasomotor control
An important feature of PAG vasomotor control is its longitudinal columnar organization. Further, the vasomotor columns are viscerotopically organized, and some partly overlap. The existence of partly overlapping vasomotor columns having different viscerotopic organizations provides a basis for evoking patterns of blood flow changes from restricted portions of the PAG.

As shown on Figure 15, activation of neurons in the intermediate part of the lateral PAG elicits a pattern of blood flow redistribution that favours the extracranial territory at the expense of the hindlimb, while activation of neurons in the caudal part of the lateral PAG evokes a pattern of blood flow redistribution that favors the hindlimb at the expense of the extracranial territories and the viscera (e.g., the renal vasculature). Experiments performed in unparalyzed decerebrate animals (Carrive et al., 1987; Carrive et al., 1988b) and in intact freely moving animals (Bandler and Carrive, 1988; Zhang et al., 1990) have shown that

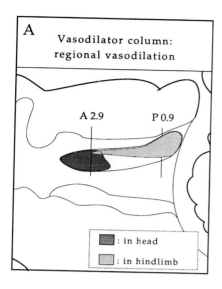

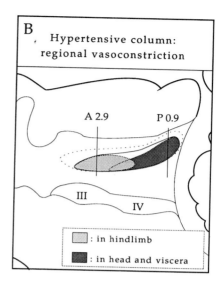

In the areas of overlap,
specific patterns of blood flow redistribution are put together

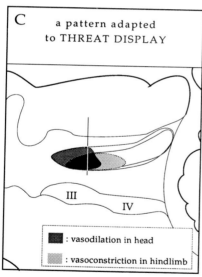

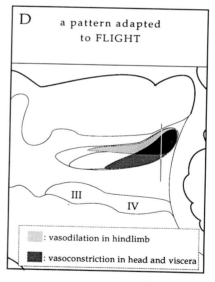

Figure 18. Schematic parasagittal representation of lateral PAG control of the regional vasculature and its relation to defensive behavior. Panel A represents the vasodilator column and its viscerotopic organization. Panel B represents the vasoconstrictor column and its viscerotopic organization, different to the one in the vasodilator column. Panel C and D show that in the area of overlap of the two vasomotor columns, specific patterns of blood flow redistribution can be evoked. Two patterns have been identified, one that is adapted to a threat display reaction, and another which is adapted to a flight reaction.

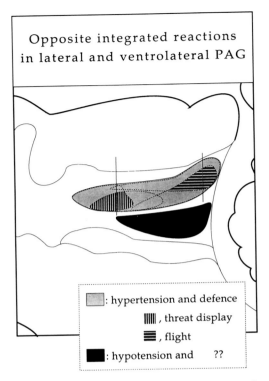

Figure 19. Schematic parasagittal representation of the PAG areas whose role in the mediation of integrated reactions have been identified. Lateral PAG plays a role in the mediation of defensive reactions such as threat display and flight. On the opposite, the ventrolateral PAG mediate an integrated reaction characterized by decreased somatomotor and autonomic (sympathetic) activity.

activation of neurons in the same two regions of the lateral PAG evoke two distinct types of defense reactions, threat display and flight (Fig. 19) (Bandler and Depaulis, this volume). Threat display, evoked from intermediate lateral PAG is characterized by intense facio-vocal changes (hissing, howling) with little phasic limb activity, whereas the flight reaction, evoked from caudal lateral PAG is characterized by intense muscular activity in the limbs, with little to no phasic facio-vocal changes (see Bandler and Depaulis, this volume). In both cases, the blood flow redistribution always favors the vasculature of the most active muscular territory. Further, the redistribution is made at the expense of less active territories.

These observations support the original finding of Abrahams et al. (1960) that the cardiovascular effects evoked from PAG are adapted to the defense reaction. However, whereas these authors postulated that only one cardiovascular pattern characterized the defense reaction from its early alerting stages onwards (Abrahams et al., 1964), the present findings suggest instead that the lateral PAG vasomotor control is responsible for the integration of specific patterns of redistribution in blood flow which are adapted to different types of defensive reactions (Fig. 18 C, D). The adaptive value of lateral PAG vasomotor control is probably in the quick and specific redistribution of blood flow during the early stages of different types of emergency situations.

Adaptive value of ventrolateral PAG vasomotor control

The adaptive value of the hypotension and decreased vasoconstrictor tone evoked from ventrolateral PAG is not well understood (Fig. 19). Experiments performed in unparalyzed decerebrate animals (Carrive and Bandler, 1991a) and in intact freely moving animals (Zhang et al., 1990) have shown that activation of neurons in the ventrolateral PAG reduces somatomotor activity: on-going activity stops and the animal adopts an immobile posture (hyporeactive type of immobility; see Bandler and Depaulis, this volume). This reaction stands in sharp contrast to the active defensive reactions evoked from the lateral part of the PAG. The cardiovascular and somatomotor changes evoked from ventrolateral PAG are perhaps part of a restful, relaxing type of integrated reaction (see Lovick, this volume). These coordinated changes could also play an active role after emergency reactions, allowing the animal to rest and recuperate more efficiently.

Speculation: A Model of Functional Organization in the PAG

Physiological and anatomical studies presented in this chapter demonstrate the existence of a functional organization within the PAG. They form the basis for a model of PAG functional organization.

The model proposed has four major features:

1. *Longitudinally oriented functional columns.* PAG functions are represented along a rostrocaudal axis and form longitudinally oriented functional columns of various lengths.

2. *Topographic organization.* Functional columns of the PAG may be topographically organized with respect to their peripheral effectors.

3. *Columnar overlap.* Some functional columns are partly overlapping. In the regions of overlap, unique patterns of physiological changes are put together. This may well be the basis for integration of individual functional components into coordinated patterns, and ultimately into complex behavioral reactions.

4. *Lateral PAG versus ventrolateral PAG.* The lateral and ventrolateral parts of the PAG control the same physiological parameters, but in opposite directions. Functional columns in the lateral PAG have their functionally opposite counterparts in the ventrolateral PAG.

This model was originally develop to describe the functional organization of PAG vasomotor control. Perhaps it can be extended to other functions represented in the PAG, such as the control of the skeletal musculature or the modulation of sensory input (e.g., analgesia). In particular, it describes relatively well the functional organization of PAG control of somatomotor activity. For instance, a "movement" column can be identified in the lateral PAG. This column lies in the intermediate and caudal portions of the lateral PAG and mediates increases in somatomotor activity (such as those that characterize certain defensive reactions); it is somatotopically organized: head is represented rostrally and limbs caudally (threat display versus flight); it overlaps with the vasodilator and the hypertensive column; and finally, it has its functionally opposite counterpart in the ventrolateral PAG: an "immobility" column mediating decreased somatomotor activity.

Defining functional columns is a convenient way to consider PAG functions. Although it is an important step towards understanding PAG organization, it says very little about how these PAG functions are coded at the level of the neuron. Is it that every functional column has its own set of output neurons, each of them dedicated to the control of a unique function, or is it that each output neuron belongs to more than one functional column, controlling multiple functions at the same time? The answer to this question is the real challenge, because it is this code that accounts for the remarkable integrative properties of the PAG.

Acknowledgments

This research was supported by grants from the Australian National Health and Medical Research Council and from the Harry Frank Guggenheim Foundation. I wish to thank Dr R. Bandler and Dr K. Keay for their constructive comments on the manuscript.

References

Abrahams, V.C., Hilton, S.M. and Zbrozyna, A.W., Active muscle vasodilatation produced by stimulation of the brain stem: its significance in the defence reaction, J. Physiol. (Lond.), 154 (1960) 491-513.

Abrahams, V.C., Hilton, S.M. and Zbrozyna, A.W., The role of active muscle vasodilatation in the alerting stage of the defence reaction, J. Physiol. (Lond.), 171 (1964) 189-202.

Bandler, R., Brain mechanisms of aggression as revealed by electrical and chemical stimulation: suggestion of a central role for the midbrain periaqueductal grey region, In: Progress in Psychobiology and Physiological Psychology, Vol. 13, Epstein A. and Morrison A. (Eds.), Academic Press, New York, 1988, pp 67-154

Bandler, R. and Carrive, P., Integrated defence reaction elicited by excitatory amino acid microinjection in the midbrain periaqueductal grey region of the unrestrained cat, Brain Res., 439 (1988) 95-106.

Berman, A.L., The Brain Stem of the Cat. A Cytoarchitectonic Atlas with Stereotaxic Coordinates, The University of Wisconsin Press, Madison, 1968.

Calaresu, F.R. and Yardley, C.P., Medullary basal sympathetic tone, Ann. Rev. Physiol., 50 (1988) 511-524.

Carrive, P. and Bandler, R., Viscerotopic organization of neurons subserving hypotensive reactions within the midbrain periaqueductal grey: a correlative functional and anatomical study, Brain Res., 541 (1991a) 206-215.

Carrive, P. and Bandler, R., Control of extracranial and hindlimb blood flow by the midbrain periaqueductal grey of the cat, Exp. Brain Res., 84 (1991b) 599-606.

Carrive, P., Bandler, R. and Dampney, R.A.L., Anatomical evidence that hypertension associated with the defence reaction in the cat is mediated by a direct projection from a restricted portion of the midbrain periaqueductal grey to the subretrofacial nucleus of the medulla, Brain Res., 460 (1988) 339-345.

Carrive, P., Bandler, R. and Dampney, R.A.L., Somatic and autonomic integration in the midbrain of the unanaesthetised decerebrate cat: A distinctive pattern evoked by excitation of neurones in the subtentorial portion of the midbrain periaqueductal grey, Brain Res., 483 (1989a) 251-258.

Carrive, P., Bandler, R. and Dampney, R.A.L., Viscerotopic control of regional vascular beds by discrete groups of neurons within the midbrain periaqueductal gray, Brain Res., 493 (1989b) 385-390.

Carrive, P., Dampney, R.A.L. and Bandler, R., Excitation of neurones in a restricted portion of the midbrain periaqueductal grey elicits both behavioural and cardiovascular components of the defence reaction in the unanaesthetised decerebrate cat, Neurosci. Lett., 81 (1987) 273-278.

Ciriello, J., Caverson, M.M. and Polosa, C., Function of the ventrolateral medulla in the control of the circulation, Brain Res. Rev., 11 (1986) 359-391.

Dampney, R.A.L. and McAllen, R.M., Differential control of sympathetic fibres supplying hindlimb skin and muscle by subretrofacial neurones in the cat, J. Physiol. (Lond.), 395 (1988) 41-56.

Dampney, R.A.L., Czachurski, J., Dembowsky, K., Goodchild, A.K. and Seller, H., Afferent connections and spinal projections of the pressor region in the rostral ventrolateral medulla of the cat, J. Auton. Nerv. Syst., 20 (1987a) 73-86.

Dampney, R.A.L., Goodchild, A.K. and McAllen, R.M., Vasomotor control by subretrofacial neurones in the rostral ventrolateral medulla, Can. J. Physiol. Pharmacol., 65 (1987b) 1572-1579.

Dampney, R.A.L., Goodchild, A.K. and Tan, E., Vasopressor neurons in the rostral ventrolateral medulla of the rabbit, J. Auton. Nerv. Syst., 14 (1985) 239-254.

Eliasson, S., Lindgren, P. and Uvnas, B., The hypothalamus, a relay station of the sympathetic vasodilator tract, Act. Physiol. Scand., 31 (1954) 290-300.

Goadsby, P.J., Lambert, G.A. and Lance, J.W., The peripheral pathway for extracranial vasodilation in the cat, J. Auton. Nerv. Syst., 10 (1984) 145-155.

Goodchild, A.K., Dampney, R.A.L. and Bandler, R., A method of evoking physiological responses by stimulation of cell bodies but not axons of passage within localized regions of the central nervous system, J. Neurosci. Meth., 6 (1982) 351-363.

Hilton, S.M., Marshall, J.M. and Timms, R.J., Ventral medullary relay neurones in the pathway from the defence areas of the cat and their effect on blood pressure, J. Physiol. (Lond.), 345 (1983) 149-166.

Jones, R.O., Kirkman, E. and Little, R.A., The involvement of the midbrain periaqueductal grey in the cardiovascular response to injury in the conscious and anaesthetized rat, Exp. Physiol., 75 (1990) 483-495.

Kabat, H., Magoun, H.W. and Ranson, J.W., Electrical stimulation of points in the forebrain and midbrain: The resultant alterations in blood pressure, Arch. Neurol., 34 (1935) 931-955.

Lindgren, P., The mesencephalon and the vasomotor system, Act. Physiol. Scand., 35 (Suppl 121) (1955) 1-183.

Lindgren, P., Rosen, A., Strandberg, P. and Uvnas, B., The sympathetic vasodilator outflow - a cortico-spinal autonomic pathway, J. Comp. Neurol., 105 (1956) 95-109.

Lipski, J., Bellingham, M.C., West, M.J. and Pilowsky, P., Limitations of the technique of pressure microinjection of excitatory amino acids for evoking responses from localized regions of the CNS, J. Neurosci. Meth., 26 (1988) 169-179.

Lovick, T.A, Ventrolateral medullary lesions block the antinociceptive and cardiovascular responses elicited by stimulating the dorsal periaqueductal grey matter in rats, Pain, 21 (1985) 241-252.

Lovick, T.A., Differential control of cardiac and vasomotor activity by neurones in nucleus paragigantocellularis lateralis in the cat, J. Physiol. (Lond.), 389 (1987) 23-35.

Lovick, T.A. and Hilton, S.M., Vasodilator and vasoconstrictor neurones of the ventrolateral medulla in the cat, Brain Res., 331 (1985) 353-357.

Luiten, P.G.M., ter Horst, G.J. and Steffens, A.B., The hypothalamus, intrinsic connections and outflow pathways to the endocrine system in relation to the control of feeding and metabolism, Prog. Neurol., 28 (1987) 1-54.

McAllen, R.M., Location of neurones with cardiovascular and respiratory function at the surface of the cat's medulla, Neuroscience, 18 (1986a) 43-49.

McAllen, R.M., Action and specificity of ventral medullary vasopressor neurones in the cat, Neuroscience, 18 (1986b) 51-59.

McAllen, R.M., Identification and properties of subretrofacial bulbospinal neurones: a descending cardiovascular pathway in the cat, J. Auton. Nerv. Syst., 17 (1986c) 151-164.

McAllen, R.M., Dampney, R.A.L. and Goodchild, A.K., The sub-retrofacial nucleus and cardiovascular control, In: Organization of the Autonomic Nervous System: Central and Peripheral Mechanisms, Calaresu F.R., Ciriello J., Polosa C. and Renaud L.P. (Eds.), A. Liss, New York, 1987, pp. 215-225.

McAllen, R.M. and Dampney, R.A.L., Vasomotor neurons in the rostral ventrolateral medulla are organized topographically with respect to type of vascular bed but not body region, Neurosci. Lett., 110 (1990) 91-96.

Reis, D.J., Morrison, S. and Ruggiero, D.A.,The C1 area of the brainstem in tonic and reflex control of blood pressure, Hypertension, 11 (Suppl.) (1988) I8-I13.

Ross, C.A., Ruggiero, D.A., Joh, T.H., Park, D.H. and Reis, D.J., Rostral ventrolateral medulla: selective projections to the thoracic autonomic cell column from the region containing C1 adrenaline neurons, J. Comp. Neurol., 228 (1984a) 168-185.

Ross, C.A., Ruggiero, D.A., Park, D.H., Joh, T.H., Sved, A.F., Fernandez-Pardal, J., Saavedra, J.M. and Reis, D.J., Tonic vasomotor control by the rostral ventrolateral medulla: effect of electrical or chemical stimulation of the area containing C1 adrenaline neurons on arterial pressure, heart rate and plasma catecholamines and vasopressin, J. Neurosci., 4 (1984b) 474-494.

Schramm, L.P. and Bignall, K.E., Central neural pathway mediating active sympathetic muscle vasodilation in cats, Am. J. Physiol., 221 (1971) 754-767.

Spencer, S.E., Sawyer, W.B., Wada, H., Platt, K.B. and Loewy, A.D., CNS projections to the pterygopalatine parasympathetic preganglionic neurons in the rat: a retrograde transneuronal viral cell body labeling study, Brain Res., 534 (1990) 149-169.

Strack, A.M., Sawyer, W.B., Hughes, J.H., Platt, K.B. and Loewy, A.D., A general pattern of CNS innervation of the sympathetic outflow demonstrated by transneuronal pseudorabies viral infection, Brain Res., 491 (1989) 156-162.

Strack, A.M., Sawyer, W.B., Platt, K.B. and Loewy, A.D., CNS cell groups regulating the sympathetic outflow to adrenal gland as revealed by transneuronal cell body labeling with pseudorabies virus, Brain Res., 491 (1989) 274-276.

Zhang, S.P., Bandler, R. and Carrive, P., Flight and immobility evoked by excitatory amino acid microinjection within distinct parts of the subtentorial midbrain periaqueductal gray of the cat, Brain Res., 520 (1990) 73-82.

Interactions Between Descending Pathways from the Dorsal and Ventrolateral Periaqueductal Gray Matter in the Rat

T.A. Lovick

Department of Physiology
The Medical School
Birmingham, U.K.

Introduction

Many psychological and environmental factors are known to influence responsiveness to noxious sensory stimuli. Until recently, the physiological mechanisms which enable such changes to take place were not understood. Current knowledge of endogenous analgesic mechanisms stems largely from the observation by Reynolds (1969) that analgesia could be produced in rats by electrical stimulation in the midbrain periaqueductal gray matter (PAG). This important paper was the first to suggest a neurophysiological basis for altered pain perception and provided a major impetus for further studies of pain control from the PAG. Although there is now a wealth of data concerning the neurophysiological and neuropharmacological mechanisms which underly the stimulation-produced analgesia, the physiological role of pain control from the PAG and the conditions under which this system becomes activated are less well understood. The PAG is known to be involved in initiating and integrating many aspects of behavioral, somatomotor and autonomic activity of which analgesia is but one component. In the past the tendancy has been to study the individual aspects of PAG function in isolation. This approach has yielded much information regarding the physiological mechanisms by which such changes are mediated but has been less profitable in terms of our understanding of the functional significance of the responses. In the search for a functional significance for pain

The Midbrain Periaqueductal Gray Matter, Edited by A. Depaulis and
R. Bandler, Plenum Press, New York, 1991

control from the PAG, it may perhaps be more appropriate to consider the analgesia as a component of a much wider response of the whole organism, taking into account autonomic and motor changes as well as alterations in sensory responsiveness. In the current investigation the analgesia and accompanying cardiovascular changes evoked by stimulation in the PAG have been studied together in the belief that an investigation of the whole response may be more revealing than a study of its individual components in isolation.

Cardiovascular Changes Accompanying Stimulation-Evoked Analgesia

Responses to electrical stimulation

In initial experiments on rats anesthetised with the steroidal anesthetic mixture, Saffan, electrical stimulation throughout the PAG, using short trains of pulses at fixed intensity, produced analgesia as judged by an increase in the latency of the tail flick reflex (Lovick, 1985). Indeed, the PAG could be divided into two functional regions on the basis of the cardiovascular change which accompanied the analgesia. One region, which will be referred to as the "dorsal" PAG encompassed the area dorsal to the level of the dorsal raphe nucleus (DRN) and included the dorsomedial, dorsolateral and lateral parts of the PAG. The second region, the "ventrolateral" PAG, included the area ventral to the level of the aqueduct and lateral to the DRN. In the dorsal PAG, the analgesia was accompanied by a clearcut pattern of cardiovascular change (Fig. 1). The response included an increase in blood pressure and heart rate and vasodilatation in hindlimb muscle together with an increase in the rate and depth of respiration. Together with these cardiorespiratory changes were other autonomic signs of arousal such as pupillodilatation and exophthalamus and at some sites, particularly in the caudal half of the PAG, movements of the jaws and vibrissae occurred, with coordinated movements of the hindlimbs and dorsiflexion of the tail. This striking pattern of autonomic and motor changes disappeared as the electrode was advanced to stimulate deeper sites. However, although analgesia could still be evoked from stimulation at sites ventral to the level of the aqueduct (Fig. 1), there was no consistent change in the cardiorespiratory parameters monitored.

Chemical stimulation

In agreement with previous studies (Hilton and Redfern, 1986), the cardiovascular changes evoked by electrical stimulation in the dorsal PAG could be reproduced by microinjection of an excitatory amino acid (D,L-homocysteic acid, DLH 65 nmol in 350 nl) into this region (Figs. 2 and 4). The cardiovascular response was accompanied by analgesia (Fig. 2). Thus the response to electrical stimulation in the dorsal PAG can be assumed to be due mainly to activation of cell bodies in the vicinity of the electrode tip, rather than to stimulation of fibers of passage.

In contrast to its effects in the dorsal PAG, injection of DLH into the ven-

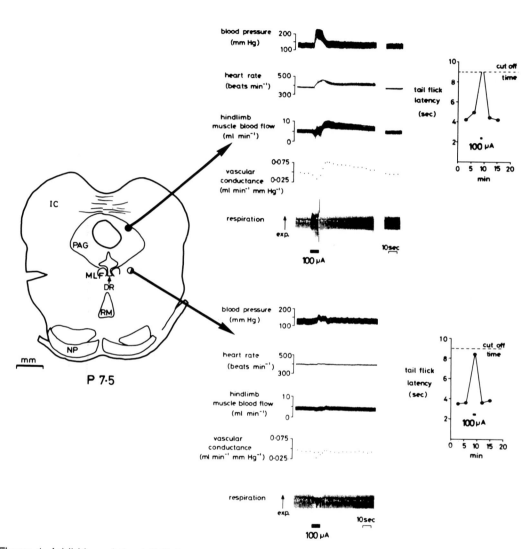

Figure 1. Inhibition of the tail flick response produced by electrical stimulation in the dorsal PAG (50 µA, 80Hz, 0.5msec pulses) is accompanied by a "defense" pattern of cardiovascular change. The analgesia evoked from stimulation further ventrally is not accompanied by any clearcut pattern of cardiovascular change. Rat anesthetised with Saffan. Reprinted with permission from Lovick (1985) Pain.

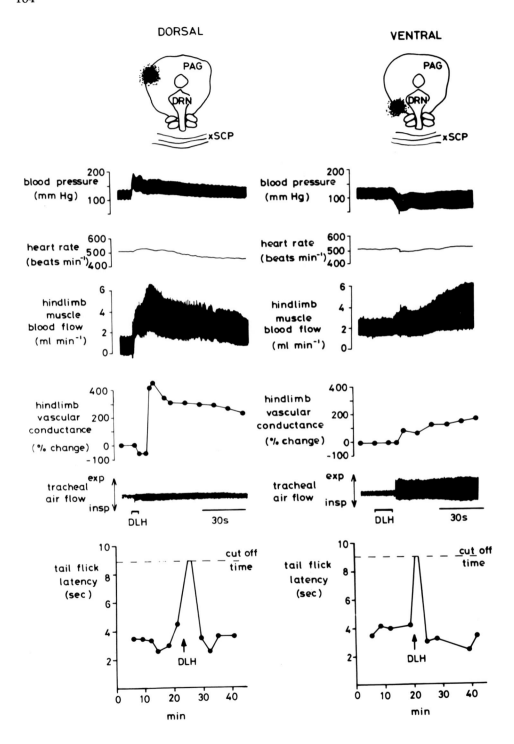

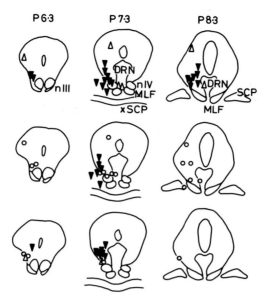

Figure 3. Top row: Outline drawings of representative sections through the PAG to show sites where injection of DLH produced an increase (open triangles) or a decrease (filled triangles) in resting blood pressure. Middle row: Sites where injection of DLH produced a decrease (filled triangles) or had no effect (open circles) on the cardiovascular components of the defense response evoked by stimulation in the dorsal PAG (not shown). Bottom row: Sites where injection of DLH excited (open triangles), inhibited (filled triangles) or had no effect (open circles) on the activity of spinally-projecting neurones in the RVLM. Abbreviations: xSCP: decussation of the superior cerebellar peduncle; SCP: superior cerebellar peduncle; MLF: medial longitudinal fasciculus; DRN: dorsal raphe nucleus; nIII: oculomotor nucleus; nIV: trochlear nucleus. Numbers above sections indicate mm caudal to Bregma.

◀━━━

Figure 2. Examples of the patterns of cardiovascular response which accompany inhibition of the tail flick response evoked by selective activation of cell bodies in the dorsal (left side of figure) and ventrolateral (right side of figure) PAG by microinjection of D,L-homocysteic acid (DLH 65nmol). Rats anesthetised with Saffan. Abbreviations: DRN: dorsal raphe nucleus; IC: inferior colliculus; PAG: periaqueductal gray matter; xSCP: decussation of the superior cerebellar peduncle. Modified from Lovick (1991) News in Physiological Sciences.

trolateral part of the nucleus did not reproduce completely the effects of electrical stimulation. Whilst both electrical and chemical stimulation produced analgesia, injection of DLH into the ventrolateral PAG evoked a clearcut fall in blood pressure with bradycardia and vasodilatation in the hindlimb (Figs. 1 and 2). This striking pattern of cardiovascular response was quite distinct from the small and inconsistent cardiovascular changes evoked by electrical stimulation. The failure of chemical stimulation to reproduce the effects of stimulating electrically has important implications for the interpretation of effects produced by electrical stimulation in the ventrolateral PAG. In terms of the cardiovascular changes, it now seems likely that the effects of electrical stimulation represent the net effect of simultaneous activation of both cell bodies and fibers of passage, or perhaps even the fibers alone. The interpretation of the analgesic effects induced by electrical stimulation must also be viewed with caution since superficial similarities in the inhibition of an analgesic test such as the tail flick reflex could mask more subtle differences in the underlying mechanisms which produce the analgesia (see also Morgan, this volume).

Efferent Pathways to the Spinal Cord

Dorsal PAG

There is anatomical and electrophysiological evidence to show that the efferent pathway for both the analgesia and the cardiovascular components of the response to stimulating in the dorsal half of the PAG relay on spinally-projecting neurons in nucleus paragigantocellularis lateralis of the rostral ventrolateral medulla (RVLM). Electrolytic lesions of the RVLM or application of inhibitory amino acids to the neuropil attenuate both the cardiovascular "defense" response and the analgesia evoked from the dorsal PAG (Hilton et al., 1983; Lovick, 1985; Foong and Duggan, 1986). The lesions of the RVLM produced a fall in resting blood pressure. However, in the rat anesthetised with Saffan, this effect is only transient and the attenuation of the defense response persisted when the blood pressure had returned to control levels. Thus the effect on stimulation in the dorsal PAG is unlikely to be secondary to the fall in resting blood pressure.

Spinally-projecting neurons in the RVLM terminate in the ventral and dorsal horns of the cord as well as in the intermediolateral cell column (Loewy and McKellar, 1981; Martin et al., 1981). Stimulating the RVLM evokes analgesia, facilitates the activity of motoneurons and induces excitatory effects in the spinal sympathetic outflow (Gray and Dostrovsky, 1985; Lovick, 1986a; Lovick and Li, 1989; Sidall and Dampney, 1989). Anatomical and electrophysiological studies have shown that neurons in the RVLM receive direct projections from the dorsal PAG and that these inputs are mainly excitatory (Andrezik et al., 1981; Lovick, 1986b; 1988; Li and Lovick, 1985; Carrive et al. 1989; Holstege, this volume). Measurements of conduction velocities in this pathway in the rat revealed a slowly conducting projection (0.3-4.4 m/sec) (Li and Lovick, 1985) which is indicative of cells with thinly myelinated or unmyelinated axons. Similar values were calculated for conduction in this pathway in the cat (Lovick, 1988). In both

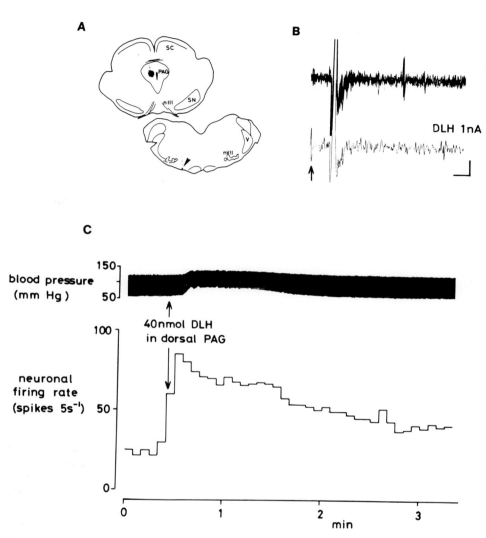

Figure 4. Effect of microinjection of D,L-homocysteic acid (DLH 40 nmol) into the lateral PAG on the activity of a medullo-spinal neuron. Urethane-anesthetised preparation. A - Locations of injection site in the dorsal PAG (top section) and recording site in the ventro-lateral medulla ventromedial to the caudal tip of the facial nucleus (bottom section). B - Top trace: superimposed sweeps of response to stimulation in the spinal cord at T9. Bottom trace: spontaneous action potential (arrow) cancels the antidromic spike. Firing rate of the neuron was raised by iontophoretic application of 1nA DLH. Scale bars: 10msec, 100 mV. C: Effect of microinjection of DLH into the dorsal PAG on the firing rate of the neuron. Abbreviations: PAG: periaqueductal gray mater, SC: superior colliculus, SN: substantia nigra, nIII: third cranial nerve, nVII: facial nucleus, V: trigeminal tract.

species stimulation in the dorsal PAG evoked mainly excitatory responses in ventrolateral medullary neurons including cells with spinally-projecting axons, identified antidromically by stimulation in the spinal cord (Fig. 4) (Lovick et al., 1984; Li and Lovick, 1985; Lovick, 1988 and unpublished work). The conduction velocities of these medullo-cells ranged from 5-120 m/sec in the cat although lower values (0.55-8.4 m/sec) were measured in the rat. Since cells with slow, medium and fast conduction velocities are thought to project respectively to the intermediolateral, dorsal and ventral horns of the spinal cord (Lovick and Li, 1989), stimulation in the dorsal PAG would appear to activate all three cell types.

Ventrolateral PAG

The descending pathway which mediates antinociception evoked from the ventrolateral PAG relays in the medullary raphe in nucleus raphe magnus (NRM). Lesions in NRM have been shown to selectively block analgesia evoked from ventral but not dorsal midbrain sites (Prieto et al., 1983) and activation of neurons in NRM evokes a powerful analgesia (see Besson and Chaouch, 1987 for review). Electrophysiological studies have demonstrated that raphe-spinal neurons can be excited by stimulation of the ventrolateral PAG (e.g., Lovick et al., 1978) and there is now compelling evidence for the involvement of excitatory amino acids as neurotransmitters in this projection (Wiklund et al., 1988; Beitz and Williams, this volume). In addition to this ventromedial synapse in NRM, other evidence suggests that the descending pathway from the ventrolateral PAG may also relay laterally in the ventral medulla. For example, Sandkühler and Gebhart (1984) have shown that a combined local anesthetic block of both ventromedial and ventrolateral medulla was more effective in reducing the effectiveness of stimulation in the PAG than when the block was confined to either ventromedial or ventrolateral sites.

In contrast to the extensive studies of the ventrolateral PAG-raphé spinal pathway with regard to analgesia, the pathways which mediate the cardiovascular and motor changes have received little attention. The ventrolateral PAG, particularly at caudal levels, projects densely to the medullary raphe nuclei magnus, pallidus and obscurus (Lakos and Basbaum, 1988; Holstege, this volume but see Mantyh, 1983). NRM sends projections to the dorsal horn and the intermediolateral cell column (IML) while the more caudal raphe nuclei have terminations in the IML and ventral horn (Holstege and Kuypers, 1982; Holstege, this volume). Recent studies have shown that injection of a glutaminergic agonist into the ventromedial medulla produces hypotension, bradycardia and a reduction in muscle tone (Lai and Seigel, 1990). While further unequivocal evidence is still needed, circumstantial evidence suggests that the analgesia, sympathoinhibition and depression of motor activity which make up the whole response evoked from the ventrolateral PAG, may all be mediated via ventromedial relays in the medullary raphe.

On the basis of the cardiovascular changes which accompany the analgesia evoked from the PAG, it has been possible to distinguish two control systems: one originating from neurons dorsal and lateral to the level of the aqueduct, and the other from cells situated deeper, in the ventrolateral PAG. It is interesting to note that while these divisions have been made on the basis of association with a particular pattern of cardiovascular change, similar conclusions have been drawn by other workers, using behavioral and pharmacological criteria (Morgan, this volume). Lesion studies suggest that the descending efferent pathways from the dorsal and ventrolateral PAG control systems respectively favor ventrolateral and ventromedial relays in the medulla (Prieto et al., 1983; Lovick, 1985). Anatomical studies have described dense projections from the dorsal PAG to the RVLM and from the ventrolateral PAG to NRM (Abols and Basbaum, 1981; Andrezik et al., 1981; Beitz et al., 1983; Lovick, 1986). However, descending fibers from the ventrolateral PAG have recently been found to course through the ventrolateral medulla (Carrive and Bandler, 1991) and a projection from the lateral and dorsolateral PAG to the medullary raphe has also been reported (Abols and Basbaum, 1981; Beitz et al., 1983). Although there is a clear functional distinction between the dorsal and ventrolateral control systems in the PAG, there may be considerable overlap between the course of their efferent pathways to the spinal cord.

Interactions between Descending Control from the Dorsal and Ventrolateral Parts of the PAG

Although functionally distinct, the two descending systems from the PAG do not appear to work in isolation. Evidence is now beginning to accumulate which suggests that there may be considerable interaction between the two descending pathways.

Cardiovascular control

In the Saffan-anesthetised rat, microinjection of DLH (65 nmol in 350 nl) into the caudal half of the ventrolateral PAG attenuated the cardiovascular components of the defense response evoked by stimulation in the dorsal PAG (Fig. 5). Stimulation in the ventrolateral PAG produced a decrease in resting blood pressure and heart rate (see above). However, it is unlikely that the reduction in the sympathoexcitatory defense response evoked from the dorsal PAG was simply due to summation of sympathoinhibitory and sympathoexcitatory influences at the level of the spinal sympathetic outflow. When pressor responses of similar magnitude to those elicited from the dorsal PAG were evoked by stimulation in the RVLM (the main relay in the descending efferent pathway from the dorsal PAG) activation of cells in the ventrolateral PAG had no effect on this pressor response, despite the fact that resting blood pressure fell. This indicates firstly, that inhibition of the sympathoexcitatory defense response evoked from the PAG could not be a consequence of the fall in resting blood pressure and secondly, that the inhibitory interaction must have occurred either within the PAG itself or more likely, at the synapse in the RVLM.

The effect of stimulating neurons in the ventrolateral PAG was tested on 12 antidromically-identified spinally-projecting neurons in the RVLM (Fig. 6).

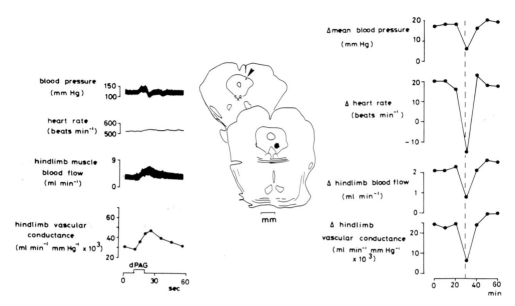

Figure 5. Effect of microinjection of 65 nmol D,L-homocysteic acid into the ventrolateral PAG on the cardiovascular response evoked from the dorsal PAG. Left panel: control response to stimulation in the dorsal PAG. Middle panel: location of stimulation site in the dorsal PAG (arrow, top section) and site of microinjection of DLH into the ventrolateral PAG (bottom section). Right panel: size of the increase in mean blood pressure, heart rate, hindlimb muscle blood flow and conductance evoked by electrical stimulation in the dorsal PAG. After injection of DLH into the ventrolateral PAG (broken line) the amplitude of all the components of the response was reduced.

Eight of these cells were barosensitive (their activity was depressed during the pressor response produced by an i.v. bolus of phenylephrine). These cells appear to correspond to the "cardiovascular" neurons of the RVLM described by Brown and Guyenet (1984). The remaining 4 cells were not classified. Injections of DLH (15-40 nmol in 75-200nl) were made at 16 sites in the ventrolateral PAG (Fig. 3). At 9 sites DLH produced a reduction in ongoing activity of each of the medullo-spinal cells tested (n=6). The effective sites were clustered in the ventrolateral PAG 7.3 - 8.3mm caudal to bregma (Fig. 3). Injections made at a further 7 sites rostral or caudal to this level had either no effect (n=3) or excited (n=4) the remaining 6 medullo-spinal neurons (Fig. 3). There was a striking overlap between the locations of the injection sites in the ventrolateral PAG which attenuated the cardiovascular components of the defense response evoked from the dorsal PAG and the sites where DLH inhibited the activity of cardiovascular neurons in the RVLM (Fig. 3).

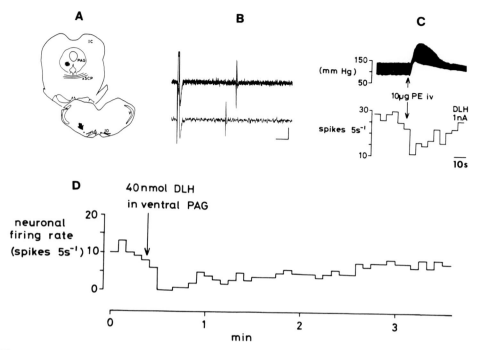

Figure 6. Effect of microinjection of D,L-homocysteic acid (DLH) into the ventrolateral PAG on the activity of a spinally-projecting cardiovascular neuron in the ventrolateral medulla in a urethane-anesthetised rat. A. Location of injection site in the PAG and recording site (arrow) in the ventrolateral medulla at the level of the rostral tip of the inferior olive. B. Top trace: superimposed record (5 sweeps) of response of neuron to stimulation in the spinal cord at T9. Lower trace: spontaneous action potential cancels the antidromic spike. Calibration bars: 10 ms, 100mV. C. Response of neuron to baroreceptor activation. Histogram of neuronal firing rate shows that activity was inhibited during the pressor response produced by injection of phenylephrine (PE). D. Histogram of neuronal firing shows depression of activity following microinjection of DLH into the ventrolateral PAG. Abbreviations: IC: inferior colliculus; IO: inferior olive; PAG: periaqueductal gray matter; xSCP: decussation of the superior cerebellar peduncle.

The pathway to the RVLM which mediates the inhibitory effects evoked from the ventrolateral PAG is still not clear. Recently, retrograde labeling studies in the cat described a direct projection to the RVLM from cells in the "hypotensive" column of the ventrolateral PAG, particularly in the subtentorial region (Bandler et al., 1991; Carrive, this volume). Such a direct projection, if inhibitory, may mediate the earliest phase of the inhibition of neurons in the RVLM. Other evidence suggests that an indirect pathway via the medullary raphe nuclei may also be involved. Retrograde transport and intracellular labeling studies have shown that neurons in both nucleus raphe magnus and obscurus send direct

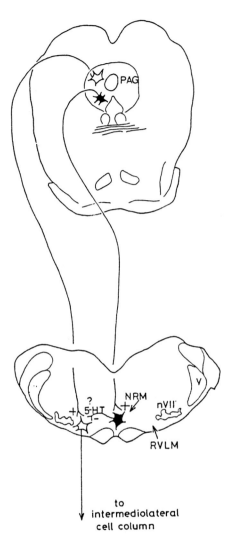

Figure 7. Schematic diagram to show possible pathway through which neurons in the ventrolateral PAG could exert inhibitory modulation of activity in the pathway descending from the dorsal PAG. Abbreviations: NRM: nucleus raphe magnus, PAG: periaqueductal gray matter, RVLM: rostral ventrolateral medulla, V: trigeminal tract, nVII: facial nucleus.

projections to the RVLM (Andrezik et al., 1981; Lovick, 1986; Nicholas and Hancock, 1990; Mason, this volume). Stimulation in NRM exerts predominantly inhibitory effects on neurons in the RVLM (Lovick, 1988). Moreover, Gong et al. (1989) recently reported that injection of DLH into the nucleus raphe obscurus inhibited the pressor response evoked from the dorsal PAG. Given that these medullary raphé nuclei receive direct excitatory inputs from the ventrolateral PAG (see above), it seems likely that their inhibitory projections to the RVLM may play an important role in modulation by the ventrolateral PAG, of activity in the pathway descending from the dorsal PAG (Fig. 7).

The NRM and NRO are rich in 5-HT-containing neurons (Steinbusch, 1981). There is also a high density of 5-HT-containing nerve terminals in the RVLM which suggests that synaptically released 5-HT might influence the activity of ventrolateral medullary cells. A direct relationship between these serotonergic terminals and the 5-HT-containing perikarya of the raphe has not yet been established but it is perhaps significant to note that injection of 5-HT into the RVLM can attenuate the cardiovascular components of the response evoked from the dorsal PAG, effectively mimicking the effects evoked from the ventrolateral PAG (Fig. 8) (Lovick, 1991). An inhibitory serotonergic projection to the RVLM from the medullary raphe is therefore a likely mediator of modulatory effects on the descending pathway from the dorsal PAG (Fig. 7).

Somatomotor control neurons

While it is clear that the cardiovascular components of the response evoked from the dorsal PAG can be modulated from the ventrolateral PAG, it is not known whether similar interactions occur with respect to the analgesia and motor components of the response. In the Saffan-anesthetised rat, there was no sign of any summation of the analgesic effects evoked from the dorsal and ventral PAG, even at times when the cardiovascular response to stimulation in the dorsal PAG was attenuated. However, it is likely that the tail flick response is not a sufficiently sensitive test with which to assess such interactions and that studies at the single unit level will be necessary to resolve this question.

Despite the lack of adequate data regarding modulation of the analgesic component of the response from the dorsal PAG, it is nevertheless interesting to note that injections of 5-HT into the RVLM which attenuated the cardiovascular components of the response evoked from the dorsal PAG, had no effect on the analgesia, at least as judged by inhibition of the tail flick response (Fig. 8) (Lovick, 1990). Although the pain control and cardiovascular control cells in the RVLM are anatomically separate (Sidall and Dampney, 1989), the relatively large injection volumes used in this study should have ensured that 5-HT spread to reach both types of cell. It is likely therefore, that compared to the cardiovascular cells, the pain control neurons are much less sensitive or even insensitive to 5-HT. If interactions do occur between the pathways which mediate analgesia from the dorsal and ventrolateral PAG, the mechanism may be pharmacologically distinct from that which modulates the cardiovascular components of the response.

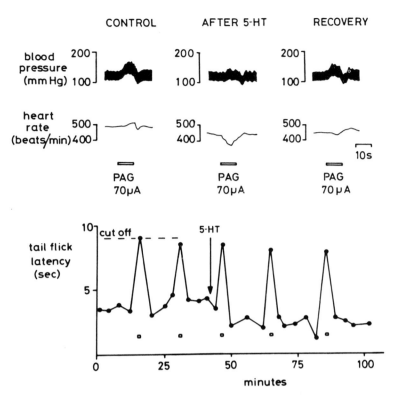

Figure 8. Effect of microinjection of 5-HT into the rostral ventrolateral medulla (RVLM) on the cardiovascular response and analgesia evoked by stimulation in the dorsal PAG. Top records. left: stimulation in the dorsal PAG (open bar) evoked a pressor response and tachycardia. Middle record: 5 min after injection of 5-HT into RVLM the pressor component of the response was reduced and the tachycardia replaced by a bradycardia. Right record: recovery of response. Bottom part of figure shows increase in latency of the tail flick response produced by stimulation in the dorsal PAG (open bar). The analgesia was still present following injection of 5-HT (arrow) into the RVLM.

Functional Significance of Descending Control from the PAG

Dorsal PAG

In order to survive animals must be able to react to novel stimuli or other potential environmental dangers with an appropriate behavioral response. Stimulation in the PAG, particularly within the region lateral to the aqueduct produces many of the motor patterns of behavioral activity which are characteristic of the animals natural response to danger (Bandler and Depaulis, this volume). The same region of the PAG has also been shown to initiate and integrate a complex series of cardiovascular and somatomotor changes and it has been suggested that these circulatory changes help to redistribute blood to appropriate vascular beds according to the metabolic demands of a particular pattern of behavior (Bandler et al., 1991).

In the anesthetised animal the pattern of cardiovascular change evoked from the dorsal half of the PAG is characterised by the redistribution of a raised cardiac output to skeletal muscle at the expense of circulation to the skin and viscera (Lovick, 1985; Hilton and Redfern, 1986). This pattern of response appears to be a fundamental reaction to many forms of threat. In the absence of anesthesia it is evoked as part of the orienting response to novel stimuli but it also occurs in response to stimuli which elicit active defensive behavior, apparently overiding other patterns of cardiovascular activity (Mancia et al., 1972 ; Martin et al., 1976). Interestingly, the analgesia evoked from the dorsal PAG was always accompanied by this same pattern of cardiovascular response. Under the extreme or emergency conditions which evoke attack or defensive behavior, an acute period of analgesia may have considerable survival value. During the analgesic phase the normal protective reflexes to a noxious stimulus (e.g., flexor withdrawal, immobilisation of the affected part) would be suppressed, thus giving priority to the execution of movements directed towards the survival of the whole animal.

Studies at the single unit level in anesthetised cats have shown that activation of the descending pathway from the dorsal PAG produces selective inhibition of peripheral input to the dorsal horn from Ad and C fibers while the response to stimulation of Ab fibers remains unchanged (Duggan and Morton, 1983). In conscious animals, defensive behavioral responses to tactile stimulation, particularly of the head region, appear to be enhanced during activation of the lateral PAG (Bandler and Depaulis, this volume). It is possible therefore, that enhanced central processing of tactile input may be yet another strategy which primes the body to respond optimally to environmental cues which trigger survival behavior.

Ventrolateral PAG

Stimulation in the ventrolateral PAG appears to produce a response which is almost the antithesis of that evoked from more dorsal sites. The sympathoexcitation and facilitation of motor activity evoked from the dorsal PAG are replaced by sympathoinhibition and immobility while the aversive nature of dorsal stimulation is reversed: electrode placements in the ventrolateral PAG become positively reinforcing and support self-stimulation behavior (Dennis et al., 1980). Although stimulation both dorsally and ventrally in the PAG produces analgesia, there are significant differences between the properties of the antinociception evoked from dorsal and ventral sites, suggesting that the quality of the sensory loss in each case may not be the same (see Morgan, this volume).

Compared to the dorsal pathway, the functional significance of the descending control pathway from the ventrolateral PAG is less clear. However, one clue may arise from the observation that in both man and experimental animals, the whole pattern of response evoked by stimulation in the ventrolateral PAG can be reproduced by electrical stimulation of peripheral nerves or by acupuncture (Omura, 1975; Kaada, 1982; Yao et al., 1982; Ernst and Lee, 1986; Olausson et al., 1986; Wang et al., 1990). Traditional acupuncture activates Group II and III afferents from skeletal muscle (Lu et al., 1981) and a characteristic feature of the response is its slow onset and long duration (Omura, 1975). An increase in small diameter afferent input to the CNS from muscle, joints and tendons would also be expected to occur during periods of intense physical exertion. Such physical activity is usually followed by a period of sympathoinhibition, immobility and reduced responsiveness to sensory stimuli which in its extreme form is manifested as sleep. The similarity between the effects of ventrolateral PAG- and muscle afferent-induced changes raises the possibility that the ventrolateral PAG may in some way be involved in mediating the longer term or recuperative phase of the response to a stressful or life-threatening encounter. This concept of a dual role for the PAG in integrating and initiating the acute and recuperative phases of the response to stressful or fear-inducing situations bears some resemblance to the Perceptual-Defensive-Recuperative model of fear and pain proposed by Bolles and Fanselow (1980). In their model the fear motivation system functions to defend the animal against natural dangers such as predation and activates appropriate defensive behaviors such as flight/freezing as well as suppressing responses to pain. Once the acute danger period has passed, the presence of pain initiates recuperative behavior characterised by immobility, sleep, etc. The results of the present physiological study point to the dorsal part of the PAG as mediator of acute defensive behavior (see also Fanselow, this volume) and the ventrolateral PAG as being involved in the recuperative phase. Bolles and Fanselow (1980) have proposed that pain acts as the initiator of recuperative behavior. This idea gains support from anatomical studies which have shown a collateral input to the PAG from spinothalamic tract cells (Harmann et al., 1988). Other lines of investigation suggest that the recuperative ventrolateral PAG control system may also be activated in other ways. Low frequency stimulation of high threshold afferents in a muscle nerve has been shown to attenuate the cardiovascular components of the defense response evoked from the dorsal PAG (Huangfu and Li, 1985, Lovick and Li, unpublished work). This effect can be abolished by lesions of the arcuate nucleus of the hypothalamus (Huangfu and Li, 1987). It has been suggested that these long lasting effects evoked by muscle afferent stimulation are mediated by activation of an endorphinergic pathway from the arcuate nucleus which excites the descending pathway from the ventrolateral PAG (Huangfu and Li, 1988). It is interesting to note that lesions of the arcuate have also been shown to abolish the analgesia produced by low frequency acupuncture (Wang et al., 1990).

As a working hypothesis it may be useful to consider the descending systems from the dorsal and ventrolateral PAG as mediating two different aspects of defensive or stress-induced behavior. The dorsal pathway appears to be activated rapidly in emergency or life-threatening situations to initiate and integrate acute physiological changes which enable the animal to respond appropriately

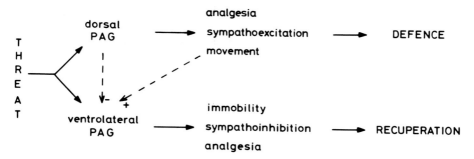

Figure 9. Schematic diagram to show possible involvement of pathways from the dorsal and ventrolateral PAG in mediating defensive responses to an external threat.

to a threat. Active defense behavior would lead to an increase in high threshold muscle afferent input to the CNS and activate the ventrolateral pathway from the PAG. Sympathetically-mediated release of β-endorphin from the adrenal medulla may also contribute to activation of the ventrolateral system by stimulation of opiate receptors in the ventrolateral PAG. A possible schema for this sequence of events is presented in Figure 9. In this model, threatening external stimuli activate neurons in both the dorsal and ventrolateral PAG. The descending pathway from the dorsal PAG initiates the autonomic, sensory and motor changes which underly the behavioral manifestation of defensive behavior. Effective defense or attack behaviour, by removing the external threat, will then lead to a reduction in activity in the descending pathway from the dorsal PAG. As activity increases in the ventrolateral pathway due to the exercize-induced increase in high threshold muscle afferent input, it will further inhibit the dorsal pathway via an inhibitory interaction in the RVLM. Eventually, as activity in the dorsal pathway wanes, the activity of the ventrolateral pathway will start to predominate, leading finally to the establishment of a recuperative phase which is characterised behaviorally by immobility associated with sympathoinhibition and reduced responsiveness to sensory stimuli.

References

Abols, I.A. and Basbaum, A.I., Afferent connections of the rostral medulla of the cat: a neural substrate for midbrain-medullary interactions in the modulation of pain, J. Comp. Neurol., 201 (1981) 285-297.

Andrezik, J.A., Chan-Palay, V. and Palay, S.L., The nucleus paragigantocellularis lateralis in the rat: demonstration of afferents by retrograde transport of horseradish peroxidase, Anat. Embryol., 161 (1981) 373-390.

Bandler, R., Carrive, C. and Zhang, S.P., Integration of somatic and autonomic reactions within the midbrain periaqueductal grey: viscerotopic, somatotopic and functional organisation, Prog. Brain Res., 87 (1991) 269-305.

Beitz, A.J., Mullett, M.A. and Weiner, L.L., The periaqueductal gray connections to the rat spinal trigeminal, raphe magnus, gigantocellularis pars alpha and paragigantocellular nuclei arise from separate neurons, Brain Res., 288 (1983) 307-314.

Besson, J.M. and Chaouch, A., Peripheral and spinal mechanisms of pain, Physiol. Rev., 67 (1987) 667-186.

Bolles, R.C. and Fanselow, M.S., A perceptual-defensive-recuperative model of fear and pain, Behav. Brain Sci., 3 (1980) 291-323.

Brown, D.L. and Guyenet, P.G., Electrophysiological study of cardiovascular neurons in the rostral ventrolateral medulla in rats, Circ. Res., 56 (1985) 359-369.

Carrive, P., Bandler, R. and Dampney, R.A.L., Viscerotopic control of regional vascular beds by discrete groups of neurons within the midbrain periaqueductal gray, Brain Res., 493 (1989) 385-390.

Carrive, P. and Bandler, R. Viscerotopic organisation of neurones subserving hypotensive reactions within the midbrain periaqueductal grey: a correlative functional and anatomical study, Brain Res., 541 (1991) 206-215.

Dennis, S.G., Choiniere, M. and Melzack, R., Stimulation-produced analgesia in rats: assessment by two pain tests and correlation with self stimulation, Exp. Neurol., 68 (1980) 295-309.

Duggan, A.W. and Morton, C.R., Periaqueductal grey stimulation: an association between selective inhibition of dorsal horn neurones and changes in peripheral circulation, Pain, 15 (1983) 237-248.

Ernst, M. and Lee, M.H.M., Sympathetic effects of manual and electrical acupuncture at the Tsusanli point: comparison with the Hoku hand point sympathetic effects, Exp. Neurol., 94 (1986) 1-10.

Foong, F. and Duggan, A.W., Brainstem areas tonically inhibiting dorsal horn neurones: studies with microinjection of the GABA analogue piperidine-4-sulphonic acid, Pain, 27 (1986) 361-371.

Gong, Q.L., Lin, R.J. and Li, P., Inhibitory effect of nucleus raphe obscurus on defense pressor response, Chin. J. Physiol. Sci., 5 (1989) 311-318.

Gray, B.G. and Dostrovsky, J.O., Descending inhibitory influences from the periaqueductal gray, nucleus raphe magnus and adjacent reticular formation I. Effects on lumbar spinal cord nociceptive and non-nociceptive neurons, J. Neurophysiol., 49 (1985) 932-947.

Harmann, P.A., Carlton, S.M. and Willis, W.D., Collaterals of spinothalamic tract cells to the periaqueductal gray: a fluorescent double labelling study in the rat, Brain Res., (1988) 87-97.

Hilton, S.M., Marshall, J.M. and Timms, R.J., Ventral medullary neurones in the pathway from the defense areas of the cat and their effect on blood pressure, J. Physiol. (Lond.), 345 (1983) 149-166.

Hilton, S.M. and Redfern, W.S., A search for brain stem cell groups integrating the defense reaction in the rat, J. Physiol., 378 (1986) 213-228.

Hoffmann, P. and Thoren, P., Electrical muscle stimulation in the hindleg of the spontaneously hypertensive rat induces a long-lasting fall in blood pressure, Acta Physiol. Scand., 133 (1988) 211-219.

Holstege, G. and Kuypers, H.G.J.M., The anatomy of brainstem pathways to the spinal cord of the cat. A labelled amino acid tracing study, Brain Res., 57 (1982) 145-175.

Huangfu, D.H. and Li, P., The effects of deep peroneal nerve inputs on defense reaction elicited by brainstem stimulation, Chin. J. Physiol. Sci., 1 (1985) 176-184.

Huangfu, D.H. and Li, P., Role of nucleus arcuatus in the inhibitory effect of deep peroneal nerve inputs on defensive reaction, Chin. J. Physiol. Sci. 3 (1987) 37-46.

Huangfu, D.H. and Li, P., The inhibitory effect of ARC-PAG-NRO system on the ventrolateral medullary neurones in the rabbit, Chin. J. Physiol. Sci., 4 (1988) 115-125.

Kaada, B., Vasodilatation induced by transcutaneous nerve stimulation in peripheral ischaemia (Raynauds' phenomenon and diabetic neuropathy), Europ. Heart J., 3 (1982) 305-314.

Lai, Y.Y. and Siegel, J.M., Cardiovascular and muscle tone changes produced by microinjection of cholinergic and glutaminergic agonists in dorsolateral pons and medial medulla, Brain Res., 114 (1990) 27-36.

Lakos, S. and Basbaum, A.I., An ultrastructural study of the projections from the midbrain periaqueductal gray to spinally-projecting serotonin immunoreactive neurons of the medullary raphe magnus in the rat, Brain Res., 443 (1988) 383-388.

Li, P. and Lovick, T.A., Excitatory projections from hypothalamic and midbrain defense areas to nucleus paragigantocellularis lateralis in the rat, Exp. Neurol., 89 (1985) 543-553.

Loewy, A.D. and McKellar, S., Serotonergic projections from the ventral medulla to the intermediolateral cell column in the rat, Brain Res., 21 (1981) 146-152.

Lovick, T.A., Ventrolateral medullary lesions block the antinociceptive and cardiovascular responses elicited by stimulating the dorsal periaqueductal grey matter in rats, Pain, 21 (1985) 241-252.

Lovick, T.A., Analgesia and the cardiovascular changes evoked by stimulating neurones in nucleus paragigantocellularis lateralis in the rat, Pain, 25 (1986a) 259-268.

Lovick, T.A., Projections from brainstem nuclei to nucleus paragigantocellularis lateralis in the cat, J. Auton. Nerv. Syst., 16 (1986b) 1-11.

Lovick, T.A., Convergent afferent inputs to nucleus paragigantocellularis lateralis in the cat, Brain Res., 456 (1988) 483-487.

Lovick, T.A., Selective modulation of the cardiovascular response but not the antinociception evoked from the dorsal PAG by 5-HT in the ventrolateral medulla, Pflugers Arch., 416 (1990) 222-224.

Lovick, T.A., Central nervous integration of pain and autonomic function, News in Physiological Sciences, 6 (1991) 82-86.

Lovick, T.A. and Li, P., Integrated activity of neurones in the rostral ventrolateral medulla, Prog. Brain Res., 81 (1989) 223-232.

Lovick, T.A., Smith, P.R. and Hilton, S.M., Spinally-projecting neurones near the ventral surface of the medulla in the cat, J. Auton. Nerv. Syst., 11 (1984) 27-33.

Lovick, T.A., West, D.C. and Wolstencroft, J.H., Responses of raphespinal and other raphe neurones to stimulation of the periaqueductal grey matter in the cat, Neurosci. Lett., 8 (1978) 45-49.

Lu, G.W., Xie, J.Q., Yang, Y.N. and Wang, Q.L., Afferent nerve fiber composition at point zusanli in relation to acupuncture analgesia: a functional morphological investigation, Chin. Med. J., 95 (1981) 255-263.

Mancia, G., Baccelli, G. and Zanchetti, A., Hemodynamic responses to different emotional stimuli in cats: patterns and mechanisms, Am. J. Physiol., 223 (1972) 925-933.

Mantyh, P., Connections of the midbrain periaqueductal gray in the monkey II. Descending efferent projections, J. Neurosci., 9 (1983) 582-593.

Martin, G.F., Cabana, T., Humbertson, A.O., Laxson, L.C. and Panneton, W.M., Spinal projections from the medullary reticular formation of the North American opossum: evidence for connectional heterogeneity, J. Comp. Neurol., 196 (1981) 663-682.

Martin, J.E., Sutherland, C.J. and Zbrozyna, A.W., Habituation and conditioning of the defense reactions and their cardiovascular components in dogs and cats, Pflugers Arch., 365 (1976) 37-47.

Nicholas, A.P. and Hancock, M.B., Evidence for projections from the rostral medullary raphe onto medullary catecholamine neurons in the rat, Neurosci. Lett., 108 (1990) 22-28.

Olausson, B., Eriksson, E., Ellmarker, L., Rydenhag, B., Shyu, B.C. and Andersson, S.A., Effects of naloxone on dental pain threshold following muscle exercize and low frequency transcutaneous nerve stimulation: a comparative study in man, Acta Physiol. Scand., 126 (1986) 299-305.

Omura, Y., Patho-physiology of acupuncture treatment: effects of acupuncture on cardiovascular and nervous systems, Acupunct. Electrotherap Res., 1 (1975) 51-141.

Prieto, G.J., Cannon, J.T. and Liebeskind, J.C., N. raphe magnus lesions disrupt stimula-
    tion-produced analgesia from ventral but not dorsal midbrain sites in the rat, Brain
    Res., 261 (1983) 53-57.
Reynolds, D.V., Surgery in the rat during electrical analgesia produced by focal brain sti-
    mulation, Science, 164 (1969) 444-445.
Sankühler, J. and Gebhart, G.F., Relative contributions of nucleus raphe magnus and adja-
    cent medullary reticular formation to the inhibition by stimulation in the periaque-
    ductal gray of a spinal nociceptive reflex in the pentobarbital-anesthetised rat, Brain
    Res., 305 (1984) 77-87.
Sidall, P.J. and Dampney, R.A.L., Relationship between cardiovascular neurones and des-
    cending antinociceptive pathways in the rostral ventrolateral medulla of the cat, Pain,
    37 (1989) 357-364.
Steinbusch, H.W.M., Distribution of serotonin-immunoreactivity in the central nervous
    system of the rat - cell bodies and terminals, Neuroscience, 6 (1981) 557-618.
Wang, Q., Mao, L. and Han, J.S., The arcuate nucleus of hypothalamus mediates low but
    not high frequency electroacupuncture analgesia in rats, Brain Res., 513 (1990) 60-66.
Wiklund, L., Behzadi, G., Kalea, P., Headley, P.M., Nicolopoulos, L.S., Parsons, C.G. and
    West, D.C., Autoradiographic and electrophysiological evidence for excitatory amino
    acid transmission in the periaqueductal grey projection to nucleus raphe magnus in
    the rat, Neurosci. Lett., 93 (1988) 158-163.
Yao, T., Andersson, S. and Thoren, P., Long lasting cardiovascular depression induced by
    acupuncture-like stimulation of the sciatic nerve in unanaesthetised hypertensive rats,
    Brain Res., 240 (1982) 77-85.

Analgesia Produced by Stimulation of the Periaqueductal Gray Matter : True Antinoceptive Effects Versus Stress Effects

Jean-Marie Besson*, Véronique Fardin**
and Jean-Louis Olivéras*

*Unité de Recherches de Physiopharmacologie
du Système Nerveux de l'INSERM (U.161), Paris, France
and **Département de Biologie, Rhône-Poulenc-Santé
Centre de Recherches de Vitry, Vitry-sur-Seine, France

Introduction

Since Reynold's initial report years ago (1969) revealed that during and after periaqueductal gray matter (PAG) stimulation rats could support severe, painful interventions such as laparotomies, the "stimulation produced analgesia phenomenon" (SPA) has been extensively studied in the rat, cat and monkey (see reviews in Liebeskind et al., 1976; Mayer and Price, 1976; Basbaum and Fields, 1978; Mayer, 1979; Besson et al., 1981; Basbaum and Fields, 1984; Olivéras and Besson, 1988). Furthermore, SPA from central gray (anatomically including the caudal periventricular hypothalamic region and the PAG) has been used to relieve severe intractable pain in human patients (Adams, 1976; Gybels et al., 1976; Hosobuchi et al., 1977; 1979; Richardson and Akil, 1977a,b). However, after an initial rush of enthusiasm by neurosurgical teams involved in this pain treatment, the use of central gray deep brain stimulation became controversial (Gybels et al., 1976; Meyerson, 1988). These controversies are justifiable in light of certain behavioral observations: the PAG and the central gray are complex regions involved in diverse functions and states including not only analgesia but also "intense emotion" (very unpleasant to intolerable sensations) when electrically stimulated in both human beings and animals. The unpleasant effects of such a stimulation have often been described since the initial investigations (Magoun et al., 1937; Kelly et al., 1946; Nashold et al., 1969; 1974) but are not

The Midbrain Periaqueductal Gray Matter, Edited by A. Depaulis and
R. Bandler, Plenum Press, New York, 1991

acceptable in the case of a medical treatment for intractable pain in humans. Considering that the effective sites which produce analgesia and intense intolerable emotion are included in the same structure, the determination of the key regions for producing "pure" analgesic effects is critical for the development of reliable treatment.

On the basis of the extensive mapping that we have performed in the cat (Guilbaud et al., 1972; Liebeskind et al., 1973; Olivéras et al., 1974a; 1978; 1979), it appears that the effective zone for producing "pure" analgesia (without other noticeable behavioral change) is predominantly the Dorsal Raphé Nucleus (DRN) located in the ventral part of the PAG. Alternatively, we reported that stimulation of the dorsal PAG systematically produced strong aversive reactions such as coordinated attack and escape which could mask the potential analgesic effects during the application of central stimulation. In contrast, all the literature devoted to SPA in the rat from 1969 to 1982 (Reynolds, 1969; Mayer et al., 1971; Akil et al., 1972; Mayer and Liebeskind, 1974; Akil and Liebeskind, 1975; Basbaum et al., 1976; Giesler and Liebeskind, 1976; Morrow and Casey, 1976; Pert and Walter, 1976; Soper, 1976; Yaksh et al., 1976a; Rhodes and Liesbeskind, 1978; Dennis et al., 1980; Cannon et al., 1982) clearly indicates that the analgesic effects can be induced by stimulation of large areas throughout the whole PAG (dorsal, lateral and ventral). Unfortunately, these studies in rats have provided a paucity of information on the possible secondary behavioral effects induced by central stimulation (possibly masked by the animal contention generally used in these studies). In other words, the above cited studies have overlooked or ignored the "secondary" behavioral effects of PAG stimulation. This lack of information is surprising considering that it is well documented that the electrical stimulation of the PAG in the rat can trigger stereotyped behavioral effects which have been used as a model for aversion by numerous investigators (Olds and Olds, 1962; Kiser and Lebovitz, 1978; Schmitt and Karli, 1980; Sandner et al., 1987; Di Scala and Sandner, 1989).

As revealed by our systematic reinvestigation performed in the freely moving rat (Fardin et al., 1981; 1984a,b), the discrepancies between cat and rat appeared to result from the definition of the so-called SPA phenomenon associated or not with other behavioral manifestations produced by the central stimulation itself. The present paper will overview some of the behavioral aspects of SPA with a particular emphasis on the rat. In this species it was possible to study, with precision, the pure analgesic effects and those effects associated with other behavioral manifestations (some of which reflected a possible stress imposed on the animal). This perspective will reveal the possible pitfalls arising from deep brain electrical stimulation in terms of effectiveness of analgesia and may also give one possible explanation to the relative lack of success of this approach in human beings. The behavioral data obtained emphasize the idea that SPA elicited from PAG must be considered as a complex and heterogeneous phenomenon and demonstrate that "pure" antinociceptive effects are found in restricted PAG areas.

Marked Differences in the Distribution of SPA Effective Sites According to the Presence of Additional Behavioral Manifestations

The discrepancies between cat and rat in regard to the localization of the SPA effective sites underlined above, led us to reinvestigate SPA from PAG and adjacent structures with two major objectives: 1) to define the effective analgesic zone(s) as accurately as possible based upon a large number of stimulation sites; and 2) to consider possible central stimulation side effects in freely moving rats since, as previously stated, most of the literature studies were performed in restrained or semi-restrained animals. The precision of such an investigation was linked to the fact that bipolar concentric electrodes were used, giving a more focal PAG electrical stimulation than the twisted wires electrodes commonly used in previous studies (Bures et al., 1983). The analgesia was evaluated during central stimulation by measuring the changes of the vocalization threshold induced by electrical tail shocks (trains of 500-ms duration containing 25 rectangular shocks of 1-ms width), but also by means of the modifications of the nocifensive reactions to noxious pinch.

The most striking result of the study carried out on 129 freely moving rats was that in order to obtain antinociceptive effects (rises of the vocalization threshold) from all parts of the PAG, it was necessary to apply central stimulation intensities which triggered additional other marked behavioral modifications (Fig. 1) we will consider in details in the next section. These concomitant behavioral reactions have not been reported by previous investigators with the exception of Pert and Walter (1976) and Yeung et al. (1977; see, however, the more recent studies of Morgan et al., 1987 and Prado and Roberts, 1985). Since there is a potential interference between these secondary behavioral manifestations and the evaluation of the analgesic effects, we have designed two experimental situations: 1) analgesia was gauged during central stimulation at currents just subthreshold to those inducing an observable secondary behavioral response; and 2) analgesia was evaluated in the same animal during threshold or suprathreshold stimulation currents which triggered the other behavioral reactions. Using PAG stimulation which did not induce such side effects (Fig. 1A), very few analgesic effects (21/129 sites) were obtained notwithstanding the parameters of central stimulation used (sine waves or biphasic rectangular pulse pairs). During central stimulation, the animals were quiet and normally attentive to innocuous stimuli. The only noticeable behavioral change consisted of the appearance of minor exploratory motion. For most of the sites, no variations of the vocalization threshold were produced by stimulation: these stimulating points were widely distributed throughout the PAG and neighboring structures.

In the ventral part of the PAG, although central stimulation did not systematically produced analgesia, it is only from sites located at this level that clear analgesic effects were obtained. As can be seen in Fig. 1A, there is no particular analgesia from caudal DRN sites (P0.8). The effective sites, spreading from the frontal plane A0.1 to P0.5, are located in two areas: the dorsomedial part of the

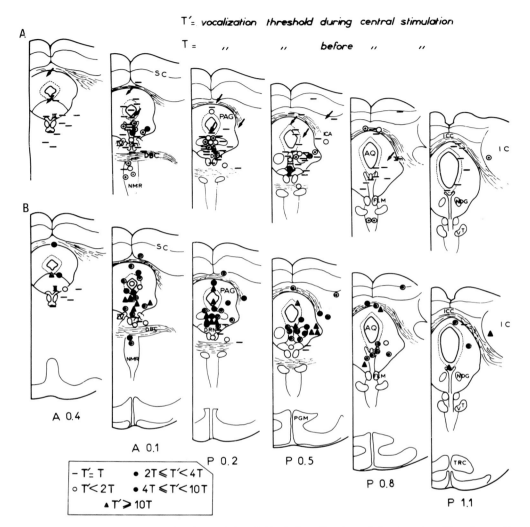

Figure 1. Rostrocaudal distribution of the Stimulation Produced Analgesia efficacious sites during PAG electrical stimulation in the freely moving rat. Analgesia was assessed by the modifications of the vocalization (squeak) threshold induced by electrical tail shocks. The arrows indicate a threshold decrease (hyperalgesia). A marked difference in the number and localization of the effective analgesic sites are observed according to the absence (A, ventral PAG) or the presence (B, all parts of the PAG) of secondary stereotyped behavioral manifestations (aversion, gnawing and rotation, see Fig. 3). Stereotaxic coordinates are according to the rat atlas of Fifkova and Marsala (1967). AQ, aqueduct of Sylvius; DBC, decussation of the brachium conjunctivum; DRN, dorsal raphé nucleus; FLM, fasciculus longitudinalis medialis; IC, inferior colliculus; ICC, intercollicular commissure; NDG, dorsalis tegmental nucleus of Gudden; NMR, nucleus raphé medianus; PAG, periaqueductal gray matter; PGM, pontine gray-medial division; SC, superior colliculus; TRC, tegmental reticular nucleus-central division; VT, ventral tegmental nucleus; III, nucleus of the oculomotor nerve.

DRN (10 sites) and a more lateral region situated just under the trigeminal mesencephalic nucleus. This latter zone contains only 4 effective sites but appears to be quite separated from the DRN: indeed, stimulation at several sites located in the lateral "wings" of the DRN did not cause any variation in the vocalization threshold (P0.2 and P0.5).

When stimulating the dorsal and lateral parts of the PAG and the ventral region surrounding the aqueduct as well as the colliculi or the adjacent intercollicular commissure, no change of vocalization threshold was obtained except in some rats in which a decrease of the threshold was observed paradoxically, indicating some "hyperalgesia". Although our observations are in good agreement with the study of Schmideck et al. (1971) who occasionally found decreases in vocalization thresholds during PAG stimulation in the awake monkey, these increases in nociceptive reactivity have not been reported in the literature. Thus, additional approaches using peripheral mechanical and thermal stimuli would be necessary to firmly established the reality of this hyperalgesia.

However, the dichotomy between the ventral and dorsal PAG disappears when the central stimulation intensity was increased (Fig. 1B). Under these conditions, analgesia was obtained from practically the entire PAG as emphasized previously, the vocalization threshold being dramatically increased on occasions. Interestingly, the greatest increases were observed primarily in the dorsomedial part of the DRN and in the ventrolateral PAG. Still under these experimental conditions, antinociceptive effects were noted from sites located outside the PAG which also produced other prominent behavioral manifestations, regardless of which area was stimulated. The extension of the "analgesic" zones to the entire PAG and surrounding areas clearly demonstrates that the production of analgesia depends considerably upon the criteria applied to define it, in terms of behavioral changes induced by the central stimulation itself. Hence, it is regrettable that, except for our study, practically no systematic investigation have considered these additional behavioral reactions as a limiting factor in the production of PAG analgesic effects. In our opinion, these facts could explain the common features and differences between cat and rat in regard to the distribution of the effective analgesic sites (Fig. 2). In the work our group performed in the cat (Guilbaud et al., 1972; Liebeskind et al., 1973; Olivéras et al., 1974a; 1979), the PAG mapping was based on gauging the gross behavioral reactions in response to strong pinches applied to extended body regions from head to tail. Any moderate (up to 100 μA) central stimulation which attenuated or suppressed all the reactions to pinch (e.g., vocalization, withdrawal, biting, flight, etc.) with no other observable behavioral modifications due to the central stimulation itself was considered to produce analgesia. With the systematic exclusion of secondary behavioral changes, we obtained in the cat a similar repartition of the analgesic sites as in the rat using the same experimental conditions, the ventral PAG, namely, the DRN and a more lateral zone. The localization to the DRN was confirmed using the increase of the threshold of the jaw-opening reflex (JOR, a nociceptive flexion reflex, Olivéras et al., 1974b).

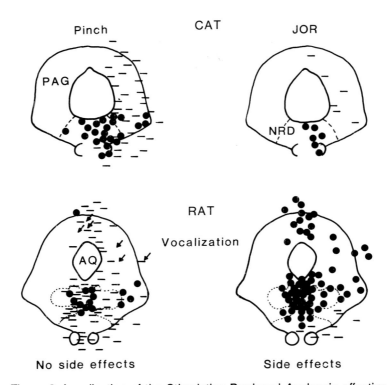

Figure 2. Localization of the Stimulation Produced Analgesia effective sites at the Periaqueductal Gray level in the freely moving cat and rat using several analgesic tests. The dots indicate the location of the efficacious analgesic sites irrespective of the strength of analgesia, the bars-no effect, the arrows-hyperalgesia. Note the similarity between the two species in the ventral PAG with no side effects due to central stimulation itself. Also note the very large distribution of the "analgesic" sites when central stimulation produces side effects together with analgesia in the rat. Abbreviations of the structures are in agreement with Fig. 1.

Figure 3 (right page). Rostrocaudal distribution of mesencephalic stimulation sites which produce stereotyped behavioral manifestations in the freely moving rat. These various coordinated behaviors are obtained from different PAG areas. The aversion (intense exploratory movements followed by flight reactions and sometimes pronounced distress vocalizations) occur within the dorsal and lateral PAG, the region just below the cerebral aqueduct and the colliculus area. Triangles: flight-jump, diamonds: flight-jump and vocalization, stars: flight-jump and motor seizure. The rotation (automatic turning movements ipsilateral to the stimulation site) is triggered from the ventral PAG (ventromedial part of the DRN, ventrolateral PAG) and the surrounding mesencephalic reticular formation. The gnawing (chewing and gnawing objects in the environment associated to intense exploration) is produced by stimulating the ventral PAG (mainly the DRN). Abbreviations are the same as in Fig. 1. Reprinted with permission from Fardin et al. (1984) Brain Research.

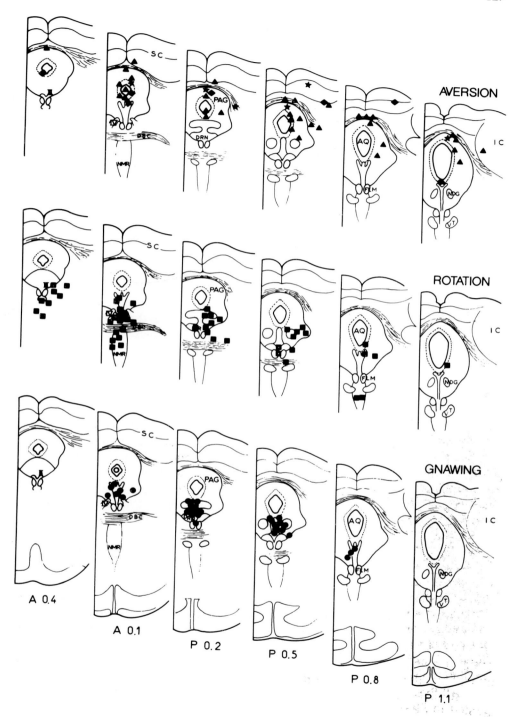

Stereotyped Behavioral Manifestations Elicited from PAG Electrical Stimulation

In the rat, these behavioral manifestations, which could be induced occasionally with low levels of central stimulation (20μA), consisted of stereotypic coordinated behaviors which clearly differed depending of the stimulation site (Fig. 3).

The gnawing behavior exclusively obtained in the ventral PAG (primarily the DRN) starts just at the onset of the central stimulation: the animal sniffs and thereafter violently gnaws and chews. If food is presented to the rats, they chew it and do not swallow it but spit it out, indicating that this behavior is different from eating.

The rotation behavior is always ipsilateral to the stimulation site located in the ventral PAG or in neighboring structures just ventral to the PAG. More precisely within the PAG, the rotation is produced by the stimulation of the ventromedial part of the DRN and a ventrolateral region including parts of the "wings" of the DRN. This behavior begins as the central stimulation is applied : a head rotation which is followed by the whole body is clearly observed. The rat then continues to rotate until the stimulation stops. This reaction follows the frequency of stimulation. When the current intensity is increased, the movements become more intense: some animals curl up completely and then whilst in this position, sometimes provoking a fall. For some stimulation sites located in the midline, there is no clear rotation but tremors of the paws and the body.

The aversive "explosive" behavior clearly contrasts with the motor reactions since it seems to be directed, solely as a means to escape. As soon as the central stimulation is applied, a "freezing" behavior associated with a polypnea is observed, rapidly followed by intense exploratory movements and a tremendous flight reaction. For some rats, central stimulation also produces distress-like vocalizations (prolonged and repeated sharp cries which did not followed the stimulation frequency). As the intensity is increased in some cases, additional complex behavioral reactions occur such as "motor seizure-like manifestations". In all cases, flight, jump and vocalizations cease at the offset of central stimulation, but some important "post-effects" such as gasping and static posture are noticed. These kinds of aversive effects are obtained in the dorsal and lateral aspects of the PAG as well as a ventral region just surrounding the cerebral aqueduct. Similar reactions are also triggered from the colliculi, the intercollicular commissure and the mesencephalic tectum just adjacent to the lateral PAG.

Our systematic mapping of these behaviors, shows that they are produced by the stimulation of clear-cut zones with a small amount of overlap (Fig. 4). Subsequent increases of the intensity of central stimulation, however, leads to more complex phenomena consisting either in an exacerbation of the initial reaction and/or the occurrence of other manifestations probably due to the spreading of current (especially for the gnawing and the rotations).

All these behavioral observations, though rarely reported in the literature devoted to SPA from PAG, are consistent with the observations of others. The

very first investigators who electrically stimulated the mesencephalic and diencephalic central gray noted the "unpleasant" nature on the induced effects
(Magoun et al., 1937; Kelly et al., 1946). Lately, it has been observed frequently
that the electrical stimulation of the dorsal and lateral parts of the PAG and the
adjacent dorsal midbrain tectum induces behavioral reactions that include marked "emotional effects" in the rat (Olds and Olds, 1962; Valenstein, 1965;
Gardner and Malmo, 1969; Wolfe et al., 1971; Schmitt et al., 1974; Kiser and
Leibovitz, 1978; Waldbillig, 1975), the cat (Spiegel et al., 1954; Hunsperger, 1956;
1959; Wada et al., 1970; Olivéras et al., 1974a; Shaikh et al., 1987) and the monkey
(Delgado, 1955; Jürgens, 1976). These emotional reactions have been even categorized as "fear-like" or "pain-like" because they are similar, to some extent, to
those induced by a peripheral noxious stimulus such as avoidance and flight,
vocalization, attack or rage and the neurovegetative pupillary dilation, piloerection, alterations of the cardiac and respiratory frequencies.

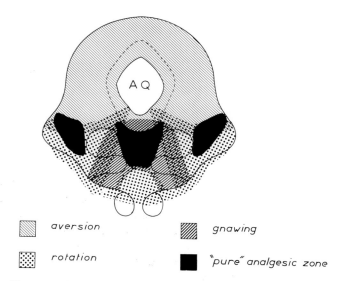

Figure 4. Summary diagram depicting the PAG areas diffe-
rentiated on the basis of the behavioral effects triggered by
central stimulation in the freely moving rat. This is done
considering the first behavioral manifestation induced by
electrical stimulation since an increase of current intensity
can produce more than one reaction according to the current
spread. These areas include all the stimulation sites from the
various frontal planes studied. This diagram illustrates the
heterogeneity of the PAG in producing behavioral changes
when electrically stimulated and emphasizes the difficulty in
producing "pure" analgesic effects. Reprinted with permission
from Fardin et al. (1984) Brain Research.

Nocifensive reactions have been also reported during PAG (Hunsperger, 1956; Olivéras et al., 1974a) and colliculi (Spiegel et al., 1954) stimulations. In the human, feelings of anxiety, distress, panic, even imminent death or precise pain sensations have been described during stimulation of these regions (Nashold et al., 1969; 1974; Cosyns and Gybels, 1979).

Gnawing and automatic motor effects have also been sometimes reported. As we emphasize in our work, these manifestations do not appear to be of an aversive nature and are produced from similar PAG areas as we found, such as the ventral part of the PAG and the subjacent mesencephalic reticular formation. More precisely, the gnawing was obtained by electrical stimulation of the DRN (Schmitt et al., 1974), the hypothalamus (Waldbillig, 1975; Roberts, 1980) which is known to project directly to the PAG (Wolf and Sutin, 1966; Saper et al., 1979; Mantyh, 1982; Holstege, 1988) or microinjection of muscimol, a GABA agonist, into the DRN (Przewlocka et al., 1979). The rotational activity, rarely described in the cat (Olivéras et al., 1974a), has been noted in the rat during ventrolateral PAG stimulation (Kiser and Lebovitz, 1978) and also after unilateral electrolytic or chemical destruction of the DRN (Jacobs et al., 1977; Giambalvo and Snodgrass, 1978; Nicolaou et al., 1979; Blackburn et al., 1980).

The existence and the importance of all these secondary stereotyped behaviors which were elicited in association with SPA, call into question the authenticity and the physiological significance of the "analgesia" triggered from large areas of the PAG. Indeed, it is not clear whether these concomitant manifestations should or should not be considered as side effects which are independent or linked with SPA. In the latter case, when the central stimulation produces complex marked behavioral reactions, analgesia may result from an inability of the animal to respond to a noxious stimulus because of the blockade of its motor functions in the case of the rotations or from a "stress" phenomenon when this stimulation is aversive. In the other case, analgesia and the secondary manifestations are strictly independent, occurring simultaneously because the neural substrates of each phenomenon is activated concomitantly by the electrical stimulation. In order to address these relevant questions, we have extended our investigations to studying the degree of both analgesia and the secondary effects as a function of the central stimulation intensity. We have also attempted to verify whether analgesics effects display different characteristics when they are induced from different regions of the PAG.

Analgesia and Side Effects as a Function of PAG Central Stimulation Intensity: Is There Interaction in the Case of Aversion?

Within the ventral PAG (Fig. 5, left), the changes in the vocalization threshold as a function of the central stimulation intensity illustrates that graded increases induce a progressive rise of the threshold.

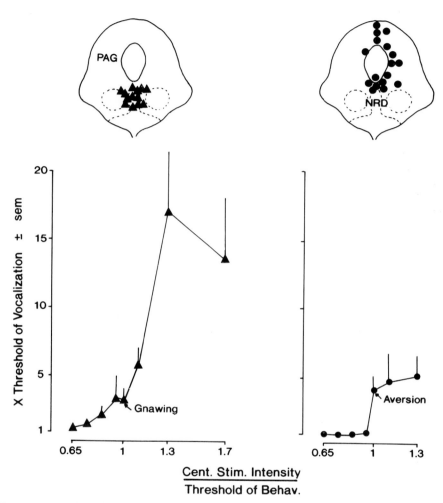

Figure 5. Strength of analgesia as a function of central stimulation intensity applied to the nucleus raphé dorsalis and the aversive zones of the PAG in the freely moving rat. For sites located in the DRN, the mean curve for analgesia illustrates that this phenomenon appears before the occurrence of gnawing (1, the abscissae indicating the ratio between the central intensity at which analgesia is tested and the threshold for the first noticeable secondary behavioral manifestation). Alternatively for the aversive zones, both analgesia and aversive effects occur concomitantly. Furthermore, increasing the current intensity does not induce further effects.

Generally, the increase of the threshold appears before the induction of any other behavioral reaction when the stimulation sites are located in the dorso-medial subdivision of the DRN and/or in the ventrolateral zone lying under the mesencephalic trigeminal nucleus (not shown in the figure). These results corroborate the existence of a "pure" distinct analgesic area. At higher intensity levels in the dorsomedial part of the DRN which produce the gnawing and a further rotation, the analgesia becomes more and more intense until a high degree. However, several observations suggest strongly that these motor reactions do not interfere with the antinociceptive effects. For example, there is no relation between the strength of analgesia and motor disturbances. Furthermore, the stimulation of the most ventral part of the PAG or subjacent mesencephalic reticular formation induces considerable motor effects without any modification of the vocalization threshold. Finally, since the ventral PAG is a small area and its subdivisions very close to each other, one can imagine that the occurrence of gnawing and motor reactions after obtaining pure analgesic effects is primarily due to a current spread to the motor zone.

This could also explain the extension of the analgesic zone to the whole ventral PAG: for example in the case of the ventrolateral motor area, the analgesia obtained after motor effects could result from a current spread to the "pure" analgesic zone. Hence, it appears that in the ventral PAG, analgesia is likely due to the activation of a neural substrate distinct from the one producing motor effects.

Within the aversive zones of the PAG (dorsal, lateral and the ventral region surrounding the aqueduct), the results are quite different (Fig. 5, right): although the magnitude of analgesia is comparable to that we obtained in the DRN before the occurrence of gnawing, the analgesic effects appear, all of a sudden, with aversion. This fact suggests that there is a possible interaction with analgesia and aversion at this level. A further increase in stimulation produces very strong aversive reactions but does not extend the degree of analgesia. Similar characteristics are also obtained from stimulation of zones located outside the PAG (i.e., colliculus, intercollicular commissure and adjacent tectum).

All of these behavioral observations are reminiscent of pharmacological and behavioral data which illustrate the dichotomy of the PAG in regard to the ventral and more dorsal parts; it emphasizes to some extent that for the aversive zones, analgesia is clearly associated with strong emotional disturbances. Indeed, although antinociceptive effects have been reported during morphine microinjections within various PAG areas (see review of Yaksh and Rudy, 1978), Sharpe et al. (1974) have shown that, in the rat, small doses of morphine were efficacious only when injected in the ventral PAG; similar results have been described by Yaksh et al. (1976b). Interestingly, using higher doses, morphine became effective at sites over the whole PAG, but the analgesia was accompanied with pronounced behavioral perturbations such as "episodes of violent uncoordinated motor activity accompanied occasionally with vocalization, ipsilateral circling behaviour and hyperreactivity as a response to a variety of stimuli" (see also the results of Jacquet and Lajtha, 1973; 1974). The dorso-ventral PAG partition is also observed when one considers the self-stimulation phenomenon: the rats never exhibit self-stimulation with electrodes placement in the dorsal and lateral PAG. Moreover, the animal will stop electrical stimulation (Olds and Olds, 1962; Valenstein, 1965; Gardner and Malmo, 1969; Schmitt et al., 1974; 1977). In contrast, the self-stimulation phenomenon has been reported in the ventral part of the PAG (Liebman et al., 1970; Schmitt et al., 1974; 1977).

Although we cannot exclude the possibility that there is a simple concomitance between analgesia and aversion, our results and the afore mentioned citations strongly suggest that antinociception is a consequence of aversion, and therefore question the "authenticity" of analgesia from aversive regions of the PAG. However, our behavioral approach only provides correlative data and no direct evidence that analgesia is a consequence of aversion. Nevertheless, one can hypothesize that violent aversive reactions trigger stress which consequently influence nociceptive reactivity. Indeed, antinociception under certain stressful conditions (Akil et al., 1976; Amir and Amit, 1978; Chance et al., 1978; Hayes et al., 1978; Bodnar et al., 1980; Lewis, 1980; Millan, 1980) have been reported. Moreover in the case of aversion, we have observed that some typical behaviors outlasted the duration of the central stimulation, for example, a rat will remain in a static posture with gasping respiration. The existence and the persistence of these behavioral modifications could explain the marked analgesic post-effects obtained by stimulation of these aversive regions. On the other hand, some studies have precisely shown that stimulation of the lateral PAG in the cat and the rat can elicit the classical autonomic variations in response to stressful conditions such as increase in respiration, tachycardia, increase in blood pressure, regional vasodilation and pupillary dilation (McDougall et al., 1985; Lovick, 1985; this volume; Carrive et al., 1989; this volume).

Conclusion

Taken as a whole, the results presented in this paper emphasize the extraordinary diversity of the behavioral effects evoked electrically from the PAG. It makes complex to some extent the data concerning analgesia. However, these results give a more realistic picture of the difficulty in obtaining analgesic effects within the rat PAG. This view contrasts greatly with the relative simplistic knowledge accumulated during more than 20 years using this species as a reference. Indeed, the majority of the studies have considered this structure as uniform in terms of the production of analgesia and very rarely consider the possibility of pitfalls due to the side effects produced by the central stimulation itself. Alternatively, our major concern since the beginning of our studies in the cat and then in the rat, was to produce SPA without other apparent behavioral modifications, hence demonstrating "pure" analgesic effects. This approach was fruitful in that we have clearly shown a dichotomy between ventral and more dorsal aversive zones of the PAG in terms of the reality of analgesia. We have demonstrated that at ventral PAG level, only the stimulation of the dorsomedial part of the DRN and a more lateral zone produce "pure" analgesia. Moreover, the motor

and gnawing side effects triggered from the ventral PAG do not appear to interact with analgesia which is not the case for the aversive regions. These data are
highly relevant since some studies demonstrated clear-cut differences between
dorsal and ventral PAG, particularly those using nucleus raphé magnus (another
"pure" analgesic zone) electrolytic lesions (Prieto et al., 1983), or pharmacology
such as the serotonin synthesis blocker, p-CPA (Akil and Liebeskind, 1975),
naloxone (Cannon et al., 1982) or diazepam (Morgan et al., 1987) administrations. All these lines of evidence demonstrate that the analgesia produced by the
ventral part and the aversive zones of the PAG are dependent on different
mechanisms, the former being more of a physiological nature, the latter likely
related to changes imposed by emotional stressful conditions.

The analgesic, motor and aversive effects that we have reported during or
even after the stimulation of different PAG areas are extremely relevant in terms
of deep brain central gray stimulation in patients suffering from chronic pain.
Indeed, the limited extent of the PAG pure analgesic zones and the presence, in
their immediate vicinity of motor and aversive regions (which also exist in
human beings) complicate this form of treatment. These facts may partly explain
some of the contradictory clinical results using PAG-PVG stimulations (see refs.
in Meyerson, 1988); particularly taking into account that the exact location of the
stimulation electrode is not known with precision in human studies. However in
patients, the region stimulated is more rostral to the PAG, primarily the periventricular gray. Furthermore, this kind of stimulation is rather used for chronic
pain and not the acute nociception generally studied in animals.

References

Adams, J.E., Naloxone reversal of analgesia produced by brain stimulation in the human,
Pain, 2 (1976) 161-166.
Akil, H. and Liebeskind, J.C., Monoaminergic mechanisms of stimulation-produced anal-
gesia, Brain Res., 94 (1975) 279-196.
Akil, H. and Mayer, D.J., Antagonism of stimulation-produced analgesia by p-CPA, a
serotonin synthesis inhibitor, Brain Res., 44 (1972) 692-697.
Akil, H., Madden, J., Patrick, R.L. and Barchas, J.D., Stress-induced increase in endoge-
nous opiate peptides: concurrent analgesia and its partial reversal by naloxone, In:
Opiate and Endogenous and Opioid Peptides, Kosterlitz H.W. (Ed.), Elsevier,
Amsterdam, 1976, pp. 63-70.
Amir, S. and Amit, Z., Endogenous opioid ligands may mediate stress-induced changes in
the affective properties of pain related behavior in rats, Life Sci., 23 (1978) 1143-1152.
Basbaum, A.I. and Fields, H.L., Endogenous pain control mechanisms: review and hypo-
thesis, Ann. Neurol., 4 (1978) 451-462.
Basbaum, A.I. and Fields, H.L., Endogenous pain control systems, brainstem spinal path-
ways and endorphin circuitry, Ann. Rev. Neurosci., 7 (1984) 309-338.
Basbaum, A.I., Marley, N.J.E. and O'Keefe, J., Spinal cord pathways involved in the production of analgesia by brain stimulation, In: Advances in Pain Research and Therapy,
Vol.1, Bonica J.J. and Albe-Fessard D. (Eds.), Raven Press, New-York, 1976, pp. 511-
515.
Besson, J.M., Olivéras, J.L., Chaouch, A. and Rivot, J.P., Role of the raphé nuclei in stimulation produced analgesia, In: Serotonin, Neurochemistry and Function, Advances in

Experimental Biology and Medicine, Vol.33, Haber B., Gabay S., Issidorides T. and Alivisatos S.G.A. (Eds.), Plenum Press, New-York, 1981, pp. 153-176.

Blackburn, T.P., Forster, G.A., Heapy, C.G. and Kemp, J.O., Unilateral lesions of the dorsal raphé nucleus (DRN) and rat rotational behaviour, Eur. J. Pharmacol., 67 (1980) 427-438.

Bodnar, R.J., Kelly, D.D., Brutus, M. and Glusman, M., Stress-induced analgesia: neural and hormonal determinants, Neurosci. Biobehav. Rev., 4 (1980) 87-100.

Bures, J., Buresova, O. and Huston, J., Techniques and basic experiments for the study of brain and behavior, Elsevier, Amsterdam, 1983.

Cannon, J.T., Prieto, G.J. and Liebeskind, J.C., Evidence for opioid and non-opioid forms of stimulation-produced analgesia in the rat, Brain Res., 243 (1982) 315-321.

Carrive, P., Bandler, R. and Dampney, R.A.C., Somatic and autonomic integration in the midbrain of the unanesthetized decerebrate cat: a distinctive pattern evoked by excitation of neurones in the subtentorial portion of the midbrain periaqueductal gray, Brain Res., 483 (1989) 251-258.

Chance, W.T., White, A.C., Krynock, G.M. and Rosecrans, J.A., Conditional fear-induced antinociception and decrease binding of [^{3}H]N-leuenkephalin to rat brain, Brain Res., 151 (1978) 371-374.

Cosyns, P. and Gybels, J., Electrical central gray stimulation for pain in man, In: Advances in Pain Research and Therapy, Vol.3, Bonica J.J., Albe-Fessard D. and Liebeskind J.C. (Eds.), Raven Press, New-York, 1979, pp. 511-514.

Delgado, J.M.R., Cerebral structures in transmission and elaboration of noxious stimulation, J. Neurophysiol., 18 (1955) 267-275.

Dennis, S.G., Choinière, M. and Melzack, R., Stimulation-produced analgesia in rats: assessement by two pain tests and correlation with self-stimulation, Exp. Neurol., 68 (1980) 295-309.

Di Scala, G. and Sandner, G., Conditionned place aversion produced by microinjection of semicarbazide into the periaqueductal gray of the rat, Brain Res., 483 (1989) 91-97.

Fardin, V., Olivéras, J.L. and Besson, J.M., Effets moteurs et aversifs induits par la stimulation électrique de la substance grise périaqueducale chez le rat, C.R. Acad. Sci. (Paris), 292 (1981) 649-652.

Fardin, V., Olivéras, J.L. and Besson, J.M., A reinvestigation of the analgesic effects induced by stimulation of the periaqueductal gray matter in the rat. I. The production of behavioral side effects together with analgesia, Brain Res., 306 (1984a) 105-123.

Fardin, V., Olivéras, J.L. and Besson, J.M., A reinvestigation of the analgesic effects induced by stimulation of the periaqueductal gray matter in the rat. II. Differential characteristics of the analgesia induced by ventral and dorsal PAG stimulation, Brain Res., 306 (1984b) 125-139.

Fifkova, E. and Marsala, J., Stereotaxic atlas for the cat, rabbit and rat, In: Electrophysiological Methods in Biological Research, Bures J., Petran M. and Zacher J. (Eds.), Academic Press, New-York, 1967, pp. 653-731.

Gardner, L. and Malmo, R.B., Effects of low-level septal stimulation on escape: significance for limbic-midbrain interactions in pain, J. Comp. Physiol. Psychol., 68 (1969) 65-73.

Giambalvo, C.T. and Snodgrass, S.R., Biochemical and behavioral effects of serotonin neurotoxins on the nigro-striatal dopamine system: comparison of injection sites, Brain Res., 152 (1978) 555-566.

Giesler, G. Jr and Liebeskind, J.C., Inhibition of visceral pain by electrical stimulation of the periaqueductal gray matter, Pain, 2 (1976) 43-48.

Guilbaud, G., Besson, J.M., Liebeskind, J.C. and Olivéras, J.L., Analgésie induite par stimulation de la substance grise périaqueducale chez le chat: données comportementales et modifications de l'activité des interneurones de la corne dorsale de la möelle, C.R. Acad. Sci. (Paris), 275 (1972) 1055-1057.

Gybels, J., Van Hees, J. and Peluso, F., Modulation of experimentally produced pain in man by electrical stimulation of some cortical, thalamic and basal ganglia structures,

In: Advances in Pain Research and Therapy, Vol.1, Bonica J.J. and Albe-Fessard D. (Eds.), Raven Press, New-York, 1976, pp. 475-478.

Hayes, R.L., Bennett, G.L., Newlon, P.G. and Mayer, D.J., Behavioral and physiological studies of non-narcotic analgesia in the rat elicited by certain environmental stimuli, Brain Res., 155 (1978) 69-90.

Holstege, G., Direct and indirect pathways to lamina I in the medulla oblongata and spinal cord of the cat, Prog. Brain Res., 77 (1988) 47-94.

Hosobuchi, Y., Adams, J.E. and Linchitz, R., Pain relief by electrical stimulation of the central gray matter in humans and its reversal by naloxone, Science, 197 (1977) 183-186.

Hosobuchi, Y., Rossier, J., Bloom, F.E. and Guillemin, R., Stimulation of human periaqueductal gray for pain relief increases immunoreactive β-endorphin in ventricular fluid, Science, 203 (1979) 279-281.

Hunsperger, R.W., Role of substantia grisea centralis mesencephali in electrically-induced rage reactions, In: Progress in Neurobiology, Ariens Kappers J. (Ed.), Elsevier, New-York, 1956, pp. 289-292.

Hunsperger, R.W., Les représentations centrales des réactions affectives dans le cerveau antérieur et dans le tronc cérébral, Neurochirurgie, 2 (1959) 207-233.

Jacobs, B.L., Simon, S.M., Ruimy, D.D. and Trulson, M.E., A quantitative rotational model for studying serotoninergic function in the rat, Brain Res., 124 (1977) 271-281.

Jacquet, Y.F. and Lajtha, A., Morphine action at central nervous system sites in rat: analgesia or hyperalgesia depending on site and dose, Science, 182 (1973) 490-492.

Jacquet, Y.F. and Lajtha, A., Paradoxical effects after microinjection of morphine in the periaqueductal gray matter in the rat, Science, 185 (1974) 1055-1057.

Jürgens, U., Reinforcing concomitants of electrically elicited vocalizations, Exp. Brain Res., 26 (1976) 203-214.

Kelly, A.H., Beaton, L.E. and Magoun, H.W., A midbrain mechanism for facio-vocal activity, J. Neurophysiol., 1 (1946) 181-189.

Kiser, S. and Lebovitz, R.M., Anatomic and pharmocologic differences between two types of aversive midbrain stimulation, Brain Res., 155 (1978) 331-342.

Lewis, J.W., Cannon, J.T. and Liebeskind, J.C., Opioid and non opioid mechanisms of stress analgesia, Science, 208 (1980) 623-625.

Liebeskind, J.C., Guilbaud, G., Besson, J.M. and Olivéras, J.L., Analgesia from electrical stimulation of the periaqueductal gray matter in the cat: behavioral observations and inhibitory effects on spinal cord interneurons, Brain Res., 50 (1973) 441-446.

Liebeskind, J.C., Giesler, G. Jr and Urca, G., Evidence pertaining to an endogenous mechanism of pain inhibition in the central nervous system, In: Sensory Functions of the Skin of Primates, Zotterman Y. (Ed.), Pergamon Press, Oxford, 1976, pp. 561-573.

Liebman, J.M., Mayer, D.J. and Liebeskind, J.C., Mesencephalic central gray lesions and fear-motivated behavior in rats, Brain Res., 23 (1970) 353-370.

Lovick, T.A., Ventrolateral medullary lesions block the antinociceptive and cardiovascular responses elicited by stimulating the dorsal periaqueductal gray matter in rats, Pain, 21 (1985) 241-252.

Magoun, H.W., Atlas, D., Ingersoll, E.H. and Ranson, S.W., Associated facial, vocal and respiratory movements of emotional expression: an experimental study, J. Neurol. Psychopathol., 17 (1937), 241-255.

Mantyh, P.W., The ascending input to the midbrain periaqueductal gray in the primate, J. Comp. Neurol., 211 (1982) 50-64.

Mayer, D.J., Wolfe, T.L., Akil, H., Carder, B. and Liebeskind, J.C., Analgesia from electrical stimulation in the midbrain of the rat, Science, 174 (1971) 1351-1354.

Mayer, D.J., Endogenous analgesia systems: neural and behavioral mechanisms, In: Advances in Pain Research and Therapy, Vol.3, Bonica J.J., Liebeskind J.C. and Albe-Fessard D. (Eds.), Raven Press, New-York, 1979, pp. 385-410.

Mayer, D.J. and Liebeskind, J.C., Pain relief by focal electrical stimulation of the brain: an anatomical and behavioral analysis, Brain Res., 68 (1974) 73-93.

Mayer, D.J. and Price, D.D., Central nervous system mechanisms of analgesia, Pain, 2 (1976) 379-404.

Mc Dougall, A., Dampney, R. and Bandler, R., Cardiovascular components of the defence reaction evoked by excitation of neuronal cell bodies in the periaqueductal grey of the rat, Neurosci. Lett., 60 (1985) 69-75.

Meyerson, B.A., Problems and controversies in PVG and sensory thalamic stimulation as treatment for pain, Prog. Brain Res., 77 (1988) 175-188.

Millan, M.J., Gramsch, C., Przewlocki, R., Höllt, V. and Herz, A., Lesions of the hypothalamic arcuate nucleus produce a temporary hyperalgesia and attenuate stress-evoked analgesia, Life Sci., 27 (1980) 1513-1523.

Morgan, M.M, Depaulis, A. and Liebeskind, J.C., Diazepam dissociates the analgesic and aversive effects of periaqueductal gray stimulation in the rat, Brain Res., 423 (1987) 395-398.

Morrow, T.J. and Casey, K.L., Analgesia produced by mesencephalic stimulation: effect on bulboreticular neurons, In: Advances in Pain Research and Therapy, Vol.1, Bonica J.J., Albe-Fessard D., Raven Press, New-York, 1976, pp. 503-510.

Nashold, B.S. Jr, Wilson, W.P. and Slaughter, D.G., Sensations evoked by stimulation in the midbrain of man, Neurosurg., 30 (1969) 14-24.

Nashold, B.S. Jr, Wilson, W.P. and Slaughter, D.G., The midbrain on pain, In: Advances in Neurology, Vol.4, Bonica J.J. (Ed.), Raven Press, New-York, 1974, pp. 157-156.

Nicolaou, N.H., Garcia-Munoz, M., Arbuthnott, G.W. and Eccleston, D., Interactions between serotoninergic and dopaminergic systems in rat brain demonstrated by small unilateral lesions of the raphé nuclei, Eur. J. Pharmacol., 57 (1979) 295-305.

Olds, M.E. and Olds, J., Approach-escape interaction in rat brain, Ann. J. Physiol., 203 (1962) 803-810.

Olivéras, J.L. and Besson, J.M., Stimulation produced analgesia in animals: behavioral investigations, Prog. Brain Res., 77 (1988) 141-157.

Olivéras, J.L., Besson, J.M., Guilbaud, G. and Liesbeskind, J.C., Behavioral and electrophysiological evidence of pain inhibition from midbrain stimulation in the cat, Exp. Brain Res., 20 (1974a) 32-44.

Olivéras, J.L., Woda, A., Guilbaud, G. and Besson, J.M., Inhibition of the Jaw Opening reflex by electrical stimulation of the periaqueductal gray matter in the awake, unrestrained cat, Brain Res., 72 (1974b) 328-331.

Olivéras, J.L., Hosobuchi, Y., Bruxelle, J., Passot, C. and Besson, J.M., Analgesic effects induced by electrical stimulation of the nucleus raphé magnus in the rat: interaction with morphine analgesia, Abstracts 7th International Congress of Pharmacology Paris, Vol.1, 1978, pp. 119.

Olivéras, J.L., Guilbaud, G. and Besson, J.M., A map of serotoninergic structures involved in stimulation producing analgesia in unrestrained, freely moving cats, Brain Res., 164 (1979) 317-322.

Pert, A. and Walter, M., Comparison between naloxone reversal of morphine and electrical stimulation-induced analgesia in the rat mesencephalon, Life Sci., 19 (1976) 1023-1032.

Prado, W.A. and Roberts, M.H.T., An assessement of the antinociceptive and aversive effects of stimulating identified sites in the rat brain, Brain Res., 340 (1985) 219-228.

Prieto, G.J., Cannon, J.T. and Liebeskind, J.C., Nucleus Raphé magnus lesions disrupt stimulation produced analgesia from ventral but not dorsal midbrain areas in the rat, Brain Res., 261 (1983) 53-57.

Przewlocka, B., Stala, L. and Scheel-Kruger, J., Evidence that GABA in the nucleus dorsalis raphe induces stimulation of locomotor activity and eating behavior, Life Sci., 25 (1979) 937-846.

Reynolds, D.V., Surgery in the rat during electrical analgesia by focal brain stimulation, Science, 164 (1969) 444-445.

Rhodes, D.L. and Liebeskind, J.C., Analgesia from rostral brainstem stimulation in the rat, Brain Res., 143 (1978) 521-532.

Richardson, D.E. and Akil, H., Pain reduction by electrical brain stimulation in man. I. Acute administration in periaqueductal and periventricular sites, J. Neurosurg., 47 (1977a) 178-183.

Richardson, D.E. and Akil, H., Pain reduction by electrical brain stimulation in man. II. Chronic self-administration in the periventricular gray matter, J. Neurosurg., 47 (1977b) 184-194.

Roberts, W.W., [¹⁴C] deoxyglucose mapping of first-order projections activated by stimulation of lateral hypothalamic sites eliciting gnawing, eating and drinking in rats, J. Comp. Neurol., 194 (1980) 617-638.

Sandner, G., Schmitt, P. and Karli, P., mapping of jumping, rearing, squealing and switch-off behaviors elicited by periaqueductal gray stimulation in the rat, Physiol. Behav., 39 (1987) 333-339.

Saper, C.B., Swanson, L.W. and Cowan, R.J., An autoradiographic study of the efferent connections of the lateral hypothalamic area in the rat, J. Comp. Neurol., 183 (1979) 689-706.

Schmideck, H.H., Fohanno, D., Ervin, F.R. and Sweet, W.H., Pain threshold alterations by brain stimulation in the monkey, J. Neurosurg., 35 (1971) 715-722.

Schmitt, P. and Karli, P., Escape induced by combined stimulation in medial hypothalamus and central gray, Physiol. Behav., 24 (1980) 111-121.

Schmitt, P., Echancher, F. and Karli, P., Etude des systèmes de renforcement négatif et de renforcement positif au niveau de la substance grise centrale chez le rat, Physiol. Behav., 12 (1974) 271-279.

Sharpe, L.G., Garnett, J.E. and Cicero, T.J., Analgesia and hypersensitivity produced by intracranial microinjections of morphine into the periaqueductal gray matter of the rat, Behav. Biol., 11 (1974) 303-313.

Shaikh, M.B., Barrett, S.A. and Siegel, A., The pathways mediating affective defense and quiet biting attack behavior from the midbrain central gray of the cat: an autoradiographic study, Brain Res., 437 (1987) 9-25.

Soper, W.Y., Effects of analgesic midbrain stimulation on reflex withdrawal and thermal escape in the rat, J. Comp. Physiol. Psychol., 90 (1976) 91-101.

Spiegel, E.A., Kletzkin, M. and Szekely, E.G., Pain reactions upon stimulation of the tectum mesencephali, Neuropathol. Exp. Neurol., 13 (1954) 212-220.

Valenstein, E.S., Independence of approach and escape reactions to electrical stimulation of the brain, J. Comp. Physiol. Behav. Psychol., 60 (1965) 120-130.

Wada, J.A., Matsuda, M., Jung, E. and Hamm, A.E., Mesencephalically induced escape behavior and avoidance performance, Exp. Neurol., 29 (1970) 215-220.

Waldbillig, R.J., Attack, eating, drinking and gnawing elicited by electrical stimulation of rat mesencephalon and pons, J. Comp. Physiol. Psychol., 89 (1975) 200-212.

Wolf, G. and Sutin, J., Fiber degeneration after lateral hypothalamic lesion in the rat, J. Comp. Neurol., 127 (1966) 137-156.

Wolfe, T.L., Mayer, D.J. and Liebeskind, J.C., Motivational effects of electrical stimulation in dorsal tegmentum of the rat, Physiol. Behav., 7 (1971) 569-574.

Yaksh, T.L. and Rudy, T.A., Narcotic analgesics: CNS sites and mechanisms of action as revealed by intrathecal injection techniques, Pain, 4 (1978) 299-359.

Yaksh, T.L., Yeung, J.C. and Rudy, T.A., An inability to antagonize with naloxone the elevated nociceptive thresholds resulting from electrical stimulation of the mesencephalic central gray, Life Sci., 18 (1976a) 1193-1198.

Yaksh, T.L., Yeung, J.C. and Rudy, T.A., Systematic examination in the rat of brain sites sensitive to the direct application of morphine: observation of differential effects within the periaqueductal gray, Brain Res., 114 (1976b) 83-103.

Yeung, J.C., Yaksh, T.L. and Rudy, T.A., Concurrent mapping of brain sites for sensitivity of the direct application of morphine and focal electrical stimulation in the production of antinociception in the rat, Pain, 4 (1977) 23-40.

Differences in Antinociception Evoked from Dorsal and Ventral Regions of the Caudal Periaqueductal Gray Matter

Michael M. Morgan

Department of Neurology
University of California, San Francisco
California, USA

Introduction

Stimulation-produced antinociception (SPA) from the periaqueductal gray matter (PAG) was first reported by Reynolds in 1969. Mayer and co-workers (1971) further characterized this phenomenon, demonstrating that stimulation at sites throughout the PAG inhibited nociception. Subsequently, the PAG has been thought of as a homogeneous structure in producing antinociception. A recurring theme throughout this book, however, is the parcellation of the PAG into discrete functional regions. Consistent with this theme, regional differences in the characteristics of SPA from the PAG have recently been described. In the preceding chapter, Besson and co-authors cite the different behavioral reactions produced by stimulation of various PAG regions as a means of distinguishing distinct nociceptive modulatory systems. Associated behavioral reactions, however, provide no information about antinociception *per se*, and are thus insufficient criteria to define different antinociceptive systems. Nonetheless, data from studies assessing the characteristics of SPA provide compelling evidence for two nociceptive modulatory systems in the caudal PAG. Although the specific boundaries for these systems have not been clearly defined, the dorsal system seems to encompass the dorsomedial, dorsolateral, and lateral subdivisions of the PAG. The ventral system seems to include the ventrolateral subdivision of the PAG and the dorsal raphe nucleus. The objectives of this chapter are: 1) to analyze further the behavioral reactions associated with SPA from different regions of the caudal PAG, and 2) to describe the characteristics of antinociception that distinguish two modulatory systems for nociception mediated by the caudal PAG.

Behavioral Reactions Associated with SPA from the PAG

The most obvious difference in SPA obtained from different regions of the PAG is in the associated stimulation-produced behavioral reactions. These behavioral reactions have been described in detail by Fardin, Olivéras and Besson (1984a and b) and in the preceding chapter by Besson et al. These authors have shown that electrical stimulation of the lateral, dorsolateral, and dorsomedial regions of the PAG produce an antinociception that is associated with aversive reactions such as running and jumping. They interpret this association as evidence that antinociception produced by stimulating these regions is a secondary response "imposed by emotional stressful conditions". In other words, these authors and others (Prado and Roberts, 1985) suggest that SPA mediated by the dorsal PAG system is merely a form of stress-induced antinociception. In contrast, stimulation of discrete regions of the ventrolateral PAG and dorsal raphe nucleus produce antinociception that is not linked to obvious behavioral reactions. Thus, antinociception from these ventral PAG regions has been described as "pure" (Fardin et al., 1984a). Characterizing SPA on the basis of associated behavioral reactions, however, is not meaningful without knowing the nature of the relationship between the stimulation-produced effects. The relationship between aversive reactions and antinociception could be causal, as postulated by Besson and co-authors, or merely coincident. In addition, the strength of the link between stimulation-produced antinociception and behavioral reactions is limited by the sensitivity of the behavioral assessment and by experimenter bias. For instance, subtle behaviors such as gnawing or freezing may be overlooked or assumed to be less relevant to SPA than more pronounced behavioral reactions such as running and jumping.

Electrical stimulation of the PAG commonly produces aversive reactions, and stimuli that elicit aversive reactions have been shown to inhibit nociception (Fanselow and Lester, 1988; Maier, 1989). In spite of the correlation between aversive reactions and antinociception produced by PAG stimulation, recent evidence suggests that these stimulation produced effects are not causally related. If antinociception is secondary to the aversive reactions produced by stimulation of dorsal PAG regions, then manipulations that attenuate the aversive reactions should lead to a concomitant attenuation of antinociception. One manipulation known to block stimulation-produced aversive reactions from the PAG is pretreatment with the anxiolytic diazepam (Audi and Graeff; 1984; Graeff and Rawlins, 1980). When rats were treated with diazepam (1 mg/kg), stimulation-produced aversive reactions were prevented in 12 of 20 rats which previously exhibited such effects (Morgan et al., 1987). This reduction in the incidence of aversive reactions did not block SPA. In fact, diazepam allowed antinociception to be measured at stimulation sites and current intensities at which aversive reactions previously had prevented assessment of nociception. These results suggest that the close association between stimulation-produced aversive reactions and antinociception is coincident. It seems likely that both effects are part of an integrated defense response. The cardiovascular changes produced by activation of dorsal PAG regions are consistent with such a hypothesis (see Carrive et al., 1989; Carrive, this volume; Duggan and Morton, 1983; Lovick, 1985).

An additional complication in describing antinociception in terms of associated behavioral reactions is that subtle behavioral reactions may be overlooked or deemed less relevant. The lack of overt behavioral reactions associated with SPA from the dorsal raphe nucleus and ventrolateral PAG has resulted in these regions being described as supporting "pure" antinociception (Fardin et al., . 1984a). Recent work, however, has shown that behavioral reactions can be produced by selectively activating cell bodies throughout the caudal PAG (Bandler and Depaulis, this volume; Gold et al., 1990; Zhang et al., 1990). Microinjection of excitatory amino acids into the lateral part of the caudal third of the PAG produces a flight reaction consisting of running and jumping. Similar injections in and around the dorsal raphe nucleus produce an immobile posture (Gold et al., 1990). This posture appears to be indistinguishable from that conditioned by aversive footshock and has been termed freezing (see Fanselow, this volume; although see also Bandler and Depaulis, this volume). Each of these PAG-mediated species-specific defense reactions (flight and freezing) occurs concomitant with antinociception (Gold et al., 1990). Thus, antinociception mediated by ventral PAG regions is no more "pure" than that produced from dorsal PAG regions. The only difference is that freezing or immobility is a rather subtle behavior compared to the flight reaction evoked from dorsal PAG regions.

It must be emphasized that although the association between behavioral reactions and antinociception from PAG stimulation suggests an integrated response, describing associated behavioral reactions provides no information about the characteristics of SPA. In other words, the existence of different forms of SPA cannot be inferred solely on the basis of different stimulation-produced behavioral reactions. A number of recent studies have shown that the characteristics of SPA do indeed differ depending on the PAG region stimulated. These differences are discussed below.

Characteristics of SPA from the Dorsal and Ventral Regions of the Caudal PAG

Involvement of Opioids

Early studies showing that brain stimulation could inhibit nociception suggested the presence of an endogenous morphine-like transmitter (Mayer and Liebeskind, 1974; Mayer et al., 1971). This hypothesis was strengthened by the discovery of endogenous opioids by Hughes and coworkers in 1975. Soon after, Akil, Mayer, and Liebeskind (1976) found evidence for endogenous opioid involvement in SPA by showing that the opiate antagonist naloxone attenuated SPA from the PAG. Although it is now well known that endogenous opioids mediate some forms of antinociception (Akil et al., 1984), the involvement of opioids in SPA from the PAG is controversial. A number of researchers have failed to see an effect of naloxone on SPA (Aimone et al., 1987; Pert and Walter; 1976; Yaksh et al.,

1976a), and even when effective, naloxone only causes a slight attenuation of SPA (Akil et al., 1976; Cannon et al., 1982).

The controversy surrounding the effect of naloxone on SPA from the PAG seems to be attributable, in part, to the site of the stimulating electrode. Cannon and co-workers (1982) found that naloxone elevated SPA thresholds in rats with stimulation sites in the dorsal raphe nucleus, but not at stimulation sites dorsal to this nucleus. This observation was the first indication that two modulatory systems for nociception might co-exist in the midbrain. A number of studies have since shown that naloxone attenuates SPA when stimulation sites are located in and around the dorsal raphe nucleus (Millan et al., 1987; Millan et al., 1986; Morozova and Zvartau, 1986; Nichols et al., 1989; Thorn et al., 1989). Antinociception from stimulation of the PAG rostral (Aimone et al., 1987) or dorsal (Thorn et al., 1989) to the dorsal raphe nucleus is not attenuated by naloxone.

Modulation of nociception from the PAG is known to occur in the spinal dorsal horn via a relay in the rostral ventromedial medulla (RVM) (Basbaum and Fields, 1984). Opioid peptides are located at all levels of this system (Akil et al., 1984; Basbaum and Fields, 1984). At the level of the PAG, a potent antinociception can be produced by direct morphine microinjection (Jacquet and Lajtha, 1974; Yaksh et al., 1976b). Consistent with the site specificity described above, the ventral quadrant of the caudal PAG shows the greatest concentration of sites sensitive to the antinociceptive effect of morphine microinjection, with the shortest latency to antinociception onset (Yaksh et al., 1976b). Work by Millan and coworkers (1986) suggests that naloxone attenuates SPA by blocking endogenous ß-endorphin in the ventral aspect of the PAG. Cells containing ß-endorphin are located in the arcuate nucleus and project to the PAG and dorsal raphe nucleus (Bloom et al., 1978). Destruction of this pathway causes a reduction in ß-endorphin levels in the PAG and an increase in the stimulation current necessary to inhibit nociception from the ventral aspect of the PAG (Millan et al., 1986). In addition, electrical or chemical stimulation of the arcuate nucleus has been shown to inhibit nociception (Fan et al., 1984; Wang et al., 1990).

The PAG circuitry underlying opioid antinociception is not known, although evidence has accumulated to suggest that PAG opioids produce antinociception by a process of disinhibition. GABA agonists have been shown to attenuate, and GABA antagonists to potentiate the antinociceptive effect of PAG morphine (Depaulis et al., 1987; Moreau and Fields, 1986). A plausible explanation for these findings is that inhibition of tonically active GABAergic neurons by opioids allows an increase in the activity of PAG output neurons. An increase in the activity of PAG output neurons, be it by electrical stimulation (Mayer et al., 1971), microinjection of excitatory amino acids (Behbehani and Fields, 1979; Urca et al., 1980), or blockade of GABA transmission (Moreau and Fields, 1986; Sandkühler et al., 1989), is known to inhibit nociception. The weak effect of opiate antagonists on SPA from the PAG presumably results from the simultaneous activation of intrinsic opioid containing terminals and PAG output neurons. Depending on stimulation parameters a sufficient number of output neurons may be activated to render recruitment of opioid neurons irrelevant.

Administration of opiates at the spinal level also is known to inhibit nociception (Yaksh and Rudy, 1976; 1977). The contribution of spinal opioids in descending inhibition from the PAG seems limited, however. Intrathecal administration of serotonergic and noradrenergic antagonists have been shown to block PAG mediated antinociception (Aimone and Gebhart, 1986; Camarata and Yaksh, 1985; Jensen and Yaksh, 1984; Tseng and Tang, 1989). Intrathecal naloxone also has been shown to attenuate descending inhibition, but only when this antinociception is evoked from the dorsal raphe nucleus (Tseng and Tang, 1989). Interestingly, induction of a unilateral hindpaw inflammation causes an increase in spinal dynorphin and enkephalin (Iadarola, et al., 1988a, b). Stimulation of the PAG of rats with unilateral hindpaw inflammation inhibits nociception in that limb and this antinociception is blocked by intrathecal administration of an opiate antagonist (Morgan et al., 1991). This attenuation of SPA by an intrathecally administered opiate antagonist occurs only with stimulation of ventral PAG regions.

Sensitivity to Tolerance
It is well known that tolerance develops to the antinociceptive effect of opiates. Consistent with the involvement of opioids in SPA, tolerance also develops to the antinociceptive effects of PAG stimulation (Mayer and Hayes, 1975; Millan et al., 1987). Moreover, rats made tolerant to the antinociceptive effects of morphine show cross-tolerance to SPA (Mayer and Hayes, 1975; Morozova and Zvartau, 1986). Recent evidence suggests that the development of tolerance to PAG mediated antinociception depends on the PAG system activated. Tolerance has been shown to occur with as little as 1 min of continuous electrical stimulation of the dorsal raphe nucleus or ventrolateral PAG (i.e., antinociception diminishes despite continuous PAG stimulation) (Morgan and Liebeskind, 1987). In contrast, no reduction in the potency of SPA is seen with as much as 20 min of continuous stimulation of the dorsal PAG system. Immediately upon termination of PAG stimulation, however, responsiveness to noxious stimuli returns. Antinociception that outlasts the stimulation period has been reported by others (Fardin et al., 1984a; Prado and Roberts, 1985), but such post-stimulation antinociception requires higher current intensities than that needed to produce an antinociceptive effect during stimulation (Cannon et al., 1982; Sandkühler and Gebhart, 1984a). Because of this difference in stimulation intensity, it is not surprising that studies which assess nociception following, as opposed to during PAG stimulation tend to report a higher incidence of aversive reactions (Fardin et al., 1984a; Prado and Roberts, 1985).

Endogenous antinociceptive systems also can be activated by exposing rats to aversive footshock. As with SPA, both opioid and non-opioid forms of footshock-induced antinociception can be elicited depending on the parameters of footshock (Lewis et al., 1980; Terman et al., 1984), and only the opioid forms of footshock-induced antinociception show the development of tolerance (Lewis et al., 1981). Moreover, rats made tolerant to opioid forms of footshock antinociception show cross-tolerance to antinociception produced by stimulation in and around the dorsal raphe nucleus, but not from stimulation at sites dorsal to this nucleus (Terman et al., 1985). Neither tolerance nor cross-tolerance to SPA occurred in rats subjected to non-opioid forms of footshock stress-induced antinociception.

Pathways for Modulation

Both the dorsal and ventral PAG systems modulate nociception at the spinal level (Basbaum and Fields, 1984; Willis, 1988). A direct projection from the PAG to the spinal cord exists (Holstege, this volume; Mantyh and Peschanski, 1982), but these fibers are few and probably have an insignificant role in modulating nociception. An abundance of evidence indicates that descending modulation of nociception from the PAG relays in the RVM. PAG stimulation sufficient to inhibit nociception activates neurons in the RVM (Fields and Anderson, 1978; Vanegas et al., 1984), and lesions of the RVM block SPA from the PAG (Aimone and Gebhart, 1986; Prieto et al., 1983; Lovick, 1985; Morton et al., 1984; Sandkühler and Gebhart, 1984b). There is some evidence that the location of the RVM lesion necessary to block SPA depends on the PAG region stimulated. Lesion of the medial RVM, specifically the nucleus raphe magnus, has been reported to block SPA from the dorsal raphe nucleus (Prieto et al., 1983), whereas lesion of the lateral RVM has been reported to block SPA in five rats with stimulation sites located in the dorsal and lateral PAG (Lovick, 1985). Others, however, have failed to see a clearly segregated descending nociceptive modulatory output from the PAG. For example, lidocaine injections that encompassed the medial and lateral aspects of the RVM disrupted SPA evoked from sites in the dorsal, lateral and ventrolateral PAG, more limited lidocaine injections (i.e., involving either only the medial or lateral RVM) were without significant effect on PAG-evoked SPA (Aimone and Gebhart, 1986; Sandkühler and Gebhart, 1984b).

Disruption of descending pathways by placing a lesion immediately caudal to the PAG prevents PAG stimulation from inhibiting spinally mediated nocifensive reflexes (Morgan et al., 1989). Despite disruption of this descending system, supraspinally mediated nocifensive reactions, such as are produced with the hot-plate test, can still be inhibited by PAG stimulation. Consistent with the other differences in SPA reported above, this supraspinal modulation of nociception occurs with stimulation of the ventral, but not the dorsal PAG system (Morgan et al., 1989). These results demonstrate that the ventral PAG system modulates nociception via both ascending and descending pathways.

Similar results have been shown by pharmacologically blocking descending modulation with intrathecal administration of serotonergic and noradrenergic antagonists. These antagonists block antinociception from the PAG when tested using the tail-flick test (i.e., a spinally mediated reflex), but do not block antinociception when tested using the hot-plate test (Camarata and Yaksh, 1985; Jensen and Yaksh, 1984). PAG stimulation and morphine microinjection have also been shown to inhibit noxious evoked activity in thalamic neurons (Anderson and Dafny, 1983; Dafny et al., 1990; Kayser et al., 1983; Qiao and

Dafny, 1988) although it is difficult to determine the site of modulation since these latter effects could be mediated by descending pathways.

Conclusion

Antinociception can be produced by electrical stimulation throughout the PAG. Relatively little is known about the characteristics of SPA mediated by regions of the PAG rostral to the level of the dorsal raphe nucleus. Within the caudal PAG, the characteristics of antinociception differ depending on the site of stimulation. Antinociception produced by stimulation of the dorsal PAG system is not attenuated by naloxone, is resistant to the development of tolerance, and acts via a descending pathway that appears to relay in the lateral RVM. In contrast, antinociception mediated by the ventral PAG system is attenuated by naloxone, shows tolerance with continuous stimulation and cross-tolerance to opioid forms of stress-induced antinociception, and acts by both an ascending pathway and a descending pathway that relays in the nucleus raphe magnus (see Table I).

Table I: Comparison of SPA from the dorsal and ventral PAG systems.

	Dorsal System	Ventral System	Reference
SPA attenuated by naloxone	No	Yes	Cannon et al., 1982 Thorn et al., 1989
Tolerance to SPA with continuous stimulation	No	Yes	Morgan & Liebeskind, 1987
Cross tolerance between SPA and opioid footshock-induced antinociception	No	Yes	Terman et al., 1985
Modulates nociception via ascending pathway	No	Yes	Morgan et al.,1989
Behavioral reaction associated with SPA	Running & Jumping	Freezing	Gold et al., 1990 Fardin et al., 1984a

The behavioral effects associated with antinociception mediated by the dorsal or ventral PAG system also differ. These behavioral effects provide an insight into the possible functions of the two PAG modulatory systems. Antinociception from the dorsal PAG system is associated with running and jumping, responses characteristic of species-specific defense reactions to life-threatening situations (Fanselow, this volume). Stimulation of the dorsal and lateral aspects of the PAG also produces autonomic effects (e.g. tachycardia, increased blood pressure, increased respiration, muscle vasodilation, pupillary dilation) consistent with a fight or flight reaction (Carrive et al., 1989; Carrive, this volume; Duggan and Morton 1983; Lovick, 1985; this volume). Inhibition of nociception seems to be merely another component of this integrated defense reaction. The characteristics of SPA from this region of the PAG suggest a relatively rigid system that seems especially suited to defense in life threatening situations. This non-opioid antinociceptive system is resistant to tolerance and appears to inhibit nociception in an all or none manner (see Besson et al., this volume) by a descending pathway to the spinal cord.

The behavioral reaction associated with SPA from the ventrolateral PAG and dorsal raphe nucleus is less pronounced than the flight reaction associated with more dorsal PAG stimulation. Nonetheless, the immobile posture assumed by rats following activation of the ventral aspect of the PAG is an equally important species-specific defense reaction that allows a rat to avoid detection by potential predators (Fanselow, this volume). This freezing response is also produced by the presence of a natural predator or following aversive footshock (Blanchard and Blanchard, 1971; Fanselow and Sigmundi, 1986). Moreover, this shock-induced freezing response is attenuated by lesions encompassing the ventral aspect of the PAG (LeDoux et al., 1988; Liebman et al., 1970). Antinociception is known to accompany both footshock-induced (Fanselow and Lester, 1988) and PAG-produced freezing (Gold et al., 1990), although the utility of inhibiting nociception is less obvious as part of a defense reaction to potentially threatening situations compared to life-threatening situations. Antinociception in these circumstances may simply allow the rat to maintain an immobile posture over an extended period of time.

The antinociceptive effect of stimulating the ventral aspect of the caudal PAG seems to be one part of a bidirectional modulatory system for nociception. Such a modulatory system would both facilitate and inhibit nociception. This modulation is evident in studies showing that tolerance to antinociception readily develops with stimulation of ventral PAG regions and cross-tolerance develops to opioid forms of footshock antinociception. Modulation is also suggested by the multiple pathways, ascending and descending, from the ventrolateral PAG and dorsal raphe nucleus.

Acknowledgements

I extend my thanks to Mary M. Heinricher for her critical review of this paper.

References

Aimone, L.D. and Gebhart, G.F., Stimulation-produced spinal inhibition from the midbrain in the rat is mediated by an excitatory amino acid neurotransmitter in the medial medulla, J. Neurosci., 6 (1986) 1803-1813.
Aimone, L.D., Jones, S.L. and Gebhart, G.F., Stimulation-produced descending inhibition

from the periaqueductal gray and nucleus raphe magnus in the rat: Mediation by spinal monoamines but not opioids, Pain, 31 (1987) 123-136.

Akil, H., Mayer, D.J. and Liebeskind, J.C., Antagonism of stimulation-produced analgesia by naloxone, a narcotic antagonist, Science, 191 (1976) 961-962.

Akil, H., Watson, S.J., Young, E., Lewis, M.E., Khachaturian, H. and Walker, J.M., Endogenous opioids: Biology and function, Ann. Rev. Neurosci., 7 (1984) 223-255.

Anderson, E. and Dafny, N., An ascending serotonergic pain modulation pathway from the dorsal raphe nucleus to the parafascicularis nucleus of the thalamus, Brain Res., 269 (1983) 57-67.

Audi, E.A. and Graeff, F.G., Benzodiazepine receptors in the periaqueductal grey mediate anti-aversive drug action, Eur. J. Pharmacol., 103 (1984) 279-285.

Basbaum, A.I. and Fields, H.L., Endogenous pain control systems: Brainstem spinal pathways and endorphin circuitry, Ann. Rev. Neurosci., 7 (1984) 309-338.

Behbehani, M.M. and Fields, H.L., Evidence that an excitatory connection between the periaqueductal gray and nucleus raphe magnus mediates stimulation produced analgesia, Brain Res., 170 (1979) 85-93.

Blanchard, R.J. and Blanchard, D.C., Defensive reactions in the albino rat, Learn Motiv., 2 (1971) 351-362.

Bloom, F., Battenberg, E., Rossier, J., Ling, N. and Guillemin, R., Neurons containing beta endorphin in rat brain exist separately from those containing enkephalin: Immunocytochemical studies, Proc. Natl. Acad. Sci., USA, 75 (1978) 1591-1595.

Camarata, P.J. and Yaksh, T.L., Characterization of the spinal adrenergic receptors mediating the spinal effects produced by the microinjection of morphine into the periaqueductal gray, Brain Res., 336 (1985) 133-142.

Cannon, J.T., Prieto, G.J., Lee, A. and Liebeskind, J.C., Evidence for opioid and nonopioid forms of stimulation-produced analgesia in the rat, Brain Res., 243 (1982) 315-321.

Carrive, P., Bandler, R. and Dampney, R.A.L., Somatic and autonomic integration in the midbrain of the unanesthetized decerebrate cat: A distinctive pattern evoked by excitation of neurones in the subtentorial portion of the midbrain periaqueductal grey, Brain Res., 483 (1989) 251-258.

Dafny, N., Reyes-Vazquez, C. and Qiao, J.T., Modification of nociceptively identified neurons in thalamic parafascicularis by chemical stimulation of dorsal raphe with glutamate, morphine, serotonin and focal dorsal raphe electrical stimulation, Brain Res. Bull., 24 (1990) 717-723.

Depaulis, A., Morgan, M.M. and Liebeskind, J.C., GABAergic modulation of the analgesic effects of morphine microinjected in the ventral periaqueductal gray matter of the rat, Brain Res., 436 (1987) 223-228.

Duggan, A.W. and Morton, C.R., Periaqueductal grey stimulation: An association between selective inhibition of dorsal horn neurones and changes in peripheral circulation, Pain, 15 (1983) 237-248.

Fan, S.G., Wang, Y.H. and Han, J.S., Analgesia induced by microinjection of monosodium glutamate into arcuate nucleus of hypothalamus, Acta Zoologica Sinica, 30 (1984) 352-358.

Fanselow, M.S. and Lester, L.S., A functional behavioristic approach to aversively motivated behavior: Predatory imminence as a determinant of the topography of defensive behavior, In: Evolution and Learning, Bolles R. C. and Beecher M. D. (Eds.), Lawrence Erlbaum Associates, Hillsdale, 1988, pp. 185-212.

Fanselow, M.S. and Sigmundi, R.A., Species specific danger signals, endogenous opioid analgesia, and defensive behavior, J. Exper. Psychol.: An. Behav. Proc., 12 (1986) 301-309.

Fardin, V., Oliveras, J.L. and Besson, J.M., A reinvestigation of the analgesic effects induced by stimulation of the periaqueductal gray matter in the rat. I The production of behavioral side effects together with analgesia, Brain Res., 306 (1984a) 105-124.

Fardin, V., Oliveras, J.L. and Besson, J.M., A reinvestigation of the analgesic effects induced by stimulation of the periaqueductal gray matter in the rat. II Differential characteristics of the analgesia induced by ventral and dorsal PAG stimulation, Brain Res., 306 (1984b) 125-139.

Fields, H.L. and Anderson, S.D., Evidence that raphe-spinal neurons mediate opiate and midbrain stimulation-produced analgesias, Pain, 5 (1978) 333-349.

Gold, M.S., Morgan, M.M. and Liebeskind, J.C., Antinociceptive and behavioral effects of low dose kainic acid injections into the PAG of the rat, Pain, Suppl. 5 (1990) S441.

Graeff, F.G. and Rawlins, J.N.P., Dorsal periaqueductal gray punishment, septal lesions and the mode of action of minor tranquilizers, Pharmacol. Biochem. Behav., 12 (1980) 41-45.

Hughes, J., Smith, T.W., Kosterlitz, H., Fothergill, L.A., Morgan, B.A. and Morris, H.R., Identification of two related pentapeptides from the brain with potent agonist activity, Nature, 258 (1975) 577-579.

Iadarola, M.J., Brady, L.S., Draisci, G. and Dubner, R., Enhancement of dynorphin gene expression in spinal cord following experimental inflammation: Stimulus specificity, behavioral parameters and opioid receptor binding, Pain, 35 (1988a) 313-326.

Iadarola, M.J., Douglass, J., Civelli, O. and Naranjo, J.R., Differential activation of spinal cord dynorphin and enkephalin neurons during hyperalgesia: Evidence using cDNA hybridization, Brain Res., 455 (1988b) 205-212.

Jacquet, Y.F. and Lajtha, A., Paradoxical effects after microinjection of morphine in the periaqueductal gray matter in the rat, Science, 185 (1974) 1055-1057.

Jensen, T.S. and Yaksh, T.L., Spinal monoamine and opiate systems partly mediate the antinociceptive effects produced by glutamate at brainstem sites, Brain Res., 321 (1984) 287-297.

Kayser, V., Benoist, J.-M. and Guilbaud, G., Low dose of morphine microinjected in the ventral periaqueductal gray matter of the rat depresses responses of nociceptive ventrobasal thalamic neurons, Neurosci. Lett., 37 (1983) 193-198.

LeDoux, J.E., Iwata, J., Cicchetti, P. and Reis, D.J., Different projections of the central amygdaloid nucleus mediate autonomic and behavioral correlates of conditioned fear, J. Neurosci., 8 (1988) 2517-2529.

Lewis, J.W., Cannon, J.T. and Liebeskind, J.C., Opioid and non-opioid mechanisms of stress analgesia, Science, 208 (1980) 623-625.

Lewis, J.W., Sherman, J.E. and Liebeskind, J.C., Opioid and non-opioid stress analgesia: Assessment of tolerance and cross-tolerance with morphine, J. Neurosci., 1 (1981) 358-363.

Liebman, J.M., Mayer, D.J. and Liebeskind, J.C., Mesencephalic central gray lesions and fear-motivated behavior in rats, Brain Res., 23 (1970) 353-370.

Lovick, T.A., Ventrolateral medullary lesions block the antinociceptive and cardiovascular responses elicited by stimulation of the dorsal periaqueductal grey matter in rats, Pain, 21 (1985) 241-252.

Maier, S.F., Determinants of the nature of environmentally induced hypoalgesia, Behav. Neurosci., 103 (1989) 131-143.

Mantyh, P.W. and Peschanski, M., Spinal projections from the periaqueductal grey and dorsal raphe in the rat, cat and monkey, Neurosci., 7 (1982) 2769-2776.

Mayer, D.J. and Hayes, R.L., Stimulation-produced analgesia: Development of tolerance and cross-tolerance to morphine, Science, 188 (1975) 941-943.

Mayer, D.J. and Liebeskind, J.C., Pain reduction by focal electrical stimulation of the brain: An anatomical and behavioral analysis, Brain Res., 68 (1974) 73-93.

Mayer, D.J., Wolfle, T.L., Akil, H., Carder, B. and Liebeskind, J.C., Analgesia from electrical stimulation in the brainstem of the rat, Science, 174 (1971) 1351-1354.

Millan, M.H., Czlonkowski, A. and Herz, A., Evidence that μ-opioid receptors mediate midbrain "stimulation-produced analgesia" in the freely moving rat, Neurosci., 22 (1987) 885-896.

Millan, M.H., Millan, M.J. and Herz, A., Depletion of central ß-endorphin blocks midbrain stimulation-produced analgesia in the rat, Neurosci., 18 (1986) 641-649.

Moreau, J.L. and Fields, H.L., Evidence for GABA involvement in midbrain control of medullary neurons that modulate nociceptive transmission, Brain Res., 397 (1986) 37-46.

Morgan, M.M., Depaulis, A. and Liebeskind, J.C., Diazepam dissociates the analgesic and aversive effects of periaqueductal gray stimulation in the rat, Brain Res., 423 (1987) 395-398.

Morgan, M.M., Gold, M.S., Liebeskind, J.C. and Stein, C., Periaqueductal gray stimulation produces a spinally mediated, opioid antinociception for the inflamed hindpaw of the rat, Brain Res., 545 (1991) 17-23.

Morgan, M.M. and Liebeskind, J.C., Site specificity in the development of tolerance to stimulation-produced analgesia from the periaqueductal gray matter of the rat, Brain Res., 425 (1987) 356-359.

Morgan, M.M., Sohn, J.-H. and Liebeskind, J.C., Stimulation of the periaqueductal gray matter inhibits nociception at the supraspinal as well as spinal level, Brain Res., 502 (1989) 61-66.

Morozova, A.S. and Zvartau, E.E., Stimulation-produced analgesia under repeated morphine treatment in rats, Pharmacol. Biochem. Behav., 25 (1986) 533-536.

Morton, C.R., Duggan, A.W. and Zhao, Z.Q., The effects of lesions of medullary midline and lateral reticular areas on inhibition in the dorsal horn produced by periaqueductal grey stimulation in the cat, Brain Res., 301 (1984) 121-130.

Nichols, D.S., Thorn, B.E. and Berntson, G.G., Opiate and serotonergic mechanisms of stimulation-produced analgesia within the periaqueductal gray, Brain Res. Bull., 22 (1989) 717-724.

Pert, A. and Walter, M., Comparison between naloxone reversal of morphine and electrical stimulation induced analgesia in the rat mesencephalon, Life Sci., 19 (1976) 1023-1032.

Prado, W.A. and Roberts, M.H.T., An assessment of the antinociceptive and aversive effects of stimulating identified sites in the rat brain, Brain Res., 340 (1985) 219-228.

Prieto, G.J., Cannon, J.T. and Liebeskind, J.C., N. raphe magnus lesions disrupt stimulation-produced analgesia from ventral but not dorsal midbrain areas in the rat, Brain Res., 261 (1983) 53-57.

Qiao, J.-T. and Dafny, N., Dorsal raphe stimulation modulates nociceptive responses in thalamic parafascicular neurons via an ascending pathway: Further studies on ascending pain modulation pathways, Pain, 34 (1988) 65-74.

Reynolds, D.V., Surgery in the rat during electrical analgesia induced by focal brain stimulation, Science, 164 (1969) 444-445.

Sandkühler, J. and Gebhart, G.F., Characterization of inhibition of a spinal nociceptive reflex by stimulation medially and laterally in the midbrain and medulla in the pentobarbital-anesthetized rat, Brain Res., 305 (1984a) 67-76.

Sandkühler, J. and Gebhart, G.F., Relative contributions of the nucleus raphe magnus and adjacent medullary reticular formation to the inhibition by stimulation in the periaqueductal gray of a spinal nociceptive reflex in the pentobarbital-anesthetized rat, Brain Res., 305 (1984b) 77-87.

Sandkühler, J., Willmann, E. and Fu, Q.G., Blockade of GABA$_A$ receptors in the midbrain periaqueductal gray abolishes nociceptive spinal dorsal horn neuronal activity, Eur. J. Pharmacol., 160 (1989) 163-166.

Terman, G.W., Penner, E.R. and Liebeskind, J.C., Stimulation-produced and stress-induced analgesia: Cross-tolerance between opioid forms, Brain Res., 360 (1985) 374-378.

Terman, G.W., Shavit, Y., Lewis, J.W., Cannon, J.T. and Liebeskind, J.C., Intrinsic mechanisms of pain inhibition: Activation by stress, Science, 226 (1984) 1270-1277.

Thorn, B.E., Applegate, L. and Johnson, S.W., Ability of periaqueductal gray subdivisions and adjacent loci to elicit analgesia and ability of naloxone to reverse analgesia, Behav. Neurosci., 103 (1989) 1335-1339.

Tseng, L.L.F. and Tang, R., Differential actions of the blockade of spinal opioid, adrenergic and serotonergic receptors on the tail-flick inhibition induced by morphine microinjected into dorsal raphe and central gray in rats, Neurosci., 33 (1989) 93-100.

Urca, G., Nahin, R.L. and Liebeskind, J.C., Glutamate-induced analgesia: Blockade and potentiation by naloxone, Brain Res., 192 (1980) 523-530.

Vanegas, H., Barbaro, N.M. and Fields, H.L., Midbrain stimulation inhibits tail-flick only at currents sufficient to excite rostral medullary neurons, Brain Res., 321 (1984) 127-133.

Wang, Q., Mao, L. and Han, J., Analgesia from electrical stimulation of the hypothalamic arcuate nucleus in pentobarbital-anesthetized rats, Brain Res., 526 (1990) 221-227.

Williams, F.G. and Beitz, A.J., Ultrastructural morphometric analysis of enkephalin-immunoreactive terminals in the ventrocaudal periaqueductal gray: Analysis of their relationship to periaqueductal gray-raphe magnus projection neurons, Neurosci., 38 (1990) 381-394.

Willis, W.D., Anatomy and physiology of descending control of nociceptive responses of dorsal horn neurons: Comprehensive review, Prog. Brain Res., 77 (1988) 1-29.

Yaksh, T.L. and Rudy, T.A., Analgesia mediated by a direct spinal action of narcotics, Science, 192 (1976) 1357-1358.

Yaksh, T.L. and Rudy, T.A., Studies on the direct spinal action of narcotics in the production of analgesia in the rat, J. Pharmacol. Exp. Ther., 202 (1977) 411-428.

Yaksh, T.L., Yeung, J.C. and Rudy, T.A., An inability to antagonize with naloxone the elevated thresholds resulting from electrical stimulation of the mesencephalic central gray, Life Sci., 18 (1976a) 1193-1198.

Yaksh, T.L., Yeung, J.C. and Rudy, T.A., Systematic examination in the rat of brain sites sensitive to the direct application of morphine: Observation of differential effects within the periaqueductal gray, Brain Res., 114 (1976b) 83-103.

Zhang, S.P., Bandler, R. and Carrive, P., Flight and immobilitiy evoked by excitatory amino acid microinjection within distinct parts of the subtentorial midbrain periaqueductal gray of the cat, Brain Res., 520 (1990) 73-82.

The Midbrain Periaqueductal Gray as a Coordinator of Action in Response to Fear and Anxiety

Michael S. Fanselow

Department of Psychology
and Brain Research Institute
University of California-Los Angeles
California, USA

Introduction

This paper describes a current psychobiological approach to motivated behavior using fear as an illustration. The first part describes the properties of fear motivation and then argues that an amygdala-periaqueductal gray system displays many of these properties. The second half of the paper details the organization of fear related behavior and then presents a model suggesting how the periaqueductal gray may be responsible for much of this organization.

Functional Behavior Systems and Fear

The major point I want to make in this chapter is that the array of phenomena in which the midbrain periaqueductal gray matter (PAG) appears to be involved, can be viewed in an integrated and cohesive fashion. The integrating theme I wish to suggest is fear (or anxiety). In order to talk about fear, and especially to make my point, we must start out with a definition. My view of fear may differ somewhat from that commonly employed. I view fear as a functional behavior system. According to Timberlake and Lucas, a functional "behavior system is a complex control structure related to a particular function or need of the organism, such as feeding, reproduction, defense or body care" (1989, p. 241). There are four components to the definition of a functional behavior system (Fanselow and Sigmundi, 1987): 1) The function or evolutionary benefit of the behavior system must be specified 2) The observable behavioral consequences must be detailed 3) The environmental antecedent stimuli must be indicated, and 4) testable mechanisms such as the neural substrate must be proposed. Each of these components of the functional behavior system related to fear is specified below.

The Midbrain Periaqueductal Gray Matter, Edited by A. Depaulis and
R. Bandler, Plenum Press, New York, 1991

1) Evolutionary benefit:

Functional behavior systems are derived from specific species-survival-related needs that the organism can only satisfy by appropriate interaction with its environment. Therefore, the purpose or environmental problem the system deals with needs to be known. In the case of fear, this function is to protect individuals of the species against environmental dangers. This paper focuses on research using rats as subjects. For the rat, predation is a major source of environmental danger that exerts considerable selection pressure on the species. Thus, fear is considered as an antipredator defensive behavioral system (Fanselow, 1984a).

2) Behavioral consequences:

The second aspect of the definition is to specify the observable behavioral consequences of activation of the behavioral system. Specifically, what responses are observed in a frightened rat. The most salient aspect of a frightened rat's behavior is the freezing response. Operationally, freezing is defined as a cessation of all observable movements except those associated with respiration. It typically, although not always, occurs in a crouching posture next to some available object such as the corners of an observation chamber (thigmotaxis). Freezing has long been recognized as a response to fear (e.g., Small, 1899). For example, in 1872 Darwin observed that the frightened subject "stands like a statue motionless and breathless, or crouches down as if instinctively to escape observation" (Darwin, 1965, p290). The freezing response should not be confused with catatonia or tonic immobility for the freezing rat is neither pliant nor flaccid. Rather, while immobile, it is highly alert with considerable muscle tone. This is illustrated by Griffith's (1920) description of a rat exploring the ceiling of its cage. Upon presentation of a cat, the rat "immediately froze and hung for 22 minutes (p23)." Additionally, freezing rats show a potentiated startle response to a sudden loud noise (e.g., Leaton and Borszcz, 1985). Besides freezing, the frightened rat becomes analgesic. This loss of sensitivity to pain has been demonstrated with both reflexive and integrative indices of pain (see Fanselow, 1991 for a review). This fear-induced analgesia prevents injury related behaviors from interfering with defensive efforts (Bolles and Fanselow, 1980). A third index of activation of the defensive system is the various autonomic changes (heart rate, blood pressure, etc) that accompany fear (e.g., Hofer, 1970). These changes, like analgesia, may support overt defensive behaviors; changes in autonomic tone may better enable the animal to defend. For example, redirection of blood flow supplies energy to skeletal muscle groups utilized during defensive action. These fear related behaviors can be taken as examples of a class of behaviors labeled species specific defense reactions (SSDRs) by Bolles (1970).

A rat's reactions to methyl 6,7-dimethoxy-4-ethyl-ß-carboline-3-carboxylic acid methyl ester (DMCM) illustrate this constellation of behaviors. This ß-carboline is a potent inverse agonist at benzodiazepine receptors and is purported to be anxiogenic unlike benzodiazepine agonists (e.g., diazepam) which are

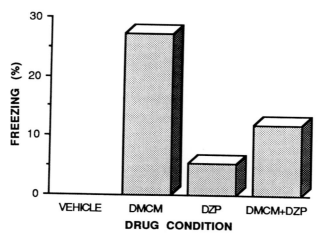

Figure 1. Four groups of 6 Long-Evans (hooded), female, rats were given an injection of either vehicle, DMCM (1.25 mg/kg ip), diazepam (2.5 mg/kg) or DMCM + diazepam shortly before placement in a novel observation chamber. The percentage of the 8 min test period spent freezing is presented. DMCM produced statistically reliable (p<.03) increases in freezing relative to the other 3 groups which did not differ. This effect was blocked by diazepam, which by itself had no reliable effect. Based on Fanselow et al., 1991.

anxiolytic (Stephens and Kehr, 1985). Figure 1 shows a study in which rats were given either DMCM or its vehicle in a factorial combination with diazepam or its vehicle. The percentage of time rats spent freezing during an 8 min test period was recorded. The figure shows that DMCM treatment resulted in a freezing response that was reversed by diazepam. The results of another study with DMCM are presented in Figure 2. In that experiment, pain sensitivity was assessed with the formalin test. The rats were injected subcutaneously into the paw with a small amount of dilute formalin solution and the percentage of time they spent reacting to the injection with either paw licking or paw lifting was recorded. Defecation and urination which might be considered as autonomic indices of fear were recorded as well. DMCM's ability to dose dependently suppress the reactions to formalin indicates an analgesia. Other studies, using the hot plate (Helmstetter et al., 1990) and tail-flick (Rodgers and Randall, 1987) tests of pain sensitivity also indicate that this anxiogenic drug produces analgesia. The lower two panels of Figure 2 show that DMCM increased defecation and urination suggesting the autonomic arousal associated with fear. Based on both behavior theory and data, overt defensive behaviors (e.g., freezing), analgesia and autonomic arousal are all linked with a defensive behavioral system. There is an impressive correspondence between these behaviors and the behavioral indices employed by the other authors in this volume to analyze PAG function. This is most clearly represented in chapters on defensive behavior (Bandler and Depaulis, this volume), analgesia (Besson et al., this volume) and autonomic function (Carrive; Lovick, this volume).

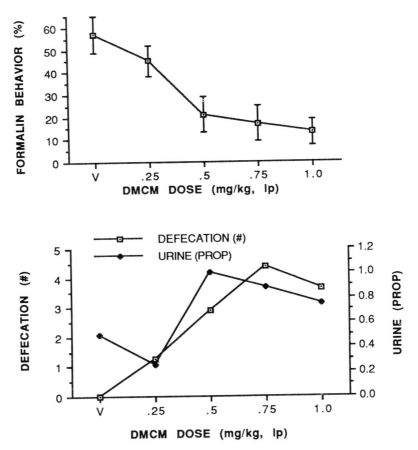

Figure 2. DMCM caused a dose dependent analgesia as indicated by a suppression of pain-related recuperative behavior caused by s.c. injection of .05 ml of 15% formalin under the dorsal surface of a hindpaw (p<.01, upper panel). This analgesia was a statistically reliable (p<.01) positive linear function of dose. As indicated in the lower panel, the ß-carboline also caused a dose dependent and reliable increase in the number of fecal boluses excreted (p<.02, defecation) and the probability that an animal would urinate (p<.01). These effects were also a positive linear function of dose (p<.01). DMCM was administered i.p. in a Molecusol vehicle. There were 5 independent groups of 8 female hooded rats.

3) Antecedent stimuli:

Having defined the response consequences of a functional behavior system one must then specify the antecedent stimulus conditions that activate the system. The question is, what environmental stimuli normally serve to produce fear? There seem to be two general classes: innately recognized threats and learned danger stimuli. As an example of the former, rats react with freezing and analgesic responses to a cat on their very first experience with this predator (Blanchard and Blanchard, 1971; Blanchard et al., 1976; Griffith, 1920; Lester and Fanselow, 1985). Results of a study by Lichtman and I are presented in Figure 3 (Lichtman and Fanselow, 1990). We compared the latency of the radiant heat-induced tail-flick of a rat in the presence or absence of a cat. In the presence of the cat, the rat's tail-flick latency was elevated indicating an analgesic response. Pretreatment with the opiate antagonist naltrexone reversed that analgesia suggesting that it was mediated by endogenous opioids. Other stimuli of biological relevance to danger such as odors of stressed conspecifics and tactile stimulation to the head and back also act as innate danger stimuli producing similar reactions (Fanselow, 1985; Fanselow and Sigmundi, 1986). This phenomena is not

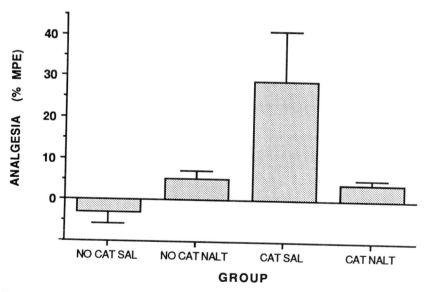

Figure 3. Latency for four groups of 8 hooded male rats to tail-flick to radiant heat is expressed as a percentage of the maximum possible effect (MPE=[Test latency-baseline latency/ cut off latency-baseline latency] X 100). Tests were made in the presence or absence of a cat. The presence of this predator caused an analgesia in the rat (p<.05). The opiate antagonist naltrexone (NALT-7mg/kg ip) had no effect on pain reactivity in the absence of the cat but reversed the analgesia produced by the cat (p<.05). Based on Lichtman and Fanselow, 1990.

confined to rats as deermice also react with freezing (Hirsch and Bolles, 1980) and analgesia (Kavaliers, 1988) to the presence of weasels. Since the ability of these stimuli to produce SSDRs is genetically transferred they can be referred to as species specific danger signals (Fanselow and Sigmundi, 1986).

The other class of antecedent stimuli are initially neutral but acquire the ability to produce fear by Pavlovian conditional association with painful stimuli such as electric shock. The results of an illustrative demonstration are presented in Figure 4. In this experiment, rats were placed in each of two discriminable observation chambers daily for 10 days. Four min into placement in one of the chambers (S+) they received a brief mild electric shock but there were no consequences of being placed in the other chamber (S-). The rats were given a shock-free test in each chamber on the eleventh day. Prior to this test they were given an injection of formalin s.c. into a hindpaw so that analgesia could be assessed. Additionally, half the rats were given an injection of the anxiolytic benzodiazepine, midazolam (2mg/kg) and the other half were given a placebo. Figure 4 shows that there was considerable freezing but little formalin-related behavior in the dangerous place (S+) compared to the safe place (S-). When fear was inhibited by midazolam both the freezing and analgesia were attenuated. As with the analgesia that occurred in response to the cat, the analgesia produced by the S+ was reversed by naltrexone (Fanselow and Baackes, 1982). Additional studies indicated that this analgesia involves a μ-opioid component (Fanselow, et al., 1989). What is important here is that these responses were not produced by the shock but rather, by the situational cues that were present at the time of shock. Indeed, shock itself does not directly produce freezing. Rather, freezing following shock is a Pavlovian conditional reaction to the contextual cues that were contiguous with shock. Several findings illustrate the conditional nature of freezing (e.g., Fanselow, 1980; 1982; 1986). For example, in the study depicted in Figure 4, the data were obtained on a shock-free test day. Additionally, animals tested immediately after shock in a manner that precludes conditioning to contextual cues do not freeze (Fanselow, 1981; 1986). Since these situational cues were initially neutral and acquired their fear provoking ability through experience they are learned danger signals.

In this section fear was made a testable psychological construct because it was anchored to, and brought about by, manipulable environmental stimuli and resulted in observable behavioral consequences. Important, for a functional behavior system, is that both the stimuli and responses are tied to the proposed adaptive function of the behavior system. Fear is activated by a variety of stimuli, all of which signal danger, and in turn there are a host of responses, all of which are directed at dealing with the danger.

4) Mechanisms:

The fourth and most difficult aspect of defining a functional behavior system is to characterize the mechanisms that mediate between the antecedent stimuli and behavioral consequences. The neural mechanisms devoted to serving this behavioral system have two important jobs. One is to detect, gather and

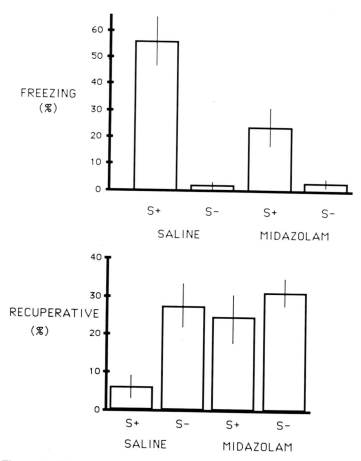

Figure 4. These data are from a shock-free test day taken after hooded female rats were trained to expect a mild electric shock in one context (S+) but were safe in another (S-). Each drug condition is represented by 14 rats. The upper panel displays freezing during this 8 min test; the lower panel shows the percent of time spent in formalin-induced recuperative (paw licking and lifting) behavior. The S+ produced reliable freezing and analgesia relative to the S-. All comparisons were made with Scheffé's test (p<.05). The anxiolytic benzodiazepine agonist, midazolam (2mg/kg ip), attenuated both freezing and analgesia. Based on Fanselow and Helmstetter, 1988.

organize the various activating stimuli that signal danger to the organism. That is, these disparate stimuli must converge on a single neural mechanism. The second is to coordinate the various defensive activities. Part of this coordination would be to select the particular defensive responses appropriate for the situation. The rat has several SSDRs available and the most appropriate one for the situation must be selected. Another part of this coordination is to determine what responses go together. A freezing rat has both a characteristic autonomic pattern (e.g., Hofer, 1970) and shows thigmotaxis (Grossen and Kelley, 1972). On the other hand, a fighting rat may show a different pattern of autonomic responses and is obviously not concerned about thigmotaxis. The section that follows provides a framework that relates the various aspects of this definition of fear to potential underlying mechanisms.

Neural Mechanisms Serving the Defensive Behavior System: An Overview

Current evidence points to the amygdala as the point of convergence of the various danger stimuli. For example, lesions of the amygdala attenuate SSDRs such as freezing and analgesia in response to both innate and learned danger stimuli (Blanchard and Blanchard, 1972; Helmstetter et al., 1988). On the other hand, the PAG may serve as coordinator of the various defensive activities. The central nucleus of the amygdala projects to the PAG (de Olmos et al., 1985; Gloor, 1978; Hopkins and Holstege, 1978; Krettek and Price, 1978; Rizvi et al., 1991; Shipley et al., this volume), which has long been associated with defensive behaviors and analgesia. Electrical and chemical stimulation of the PAG produces defensive behavior including freezing (Bandler and Depaulis, 1988; Depaulis et al., 1989). Lesions of PAG reduce freezing (Borszcz et al., 1989; LeDoux et al., 1988; Liebman et al., 1970). Indeed, PAG lesions block the high

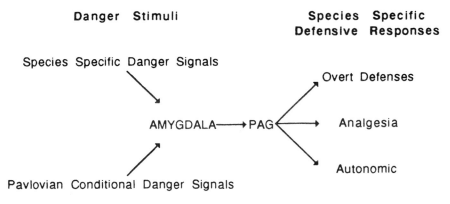

Figure 5. A framework for the defensive behavior system.

level of defensiveness seen in wild rats (Blanchard et al., 1981). The PAG is the prototypical site for brain stimulation produced analgesia (Mayer and Liebeskind, 1974; Oliveras and Besson, 1988; Reynolds, 1969). Both excitatory amino acids and opiates infused into PAG are profoundly analgesic (Criswell, 1975; Jacquet and Lajtha, 1974; Urca et al., 1980). Additionally, autonomic changes associated with defensive behavior are provoked by local injections of excitatory amino acids into the PAG (Bandler, 1988; Carrive et al., 1989; Bandler et al., 1991; Carrive, this volume). This framework is diagrammed in Figure 5.

The amygdala-PAG system seems to correspond with the Defensive Behavioral System in that it represents a convergence of the sensory inputs producing fear-like behavior and is also where the various fear responses begin to diverge. The various fear related responses can be separated within the PAG.

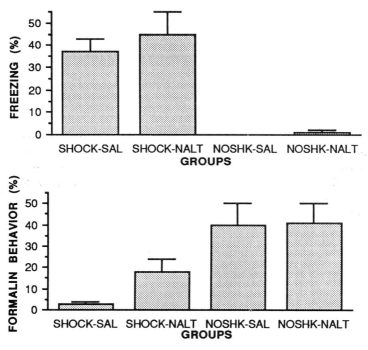

Figure 6. Rats (hooded females) either received shock or not (NOSHK) on a training day. The next day they were tested for formalin-induced recuperative behavior (upper panel) or freezing (lower panel) in the training context. Just prior to this test, the rats received either naltrexone (NALT—5 μg/rat) or saline administered through chronic indwelling cannulae in the ventral PAG contralateral to formalin treated paw. Prior shock experience resulted in conditional analgesia (suppressed reactivity to formalin) and freezing. While naltrexone attenuated the analgesia it had no effect on freezing. Based on Helmstetter and Landeira-Fernandez, 1991.

For example, in the fear conditioning preparation described above, if naltrexone is injected into ventral PAG (v-PAG) during testing, it reduces the fear-induced analgesia as assessed with formalin but not performance of the freezing response (see Fig. 6). It appears that within the v-PAG, analgesia is initiated by release of endogenous opioids. However, freezing, and perhaps other SSDRs, are independent of this opioid mechanism. Therefore, the present form of the model suggests that the PAG is stimulated by the amygdala, possibly by excitatory amino acids, which in turn generate a number of SSDRs via several other neurochemical systems.[1] This explains the fact that while electrical or kainic acid stimulation of the PAG produces both freezing and analgesia, opiate antagonists block the analgesia but not the freezing (Helmstetter and Landeira-Fernandez, 1991).

The Organization of Defensive Behavioral Topography

The final goal of this paper is to present a preliminary outline of how the PAG may act as a coordinator of various defensive behavioral topographies. In order to do this we must first consider how defensive behavior is organized and what leads to the selection of one defensive behavioral strategy over another. Elsewhere, I argued that the overt defensive behavior observed in a particular situation is determined by the spatial and temporal relationship of the rat to danger (Fanselow and Lester, 1988; Fanselow, 1989). Defensive behavior seems to change, both quantitatively and qualitatively, as the risk of a danger such as predation increases; the form or topography of defensive reactions varies along a continuum of predatory imminence. This continuum can be broken into three general stages that correspond to the qualitative changes in defensive behavioral topography. The first stage of defense is labeled, "pre-encounter defense." This occurs when a rat must leave a safe area to forage in a slightly more dangerous area. The rat may do so in a cautious manner (Blanchard and Blanchard, 1988) characterized by stretched-approach behaviors (Pinel and Mana, 1989). Indeed, even foraging and feeding patterns may be reorganized if there is some moderate risk in the foraging/feeding area (Fanselow et al., 1988). When an actual predator is detected the rat goes into the next stage of defense, "post-encounter defense". For the rat, this stage of defense is characterized by the freezing response. If the predator is about to make contact the freezing pattern is disrupted and the rat engages in "circa-strike defense" such as vigorous escape attempts or jump attacks (e.g., Blanchard et al., 1989). An overview of the predatory imminence continuum is presented in Figure 7 (see Fanselow and Lester, 1988; Fanselow, 1989; for more detail).

---

1 This is not to deny that other brain structures may also signal the PAG of threat; however, the amygdala is the structure that has received the greatest degree of experimental attention in this regard. Additionally, information concerning danger may be communicated to the PAG via more direct routes such as the superior colliculus (Rhoades et al., 1989) and afferents from spinal cord (Blomqvist and Craig, this volume).

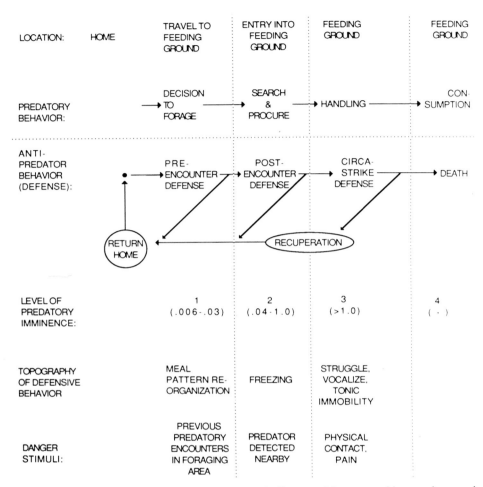

Figure 7. The Predatory Imminence Continuum is illustrated in terms of increasing preda-tory imminence from left to right. It suggests that the type of defensive behavior in a situa-tion varies both quantitatively (within a stage of antipredator defense) and qualitatively (between stages) with the prey's perception of the degree of danger in that situation. Reprinted with permission from Fanselow (1989) Kluwer Academic Publishers, Boston.

In the above examples I used the physical distance of the predator to illus-trate how behavior changes with predatory imminence. However, there are also important temporal aspects to the continuum. This temporal quality of predato-ry imminence is perhaps best demonstrated by the fact that varying the density of a single stimulus (e.g., electric shock) produces behaviors belonging to each of the three stages on the continuum. The row labeled "level of predatory immi-nence" of Figure 7 provides ranges of the shock densities (shocks per minute) associated with the various stages of "antipredator behavior" and "topography of defensive behavior".

Another important temporal aspect to the continuum is that the various stages of defense can be seen following a single stimulus presentation. The behavioral changes seem to work backwards from stimulus presentation with predatory imminence being highest at the time of exposure and progressively decreasing as time continues. For example, Blanchard et al. (1989) suggest that following a single cat presentation, rat's living in a laboratory burrow system show orderly changes in defensive behaviors with high imminence (circa-strike) behaviors following immediately after cat exposure with freezing following shortly thereafter. Finally, pre-encounter behaviors replace freezing at even longer intervals. This is quite similar to the pattern produced by electric shock. Shock causes an immediate activity burst that persists for a brief time beyond shock presentation (Fanselow, 1982). This activity is an unconditional reaction to the shock and may represent circa-strike behaviors. This activity burst is gradually replaced by freezing (the dominant post-encounter defense) which is a conditional reaction to the apparatus cues present at the time of shock. After a more prolonged period of time the freezing begins to dissipate. Figure 8 illustrates this biphasic pattern of an activity burst followed by freezing.

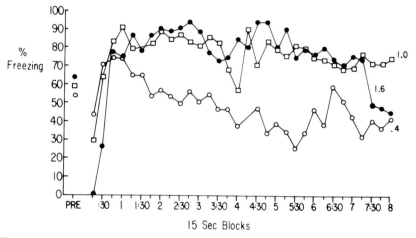

Figure 8. Rats (hooded, females) received 4 brief (0.75 sec) shocks (1 mA) spaced 20 sec apart on a training day. The next day they were returned to the chamber. PRE indicates the percentage of time spent freezing during a 3 min baseline period. All the rats then received a single 0.75 sec shock that varied in intensity for different groups. The curve presents percentage of time during 15 sec blocks the 8 rats in each group spent freezing beginning immediately with shock termination. Note that the shock disrupted freezing and that the duration of this disruption varied positively with shock intensity (p<.0001). This disruption is an unconditional reaction to shock and is labelled the activity burst. The rats resume freezing, a conditional reaction to shock associated stimuli, as the activity burst dissipates. Note that freezing is weakest in the rats that received the mildest shock(p<.004). Reprinted with permission from Fanselow (1982) Psychonomic Society.

   This pattern can be analyzed by applying the predatory imminence continuum to Pavlovian conditioning (Fanselow, 1989). According to this application of the model, a conditional stimulus elicits a response one step lower on the predatory imminence continuum than the response elicited by the unconditional stimulus with which it was paired. The shock is treated by the rat as a painful contact stimulus eliciting unconditional circa-strike behavior (see Fig. 7). The conditional response to the apparatus cues paired with such a contact stimulus are not circa-strike behaviors but rather they are responses one step reduced in predatory imminence, post-encounter responses such as freezing (see Fanselow, 1989 for an elaboration of this response determination rule). Immediately after shock the unconditional responses dominate but as they dissipate the conditional responses to the situational cues (freezing) are observed. As time goes on, since there is no additional shock, the conditional freezing extinguishes and hence the reduction in this behavior.

Relationship between Anatomical Subdivisions of the PAG and Different Stages of Defense

   The data depicted in Figure 8 and the interpretation given to them above, fit quite nicely with data relevant to PAG function. Together with Michael M. Morgan and John Liebeskind, I examined responses of rats to electrical stimulation of either v-PAG sites or sites more dorsal and lateral (lateral-PAG). Ventral electrode placements were directed at areas known to support opioid mediated analgesia; lateral-PAG (l-PAG) placements were targeted at areas known to produce nonopioid analgesia (Cannon et al., 1982). Stimulation of l-PAG sites produced analgesia and vigorous motor behaviors (e.g., running, turning) that looked identical to a footshock-induced activity burst. These are probably the sort of reactions labeled as aversive reactions by Besson and others (e.g., Besson et al., this volume). This activity burst barely outlasted the stimulation and the rat began to freeze shortly thereafter. This reaction was remarkably similar to the reaction to shock: an activity burst during the stimulation that rapidly gave way to freezing. While ventral stimulation also produced analgesia the overt behaviors were quite different. There was no activity burst to the current. Rather, the rats immediately froze during stimulation. This freezing response was indistinguishable from that conditioned by foot shock.
   In a second study, conducted with J. J. Kim and Michael Gold, we made lesions of either dorsal/lateral PAG or ventral PAG. There were also sham surgery controls. After recovery from surgery, the rats were placed in a chamber and given 3 electric footshocks (.75 s, 1 mA) spaced 1 min apart. We measured the activity burst during this session. The next day they were returned to the chamber and observed for conditional freezing in the absence of shock. Figure 9 displays the activity burst data, while freezing data are in Figure 10. Dorsal/lateral lesions completely eliminated the activity burst. However, conditional freezing was not significantly affected. Note that even though the unconditional reaction to shock was eliminated, the conditional response remained, and the rats sho-

wed near normal freezing on the shock-free test day. The effective lesions were large and included most of the dorsal and lateral areas of the PAG (see Fig. 11B). Smaller lesions that only damaged the dorsal medial PAG (see Fig. 11A) did not produce the effect. Therefore, it seems likely that the l-PAG (Fig. 11D) is critical for circa-strike defensive behaviors. It should be noted that what is referred to as lateral (l-PAG) here corresponds closely to what has been referred to as dorsal-PAG in the stimulation produced analgesia literature (e.g., Morgan, this volume; Morgan and Liebeskind, 1987). The lateral terminology is adopted to be consistent with the other chapters in this volume.

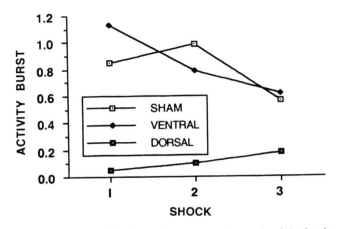

Figure 9. The activity burst in response to each of 3 shocks spaced 1 min apart is presented. Rats (12 Sprague-Dawley males) had lesions of either ventral (see Fig. 11C for histology) or the dorsal/lateral (labeled dorsal on figure, see Fig. 11B) areas of the PAG. Dorsal/lateral PAG but not ventral PAG lesions attenuated the burst relative to sham operated controls. The activity burst was measured by digitizing the output of a videocamera and calculating the number of pixels that changed every 30 msec with the aid of a Videotrack (Lyon Electronique) image analysis system. The animal was considered to be showing an activity burst if over 1500 pixels changed in 30 msec. The activity burst measure reflects a difference score between shock and baseline. The plotted measure is a difference score between seconds scored as bursting following shock and an equivalent preshock baseline.

On the other hand, it is clear from Figure 9 that ventral lesions did not alter the activity burst. However, when these rats were returned to the chamber the next day they showed very little freezing (Fig. 10). Ventral lesions are displayed in Figure 11C. These results confirm the many previous reports that v-PAG lesions block the fear-induced freezing response described in the second

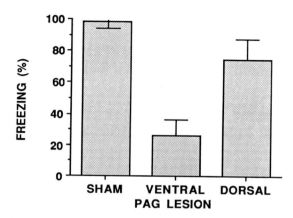

Figure 10. These data are from a shock-free test day that occurred 24 h after the shock session displayed in Fig. 9. Ventral PAG lesions dramatically reduced the percentage of time spent freezing during an 8 min test. Freezing was scored automatically by the same system described in Fig. 9. A rat was considered freezing if less than 5 pixels changed in 30 msec. The rats were Sprague-Dawley males.

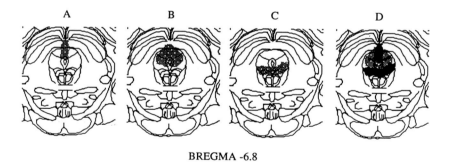

BREGMA -6.8

Figure 11. Examples of the electrolytic lesions of PAG examined by Kim, Gold and Fanselow. Panel A: the shaded portion indicates a lesion that affected neither freezing nor the activity burst. Panel B: The shaded portion indicates a lesion that completely eliminated the activity burst but had no effect on freezing. Panel C: The shaded portion represents a lesion that eliminated freezing but did not affect the activity burst. Panel D: The lesions which did not effect the activity burst (A and C) are superimposed in black on the lesion that eliminated the activity burst (B). It is suggested that the shaded portion on this section (D) corresponds to the areas of the PAG mediating circa-strike behavior. This region is referred to as l-PAG in this chapter.

paragraph of this paper (e.g., Borszcz et al., 1989; Liebman et al., 1970). These data implicate l-PAG with circa-strike behaviors such as the activity burst and v-PAG with the post-encounter defenses such as freezing. A simplistic and preliminary circuit using this distinction is shown in Figure 12.

Ventral PAG: The Figure suggests that the presence of an environmental danger signal is communicated to the v-PAG via the amygdala. The v-PAG in turn initiates post-encounter defenses like freezing and opioid analgesia. However, the autonomic changes (e.g., increased arterial pressure) associated with conditional freezing seem to be mediated by projections from the amygdala to the lateral hypothalamus (LeDoux et al., 1988).

Thus the v-PAG is seen as the organizer or coordinator of post-encounter defensive responses. The linkage of this structure with both freezing and opioid analgesia corresponds nicely with the fact that stimuli such as innately recognized predators and stimuli associated with shock produce freezing accompanied by opioid analgesia (see Fig. 3; Lester and Fanselow, 1985; Lichtman and Fanselow, 1990; Fanselow, 1984b; Fanselow and Baackes, 1982).

Lateral PAG: The l-PAG mediates circa-strike defenses such as the activity burst, jump attack, and vigorous escape behaviors. These active defensive behaviors have different autonomic requirements than freezing and the appropriate adjustments in autonomic function to support these behaviors are also generated by l-PAG (Bandler et al., 1991; Bandler and Depaulis; Carrive, this volume). The circa-strike pattern may be triggered directly by a nociceptive stimulus acting at the l-PAG. However, if exceptionally strong danger stimuli are activating the v-PAG this intense activation may be communicated to the l-PAG resulting in circa-strike defenses. The excitatory connection between the ventral and l-PAG is indicated by a dotted line in Figure 12 because this input is only active in cases of extreme fear (high predatory imminence). In such cases, the l-PAG suppresses post-encounter defenses by inhibiting v-PAG and activating the appropriate circa-strike motor patterns. These motor patterns would be supported by nonopioid analgesia and appropriate autonomic patterns which are also mediated by this subdivision of the PAG.

These vigorous circa-strike defenses must be well coordinated with the location of stimuli in the environment. A rat jumping away from a predator about to make contact must know where the predator is, as well as the location of any available cover, and must make the jump accordingly. This sensory-motor coordination must be performed with the utmost rapidity. The superior colliculus seems ideal, not only because of its anatomical proximity to the l-PAG, but also because it has all the necessary sensory information. Evidence suggesting the superior colliculus is involved with defensive behavior (Dean et al., 1989) and that it projects to the PAG (Rhoades et al., 1989) is consistent with this suggestion.

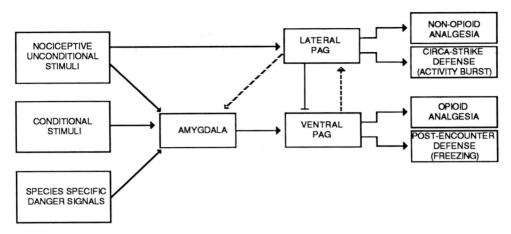

Figure 12. A theoretical model of the defensive behavior system indicating how the amygdala may organize various stimulus inputs to the system and how subdivisions of the PAG may in turn coordinate various defensive activities.

Preliminary Evaluation of the Model

The model presented in Figure 12, seems to handle the available data quite well. Of course, it was designed specifically with the differing effects of lateral versus ventral stimulation in mind: that v-PAG stimulation seems to produce an analgesia that is reversible by opioid antagonists while the analgesia produced by l-PAG stimulation typically is not (e.g., Morgan, this volume). It also was designed to accommodate our observation that either ventral stimulation or lesions affected freezing while l-PAG stimulation or lesions affected the activity burst.

The linkage between the l-PAG, the activity burst and nonopioid analgesia combined with the proposal that this structure is activated directly by nociceptive input accounts for the observation that electric shock produces an immediate but short lived activity burst and nonopioid analgesia (Fanselow, 1982; Grau, 1987; Maier, 1989; 1990). The ability for very strong activation of v-PAG circuits to in turn activate l-PAG circuits fits well with the finding that more severe stress seems to produce a nonopioid rather than an opioid form of analgesia. For example, increasing the intensity or duration of shock leads to nonopioid analgesia (Fanselow, 1984b; Terman et al., 1984) as does increasing the number of shocks received (Lichtman and Fanselow, 1989; Maier, 1989). It is also consistent with reports that very strong shocks tend to condition freezing less than moderate shock (Leaton and Borszcz, 1985). Contrary to theoretical views of stress-indu-

ced analgesia that suggest that the analgesia produced by fear conditioning preparations is always opioid mediated (Grau, 1987; Watkins and Mayer, 1986) the
present approach predicts that very potent conditional fear stimuli will produce
a nonopioid mediated analgesia. This latter view appears far more consistent
with the available data (Chance, 1980; Fanselow, 1984b; Lichtman and Fanselow,
1989; Hagen and Green, 1988)

The finding that shock unconditionally elicits activity but never unconditionally produces freezing even though the conditional response to stimuli associated with shock is freezing has always been something of a paradox to conditioning theory (Fanselow, 1989). The present proposal reckons with the paradox
by making clear that a nociceptive stimulus has two independent functions. The
unconditional response is to provoke l-PAG mediated defenses which are active
responses. This stimulus' ability to support conditioning is mediated by different pathways which potentiate a neutral stimulus' ability to activate the amygdala and v-PAG. The inhibitory connection from l-PAG to v-PAG insures that
such a nociceptive stimulus never results in freezing behavior. When a rat is freezing and showing opioid analgesia, presenting it with an electric shock immediately disrupts the freezing (Fanselow, 1982) and replaces the opioid analgesia
with a nonopioid one (Maier, 1990).

There are monosynaptic projections from l-PAG to amygdala (Rizvi et al.,
1991). This finding is added to Figure 12 because it allows the model to assimila-
te two potentially important findings. One is that stimulation of the l-PAG can
support fear conditioning (Di Scala et al., 1987). This would also account for our
finding of delayed freezing following l-PAG stimulation. Accordingly, this
delayed freezing would be interpreted as a conditional response to apparatus
cues contiguous with stimulation. It was delayed for the same reason post-shock
freezing is delayed; the inhibition of the v-PAG by the l-PAG must dissipate first.
This ability of l-PAG activity to support conditioning may also play a role in
second-order conditioning. In that phenomenon a very potent conditional stimulus acts as an unconditional stimulus, conditioning neutral stimuli with
which it is paired (Rescorla, 1980). For example, a light may be paired with a
shock to produce first-order fear conditioning to the light. Then a tone may be
paired with the light to produce second-order conditioning to the tone. Freezing
and analgesia can be produced by second-order conditioning (Helmstetter and
Fanselow, 1989; Ross, 1988). Importantly, second-order conditioning often
appears to be mediated differently than first order conditioning (Rescorla, 1974;
Rizley and Rescorla, 1972). The first order conditional response appears related
to a memory of the sensory events with which it is associated. The second-order
conditional response seems to be characterized by the subjects emotional reaction to the first-order stimulus. This may be because the neural input supporting
first order conditioning is coming from the thalamus (LeDoux et al., 1986), which
encodes relatively specific sensory information, while the neural input that supports second-order conditioning arises from the l-PAG which is associated with
strong emotional reactivity. It should be noted that while l-PAG stimulation can
support conditioning our lesion data clearly indicate that this structure is not
necessary for first-order conditioning.

Summary and Conclusion

In order to survive, animals must be able to react to a variety of environmental dangers with swift but effective patterns of defensive behavior. For example, without the benefit of experience, laboratory reared rodents show defensive reactions to predators from their ecological niche. Besides these innately recognized dangers, initially neutral environmental events can acquire rapidly the ability to produce similar fear reactions if they are associated with noxious stimulation. Thus a number of learned and innate danger stimuli appear to converge on a common fear system. Not only are the environmental activators of this fear system complex but so too are the responses. The animal may react with any one of a variety of motor behaviors such as changes in feeding pattern, freezing, or vigorous escape, with the particular behavior selected being a function of the degree of threat. Besides motor behaviors the fear reaction is accompanied by autonomic modulation and analgesia that support the motor behaviors. This coordination of environmental stimuli and behavior serving the function of defense against environmental dangers is referred to as the defensive behavior system. Within this framework, fear and anxiety are convenient labels to refer to the activation of this functional behavior system.

Obviously, this elaborate orchestration of stimuli and responses must be performed by an efficient and well integrated neural circuitry. The evidence reviewed suggests that the amygdala synthesizes the various stimulus inputs and then signals the periaqueductal gray of the degree of threat. The PAG, with its well established influence on defensive behavioral topography, descending analgesic system, autonomic and somatic responses, is in an ideal position to coordinate these various response components with the degree of threat. Thus the PAG may be the organ of defensive response selection and generation. When the threat is immediate, such as when a predator is making or about to make contact, the vigorous active defensive behaviors and nonopioid analgesia associated with the lateral subdivision of the PAG dominate. Less immediate threats (e.g., a cat in the vicinity) or conditional stimuli predicting danger activate freezing responses and opioid analgesia mediated by the ventral subdivision of the PAG. Recognizing these various features of defensive behavior organizes the many behavioral aspects of the PAG under a unitary functional role.

Acknowledgements

This research was supported by grants from the National Institute of Mental Health (MH39786) and National Science Foundation (BNS-9008820). I thank S. L. Young for helpful comments on an earlier draft.

References

Bandler, R., Brain mechanisms of aggression as revealed by electrical and chemical stimulation: Suggestion of a central role for the midbrain periaqueductal grey region, Prog. psychobiol. physiol. psychol., 13 (1988) 67-154.

Bandler, R., Carrive, P., and Zhang, S. P., Integration of somatic and autonomic reactions within the midbrain periaqueductal grey: Viscerotopic, somatotopic and functional organization, Prog. Brain Res., 87 (1991) 269-305.

Bandler, R. and Depaulis, A., Elicitation of intraspecific defence reactions in the rat from midbrain periaqueductal grey by microinjection of kainic acid, without neurotoxic effects, Neurosci. Lett., 88 (1988) 291-296.

Blanchard, R.J. and Blanchard, D.C., Defensive reactions in the albino rat, Learn. Motiv., 2 (1971) 351-362.

Blanchard, D. C. and Blanchard, R. J., Innate and conditioned reactions to threat in rats with amygdaloid lesions, J. Comp. Physiol. Psychol., 81 (1972) 281-290.

Blanchard, R.J. and Blanchard, D.C., Ethoexperimental approaches to the biology of emotion, Ann. Rev. Psychol., 39 (1988) 43-68.

Blanchard, R.J., Blanchard, D.C. and Hori, K., An ethoexperimental approach to the study of defense, In: Ethoexperimental Approaches to the study of Behavior, Blanchard R. J., Brain P. F., Blanchard D.C. and Parmigiani S. (Eds.), Kluwer, Dordrecht, 1989, pp. 114-136.

Blanchard, R. J., Fukunaga, K. K. and Blanchard, D. C., Environmental control of defensive reactions to a cat, Bull. Psychon. Soc., 8 (1976) 179-181.

Blanchard, D. C., Williams, G., Lee, E. M. C. and Blanchard, R. J., Taming of wild Rattus norvegicus by lesions of the mesencephalic central gray, Physiol. Psychol., 9 (1981) 157-163.

Bolles, R. C., Species-specific defense reactions and avoidance learning, Psychol. Rev., 77 (1970) 32-48.

Bolles, R.C. and Fanselow, M.S., A perceptual-defensive-recuperative model of fear and pain, Behav. Brain Sci., 3 (1980) 291-301.

Borszcz, G. S., Cranney, J. and Leaton, R. N., Influence of long-term sensitization on long-term habituation of the acoustic startle response in rats: Central gray lesions, preexposure, and extinction, J. Exper. Psychol.: An. Behav. Pr., 15 (1989) 54-64.

Cannon, J T., Prieto, G. J., Lee, A. and Liebeskind, J. C., Evidence for opioid and non-opioid forms of stimulation-produced analgesia in the rat, Brain Res., 243 (1982) 315-321.

Carrive, P., Bandler, R. and Dampney, A.L., Somatic and autonomic integration in the midbrain of the unanesthetized decerebrate cat: a distinctive pattern evoked by excitation of neurones in the subtentorial portion of the midbrain periaqueductal grey, Brain Res., 483 (1989) 251-258.

Chance, W.T., Autoanalgesia: Opiate and nonopiate mechanisms, Neurosci. Biobehav. Rev., 4 (1980) 55-67.

Criswell, H.E., Analgesia and Hyperreactivity Following Morphine Microinjection into Mouse Brain, Pharmacol. Biochem. Behav., 4 (1975) 23-26.

Darwin, C., The expression of the emotions in man and animals, University of Chicago Press, Chicago,1965.

Dean, P., Redgrave, P. and Westby, G.W.M., Event or emergency? Two response systems in the mammalian superior colliculus, Trends Neurosci., 12 (1989) 137-147.

de Olmos, J., Alheid, G. F. and Beltramino, C. A., Amygdala, In: The rat nervous system. I. Forebrain and midbrain, Paxinos G. (Ed.), Academic Press, New York, 1985, pp. 223-334.

Depaulis, A., Bandler, R. and Vergnes, M., Characterization of pretentorial periaqueductal gray matter neurons mediating intraspecific defensive behaviors in the rat by microinjections of kainic acid, Brain Res., 486 (1989) 121-132.

Di Scala, G., Mana, M. J., Jacobs, W. J. and Phillips, A. G., Evidence of Pavlovian conditioned fear following electrical stimulation of the periaqueductal grey in the rat, Physiol. Behav., 40 (1987) 55-63.

Fanselow, M.S., Conditional and unconditional components of post-shock freezing, Pavlovian J. Biol. Sci., 15 (1980) 177-182.

Fanselow, M.S., Naloxone and Pavlovian fear conditioning, Learn. Motiv., 12 (1981) 398-419.

Fanselow, M.S., The post-shock activity burst, Anim. Learn. Behav., 10 (1982) 448-454.

Fanselow, M.S., What is conditioned fear?, Trends Neurosci., 7 (1984a) 460-462.

Fanselow, M.S., Shock-induced analgesia on the Formalin Test: Effects of shock severity, naloxone, hypophysectomy and associative variables, Behav. Neurosci., 98 (1984b) 79-95.

Fanselow, M. S., Odors released by stressed rats produce opioid analgesia in unstressed rats, Behav. Neurosci., 99 (1985) 589-592.

Fanselow, M. S., Associative vs topographical accounts of the immediate shock freezing deficit in rats: Implications for the response selection rules governing species-specific defensive reactions, Learn. Motiv., 17 (1986) 16-39.

Fanselow, M. S., The Adaptive Function of Conditioned Defensive Behavior: An Ecological Approach to Pavlovian Stimulus Substitution Theory, In: Ethoexperimental Approaches to the study of Behavior, Blanchard R.J., Brain P.F., Blanchard D.C., and Parmigiani S. (Eds.), Kluwer, Dordrecht, 1989, pp.151-166.

Fanselow, M. S., Analgesia as a response to aversive Pavlovian conditional stimuli: Cognitive and emotional mediators, In: Fear, Avoidance, and Phobias: A fundamental analysis, Denny M.R. (Ed.), Erlbaum, Hillsdale, 1991, pp. 61-86.

Fanselow, M. and Baackes, M.P., Conditioned Fear-Induced Opiate Analgesia on the Formalin Test: Evidence for Two Aversive Motivational Systems, Learn. Motiv., 13 (1982) 200-221.

Fanselow, M. S., Calcagnetti, D. J. and Helmstetter, F. J., The Role of μ and κ Opioid Receptors in Conditional-Fear Induced Analgesia: The Antagonistic Actions of Nor-Binaltorphimine and the Cyclic Somatostatin Octapeptide, Cys^{2}Tyr^{3}Orn^{5}Pen^7-Amide, J. Pharmacol. Exp. Ther., 250 (1989) 825-830.

Fanselow, M.S. and Helmstetter, F.J., Conditional Analgesia, Defensive Freezing and Benzodiazepines, Behav. Neurosci., 102 (1988) 233-243.

Fanselow, M. S., Helmstetter, F. J. and Calcagnetti, D. J., Parallels between the behavioral effects of Dimethoxy-ß-Carboline (DMCM) and conditional fear stimuli, In: Current Topics in Animal Learning: Brain, Emotion, and Cognition, Dachowski L. and Flaherty C.F. (Eds.), Erlbaum, Hillsdale, 1991, pp. 187-206.

Fanselow, M. S. and Lester, L. S., A functional behavioristic approach to aversively motivated behavior: Predatory imminence as a determinant of the topography of defensive behavior, In: Evolution and Learning, Bolles R. C. and Beecher M. D. (Eds.), Erlbaum, Hillsdale,1988, pp. 185-212.

Fanselow, M. S., Lester, L. S. and Helmstetter, F. J., Changes in Feeding and Foraging Patterns as an Antipredator Defensive Strategy: A Laboratory Simulation using aversive stimulation in a Closed Economy, J. Exper. Anal. Behav., 50 (1988) 361-374.

Fanselow, M. S. and Sigmundi, R. A., Species specific danger signals, endogenous opioid analgesia, and defensive behavior, J. Exper. Psychol.: An. Behav. Proc., 12 (1986) 301-309.

Fanselow, M. S. and Sigmundi, R. A., Functional behaviorism and aversively motivated behavior: A role for endogenous opioids in the defensive behavior of the rat, Psychol. Rec., 37 (1987) 317-334.

Gloor, P., Inputs and outputs of the amygdala: What the amygdala is trying to tell to the rest of the brain, In: Limbic mechanisms: The continuing evolution of the limbic system concept, Livingston K. and Hornykiewicz O. (Eds.), Plenum, New York, 1978, pp. 196-206.

Grau, J. W., The Central Representation of an Aversive Event Maintains the Opioid and

Nonopioid Forms of Analgesia, Behav. Neurosci., 101 (1987) 272-288.

Griffith, C., The behavior of white rats in the presence of cats, Psychobiol., 2 (1920) 19-28.

Grossen, N.E. and Kelley, M.J., Species-specific behavior and acquisition of avoidance behavior in rats, J. Comp. Physiol. Psychol., 81 (1972) 307-310.

Hagen, H. S. and Green, K. F., Effects of time of testing, stress level, and number of conditioning days on naloxone sensitivity of conditioned stress-induced analgesia in rats, Behav. Neurosci., 102 (1988) 906-914.

Helmstetter, F. J., Calcagnetti, D. J. and Fanselow, M. S., The beta-carboline DMCM produces hypoalgesia after central administration, Psychobiol., 18 (1990) 293-297.

Helmstetter, F. J. and Fanselow, M. S., Differential second-order aversive conditioning using contextual stimuli, An. Learn. Behav., 17 (1989) 205-212.

Helmstetter, F. J. and Landeira-Fernandez, J., Conditional hypoalgesia is attenuated by naltrexone applied to the periaqueductal gray, Brain Res., 537 (1991) 88-92.

Helmstetter, F. J., Leaton, R. N., Fanselow, M. S. and Calcagnetti, D. J., The amygdala is involved in the expression of conditional analgesia, Soc. Neurosci. Abst., 14 (1988) 1227.

Hirsch, S.M. and Bolles, R.C., On the ability of prey to recognize predators, Z. Tierpsychol., 54 (1980) 71-84.

Hofer, M. A., Cardiac and respiratory function during sudden prolonged immobility in wild rodents, Psychosom. Med., 32 (1980) 633-647.

Hopkins, A. and Holstege, G., Amygdaloid projections to the mesencephalon, pons and medulla oblongata in the cat, Exp. Brain Res., 32 (1978) 529-547.

Jacquet, Y. F. and Lajtha, A., Paradoxical Effects after Microinjection of Morphine in the Periaqueductal Gray Matter in the Rat, Science, 185 (1974) 1055-1057.

Kavaliers, M., Brief Exposure to a Natural Predator, the Short-tail Weasel, Induces Benzodiazepine Sensitive Analgesia in White-footed Mice, Physiol. Behav., 43 (1988) 187-193.

Krettek, J.E. and Price, J. L., Amygdaloid projections to subcortical structures within the basal forebrain and brainstem in the rat and cat, J. Comp. Neurol., 178 (1978) 225-254.

Leaton, R. N. and Borszcz, G. S., Potentiated startle: Its relation to freezing and shock intensity in rats, J. Exper. Psychol.: An. Behav.Proc., 2 (1985) 248-259.

LeDoux, J. E., Iwata, J., Cicchetti, P. and Reis, D. J., Different projections of the central amygdaloid nucleus mediate autonomic and behavioral correlates of conditioned fear, J. Neurosci., 8 (1988) 2517-2529.

LeDoux, J. E., Iwata, J., Pearl, D., and Reis, D. J., Disruption of auditory but not visual learning by destruction of intrinsic neurons in the rat medial geniculate body, Brain Res., 371 (1986) 395-399.

Lester, L. S. and Fanselow, M. S., Exposure to a cat produces opioid analgesia in rats, Behav. Neurosci., 99 (1985) 756-759.

Lichtman, A. H. and Fanselow, M. S., Yohimbine administered intrathecally (i.th.) reverses both opioid and nonopioid conditional antinociception (CA) in rats, Soc. Neurosci. Abst., 15 (1989) 372.

Lichtman, A. H. and Fanselow, M. S., Cats produce analgesia in rats on the tail-flick test: Naltrexone sensitivity is determined by the nociceptive test stimulus, Brain Res., 533 (1990) 91-94.

Liebman, J. M., Mayer, D. J. and Liebeskind, J. C., Mesencephalic central gray lesions and fear-motivated behavior in rats, Brain Res., 23 (1970) 353-370.

Maier, S. F., Determinants of the nature of environmentally-induced hypoalgesia, Behav. Neurosci., 103 (1989) 131-143.

Maier, S. F., Diazepam modulation of stress-induced analgesia depends on the type of analgesia, Behav. Neurosci., 104 (1990) 339-347.

Mayer, D. J. and Liebeskind, J. C., Pain reduction by focal electrical stimulation of the brain: An anatomical and behavioral analysis, Brain Res., 68 (1974) 73-93.

Morgan, M. M. and Liebeskind, J. C., Site specificity in the development of tolerance to stimulation-produced analgesia from the periaqueductal gray matter of the rat, Brain Res., 425 (1987) 356-359.

Oliveras, J. L. and Besson, J. M., Stimulation-produced analgesia in animals: behavioural investigations, Prog. Brain Res., 77 (1988) 141-157.

Pinel, J. P. J. and Mana, M. J., Adaptive interactions of rats with dangerous inanimate objects: Support for a cognitive theory of defensive behavior, In: Ethoexperimental Approaches to the study of Behavior, Blanchard R. J., Brain P. F., Blanchard D. C. and Parmigiani S. (Eds.), Kluwer, Dordrecht, 1989, pp. 137-150.

Rescorla, R.A., Effect of Inflation of the Unconditioned Stimulus Value Following Conditioning, J. Comp. Physiol. Psychol., 86 (1974) 101-106.

Rescorla, R. A., Pavlovian second-order conditioning, Erlbaum, Hillsdale, 1980.

Reynolds, D. V., Surgery in the rat during electrical analgesia induced by focal brain stimulation, Science, 164 (1969) 444-445.

Rhoades, R. W., Mooney, R. D., Rohrer, W. H., Nikoletseas, M. M. and Fish, S. E., Organization of the projection from the superficial to the deep layers of the hamster's superior colliculus as demonstrated by anterograde transport of Phaseolus vulgaris leucoagglutinin, J. Comp. Neurol., 283 (1989) 54-70.

Rizley, R. C. and Rescorla, R. A., Associations in second-order conditioning and sensory preconditioning, J. Comp. Physiol. Psychol., 81 (1972) 1-11.

Rizvi, T. A., Ennis, M., Behbehani, M. and Shipley, M. T., Connections between the central nucleus of the amygdala and the midbrain periaqueductal gray: Topography and reciprocity, J. Comp. Neurol. (1991) in press.

Rodgers, R. J. and Randall, J. I., Benzodiazepine ligands, nociception and 'defeat' analgesia, Psychopharmacology, 91 (1987) 305-315.

Ross, R. T., Pavlovian second-order conditioned analgesia, J. Exper. Psychol.: An. Behav. Proc., 12 (1988) 32-39.

Small, W., Notes on the psychic development of the young white rat, Am. J. Psychol., 11 (1899) 80-100.

Stephens, D.N. and Kehr, W., β-Carbolines can enhance or antagonize the effects of punishment in mice, Psychopharmacology, 85 (1985) 143-147.

Terman, G. W., Shavit, Y., Lewis, J. W., Cannon, J. T. and Liebeskind, J. C., Intrinsic mechanisms of pain inhibition: Activation by stress, Science, 226 (1984) 1270-1277.

Timberlake, W. and Lucas, G. A., Behavior system and learning: From misbehavior to general principles, In: Contemporary learning theories: Instrumental conditioning theory and the impact of Biological constraints on learning, Klein S. B. and Mowrer R. R. (Eds.), Erlbaum, Hillsdale,1989, pp. 237-275.

Urca, G., Nahin., R, Liebeskind, J., Glutamate-induced Analgesia: Blockade and Potentiation By Naloxone, Brain Res., 192 (1980) 523-530.

Watkins, L. R. and Mayer, D. J., Multiple endogenous opiate and non-opiate analgesia systems: Evidence of their existence and clinical implications, Ann. N. Y. Acad. Sci., 467 (1986) 273-299.

Midbrain Periaqueductal Gray Control of Defensive Behavior in the Cat and the Rat

Richard Bandler* and Antoine Depaulis**

*Department of Anatomy, The University of Sydney
Sydney, Australia
**L.N.B.C., Centre de Neurochimie du CNRS
Strasbourg, France

Introduction

In 1943, Hess and Brügger observed that electrical stimulation within the perifornical hypothalamus transformed a normally calm and placid cat into a highly defensive animal. Subsequently, Hunsperger (1956) reported that electrical stimulation within the midbrain periaqueductal gray region (PAG) or the adjacent midbrain tegmentum, of the freely moving cat, evoked defensive reactions. At about the same time, researchers in Sweden (Eliasson et al., 1954; Lindgren, 1955; Lindgren et al., 1956) found that electrical stimulation within the midbrain, of the anesthetized cat, evoked cardiovascular changes, which included increased arterial pressure, tachycardia and regional blood flow changes. It was left to Abrahams et al. (1960) to confirm that these behavioral and cardiovascular reactions were evoked at the same midbrain sites.

Support for an integral role for the midbrain, in mediating defensive behavior, came from the findings that midbrain electrical stimulation still evoked the coordinated patterns of rapid somatic and autonomic adjustments characteristic of a cat's responses to threatening or stressful stimuli: (i) in the precollicular decerebrate cat (Bard and Macht, 1958; Keller, 1932; Lindgren, 1955; Schramm and Bignall, 1971); (ii) after large hypothalamic or amygdaloid lesions (Fernandez DeMolina and Hunsperger, 1962; Kelly et al., 1946); or, (iii) after surgical isolation of the hypothalamus (Ellison and Flynn, 1962; Gellen et al., 1972). As well, lesions which extensively damaged the central core of the midbrain were found to either eliminate or attenuate defensive reactions evoked by nociceptive stimuli, or by electrical stimulation of the hypothalamus or the amygdala (Fernandez DeMolina and Hunsperger, 1962; Hunsperger, 1956; Skultety, 1963). Similar studies, in species other than the cat, also supported the view that a mid-

The Midbrain Periaqueductal Gray Matter, Edited by A. Depaulis and
R. Bandler, Plenum Press, New York, 1991

brain region, centered on the PAG, provided an essential neural substrate for the integration of an organism's responses to threatening or stressful stimuli (see Bandler, 1988 for a recent review).

Although it established an important role for the midbrain in mediating defensive behavior, a major shortcoming of this research was that since electrical stimulation excites both cell bodies and axons of passage, it was not possible to determine the exact location of the specific population(s) of midbrain neurons mediating such defensive reactions. The technique of microinjection of excitatory amino acids (EAA) provided a way around this problem as EAA selectively excite cell bodies (and their dendritic processes) but not axons of passage (Goodchild et al., 1982; Lipski et al., 1988). Using this technique, a series of studies by Bandler (Bandler, 1982a,b; 1984; Bandler et al., 1985) and others (Hilton and Redfern, 1986; Krieger and Graeff, 1985) firmly localized populations of neurons mediating these reactions within the PAG.

In this chapter we will review the results of recent studies, in the cat and the rat, undertaken to further delineate the neural organization within the PAG mediating defensive behavior.

Methods

Freely moving cat

One week prior to the start of experimental testing, cats (2.5-4.8 kg) were anesthetized (sodium pentobarbital, 40 mg/kg), secured in a stereotaxic frame and 2-4 guide tubes, through which guide cannulae could later be implanted into the midbrain, were cemented over holes drilled stereotaxically in the skull. A vertical approach was used for the positioning of guide tubes over the rostral and intermediate thirds of the PAG. However, for the caudal third of the PAG, because the bony tentorium precluded a vertical stereotaxic approach, the guide tubes were positioned at an angle of 20° from the vertical. Guide cannulae implantation and EAA microinjections (20, 40 or 80 nmol of D,L-homocysteic acid (DLH) in either 0.20 or 0.40 µl of phosphate buffer, pH 7.4), were carried out according to procedures described previously (Bandler and Carrive, 1988; Zhang et al., 1990). Prior to making a chemical injection the cat was allowed to move freely about the test cage (1.8m x 1.0m x 1.0m) for several minutes and its behavior was recorded on videotape. An analysis of this period was used to establish a control level of behavior. Each cat's behavior from the start of a DLH injection until the end of the experimental period (up to 30 min) was observed and portions (which always included the first 2-4 min) videotape recorded. The occurrence of different behavioral items were later encoded from the video recordings. Items individually encoded included: *General movement:* turning of head, arching of back, non-locomotory limb movement, nibbling, licking, scratching any part of its own body; *Locomotion:* walking, backing, circling, running; *Escape movements:* rearing, sometimes accompanied by pawing movements at the side of the test cage; *Jumps:* directed usually to the top or sides of the test cage; *Vocalization:*

hissing, howling, mewing or growling. Ear position, pupil size and the degree of piloerection were also noted.

Freely moving rat

Four month old male Wistar rats (350-400 g) were implanted, under anesthesia (40 mg/kg sodium pentobarbital) with a stainless steel guide cannula aimed at the PAG. Each animal was then allowed to recover for at least a week. Intracerebral injections were made in the awake animal, held by the experimenter (for details see Depaulis et al., 1989). Kainic acid (KA) was used in preference to EAA such as glutamate, aspartate, or DLH because of its longer duration of action (Bandler and Depaulis, 1988; Depaulis et al., 1989). A standard injection of 40 pmol of kainic acid (KA) in 0.20 µl of sterile distilled water (pH 7.4) was used. After injection the animal was placed in a test cage (25 x 30 x 35 cm) and its behavior for the next 2 min was recorded on videotape. A weight-matched untreated male rat was then introduced into the test cage for 6-8 min and the resultant social interactions were recorded on videotape. The condition of a "neutral" test cage made it possible to observe the reactions of the KA-injected animal to the "non-aggressive" approaches (mainly investigations) of the partner. The occurrence of different behavioral items (adapted from Grant and Mackintosh (1963)) were encoded from the video recordings. The following behavioral items were encoded: *Defensive behavior:* defensive upright, defensive alerting, and defensive sideways postures; backward locomotion; forward locomotion; avoidance and jumps; *Immobility:* immobile posture; *Nonsocial behavior:* exploration and self-grooming; *Social behavior:* investigation and allogrooming (for details see Depaulis et al., 1989; 1991). Ultrasounds in the 22-28 kHz range were made audible by means of a "bat" detector and also recorded. Although the partners were not muted, the rat emitting the ultrasounds could be clearly identified by respiratory movements concomitant with the vocalization.

At the completion of the social interaction test, some injected rats were placed three times, in the center of a 3 m (dia.) circle drawn on the floor of a large enclosure and the latency to leave the circle was measured. Each open field test was terminated if the rat had not left the circle within 20s.

Sensorimotor testing

Cats

The behavior of 4 cats, in which DLH injection evoked defensive behavior was observed also following injection of KA (940 pmol). Because of the longer duration of the KA-evoked reactions it was possible in these experiments to determine the cat's response: (i) to visual approach (i.e., a glove taped to the end of a stick was placed within the test cage and moved towards the cat) in the visual hemifield contralateral or ipsilateral to the injection site; and (ii) to touch (by the glove) of the contralateral or ipsilateral muzzle and upper body. Each cat was tested 10 min, 5 min and immediately prior to a KA injection and then at 5 min intervals until the end of the experiment (30-40 min). The results of each experiment were recorded on videotape for later analysis.

Rats

In these experiments, 13 rats in which KA injections evoked defensive behavior were tested for their reaction to tactile stimulation of different parts of the body. Light tactile stimulation with a 1 cm dia. nylon brush was applied to the hindleg, the flank, the foreleg and the head on each side of the body. The side ipsilateral to the injection site was tested first. Tactile stimulation was also applied along the midline of the back. As well, the rat's behavior as the brush was moved toward its head and forelimb provided an indication of its reactivity to visual stimulation. In later experiments the rat's responses to the approach of a glove were also examined. The results of each experiment were recorded on videotape for later analysis. The reaction of each rat was scored as follows: 0, no response; 1, orientation of the head in the direction of the stimulus; 2, locomotion (forward or backward); 3, defensive alerting/sideways posture or running; 4 defensive upright posture or jump. Each animal was tested 5 min before and immediately prior to a KA injection, at 3 min post-KA injection and then at 4 min intervals until the end of the experimental period (31 min) (for details see Depaulis et al., 1989; 1991).

Histology

At the completion of each set of experiments the cat or rat was anesthetized and perfused intracardially with physiological saline followed by 4% formalin. The brain was then removed and 50 μm frozen coronal sections (cat) or 20 μm paraffin embedded sections (rat) were cut and stained with Cresyl Violet. The track and tip of each injection cannula, as indicated by damage and gliosis, were determined by light microscopic examination.

Results

I. Behavioral effects

Cat

THREAT DISPLAY

DLH injections made at sites located within a longitudinal column, lateral to the midbrain aqueduct, extending from approximately A4.1 to A1.0, elicited a threat display, which consisted of a moderate degree of pupillary dilation and piloerection; vocalization (howling usually mixed with hissing; or hissing alone); and dependent on the specific site, retraction of the ears and/or arching of the back (Bandler and Carrive, 1988). At sites at which a "strong" threat display (characterized by howling) was evoked, the cat sometimes turned sideways and backed away from the experimenter. The threat displays characterized by howling were the most intense and had the longest mean duration (howl 97.7s vs hiss alone 61.2s, t-test, p<0.05). As seen in Figure 1, the sites at which DLH injection evoked a strong threat display were localized, almost exclusively, to a restricted part of the intermediate third of the PAG, lateral to the aqueduct, between A2.9 to A1.6. "Moderate" threat displays (characterized by hissing alone) were

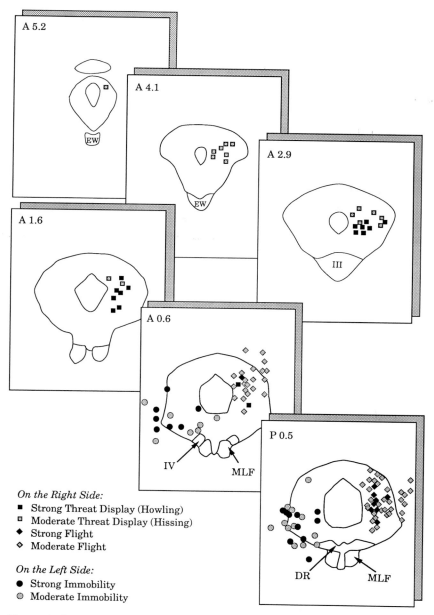

On the Right Side:
- ■ Strong Threat Display (Howling)
- ▫ Moderate Threat Display (Hissing)
- ◆ Strong Flight
- ◇ Moderate Flight

On the Left Side:
- ● Strong Immobility
- ◎ Moderate Immobility

Figure 1. Coronal sections through the PAG of the cat (modified from the atlas of Berman, 1968) showing the distribution of sites at which DLH microinjection evoked threat display, flight or immobility reactions. The centers of the sites evoking a strong or moderate threat display (squares) or a strong or moderate flight reaction (diamonds) are represented on the right side. Sites at which strong or moderate immobility (circles) was evoked are represented on the left. Abbreviations: DR, dorsal raphe nucleus; EW, Edinger-Westphal nucleus; MLF, medial longitudinal fasciculus; III, oculomotor nucleus; IV, trochlear nucleus.

evoked from sites located generally more dorsal and rostral (A4.1 and A2.9), but still within the intermediate third of the lateral PAG.

FLIGHT

DLH injections made at sites in the caudal third of the lateral PAG, evoked a flight reaction, which was characterized by rapid running about the cage (usually in a direction contralateral to the side of injection) multiple jumps and attempts to escape from the test cage (Fig. 2). Additional components of the flight reaction included moderate pupil dilation, piloerection, and occasional mewing. The reaction lasted approximately 120s, the individual components of the reaction declining in parallel (for details see Zhang et al., 1990). The sites from which a strong flight reaction was evoked were localized to a restricted part of the caudal third of the lateral PAG, centered at approximately P0.5 (Fig. 1). Sites at which a moderate flight reaction was evoked included a more extensive portion of the caudal lateral PAG, as well as a number of sites in the tegmentum laterally and dorsolaterally adjacent to the PAG (Fig. 1).

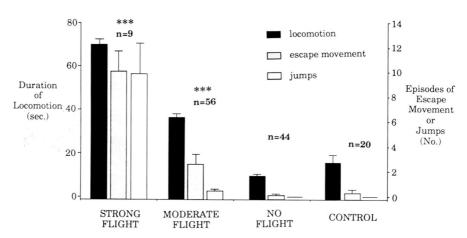

Figure 2. Histogram showing the DLH evoked changes (mean ± SEM) in duration of locomotion, number of episodes of escape movements and number of jumps. Sites were grouped on the basis of the duration of locomotion during the first 2 min post-injection (strong, >61s locomotion; moderate, 21-60s locomotion; no, 4-20s locomotion). Control levels of behavior in each category were determined from analysis of videotape records of the behavior of cats (n=20) during the pre-injection, post-handling period. Significance between groups was assessed by the Mann-Whitney U-test. ***P<0.001. Reprinted with permission from Zhang et al. (1990) Brain Research.

IMMOBILITY

In contrast to the defensive behaviors (threat display, flight) which were evoked from the lateral PAG, DLH injections made at sites located ventrolateral to the aqueduct in the caudal third of the PAG evoked a significant suppression of locomotion/cage exploration. At some of these same sites the suppression of locomotion was accompanied by a period of profound inactivity, during which time the cat showed a dramatic reduction in other spontaneous behaviors such as turning of the head, licking, grooming, stretching, vocalization, etc. At other sites, although the DLH injection suppressed locomotion, there was no change in the level of other spontaneous activity (Fig. 3). The absence of other spontaneous behavior has been called inactivity and we have used the term immobility to refer to the combination of the suppression of locomotion and increased inactivity.

As seen in Figure 1, the sites at which immobility was evoked were confined to the region ventrolateral to the aqueduct, within the caudal third of the PAG and adjacent tegmentum (A0.6 and P0.5). Although the pooled data suggest a degree of "apparent" dorso-ventral overlap of sites at which flight or immobility are evoked, this was not true in the individual animal. Figure 4 illus-

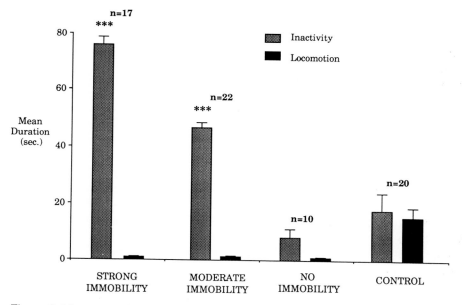

Figure 3. Histogram showing the DLH evoked changes (mean ± SEM) in duration of inactivity and locomotion. Sites were divided into 3 groups on the basis of the duration of inactivity (strong, > 61 s inactivity; moderate, 30-60 s inactivity; no immobility, < 29 s inactivity). Control levels of locomotion and inactivity were determined from analysis of the behavior of cats (n=20) during the control period. ***P<0.001, Mann-Whitney U-test. Reprinted with permission from Zhang et al. (1990) Brain Research.

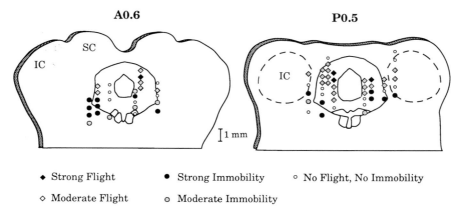

• Strong Flight   • Strong Immobility   ○ No Flight, No Immobility

◇ Moderate Flight   ◎ Moderate Immobility

Figure 4. Representative coronal sections showing the sites of DLH injections made along 14 tracks in a series of steps, passing through the midbrain from dorsal to ventral. Coordinates of the coronal sections (A0.6, P0.5) are in accordance with those of the atlas of the cat of Berman (1968). The symbols indicate the injection sites at which DLH evoked the various reactions. Reprinted with permission from Zhang et al. (1990) Brain Research.

trates the effects of DLH injections made along 14 tracks, in a series of steps, passing through the caudal third of the PAG from dorsal to ventral. It can be seen: (i) that along any individual track, sites from which flight was evoked always lay dorsal to sites from which immobility was evoked; (ii) that different behavioral reactions were evoked in the same cat, at sites separated dorso-ventrally by 0.50-1.0 mm. Thus, the quite different reactions are due to the precise site of DLH injection within the caudal PAG and not to either the past history or the personality of the individual cat.

Rat

The effects evoked by unilateral injection of KA (40 pmol) in the midbrain were examined in two sets of experiments using similar test conditions. In the first set of experiments (Depaulis et al., 1989) the effects of KA injection at 88 midbrain sites, located between A1.7 and A 3.7, were studied when the rat was alone in the test cage (the first 2 min post-injection), and then during a social interaction test which lasted for 8 min. In a second set of experiments (Depaulis et al., 1991) the effects of KA injection at 120 midbrain sites, located between P0.2 and A3.2 were studied while the animal was alone (the first 2 min post-injection) and then during a social interaction test, which lasted for 6 min. In the second set of experiments, 22-28 kHz ultrasound was recorded (N=81 sites) during both the isolation and social interaction tests. At the completion of the social interaction test, the latency to leave the 3m dia. circle was measured (N=115 sites).

Following midbrain injections of KA three major categories of behavioral effect could be distinguished during the social interaction test: backward defense; forward avoidance; and immobility. They were defined as follows:

(i) *Backward Defense:* a *strong* reaction was characterized by from 30 to 60% of the observation time spent in defensive postures (uprights, alerting, backing); a *moderate* reaction, when the time spent in defensive postures included from 10 to 30% of the observation period. (ii) *Forward Avoidance:* a *strong* reaction was characterized by from 20 to 40% of the observation time spent in forward locomotion; a *moderate* reaction when this time spent in forward locomotion included from 5 to 20% of the observation period. *Immobility:* was characterized by the injected rat remaining immobile for more than 95% of the observation time. Sites which did not fall into any of these groups were classified as without significant behavioral effect.

BACKWARD DEFENSE

This behavioral pattern was observed after KA injection at a total of 51 sites. During the isolation test (initial 2 min post-KA injection), these rats remained immobile, i.e., they did not locomote or explore the cage. There was, however, an observable increase in the rate of respiration, and any noise often evoked upright postures and/or backing. Twenty two to 28 kHz ultrasounds were measured following injection at 5 sites, but no emissions were recorded (Fig. 5).

During the social interaction test, a significant increase in both the frequency and duration of defensive postures and immobility was observed, compared to control injections of saline at the same sites (Fig. 6). The KA injected rat usually assumed a posture such that there was a contralateral concavity to its body as it faced the partner. It is important to note that the increased incidence of defensive postures (upright, alerting, backing) was in response to the "non-aggressive" investigations of the injected rat by the partner. There were some ultrasound emissions by the KA injected rats during the social interaction test, but the durations were not significantly greater than controls (Fig. 5). At the end of the social interaction test, when these animals were placed in the open field, they either remained immobile in the center of the circle, or backward locomotion toward the ipsilateral side was observed.

FORWARD AVOIDANCE

This behavioral pattern was observed at 32 midbrain sites (Depaulis et al., 1991). During the 2 min isolation test, these rats showed bursts of forward locomotion (and occasional jumps) which alternated with periods of immobility. Compared to control injections, there was a significant increase in the emission of 22-28 kHz vocalization (Fig. 5).

During the social interaction test, these rats usually assumed a posture with a slight contralateral concavity of the body, although it was less marked than in the backward defense animals. There was a significant increase in: (i) the frequency and duration of forward avoidance reactions (forward locomotion and jumps); and (ii) the duration of immobility (Fig. 7). Often a burst of forward locomotion (with an occasional jump), evoked by the approach of the partner, was followed by a phase of immobility. During the phases of immobility there was a significant increase in the emission of ultrasonic vocalization for the KA-

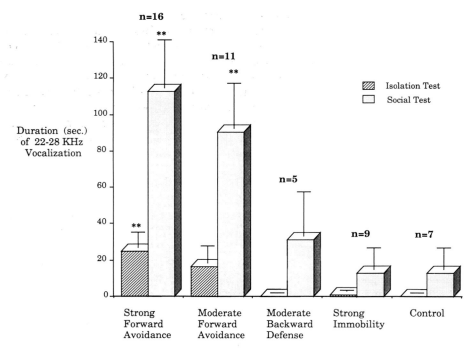

Figure 5. Mean duration (± SEM) of 22-28 kHz vocalization evoked by microinjection of KA within the PAG, during the initial 2 min period (isolation test) and
during a 6 min social interaction test (N=48). Differences between each behavioral group and the control group. ** p< 0.01 (Mann-Whitney U test).

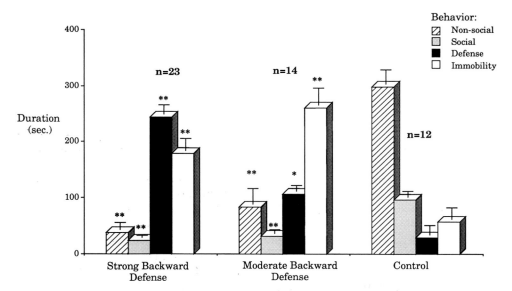

Figure 6. Mean duration (± SEM) of non-social and social behaviors, defense (backward
defensive reactions) and immobility, evoked by KA microinjection during an 8 min social
interaction test (N=45). * p< 0.05, ** p< 0.01 vs control group (Mann-Whitney U test).

injected rats (Fig. 5). When rats showing strong forward avoidance were placed in the open field, they immediately ran forward and escaped from the circle with a significantly shorter latency than controls (Table I).

Table I. Mean latency (± SEM) to leave the 3 m dia. circle after being placed in the center (N=59). (Maximum latency = 20 sec ** p< 0.01 vs control group (Mann-Whitney U test).

Group (N)	Latency
Control (6)	9.7 ± 1.5
Strong Backward (4)	16.5 ± 3.0
Moderate Backward (7)	15.4 ± 2.7
Strong Forward (17)	4.4 ± 0.3**
Moderate Forward (13)	7.1 ± 1.5
Immobile (12)	18.6 ± 1.4**

IMMOBILITY

This behavioral pattern was evoked following KA injection at 18 sites. During the initial 2 min social isolation period, although their respiration rate was increased, the rats of this group remained totally immobile. Ultrasounds were measured at 9 sites, but emissions were detected at only one site (Fig. 5).

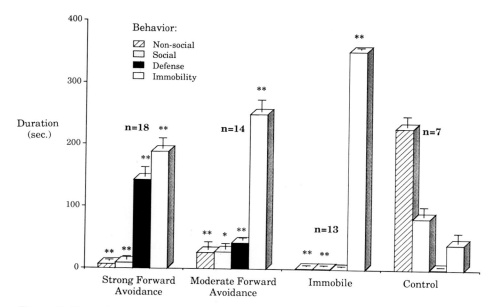

Figure 7. Mean duration (± SEM) of non-social and social behaviors, defense (forward avoidance reactions) and immobility evoked by KA injection during a 6 min social interaction test (N=52). * p< 0.05, ** p< 0.01 vs control group (Mann-Whitney U test).

During the social interaction test these rats remained immobile for almost the entire test period (mean, 352.5 sec of a possible 360 sec) (Fig. 7). No obvious postural asymmetries could be observed. An absence of any reaction to even vigorous investigation by the partner was observed at 11 of the 18 immobility sites. At the 7 other sites although occasional defensive responses were evoked by the investigations of the partner, such responses were slower and less vigorous than normal. At the 9 sites at which ultrasounds were measured, emissions were detected at only one site (Fig. 5). Following the social interaction test, when these rats were placed in the center of the circle, they usually remained in the same spot until the end of the observation period (Table I).

PAG LOCALIZATION

It can be seen in Figure 8 that the majority of sites at which *strong backward defense* was evoked, were localized to a region (centered at approximately A 2.7), lateral to the aqueduct, within the *intermediate third of the lateral PAG*. Strong backward defense was observed at only 1 site located posterior to A1.7. In contrast, the majority of sites (16 of 18) at which *strong forward avoidance* was evoked, were localized to a region (centered at approximately A1.0) lateral to the aqueduct within the *caudal third of the lateral PAG*. Strong forward avoidance was never evoked at sites rostral to A1.7. Moderate backward defense and moderate forward avoidance sites were similarly located within the intermediate and caudal parts of the lateral PAG respectively.

In contrast to the localization of backward defense and forward avoidance sites within the lateral PAG, the majority of sites (13 of 18) at which *immobility* was evoked were located *ventrolateral to the aqueduct, in the caudal one-half of the PAG* (A1.7-P0.2) or the immediately adjacent tegmentum (Fig. 8). The 5 other immobility sites were located primarily dorsolateral to the aqueduct within the rostral one-half of the PAG.

II. Sensorimotor Testing

Following KA injection into the "threat display region" of the lateral PAG of the cat (Bandler and Carrive, 1988) it was observed that approach, in the *contralateral* visual hemifield or light touch on the *contralateral* muzzle or upper body: (i) increased the intensity of the threat display (e.g., louder vocalization, increased piloerection, increased ear retraction, and backing); and (ii) often evoked striking or biting (Fig. 9). In contrast, approach in the ipsilateral visual hemifield or tactile stimulation of the ipsilateral muzzle or upper body did not reliably evoked such behavior.

A comparable asymmetry in the response to tactile stimulation was observed in experiments undertaken with rats in which KA microinjections evoked either strong backward (n=7) or strong forward avoidance (N=6) behavior. As seen in Figure 10, in these rats, touch on the *contralateral, but generally not the ipsilateral*, side of the body evoked respectively, backward defense or forward avoidance behavior (Depaulis et al., 1989; 1991). Interestingly, for those PAG sites at which backward defense was evoked: (i) touch in the region of the *contralateral*

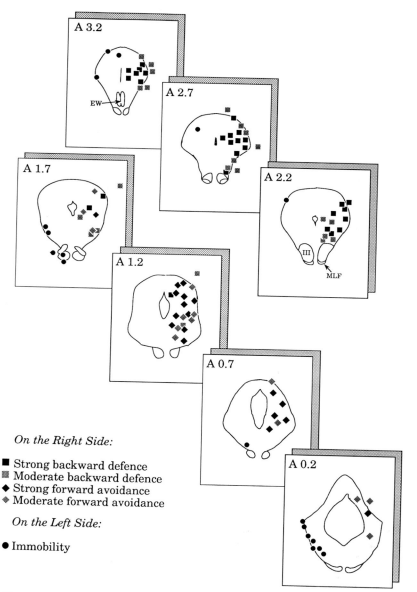

On the Right Side:

■ Strong backward defence
▦ Moderate backward defence
◆ Strong forward avoidance
◈ Moderate forward avoidance

On the Left Side:

● Immobility

Figure 8. Coronal sections through the PAG of the rat (modified from Paxinos and Watson, 1986) showing the distribution of sites at which KA microinjections evoked backward defense, forward avoidance and immobility reactions. The centers of the sites evoking a strong or moderate backward defense reaction (squares) or a strong or moderate forward avoidance reaction (diamonds) are represented on the right side. Sites at which an immobility reaction was evoked are represented on the left (circles). Abbreviations: EW, Edinger Westphal nucleus; MLF, medial longitudinal fasciculus; III, oculomotor nucleus.

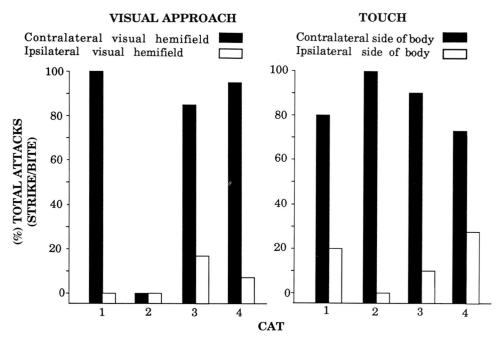

Figure 9. Each histogram compares the total number of strikes and bites which were elicited by visual and tactile stimuli. Ipsilateral and contralateral are defined with respect to the PAG injection site. For each site an equal number of ipsilateral and contralateral stimulus presentations were made. The analyses are based on a minimum of 20 attacks to visual approach or touch. Note that Cat 2 never attacked the visual stimulus. Reprinted with permission from Bandler and Carrive (1988) Brain Research.

head, elicited the most intense and longest duration backward defense reaction; (ii) touch in the region of the *contralateral forelimb*, elicited a somewhat lessened backward defense reaction; and (iii) touch of either the contralateral flank or hindlimb failed to elicit backward defense (Fig. 10). However, following KA injections at PAG sites which evoked forward avoidance reactions, such reactions were generally evoked by tactile stimulation of all regions of the contralateral body (head, forelimb, flank, hindlimb) as well as along the midline of the back (data not shown).

Although it was not systematically studied it was observed that the responsiveness to visual stimuli, of the KA-injected rat was different from that of the KA-injected cat. Thus, it was only tactile stimulation by the brush, and not its visual approach, which evoked backward defensive or forward avoidance reactions in the KA-injected rat; whereas, the presence of a stimulus in the contralateral visual hemifield readily evoked defensive behavior for the KA-injected cat (Fig. 9).

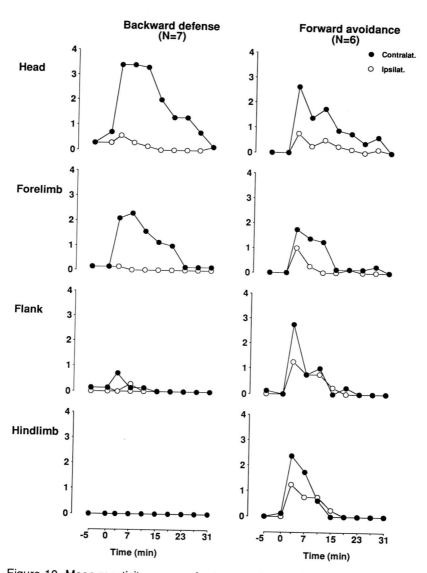

Figure 10. Mean reactivity scores of rats to tactile stimulation applied at four different body sites on the contralateral and ipsilateral side of the body. Rats were injected with KA at sites in the PAG at which strong backward defense (N= 7) or strong forward avoidance (N=6) reactions were evoked.

Discussion

Distinct defense reactions are evoked from different parts of the lateral PAG

These experiments clearly indicate: (i) that integrated defense reactions are evoked by excitation of neurons found lateral to the aqueduct within the caudal two-thirds of the PAG; and (ii) that neurons within the intermediate and caudal thirds of the lateral PAG mediate different types of defensive behavior. It is most important to note also, that although experiments were carried out in different species, remarkably similar patterns of defensive behavior were evoked from "homologous" regions of the PAG. Thus, whether in the cat or the rat, the pattern of defense evoked from the intermediate lateral PAG was characterized predominantly by facing and backing away from a "threatening" stimulus; whereas, the defensive response evoked from the caudal lateral PAG took the form of a forward avoidance reaction, which in its strongest form became flight. In the area transitional between the intermediate and the caudal PAG mixed effects were sometimes evoked. These findings are illustrated schematically in Figure 11.

It has been reported previously in the rat that microinjections of high doses of other EAA (glutamate, aspartate), high doses of GABA antagonists (bicuculline, picrotoxin) or high doses of morphine, within the intermediate and caudal thirds of the PAG, evoke vigorous, undirected jumping which has been referred to as "explosive motor behavior" (Bandler et al., 1985; Di Scala et al., 1984; Hilton and Redfern, 1986; Jacquet and Lajtha, 1974; Krieger and Graeff, 1985; Schmitt et al., 1984). In recent experiments we have observed that injections of higher doses of KA (i.e., 200 pmol rather than the usual 40 pmol) in the intermediate or caudal lateral PAG similarly evoked many jumps, but always superimposed on the basic backward defense or forward avoidance patterns of behavior (Depaulis et al., 1989; Depaulis et al., 1991). Microinjection of low-moderate doses of the GABA antagonist picrotoxin, into the intermediate lateral PAG of the rat, has also been reported to evoke backward defensive, rather than explosive motor behavior (Depaulis and Vergnes, 1986). We interpret these findings to mean that "explosive motor behavior" represents not a single reaction, but rather the strong form of two distinct behavioral patterns, namely the backward defensive reaction associated with the excitation of neurons within the intermediate lateral PAG and the forward avoidance pattern associated with excitation of neurons in the caudal lateral PAG.

Distinct immobility reactions are evoked from the ventrolateral and lateral PAG

In contrast to the defensive reactions evoked from the lateral PAG, microinjections of EAA made in the ventrolateral PAG, either in the cat or rat, evoked an immobility response. It is important to note, however, that this response usually took the form of a *"hyporeactive immobility"*. Rats, for example, injected in the ventrolateral PAG were not only immobile but also generally

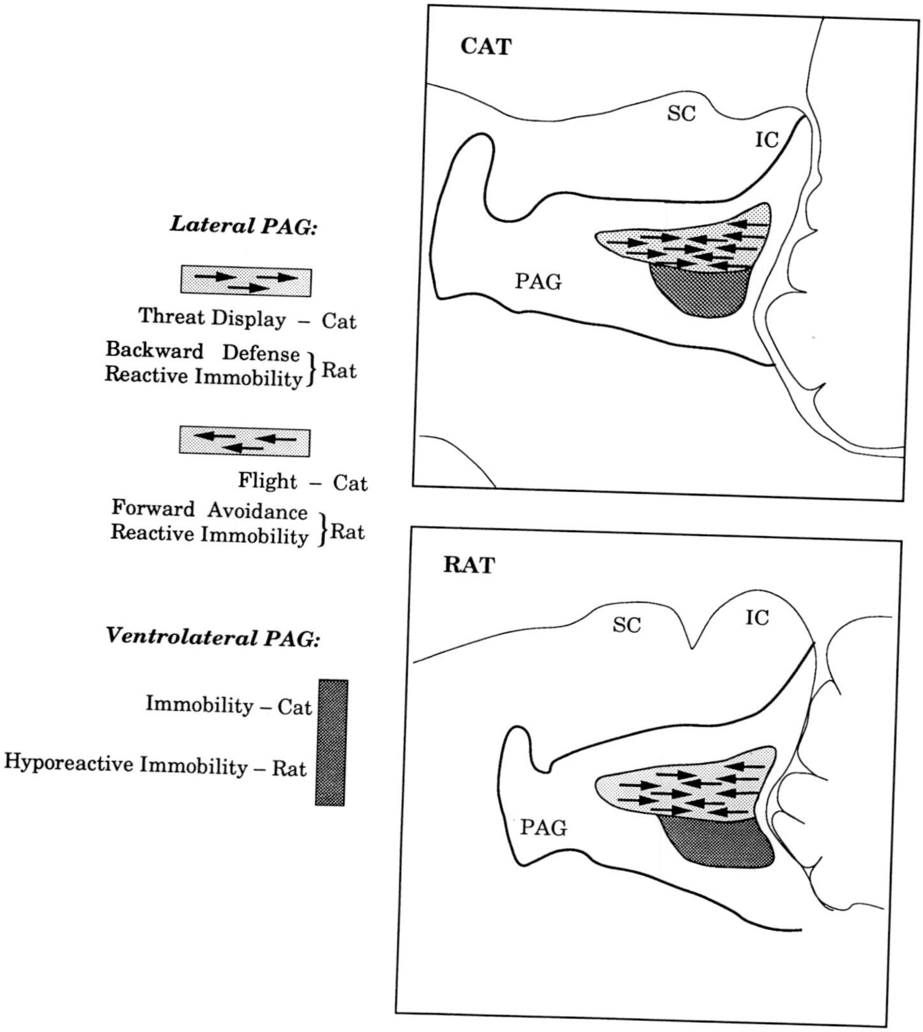

Figure 11. Schematic sagittal representation of regions in the PAG of the cat and the rat mediating defense and immobility reactions. Abbreviations: IC, inferior colliculus; PAG, periaqueductal gray; SC, superior colliculus.

unresponsive to external stimulation (e.g., the approach/investigation by a partner, touch on their body).

Periods of immobility reactions were evoked also from the lateral PAG of the rat, but always as components of defensive reactions (Figs. 6 and 7). That is, these were *"reactive immobility"* responses, in the sense that the rats although immobile showed a heightened defensive reactivity to relatively innocuous external stimuli (e.g., investigation by a partner). It may be recalled also that the reactive immobility associated with forward avoidance was characterized by the frequent emission of ultrasonic vocalization (Fig. 5).

To summarize, then, neurons in the lateral PAG mediate reactive immobility responses which occur as components of defensive reactions. In contrast, the immobility reaction evoked from the ventrolateral PAG is associated with a decreased responsiveness to external stimulation. The functional significance of this hyporeactive immobility is unknown. It is interesting to note, however, that EAA injections made in the ventrolateral region of the caudal PAG have been reported to evoke a hypotensive reaction (Bandler et al., 1991; Carrive and Bandler, 1991a; Carrive, this volume; Lovick, this volume) as well as an opioid-mediated analgesia (Morgan, this volume; Yaksh et al., 1976). It is possible that a coordinated behavioral response comprising, in part, a hyporeactive immobility, a fall in blood pressure and an opioid-mediated analgesia, is integrated within the caudal ventrolateral PAG. Such a coordinated response might be biologically adaptive, perhaps as an aid in the healing process following an injury received during a stressful encounter, or if lethally grasped by a predator (see also Duggan, 1983; Fleischmann and Urca, 1988; Lovick, this volume).

Elsewhere in this volume both Fanselow and Morgan have used the term "freezing" to describe "reactive immobility" responses. However, from their experiments, in which rats were tested only in conditions of social isolation, they concluded that "freezing" was mediated not by the lateral PAG, but rather by the ventrolateral PAG. Clearly, however, the hyporeactive immobility, which was characteristic of the social behavior of our rats following KA injections made in the ventrolateral PAG does not correspond to freezing.

Afferent regulation of the lateral PAG

The sensorimotor experiments indicate that, following unilateral injection of EAA within the lateral PAG of the cat and the rat, contralateral but not ipsilateral tactile stimuli are likely to evoke defensive reactions (Figs. 9 and 10). A similar lateralization of defensive reactions has also been observed after PAG injections of other EAA or GABA antagonists (Bandler, 1988; Bandler et al., 1985; Depaulis and Vergnes, 1986; Schmitt et al., 1984). Such findings lend support to the previous suggestions that the coordinated behavior patterns evoked from the PAG, and other parts of the central nervous system (e.g., hypothalamus), may be associated with significant changes in the processing of sensory information (Bandler, 1982; 1984;1988; Flynn, 1972).

Although it needs still to be systematically examined, our preliminary observations suggest that following KA injection in the PAG, visual stimuli (i.e., the approach of the nylon brush or glove) do not readily evoke defensive reactions in the rat. These results contrasts dramatically with the findings that visual stimuli readily evoke defensive reactions, such as running and freezing in the rat, following EAA injections made in the medial part of the superior colliculus (Redgrave and Dean, this volume). Recently it has been reported that the visually responsive cells in the medial part of the intermediate and superficial layers of the superior colliculus project, via an ipsilateral pathway, to the cuneiform nucleus. The cuneiform nucleus is also a region in which injections of EAA have been found to evoke defensive responses such as running and freezing (Mitchell et al., 1988; Redgrave et al., 1988; Redgrave and Dean, this volume; Westby et al., 1990). Although quite speculative, these findings suggest that, for the rat (and perhaps other species that are "preyed upon"): (i) the *superior colliculus - cuneiform pathway* provides an important neural substrate for defensive reactions, such as flight or freezing, when they are *evoked by **distant** (visual) threatening stimuli* (e.g., if a predator is in the vicinity); whereas (ii) the *lateral PAG* provides an integral neural substrate mediating defensive reactions, such as backward defense or forward avoidance, when they are *evoked by **nearby** (tactile) threatening stimuli* (e.g., if attacked by a predator or during a social encounter with another rat). In continuing this speculation, it is interesting to note: (i) that somatosensory afferents arising from the laminar spinal trigeminal nucleus, and the cervical and lumbar enlargements of the spinal cord, project specifically and quite heavily to the intermediate and caudal lateral PAG (Blomqvist and Craig, this volume; Blomqvist and Wiberg, 1985); and, in contrast (ii) the absence of any significant somatosensory input to neurons in the superior colliculus which project to the cuneiform nucleus (Westby et al., 1990). Finally, our findings that contralateral visual stimuli evoke strong defensive responses following EAA injections in the lateral PAG of the cat, suggest that visual stimuli have either a direct input to, or an influence upon, the lateral PAG of the cat, which is largely lacking in the rat (although see Redgrave and Dean, this volume).

The integration of somatic and autonomic response patterns in the lateral PAG

In the unanesthetized pre-collicular decerebrate rat, microinjections of DLH (10 nmol) made in the lateral PAG evoke large pressor responses (>20%). Consistent with our results in the freely moving rat, the pressor responses evoked from the intermediate, lateral PAG were associated with dorsiflexion of the back and extension of the forelimbs so as to push the animal backwards; whereas, the pressor responses evoked from the caudal, lateral PAG were usually accompanied by vigorous running movements (Keay et al., 1990).

In the unanesthetized pre-collicular decerebrate cat, both arterial pressure and regional blood flow have been monitored during PAG stimulation. As summarized in Figure 12, DLH microinjections made in the intermediate, lateral PAG, A2.5 (the same region in which a threat display is evoked in the freely moving cat) evoked howling and a hypertensive reaction associated with increased extracranial blood flow and decreased blood flow to skeletal muscle in the hindlimb (Carrive, this volume; Carrive et al., 1987; 1991b). In contrast, DLH microinjections made in the caudal lateral PAG, P0.9 (the same region in which a

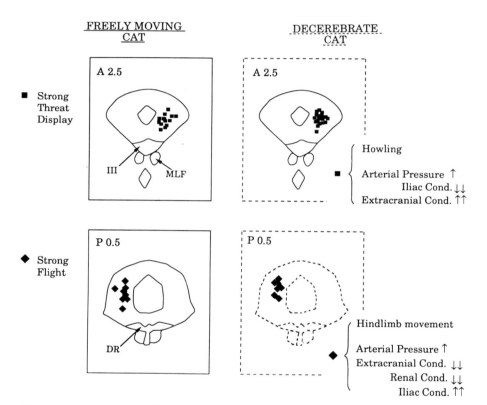

Figure 12. Left column: Representative coronal sections through the midbrain of the cat showing the centers of the sites of DLH microinjections, within the midbrain PAG, which evoked in the freely moving cat: strong threat display (top row) or strong flight (bottom row). Right column: Representative coronal sections through the midbrain of the cat showing the centers of the sites of DLH microinjections within the midbrain PAG, which in the unanesthetized decerebrate cat evoked: a hypertensive reaction associated with howling (top row) or a hypertensive reaction associated with strong hindlimb movement (bottom row). Modified from Bandler et al. (1991) Progress in Brain Research.

strong flight reaction is evoked in the freely moving cat) evoked repetitive bursts of hindlimb movement and a hypertensive reaction characterized by increased blood flow to the limbs and decreased blood flow to the head and viscera (Fig. 12) (Carrive, this volume; Carrive et al., 1989;1991b).

The findings in the unanesthetized, pre-collicular decerebrate cat, that different patterns of cardiovascular changes are associated with different somato-motor "defense" patterns, correlate well with the results of earlier studies of Zanchetti and co-workers, carried out in the conscious freely moving cat. They reported also that distinct patterns of cardiovascular changes were evoked in specific association with different kinds of defensive reactions (Adams et al., 1969; 1971; Mancia and Zanchetti, 1981; Mancia et al., 1972; 1974; see Bandler et al., 1991 or Carrive and Bandler, 1991b for further discussion of these data). Although anecdotal, the defensive reactions evoked from the lateral PAG of the cat possess, at least, a superficial correspondence to certain basic patterns of emotional expression in man. Thus, the threat display evoked from the interme-diate lateral PAG of the cat, which is characterized by both *increased blood flow and intense muscular activity in the region of the face* (Fig. 12), has many of the cha-racteristics of the human emotional reaction of "red-faced with anger"; whereas, the flight reaction evoked from the caudal lateral PAG of the cat, which is charac-terized by increased blood flow and intense muscular activity in the limbs and *decreased blood flow in the face* and viscera (Fig. 12), has certain features in com-mon with the human emotional reaction of "white-faced with fear".

The role of distinct lateral PAG regions in mediating basic patterns of human emotional expression remains highly speculative, although it has been reported that electrical stimulation of the PAG in man evokes intense feelings of fear and anxiety (see Bandler, 1988 for a recent review). In the future, immuno-histochemical and receptor binding studies of the human PAG may provide evi-dence for the existence of subdivisions, within the PAG of man, which are simi-lar to the functional/anatomical divisions suggested from our studies in the cat and the rat.

To summarize, we have found in both the cat and the rat: (i) that longitu-dinally-organized pools of neurons in the caudal two-thirds of the lateral PAG provide an integral substrate for the coordination of defensive reactions; and (ii) that two fundamentally different kinds of defensive reactions are mediated within different parts of this lateral PAG neuronal column (see Figs. 11 and 12). The important questions about how the integration of the somatic and autono-mic adjustments characteristic of these defensive reactions is achieved, remains unanswered. In attempting to answer this question, the study of the PAG will provide, in the future, an important model for understanding how basic integra-ted behavioral patterns are organized within the central nervous system.

Acknowledgements

This research was supported by grants from the Australian National Health and Medical Research Council and the Harry Frank Guggenheim

Foundation. We wish to thank Mme Any Boehrer for her excellent technical assistance and Dr P. Carrive, Dr K.A. Keay and Dr S.P. Zhang for their help in the preparation of figures and their critical comments on earlier versions of this chapter.

References

Abrahams, V.C., Hilton, S.M. and Zbrozyna, A.W., Active muscle vasodilatation produced by stimulation of the brain stem: its significance in the defence reaction, J. Physiol. (Lond.), 154 (1960) 491-513.

Adams, D.B., Baccelli, G., Mancia and G. Zanchetti, A., Cardiovascular changes during naturally elicited fighting behavior in the cat, Am. J. Physiol., 216 (1969) 1226-1235.

Adams, D.B., Baccelli, G., Mancia, G. and Zanchetti, A., Relation of cardiovascular changes in fighting to emotion and exercise, J. Physiol. (Lond.), 212 (1971) 321-326.

Bandler, R., Induction of rage following microinjections of glutamate into midbrain but not hypothalamus of cats, Neurosci. Lett., 30 (1982a) 183-188.

Bandler, R., Neural control of aggressive behaviour, Trends Neurosci., 5 (1982b) 390-394.

Bandler, R., Identification of hypothalamic and midbrain neurones mediating aggressive and defensive behaviour by intracerebral microinjections of excitatory amino acids, In: Modulation of sensorimotor activity during alterations in behavioural states, Bandler R. (Ed.), Alan R. Liss, New York, 1984, pp. 369-392.

Bandler, R., Brain mechanisms of aggression as revealed by electrical and chemical stimulation: Suggestion of a central role for the midbrain periaqueductal grey region, In: Progress in Psychobiology and Physiological Psychology, Vol. 13, Epstein A. and Morrison A. (Eds.), Academic Press, New York, 1988, pp. 67-154.

Bandler, R. and Carrive, P., Integrated defence reaction elicited by excitatory amino acid injection in the midbrain periaqueductal grey region of the unrestrained cat, Brain Res., 439 (1988) 95-106.

Bandler, R. and Depaulis, A., Elicitation of intraspecific defence reactions in the rat from midbrain periaqueductal grey by microinjection of kainic acid, without neurotoxic effects, Neurosci. Lett., 88 (1988) 291-296.

Bandler, R., Carrive, P. and Zhang, S.P., Integration of somatic and autonomic reactions within the midbrain periaqueductal grey: Viscerotopic, somatotopic and functional organization, Prog. Brain Res., 87 (1991) 269-305.

Bandler, R., Depaulis, A. and Vergnes, M., Identification of midbrain neurons mediating defensive behaviour in the rat by microinjections of excitatory amino acids, Behav. Brain Res., 15 (1985) 107-119.

Bard, P. and Macht, M.B., The behaviour of chronically decerebrate cats, In: CIBA Foundation Symposium on Neurological Basis of Behaviour, Wolstenholme G.E.W. and O'Connor C.M. (Eds.), Churchill, London, 1958, pp. 55-75.

Berman, A.L., The brainstem of the cat. A cytoarchitectonic atlas with stereotaxic coordinates, The University of Wisconsin Press, Madison, 1968.

Blomqvist, A. and Wiberg, M., Some aspects of the anatomy of somatosensory projections to the cat midbrain, In: Development, organization, and processing in somatosensory pathways, Rowe M.J. and Willis W.D. (Eds.), Alan R. Liss, New York, 1985, pp. 215-222.

Carrive, P. and Bandler, R., Viscerotopic organization of neurons subserving hypotensive reactions within the midbrain periaqueductal grey: a correlative functional and anatomical study, Brain Res., 541 (1991a) 206-215.

Carrive, P. and Bandler, R., Redistribution of blood flow in extracranial and hindlimb vascular beds evoked by excitation of neurons in the midbrain periaqueductal grey of the decerebrate cat, Exp. Brain Res., 84 (1991b) 599-606.

Carrive, P., Bandler, R. and Dampney, R.A.L., Somatic and autonomic integration in the midbrain: A distinctive pattern evoked by excitation of neurones in the subtentorial portion of the midbrain periaqueductal grey region, Brain Res., 483 (1989a) 251-258.

Carrive, P., Dampney, R.A.L. and Bandler, R., Excitation of neurones in a restricted region of the midbrain periaqueductal grey elicits both behavioural and cardiovascular components of the defence reaction in the unanaesthetised decerebrate cat, Neurosci. Lett., 81 (1987) 273-278.

Depaulis, A. and Vergnes, M., Elicitation of intraspecific defensive behaviors in the rat by microinjection of picrotoxin, a GABA antagonist, into the midbrain periaqueductal gray matter, Brain Res., 367 (1986) 87-95.

Depaulis, A., Bandler, R. and Vergnes, M., Characterization of pretentorial periaqueductal gray neurons mediating intraspecific defensive behaviors in the rat by microinjections of kainic acid, Brain Res., 486 (1989) 121-132.

Depaulis, A., Keay, K.A. and Bandler, R., Organization of defensive behavior in the midbrain periaqueductal gray matter of the rat, in preparation.

Di Scala, G., Schmitt, P. and Karli, P., Flight induced by infusion of bicuculline methiodide into periventricular structures, Brain Res., 309 (1984) 199-208.

Duggan, A. W., Injury, pain and analgesia, Proc. Australian Physiol. Pharmacol. Soc., 14 (1983) 218-240.

Eliasson, S., Lindgren, P. and Uvnas, B., The hypothalamus, a relay station of the sympathetic vasodilator tract, Act. Physiol. Scand., 31 (1954) 290-300.

Ellison, G.D. and Flynn, J.P., Organized aggressive behavior in cats after surgical isolation of the hypothalamus, Arch. ital. Biol., 106 (1968) 1-20.

Fernandez De Molina, A. and Hunsperger, R.W., Organization of the subcortical system governing defence and flight reactions in the cat, J. Physiol. (Lond.), 160 (1962) 200-213.

Fleischmann, A. and Urca, G., Different endogenous analgesia systems are activated by noxious stimulation of different body regions, Brain Res., 455 (1988) 459-57.

Flynn, J.P., Patterning mechanisms, patterned reflexes and attack behavior in cats, In: Nebraska Symposium on Motivation, Coles J.K. and Jensen D.D. (Eds.), Univ. Nebraska Press, Lincoln, 1972, pp. 125-153.

Gellen, B., Gyorgy, L. and Doda, M., Influence of the surgical isolation of the hypothalamus on oxotremorine-induced rage reaction and sympathetic response in the cat, Act. Physiol. Acad. Sci. Hung., 42 (1972) 195-202.

Goodchild, A.K., Dampney, R.A.L. and Bandler, R., A method for evoking physiological responses by stimulation of cell bodies, but not axons of passage, within localized regions of the central nervous system, J. Neurosci. Meth., 6 (1982) 351-363.

Grant, E.C. and Mackintosh, J. H., A comparison of the social postures of some common laboratory rodents, Behav., 21 (1963) 246-259.

Hess, W.R. and Brügger, M., Das subkorticale Zentrum den affektiven Abwehrreaktion, Acta Helv. Physiol., 1 (1943) 33-52.

Hilton, S.M. and Redfern, W.S., A search for brainstem cell groups integrating the defence reaction in the rat, J. Physiol., 378 (1986) 213-228.

Hunsperger, R.W., Affektreaktionen auf elektrische Reizung in Hirnstamm der Katze, Acta. Helv. Physiol. Pharmacol., 14 (1956) 70-92.

Jacquet, Y. and Lajtha, A., Paradoxical effects after microinjection of morphine in the periaqueductal gray matter in the rat, Science, 185 (1974) 1055-1057.

Keay, K.A., Depaulis, A., Breakspear, M.J. and Bandler, R., Pre- and subtentorial periaqueductal grey of the rat mediates different defense responses associated with hypertension, Neurosci. Soc. Abst., 16 (1990) 598.

Keller, A.D., Autonomic discharges elicited by physiological stimuli in mid-brain preparations, Am. J. Physiol., 100 (1932) 576-586.

Kelly, A.H., Beaton, L.E. and Magoun, H.W., A midbrain mechanism for facio-vocal activity, J. Neurophysiol., 9 (1946) 181-189.

Krieger, J. E. and Graeff, F. G., Defensive behavior and hypertension induced by glutamate in the midbrain central gray of the rat, Brazil. J. Med. Biol. Res., 18 (1985) 61-67.

Lindgren, P., The mesencephalon and the vasomotor system, Acta. Physiol. Scand., 35, Suppl 121 (1955) 1-183.

Lindgren, P., Rosen, A., Strandberg, P. and Uvnas, B., The sympathetic vasodilator outflow - a cortico-spinal autonomic pathway, J. Comp. Neurol., 105 (1956) 95-109.

Lipski, J., Bellingham, M.C., West, M.J. and Pilowsky, P., Limitations of the technique of pressure microinjection of excitatory amino acids for evoking responses from localized regions of the CNS, J. Neurosci. Meth., 26 (1988) 169-179.

Mancia, G. and Zanchetti, A., Hypothalamic control of autonomic functions, In: Handbook of the Hypothalamus, Morgane P.J. and Panksepp J. (Eds.), Marcel Dekker, New York, 1981, pp. 147-201.

Mancia, G., Baccelli, G. and Zanchetti, A., Hemodynamic responses to different emotional stimuli in the cat: Patterns and mechanisms, Am. J. Physiol., 223 (1972) 925-933.

Mancia, G., Baccelli, G. and Zanchetti, A., Regulation of renal circulation during behavioral changes in the cat, Am. J. Physiol., 227 (1974) 536-542.

Mitchell, I., Dean, P. and Redgrave, P., The projection from superior colliculus to cuneiform area in the rat. II. Defence-like responses to stimulation with glutamate in cuneiform nucleus and surrounding structures, Exp. Brain Res., 72 (1988) 626-639.

Redgrave, P., Dean, P., Mitchell, I.J., Odekunle, A. and Clark, A., The projection from superior colliculus to cuneiform area in the rat. I. Anatomical studies, Exp. Brain Res., 72 (1988) 611-625.

Schmitt, P., Di Scala, G., Jenck, F. and Sandner, G., Periventricular structures, elaboration of aversive effects and processing of sensory, In: Modulation of Sensorimotor Activity During Alterations in Behavioural States, Bandler R. (Ed.), Alan R. Liss, New York, 1984, pp. 393-414.

Schramm, L. P. and Bignall, K. E., Central neural pathway mediating active sympathetic muscle vasodilation in cats, Am. J. Physiol., 221 (1971) 754-767.

Skultety, F.M., Stimulation of periaqueductal gray and hypothalamus, Arch. Neurol., 8 (1963) 609-620.

Yaksh, T.L., Yeung, J.C. and Rudy, T.A., Systematic examination in the rat of brain sites sensitive to the direct application of morphine: observation of differential effects within the periaqueductal gray, Brain Res., 114 (1976) 83-103.

Westby, G.W.M., Keay, K.A., Redgrave, P., Dean, P. and Bannister, M., Output pathway from the rat superior colliculus mediating approach and avoidance have different sensory properties, Exp. Brain. Res., 81 (1990) 626-638.

Zhang, S.P., Bandler, R. and Carrive, P., Flight and immobility evoked by excitatory amino acid microinjection within distinct parts of the subtentorial midbrain periaqueductal grey of the cat, Brain Res., 520 (1990) 73-82.

Does the PAG Learn about Emergencies from the Superior Colliculus?

Peter Redgrave and Paul Dean

Department of Psychology
University of Sheffield, England

Visual Threat and the PAG

Recent work on the periaqueductal gray matter (PAG) has tended to focus on the organisation and functions of its outputs. There has been rather less emphasis on how the PAG is told about the emergencies which its responses enable the organism to confront. One possible reason for this lack of emphasis is the old view that the superior colliculus, neighbour to much of the PAG and an obvious source of sensory input to it, was concerned exclusively with orienting and approach, not with emergencies and defensive responding. However, it is now known that at least in rodents the superior colliculus is intimately involved in defensive responding to certain classes of emergency (Dean et al., 1989). Thus, the main purpose of this chapter is to argue that the superior colliculus is a vital link in informing the PAG that particular kinds of emergency are taking place.

The ethological studies of the Blanchards and others (for review see Blanchard and Blanchard, 1987) have provided experimental support both for the commonsense notion that visual stimuli can be extremely threatening, and also for the idea that the PAG is necessary for organising appropriate defensive responding to such threats.

(1) The distance between an approaching experimenter and a wild rat has been shown to be an important factor in determining the specific form of the

rat's defensive response. The initial response to a distant (> 3 m) threat was free-zing. At a threat distance of 2 -3 m the animals exhibited flight if an escape route were available. If an escape route were not available, in other words if the animals were cornered, then when the threat distance became less than 1 m the rats directed a wild jumping and biting attack towards the source of the threat.

(2) While it is possible to rule out the somatosensory system in the control of the above responses, the animals could have been reacting to the experimenter's sound or smell. However, previous experiments indicated that visually detected movement is a much more important cue than sound or smell in eliciting defensive behaviours (Blanchard et al., 1975).

(3) The defensive reactions of a wild rat to an approaching experimenter are abolished by destruction of the PAG (Blanchard et al., 1981). Control lesions of collicular tissue necessarily damaged by PAG lesions produced deficits in defensive reactions to visual and some tactile stimuli, however, unlike PAG lesioned animals, avoidance and escape responses once elicited appeared normal.

These studies indicate that visual stimuli can induce defensive reactions which appear to be organised at the level of the PAG, and so raise the important issue of how the PAG gets to know about visual emergencies.

Origin of Visual Signals to PAG

There is a range of experimental evidence showing that in rodents a major source of visual information for controlling defensive reactions is the superior colliculus. Here a brief summary is given; further details are available elsewhere (Dean et al., 1988a,b; 1989).

First, lesion studies have indicated that after destruction of the superior colliculus wild rats ignore the approach of the experimenter (until actual contact is made, when defensive responding occurs as usual) (Blanchard et al., 1981). Similarly, whereas normal hamsters, accustomed to living undisturbed in a simulated natural environment, respond with violent flight to sudden overhead movement when caught in the open, hamsters with collicular damage show no response at all (Merker, 1980).

Secondly, electrical and chemical stimulation of the superior colliculus can elicit a wide range of responses that resemble naturally occurring defensive reactions. These reactions include both movements and physiological reactions.

(i) Redgrave et al. (1981) mapped the dorsal midbrain with bilateral microinjections of picrotoxin (12.5ng/500 nl saline per side), a blocker of the inhibitory neurotransmitter GABA. At some collicular sites, the defense-like responses were restricted to freezing, suppression of orientation, and exaggerated startle reactions to sudden stimuli. At others this pattern of behaviour gave way to a backward shuffling movement during which mild visual, auditory or tactile stimulation evoked a strong flinching response, or a defensive posture in which the rat reared on its hind legs and oriented towards the stimulus. At the most sensitive sites microinjections of picrotoxin caused mild sensory stimulation to produce violent jumping retreat. An important point is that this dose of picrotoxin was relatively ineffective when injected directly into underlying PAG, suggesting that the behavioural effects of the collicular injections were not produced by spread to intermediate and rostral PAG.

A number of subsequent studies have produced similar results, although other neuroactive agents such as glutamate (as well as electrical stimulation) produce defense-like responses directly, that is without additional sensory stimulation. One interesting feature of the distribution of sites producing defensive responding to electrical stimulation is shown in Figure 1.

Electrical stimulation in the superficial layers of the superior colliculus induced a range of defensive movements when applied within the region corresponding to the upper visual field. These movements took the form of a back-

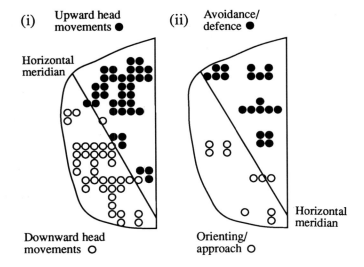

Figure 1: The threshold response to electrical stimulation of the superficial layers of the rat SC varies systematically with the location of the stimulation site. (i) Illustrates a dorsal view of the SC on which are plotted the stimulation sites from all layers that produced head movements with an upward component (filled circles) or a downward component (open circles: horizontal movements are excluded). From these data we were able to derive an approximation of the horizontal meridian in terms of movement coordinates. (ii) Illustrates a similar view of the SC on which are plotted the different types of response obtained at threshold when only the superficial layers were electrically stimulated. There is a clear association between upward head movements and defensive responses and downward movements and approach. This association suggests that visual stimuli in the upper field are more likely to elicit avoidance and defensive behaviour (references in Dean et al., 1988b).

ward roll of the body resembling cringing or flinching which, with higher stimulation intensity, developed a locomotor component giving the appearance of an animal shying away from a threatening stimulus. At some sites explosive running and jumping was observed. These electrically induced reactions may be related to behavioural observations indicating that unexpected movement overhead is more likely to produce flight than is movement at floor level (references in Dean et al., 1989), which perhaps reflects the likelihood of large predators (as opposed to, for example, conspecifics) stimulating the upper visual field.

(ii) A major problem in demonstrating physiological responses to collicular stimulation in the unanesthetised/unparalysed rat is that any responses observed could be secondary to the violent movements produced, rather than the stimulation as such. We attempted to avoid this problem by using an anesthetised preparation, and were able to produce increases in blood pressure and heart rate by microinjection of the excitatory amino acid NMDA (Keay et al., 1990b). Two areas of the dorsal midbrain contained high concentrations of active sites - rostromedial superior colliculus, and most regions of the caudal PAG. The rostro-dorsal PAG which was closest to the active collicular zone was significantly less active. This pattern of results again suggests that the collicular injections were unlikely to have altered blood pressure because of spread into underlying rostral and intermediate PAG.

Finally, electrophysiological studies in many laboratories have shown how sensitive collicular cells are to moving stimuli - precisely the visual attribute most important for eliciting defensive responses to predators in rodents. An intriguing, but still preliminary finding, is that some cells within the rat superior colliculus appear to respond specifically to suddenly expanding shadows that in natural surroundings would be produced by a particularly potent threat, an object on collision course (see below).

In conclusion then, these three lines of evidence strongly suggest that the superior colliculus is an essential source of information used in the production of defensive reactions to threatening visual stimuli in rodents. This conclusion in turn raises another question: by which route is this essential information conveyed from the superior colliculus to the PAG?

Anatomical Connections between Superior Colliculus and PAG

Current anatomical evidence suggests that there may be more than one pathway by which information concerning visual threat could be conveyed from the superior colliculus to the PAG. Figure 2 shows these connections schematically.

(i) The superior colliculus and rostral PAG appear to communicate directly and indirectly. Cells in superficial (i.e. visual) layers of the superior colliculus project directly into the dorsolateral PAG, as demonstrated by the PHAL anterograde tracing technique (Rhoades et al., 1989). A projection from the deep layers to rostral dorsolateral PAG was reported at this meeting (Holstege, this volume).

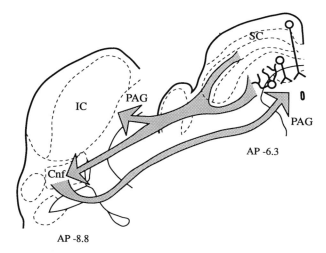

Figure 2: Connections between the SC and PAG which might relay visual threat are represented on modified coronal sections from the atlas of Paxinos and Watson (1986) and can be itemised as follows: first, direct connections from superficial, intermediate and deep SC layers to cells in dorsal PAG at both rostral and caudal levels; second, direct connections via dendrites of PAG cells extending dorsally into the overlying deep layers of the SC; third, indirect connections via a relay in the cuneiform area. AP numbers indicate the distance in mm of the coronal section caudal to bregma. Abbreviations: Cnf - cuneiform nucleus; IC - inferior colliculus; PAG - periaqueductal gray; SC - superior colliculus.

In addition, Beitz and Shepard (1985) have shown with the Golgi silver impregnation technique that the dendrites of neurons in dorsolateral PAG extend into the adjacent deep layers of the superior colliculus. Finally, anterograde and retrograde transport from a single injection of the tracer WGA-HRP into the cuneiform area suggests that this structure may act as an important relay between superior colliculus and PAG (Fig. 3). Extensive anterogradely transported terminal label was observed in dorsolateral PAG while numerous retrogradely labeled cells were seen in the overlying dorsal intermediate grey layer of the superior colliculus.

(ii) Experiments in our laboratory with anterograde and retrograde tracing techniques have identified an extensive projection from the intermediate and deep layers of the superior colliculus to the dorsal half of caudal PAG and the adjacent cuneiform area (Redgrave et al., 1987a; 1988).

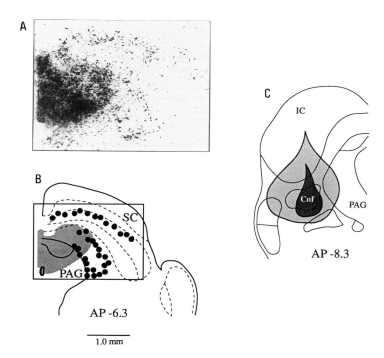

1.0 mm

Figure 3: An injection of the tracer WGA-HRP centered on the cuneiform area provides evidence for afferent connections with the SC and
efferent connections with PAG.   (A) A photomicrograph showing (i)
retrogradely labelled cells in the dorsal intermediate grey layer and
lateral deep layers of the SC; and (ii) dense anterogradely transported
terminal label in the dorsal PAG. (B) A schematic representation of
the photograph illustrated in A superimposed on a modified coronal
section from the atlas of Paxinos and Watson (1986). (C) A coronal
reconstruction of the maximum extent of the WGA-HRP injection site
which was centered on the cuneiform nucleus. The dark shading
depicts the injection's central zone while the outer 'flare zone' is represented by lighter shading. We have provided evidence elsewhere
(Redgrave et al., 1987a; 1988) suggesting that active transport of
WGA-HRP is restricted to the injection's central zone. AP numbers
indicates the distance in mm of the coronal section caudal to bregma.
Abbreviations see legend to Figure 2.

Functional Connections between Superior Colliculus and PAG

There is to our knowledge no evidence bearing on the functional status of the direct projections between the superior colliculus and rostral PAG. We have, however, begun to use a range of experimental techniques to investigate possible functions of the projections linking superior colliculus, cuneiform area, and caudal PAG.

(i) Preliminary data from studies in which the cuneiform area and adjacent caudal PAG were destroyed indicated that escape-like responses to collicular microinjections of the GABA-antagonist bicuculline were abolished (Dean et al., 1988b).

(ii) Electrophysiological recordings were made from cells in the superior colliculus, identified as projecting to the cuneiform area by antidromic stimulation (Westby et al., 1990). In tests of sensory specificity these cells were almost exclusively responsive to visual stimuli in the upper visual field. About 10% of these cells appeared to be especially responsive to a particular kind of visual threat (see above), since within the range of tests carried out, they were specifically activated by rapidly expanding or looming shadows which under natural conditions would threaten collision. Cells identified as projecting to other target regions of the superior colliculus, that are probably concerned with approach rather than defense, were more responsive to somatosensory or auditory stimuli (Keay et al., 1990a).

(iii) Microinjections into the cuneiform area of the excitatory agent sodium glutamate produced a specific subset of defense-like responses, namely either freezing or fast running (Mitchell et al., 1988). Similar responses (among others) can also be elicited by stimulation of the region of the superior colliculus that contains cells projecting to the cuneiform area.

Taken together, these data suggest that projections from the superior colliculus to caudal PAG and cuneiform area are involved in producing at least two kinds of defensive response appropriate to visual threat.

Superior Colliculus and PAG: A Speculative Model

The evidence reviewed above begins to provide clues as to how the collicular-PAG system might be used by a rat to solve the kind of problem it faces when a dangerous threat (the experimenter) appears and starts to approach (e.g., Blanchard et al., 1981). The solution adopted by the rat is to select different integrated patterns of defensive responding, depending on the distance of the threat. These responses range from freezing, to flight, to violent jumping and biting attack. The collicular-PAG system could implement this solution as follows (Fig. 4).

First, it is well established that collicular neurons are heterogeneous with respect to preferred stimulus properties (for review see e.g., Chalupa, 1984). To begin with there is a topographic map of the visual field projected from the retina onto the superior colliculus, so that individual tectal units have their own

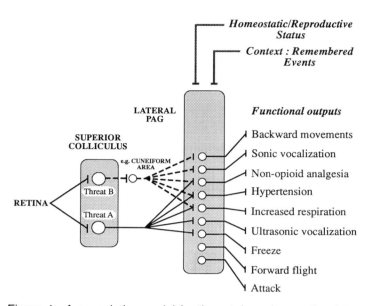

Figure 4: A speculative model for the rat, based upon the detection of specific stimulus cues, which would be able to select different integrated patterns of defensive responding . (It should be noted that certain anatomical and physiological aspects of the proposed connections are, as yet, only hypothetical). Different populations of collicular neurons are known to vary according to preferred stimulus properties; depicted here as differential sensitivity to threat A (e.g., small dark object moving overhead) and threat B (e.g., a rapid expansion of an object's image on the retina). Anatomical evidence indicates that different populations of collicular neurons have different efferent targets; in the model collicular cells responsive to the stimulus properties of threat A make direct contact with cells in lateral caudal PAG while the population sensitive to threat B make indirect contact with the intermediate lateral PAG via a relay in the cuneiform area - (a hypothetical connection between dorsolateral and lateral PAG may also be required for this circuit to be complete). There is increasing evidence for segregated functional outputs originating from different cell groups within the caudal two thirds of lateral PAG (see Bandler and Depaulis; Carrive, this volume); in the model we have arbitrarily connected the direct output of collicular units responsive to threat A to functional elements in the caudal lateral PAG associated with the freeze response. Alternatively, the indirect connections relaying information about threat B have been routed to functional units associated with backward movements in the intermediate lateral PAG . Basic collicular-PAG defense circuits could be made more 'intelligent' by input from the forebrain carrying information about internal state and external context (see text).

patch of field to which they respond. In addition, some units respond only to movement of small visual stimuli, whereas others apparently require actual expansion of the image (see above). It is therefore likely that different populations of collicular neurons will be activated by distant threats as opposed to close ones, and by movement in the upper field as opposed to the lower. A detailed mechanism for achieving defense-related discriminations of this kind, based on known electrophysiological properties of single units, has recently been modelled for the frog optic tectum (Liaw and Arbib, 1991).

Secondly, there is now a substantial body of anatomical evidence to show that different populations of collicular neurons project to different targets within the brainstem, with relatively little overlap between the projections. Using retrograde double-labeling techniques we have demonstrated in rat that largely separate populations of tectal cells project into the major contralateral and ipsilateral descending bundles (Redgrave et al., 1986). A similar separation was observed with cells projecting ipsilaterally to the dorsolateral pons and the cuneiform area (Redgrave et al., 1987b), and contralaterally to the periabducens area and caudal medulla (Redgrave et al., 1990). Insofar as these separate projections have been investigated, the sensory properties of collicular cells seem to be related to where in the brain they project (see above for reference to the work of Westby et al., 1990 and Keay et al., 1990).

Thirdly, much of the work reported at this meeting has been directed towards characterizing the multiple functional outputs of the PAG (see in particular Carrive; Bandler and Depaulis this volume). Indeed, the concluding summary diagrams of where in the PAG different reactions could be obtained showed clearly that individual defensive reactions could be associated with a more or less restricted population of cells within the PAG.

The idea of the scheme shown in Figure 4 is to link these organisational features together in an obvious way. It should be noted that this is a conceptual model, not a precise map of anatomical and functional pathways within the collicular-PAG defense system. First, it supposes that the set of neurons in the superior colliculus most responsive to distant (or upper field) threat is preferentially connected (in this case directly) with those neurons in the PAG that mediate the defensive reactions most appropriate to distant threat, namely freezing. As the threat approaches, a different population of collicular neurons takes over, which are linked (perhaps indirectly) to a different set of caudal PAG neurons responsible for flight. Although not shown in the model, as the threat comes closer still, we might expect a third set of collicular neurons to trigger PAG neurons concerned with backward defense, defensive attack, wild jumping. Separate sets of connections (direct or indirect) linking different populations of sensory neurons in the colliculus to PAG output systems which produce different defensive reactions may provide a basic level integrated defense system able to respond appropriately to a range of threatening stimuli.

These connections by themselves would produce highly predictable and stereotyped behaviour, rather than the flexible and context-dependent responding that we know characterises real animals. The required sophistication could be provided by forebrain afferents from areas known to be important in defense, such as the cortex, amygdala or hypothalamus, that carry important information concerning internal state and external context (e.g., Veening et al.; Shipley et al., this volume). For example, a hungry animal needs to be more tolerant of threat than one which is satiated. Similarly, a rat in cover needs to respond to an overhead threat by freezing, whereas one in the open needs to run: context is crucial. Moreover, such forebrain inputs could provide a substrate for learning how emergencies can be predicted. In general terms forebrain afferents would have the ability to potentiate or override the basic collicular-PAG circuitry shown in Figure 4. Such use of basic circuitry by 'higher-order' structures was proposed over fifty years ago as a fundamental principle of neural organisation (e.g., Hughlings Jackson, 1932), and has recently attracted considerable attention as 'subsumption architecture' in the design of autonomous mobile robots (Brooks 1987). Understanding exactly how the primitive defense mechanisms sketched in Figure 4 could be modulated to produce intelligent defensive behaviour may therefore be a major research issue in the future.

References

Beitz, A.J. and Shepard, R.D., The midbrain periaqueductal gray in the rat. II. A Golgi analysis, J. Comp. Neurol., 237 (1985) 460-475.

Blanchard, R.J. and Blanchard, D.C., An ethoexperimental approach to the study of fear, Psychological Record, 37 (1987) 305-316.

Blanchard, R.J., Mast, M. and Blanchard, D.C., Stimulus control of defensive reactions in the albino rat, J. Comp. Physiol. Psychol., 88 (1975) 81-88.

Blanchard, D.C., Williams, G., Lee, E.M.C. and Blanchard, R.J., Taming of wild *Rattus norvegicus* by lesions of the mesencephalic central gray, Physiol. Psychol., 9 (1981) 175-163.

Brooks, R.A., Autonomous mobile robots, In: AI in the 1980's and beyond, Grimson W.E.L. and Patil R.S. (Eds.), MIT press, Cambridge MA, 1987, pp. 343-365.

Chalupa, L.M., Visual physiology of the mammalian superior colliculus, In: Comparative neurology of the optic tectum, Vanegas H. (Ed.), Plenum Press, New York, 1984, pp. 775-818.

Dean, P., Mitchell, I.J. and Redgrave, P., Responses resembling defensive behaviour produced by microinjection of glutamate into superior colliculus of rats, Neuroscience, 24 (1988a) 501-510.

Dean, P., Redgrave, P. and Westby, G.W.M., Event or emergency ? Two response systems in the mammalian superior colliculus, Trends Neurosci., 12 (1989) 137-147.

Dean, P., Redgrave, P. and Mitchell, I.J., Organisation of efferent projections from superior colliculus to brainstem in rat: evidence for functional output channels, Prog. Brain Res., 75 (1988b) 27-36.

Hughlings Jackson, J., Relations of different divisions of the central nervous system to one another and to parts of the body, In: Selected writings of John Hughlings Jackson, Taylor J. (Ed.), Hodder and Stoughton, London, 1932, pp. 422-443.

Keay, K., Westby, G.W.M., Frankland, P., Dean, P. and Redgrave, P., Organization of the crossed tecto-reticulo-spinal projection in rat. 2. Electrophysiological evidence for separate output channels to the periabducens area and caudal medulla, Neuroscience, 37 (1990a) 585-601.

Keay, K.A., Dean, P. and Redgrave, P., N-methyl D-aspartate (NMDA) evoked changes in blood pressure and heart rate from the rat superior colliculus, Exp. Brain Res., 80 (1990b) 148-156.

Liaw, J-S. and Arbib, M.A., A neural network model for response to looming objects by frog and toad, In: Visual structures and integrated functions, Arbib M.A. and Ewert J-P. (Eds.), Research Notes in Neural Computing, Springer Verlag, Berlin., 1991, in press.

Merker, B., The sentinel hypothesis: A role for the mammalian superior colliculus, Doctoral Dissertation, M.I.T. Press, Cambridge MA, 1980.

Mitchell, I.J., Dean, P. and Redgrave, P., The projection from the superior colliculus to the cuneiform area in the rat. II. Defense-like responses to stimulation with glutamate in cuneiform nucleus and surrounding structures, Exp. Brain Res., 72 (1988) 626-639.

Paxinos, G. and Watson, C., The rat brain in stereotaxic coordinates, Academic Press, Sydney, 1986.

Redgrave, P., Dean, P., Souki, W. and Lewis, G., Gnawing and changes in reactivity produced by microinjections of picrotoxin into the superior colliculus of rats, Psychopharmacology, 75 (1981) 198-203.

Redgrave, P., Dean, P. and Westby, G.W.M., Organization of the crossed tecto-reticulo-spinal projection in rat - I. Anatomical evidence for separate output channels to the periabducens area and caudal medulla, Neuroscience, 37 (1990) 571-584.

Redgrave, P., Dean, P., Mitchell, I.J., Odekunle, A. and Clark, A., The projection from superior colliculus to cuneiform area in the rat. I. Anatomical studies, Exp. Brain Res., 72 (1988) 611-625.

Redgrave, P., Mitchell, I.J. and Dean, P., Descending projections from the superior colliculus in rat: a study using orthograde transport of wheatgerm-agglutinin conjugated horseradish peroxidase, Exp. Brain Res., 68 (1987a) 147-167.

Redgrave, P., Michell, I.J. and Dean, P., Further evidence for segregated output channels from superior colliculus in rat: ipsilateral tecto-pontine and tecto-cuneiform projections have different cells of origin, Brain Res., 413 (1987b) 170-174.

Redgrave, P., Odekunle, A. and Dean, P., Tectal cells of origin of predorsal bundle in rat: location and segregation from ipsilateral descending pathway, Exp. Brain Res., 63 (1986) 279-293.

Rhoades, R.W., Mooney, R.D., Rohrer, W.H., Nikoletseas, M.M. and Fish, S.E., Organization of the projection from the superficial to the deep layers of the hamster's superior colliculus as demonstrated by the anterograde transport of Phaseolus vulgaris leucoagglutinin, J. Comp. Neurol., 283 (1989) 54-70.

Westby, G.W.M., Keay, K.A., Redgrave, P., Dean, P. and Bannister, M., Output pathways from the rat superior colliculus mediating approach and avoidance have different sensory properties, Exp. Brain Res., 81 (1990) 626-638.

Midbrain PAG Control of Female Reproductive Behavior: *In Vitro* Electrophysiological Characterization of Actions of Lordosis-Relevant Substances

Sonoko Ogawa, Lee-Ming Kow, Margaret M. McCarthy,
Donald W. Pfaff and Susan Schwartz-Giblin

Laboratory of Neurobiology and Behavior
The Rockefeller University
New York, U.S.A.

Introduction

Among the various innate behavioral repertoires of mammals, lordosis, as a major component of female reproductive behaviors is one of the best characterized. It is a stereotyped posture consisting of dorsiflexion of the vertebral column, which causes the elevation of the head and rump, accompanied by hind limb extension (Pfaff and Lewis, 1974). Somatosensory input given by a male partner is essential for every component of naturally occurring lordosis; it can be mimicked by an experimenter's manual stimulation to the flanks, posterior rump, and perineum. The most important feature of lordosis is its steroid hormone dependency; an elevated level of estrogen is a prerequisite for the induction of lordosis both in naturally cycling and experimentally manipulated females. The nature and mechanisms of estrogen and progesterone control of lordosis have been extensively investigated and are well reviewed elsewhere (Pfaff and Schwartz-Giblin, 1988).

PAG as a Major Component of Neural Circuitry for Lordosis

It is well accepted that the PAG is one of the essential neural components for the induction of lordosis. Electrical stimulation of dorsal and lateral PAG can facilitate lordosis (see Fig. 1; Sakuma and Pfaff, 1979a) and lesions of this brain area centered at the midsuperior collicular level suppressed lordosis within 2 min in estrogen-primed rats (Sakuma and Pfaff, 1979b); lordosis was also sup-

The Midbrain Periaqueductal Gray Matter, Edited by A. Depaulis and
R. Bandler, Plenum Press, New York, 1991

pressed in rats (Riskind and Moss, 1983b) and hamsters (Floody and O'Donohue, 1980) when they were tested more than 2 weeks after PAG lesions. The most important feature of the PAG control of lordosis is that the PAG serves as a relay station between the hypothalamus and the lower brainstem. Anatomical studies (Krieger et al., 1979; Morrell et al., 1981; Beitz, 1982; Veening et al., this volume) have shown that dorsal and lateral subdivisions of the rostral and intermediate PAG, which are the most effective stimulation (or lesion) sites for the facilitation (or suppression) of lordosis, receive strong descending input from the ventromedial nucleus of the hypothalamus (VMH) where the estrogen-containing cells necessary for the behavior are located (Pfaff and Keiner, 1973), through two different pathways, one lateral (through the supraoptic commissure or the zona incerta) and the other medial (through periventricular structures). Several studies have consistently shown that the lateral pathway is especially important for facilitation of lordosis. Thus, knife-cuts of the lateral pathway, which leave the PAG completely intact, severely suppressed lordosis and lateral, but not medial, knife-cuts, interfered with estrogen's ability to increase the probability of antidromic invasion of VMH cells following electrical stimulation of the PAG (Manogue et al., 1980; Akaishi and Sakuma, 1986; Sakuma and Akaishi, 1987; Hennessey et al., 1990). Electrical stimulation of the VMH also facilitates lordosis (Pfaff and Sakuma, 1979a) and lesions of it suppress lordosis (Pfaff and Sakuma, 1979b) but the latencies of these effects were much longer in the VMH (at least 30 min, usually 2 h) than in the PAG. Also, VMH lesions did not affect the PAG-facilitation of lordosis (Sakuma and Pfaff, 1979a) whereas PAG lesions abolished the VMH-facilitated lordosis (Sakuma and Pfaff, 1979b). However, after transection at the midbrain-diencephalic junction, estrogen plus progesterone-treated females do not show lordosis to male mounts (Kow et al., 1978) but it has been reported that about half of non hormone-treated ovariectomized rats will elicit a lordosis-like response to concurrent somatosensory and vaginocervical stimulation (Rose and Flynn, 1989). These findings altogether suggest that the PAG is necessary for lordosis whereas the VMH provides a tonic hormone-related facilitation through estrogen-induced peptides which are transported to the PAG in order to prime it for the induction of lordosis (Romano et al., 1988; 1989; Mobbs et al., 1988; 1990). In fact, one third of estrogen-concentrating cells in ventrolateral VMH are known to send projections to the dorsal midbrain including the PAG (Morrell and Pfaff, 1982); other estrogen-containing forebrain and diencephalic regions also project to the PAG (Pfaff, 1980).

In turn, the PAG sends descending projections to the medulla, another important component of the lordosis neural circuit. Lesions of the gigantocellular nucleus in the medullary reticular formation or the lateral vestibular nucleus of the medulla produced significant deficits in lordosis (Modianos and Pfaff, 1976). Electrical stimulation of these medullary nuclei, through reticulospinal and vestibulospinal projections which travel in the anterolateral columns of the spinal cord (Kow et al., 1977; Zemlan et al., 1979), are known to excite the deep lumbar axial muscles (Femano et al., 1984a,b; Cottingham et al., 1988; Robbins et al., 1990, 1991) which are activated during the lordosis posture (Schwartz-Giblin

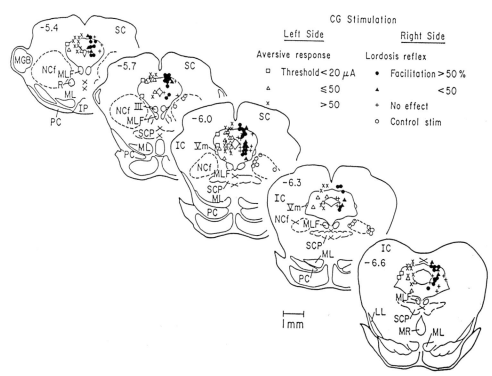

Figure 1: Localization of stimulation sites in central gray (CG) and adjacent mesencephalic structures, plotted on sections 300 μm apart. On *right side* of each section, sites of electrical stimulation are indicated according to effectiveness for lordosis facilitation (n = 82). *Filled circles* (●), facilitation exceeding 50% of prestimulation lordosis reflex score; *filled triangles* (▲), facilitation under 50%; crosses (+), no effect. Sixteen *open circles* (O) in cuneiform nucleus of mesencephalic reticular formation (NCf) indicate sites of control stimulation, which had no effects on lordosis. Numbers in *upper left* corner of diagrams indicate rostrocaudal distance in mm from bregma. On *left side* of each section, thresholds for aversive response are given for 58 points. Aversive responses are defined in the electrical stimulation study as "abrupt running and immediate attempt to escape" from manual stimulation. *Open squares* (□) and *triangles* (△), points with threshold lower than 20 μA and 50 μA, respectively; crosses (X), points with threshold higher than 50 μA. IC, inferior colliculus; IP, interpeduncular nucleus; LL, lateral lemniscus; MGB, medial geniculate body; ML, medial lemniscus; MLF, medial longitudinal fascicule; MR, median raphe nucleus; NCf, cuneiform nucleus of mesencephalic reticular formation; PC, cerebral peduncle; R, red nucleus; SC, superior colliculus; SCP, superior cerebellar peduncle; III, nucleus of oculomotor nerve; Vm, mesencephalic nucleus of trigeminal nerve. Reprinted with permission from Sakuma and Pfaff (1979a) American Journal of Physiology.

et al., 1984). Connections relevant for reproductive behavior between the PAG and the gigantocellular nucleus have been established both electrophysiologically and anatomically. For example, PAG cells antidromically activated by electrical stimulation of the gigantocellular nucleus have been identified mainly in the lateral and the ventrolateral PAG (Sakuma and Pfaff, 1980a,b), and antidromic propagation was enhanced by estrogen treatment or VMH stimulation, and suppressed by VMH lesions (Sakuma and Pfaff, 1980b). Fewer antidromically activated cells were found in the dorsal PAG. Also, deposits of a retrograde tracer into the gigantocellular nucleus, at sites where electrical stimulation activated lumbar axial muscle EMG, resulted in heavy labeling in the caudal half of PAG primarily in ventral and ventrolateral subdivisions; retrogradely labeled cells were also seen in dorsal and lateral subdivisions in the caudal half of PAG but not in the rostral half of PAG (Fig. 2; Robbins et al., 1990). Furthermore, the PAG has an excitatory influence on the medullary sites that drive the back muscles which perform the behavior. Electrical stimulation of the dorsal and particularly the ventral PAG facilitated deep back muscle EMG response to both gigantocellular nucleus and lateral vestibular nucleus stimulation (Cottingham and Pfaff, 1987; Cottingham et al., 1987), whereas PAG stimulation itself rarely evoked back muscle EMG. It has been shown also that the vestibulospinal and reticulospinal systems interact in a facilitatory way to activate back muscle EMG (Cottingham et al., 1988). Since a direct projection from the PAG to the lateral vestibular nucleus has not been reported (Hamilton and Skultety, 1970), PAG facilitation of lateral vestibular nucleus stimulation-evoked deep muscle EMG must be mediated indirectly, and a projection from the gigantocellular nucleus to the lateral vestibular nucleus has been confirmed in the cat (Corvaja et al., 1979) but not in the rat. The lordosis-relevant PAG output in the ventrolateral quadrant includes a large and unique subset of midbrain estradiol-concentrating cells that were labeled retrogradely from the dorsal medulla at the level of the obex (Corodimas and Morrell, 1990).

The PAG also receives lordosis-relevant somatosensory information (Rose, 1986; Malsbury et al., 1972) via direct ascending projections from the spinal cord (Swett et al., 1985; Menétrey et al., 1980; 1982) as well as from the gigantocellular nucleus (Morrell and Pfaff, 1983). The PAG neurons which were *not* antidromi-

Figure 2A-C (right page): Distributions of fluoro-gold (FG) labeled cells in the (A) forebrain and hypothalamus, (B) midbrain, and (C) medulla from a FG deposition at an effective stimulation site for evoking back muscle EMG responses, indicated by the X in C. In C, the black area indicates the FG core, and the gray area represents the FG halo. ● = 1 cell, ✳ = 5 cells. AMYG, amygdala; Aq, Aqueduct; ECu, External cuneate; F, Fornix; Gi, Gigantocellular reticular nucleus; GiV, Gigantocellular reticular nucleus, ventral; icp, inferior cerebellar peduncle; IO, Inferior olive; MCG, Midbrain central gray; ml, medial lemniscus; MVN, Medial vestibular nucleus; OT Optic tract; PCRt, Parvocellular reticular nucleus; Pn, Pontine nuclei; PnO, Pontine reticular nucleus; PPT, Pedunculopontine tegmental nucleus; PVN, Paraventricular nucleus; py, pyramidal tract; Sol, nucleus of the solitary tract; Sp5, Spinal trigeminal nucleus; VMN, Ventromedial nucleus. Reprinted with permission from Robbins et al. (1990) Experimental Brain Research.

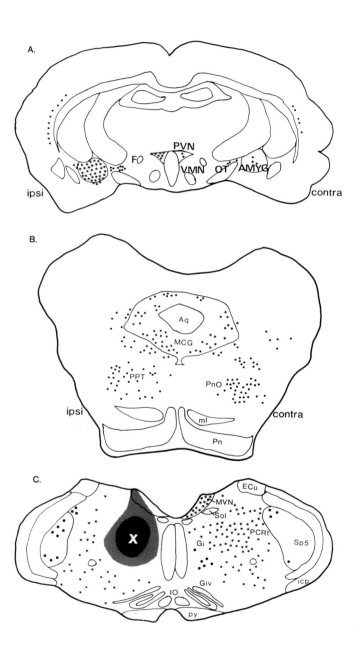

cally activated or orthodromically activated by electrical stimulation of giganto-cellular nucleus actually responded to the lordosis-relevant somatosensory stimulus and they were located in the dorsolateral and ventrolateral parts of the PAG (Sakuma and Pfaff, 1980c). In contrast, antidromically activated neurons were not responsive to this somatosensory stimulus (Sakuma and Pfaff, 1980c) and they were located mainly in the ventrolateral PAG (see Sakuma and Pfaff, 1980a,b). This indicates that the lordosis-relevant somatosensory input must innervate interneurons, rather than the PAG output neurons that project to the gigantocellular nucleus. Similarly, hormone-related descending afferents from the VMH which terminate in the dorsal and lateral subdivisions of the rostral and intermediate PAG most likely innervate interneurons, since retrogradely labeled cells from the gigantocellular nucleus are seen only in the caudal half of the PAG (see Fig. 2; Robbins et al., 1990; see also Veening et al., this volume). Thus, both ascending and descending information for reproductive behavior converge in the PAG (as well as in the medulla as proposed by Kow and Pfaff, 1982) and are processed in PAG neuronal groups separate from the output to gigantocellular nucleus.

In spite of the well defined neural pathway for lordosis and knowledge about estrogen-induced molecular changes in the VMH (Pfaff, 1989), it is still not known how the behaviorally-relevant neurotransmitters and neuropeptides affect PAG local circuitry. The next sections of the chapter describe data from recent studies that investigated the effects of three different sets of neurochemical substances which have presumptive importance both for PAG neuronal activity and reproductive behavior: GABA and its-related agents, lordosis-relevant neuropeptides, and opioid peptides.

Involvement of GABAergic Transmission in PAG Control of Lordosis

GABA as a major inhibitory neurotransmitter in the CNS is known to be involved in the regulation of various types of behavior in many different brain regions. It has been shown that in the hypothalamus, GABAergic systems may be a part of a neuronal mechanism responsible for the control of lordosis behavior (McCarthy et al., 1990). Accumulated studies on the PAG GABAergic system, such as the distribution of GABAergic cell bodies, fibers, terminals (Perez de la Mora et al., 1981; Mugnaini and Oertel, 1985; Reichling, this volume) as well as receptors (Bowery et al., 1987; Hironaka, et al., 1990; Gundlach, this volume) support the idea that GABA may be one of the most important neurotransmitters in the PAG. These data naturally led us to test the hypothesis that GABAergic transmission may participate in the PAG control of female reproductive behavior and if this is the case, to investigate the possible underlying mechanisms.

Behavioral effects of a GABA$_A$ agonist and antagonist in the PAG

GABAergic modulation of lordosis behavior in the PAG was investigated by microinjections of the GABA$_A$ agonist and antagonist, muscimol and bicuculline methiodide (BMI), respectively (McCarthy et al., 1991). Bilateral cannulae were histologically confirmed to be in the dorsal and lateral subdivisions of the rostral and intermediate thirds of the PAG (from 5.2 to 6.3 caudal to Bregma, or 3.8 to 2.7 anterior to the interaural line) in a distribution which is coextensive with the labeled fibers seen in the rostral and intermediate thirds of the PAG after PHA-L injections into the VMH (see Veening et al., Fig. 3 B,D, this volume). Behavioral tests were conducted in a two-chamber apparatus which consisted of two compartments, one large (referred to as the TEST CHAMBER) and one small (referred to as the SECOND CHAMBER), connected by a 30 cm long tunnel. Females were placed in the TEST CHAMBER at the beginning of the behavioral tests and then allowed to move freely in the entire apparatus. Males were tethered to the far wall of the large TEST CHAMBER and their movement confined to distal third of this chamber. This arrangement allowed the expression of the full complement of female sexual behaviors which include receptive, and proceptive behavior such as hopping and darting which attract the male's attention, as well as escape and defensive behaviors.

In female rats that were optimally primed for reproductive behavior with estrogen and progesterone, infusion of a total dose of 30 ng of BMI, a GABA$_A$ antagonist, into the PAG (0.25 µl per side) significantly decreased the Lordosis Ratio (# of lordosis / # of mounts and attempted mounts x 100; Wilcoxon, p< .05) and proceptive behaviors (Wilcoxon, p< .05) without affecting the Defense Index (# of rejections / # of mounts and attempted mounts x 100) at 5 min compared to pretest levels (Fig. 3A). Infusions of muscimol, a GABA$_A$ agonist, into females primed with estrogen plus progesterone had no effect, whereas in females sub-optimally primed for reproductive behavior with estrogen pretreatment only, infusion of 50 ng of muscimol (0.25 µl per side) increased the Lordosis Ratio at 5 and 60 min post-infusion compared to pretest (Fig. 3B) as well as to saline-infused controls (Fig. 3C, Mann-Whitney, p< .05). Muscimol infusion also decreased the Defense Index at 5 min post-infusion but did not significantly increase proceptive behaviors. Both muscimol and BMI had no significant effects on audible vocalization or escape behavior (number of exits from the TEST CHAMBER as well as time spent in the TEST CHAMBER).

At the completion of the behavioral experiment, six of the experimental females were infused with a 5-10% solution of the retrograde fluorescent tracer, Fluoro-gold, through the same cannulae with the same volume (0.25µl per side) as previously used for drug infusion. Fluorescent microscope analysis of cryostat-cut sections revealed heavy labeling of neurons in the VMH, anterior hypothalamus and zona incerta indicating strong descending projections from these areas to GABA effective sites in the PAG. There were also ascending projections from the lumbar spinal cord including cells in laminae I, V, X and the lateral spinal nucleus. All of these findings are consistent with those in the previous studies (see "PAG as a major component of neural circuitry for lordosis").

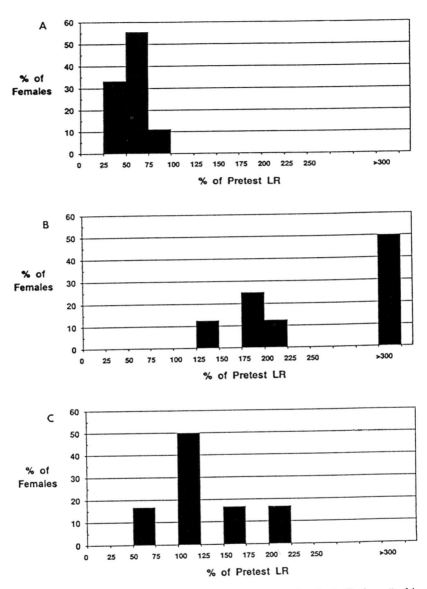

Figure 3A-C: Distributions showing the change in LR (Lordosis Ratio = # of lordosis / # of mounts and attempted mounts x 100) at 5 min post-infusion compared with the pre-infusion test levels. A: Decrease in LR after PAG infusion of 30 ng of a GABA$_A$ antagonist, bicuculline methiodide (0.25 μl per side), in estrogen plus progesterone-treated females (n = 9). B: Increase in LR after infusion of 50 ng of a GABA$_A$ agonist, muscimol (0.25 μl per side), in estrogen-treated females (n = 8). Since repeated tests tend to increase LR, the data of this group were compared with a subset of estrogen-treated females infused with saline. C: Change in LR after saline infusion in estrogen-treated females (n = 6).

These results suggest that GABAergic modulation of lordosis occurs in the area of the PAG which receives descending and ascending input from lordosis-relevant CNS regions. Also, since manipulation of GABAergic activity resulted in changes of not only lordosis but also rejection, GABAergic action in multiple neuronal pathways in the PAG may be in part responsible for the GABAergic modulation of reproductive behavior. In this regard, it is interesting that escape behavior was not significantly affected when directly measured. All of these points will be further discussed below (see Integrative hypothesis of GABAergic and neuropeptidergic modulation of lordosis).

Electrophysiological actions of GABA and GABA-related agents on PAG neurons

To determine the influences of GABAergic activity in the PAG, responsiveness of individual PAG neurons to GABA or GABA-related agents was examined *in vitro* (Ogawa et al., 1989). 350 µm thick sagittal sections containing the entire PAG along the dorso-ventral and rostro-caudal axis were prepared from ovariectomized female rats either primed with estrogen for at least 7 days or non-primed. Extracellular single-unit activity of PAG neurons randomly picked from the entire PAG was recorded from a tissue slice laid on a nylon net submerged in the recording chamber (approximately 2 ml in volume) and continuously perfused with prebubbled regular artificial cerebrospinal fluid. The firing rates (number of spikes/sec) were monitored and stored on-line with AT compatible personal computer interfaced with a 12-Bit High Speed A/D board (AD1000: Real Time Devices, INC.) for analog-to-digital conversion, and displayed both on-line and off-line as firing rate histograms. Once a steady firing rate of a neuron in the PAG was established, 50 µl or 100 µl of solutions of testing agents were applied in a bolus directly into the recording chamber with a glass microsyringe.

Mean firing rates of spontaneously active neurons were 5.8 ± 0.5 spikes/sec; the pattern of spontaneous firing was either regular or irregular. There were no regional differences in spontaneous firing rates along the rostro-caudal axis. Also, there was no difference between estrogen-primed and nonprimed preparations in mean firing rates or responsiveness to testing agents. The former finding can be compared with the *in vivo* data of Sakuma and Pfaff (1980b) which showed that the spontaneous firing rates of a specific subgroup of PAG neurons, those antidromically activated from the medullary reticular formation, were affected by estrogen pretreatment; it must be kept in mind that the neurons recorded in the present *in vitro* study were randomly selected (also see an *in vivo* study by Chan et al., 1985) and that there was no estrogen in the bath.

A total of 71% of the tested PAG neurons (122 out of 171 neurons) were inhibited by GABA (100 µM; see Fig. 4): these neurons either became silent for a few seconds to a few minutes, or slowed their firing rates. No PAG neurons were excited by GABA. There was no difference in the number of GABA-responsive neurons between the dorsal (dorsal and lateral to the upper half of the aqueduct) and ventral (lateral and ventral to the lower half of the aqueduct) parts of PAG. However, in the dorsal PAG, GABA-responsive neurons were unevenly distributed along the rostro-caudal axis, i.e., more neurons were responsive to GABA in the middle third (84%) or caudal third (68%) compared to the rostral third (57%). Similar trends appeared in the ventral PAG although the sampling numbers were small (from the rostral to caudal, 43%, 88%, and 83%). It has been shown by an autoradiographical study that both $GABA_A$ and $GABA_B$ receptors exist in the PAG (Bowery et al., 1987). Consistent with this, we found that THIP (10-100 µM), a $GABA_A$ receptor agonist, and baclofen (0.125-1.25 µM), a $GABA_B$ receptor agonist, each inhibited about 60 - 80% of PAG neurons. Out of 18 tested neurons, 13 were actually inhibited by both THIP and baclofen, indicating that many of the GABA-responsive PAG neurons have both types of receptors.

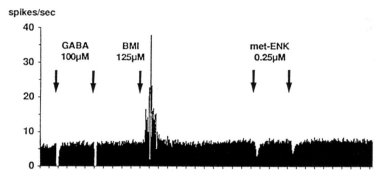

Figure 4: A firing rate histogram of a spontaneously active PAG neuron which was inhibited by bath applications of GABA and excited by bicuculline methiodide (BMI). Note BMI induced bursting. This neuron was also inhibited by met-enkephalin (met-ENK). Tick marks on the horizontal axis indicate 1 min.

To determine the proportion of PAG neurons that were tonically inhibited by GABAergic terminals, responsiveness of PAG neurons to the $GABA_A$ antagonist, bicuculline methiodide (BMI; 125 µM) was examined. Since these responses were obtained as long as 3 h after the onset of the recording session, noting in addition that the recording session started after at least 2 h of incubation time to ensure the full recovery from the trauma during preparation of slices, the data strongly suggest intrinsic GABA neurons. We can certainly rule out extrinsic GABAergic inputs because release from cut terminals would not be expected to persist for the 5 hours after which we can record very stable resting activity which increases in response to BMI. However, we cannot rule out that some intrinsic GABA neurons may be disrupted. A total of 47% of all PAG neurons tested with BMI (71 out of 152 neurons) were excited (Fig. 4), and 4% were inhibited; not all of these neurons were tested with GABA. There was no specific baseline firing pattern for the affected neurons, i.e., excited neurons included spontaneously firing neurons, both regularly firing and irregularly firing, and silent neurons. Interestingly, BMI induced bursting in 38% (27 out of 71) of BMI responsive-neurons (Fig. 4). This phenomenon was also described by Behbehani et al. (1990) and is consistent with the description by Ribas and Sanchez (NATO workshop, 1990) of burst responses in ventral PAG neurons after they had been hyperpolarized at rest. There were regional differences in the proportion of BMI-responsive neurons. The number of BMI-responsive neurons was significantly higher in the dorsal PAG (51%) than in the ventral PAG (26%). This implies that more dorsal PAG neurons were tonically inhibited by endogenous GABA than ventral neurons and is again consistent with the report by Ribas and Sanchez (NATO workshop, 1990) that ventral PAG neurons tended to have low resting membrane potentials. Moreover, since BMI is a specific $GABA_A$ antagonist, our finding of dorso-ventral differences in BMI-responsive neurons is consistent with the report showing that there is a higher density of $GABA_A$ receptors in the dorsal part of the cat PAG (Gundlach, this volume). Also, in the dorsal PAG, more neurons were BMI-responsive in the middle (64%) and caudal third (55%) than in the rostral third (31%).

Responses to both GABA and BMI were tested in some PAG neurons. Based on their increased firing with BMI, it was concluded that 61% of the GABA-responsive PAG neurons may be tonically inhibited through $GABA_A$ receptors. The proportion of this type of neuron was highest in the middle third of dorsal PAG (77%) followed by the caudal third of dorsal PAG (62%), the rostral third of dorsal PAG (47%), and lowest in the ventral PAG (36%). However, 12 out of 42 BMI-responsive neurons, which were tested with BMI and GABA, were not responsive to GABA. Given that drugs are applied to the bath, these data suggest that some GABA-nonresponsive PAG neurons may be synaptically contacted by excitatory neurons which are themselves tonically inhibited by GABAergic neurons (see also Behbehani et al., 1990).

Electrophysiological Action of Lordosis-Relevant Neuropeptides in PAG

PAG infusion of certain neuropeptides is known to affect lordosis in steroid-primed rats. Neuroanatomical and neurochemical findings on peptide-containing axon terminals and/or varicosities (the existence of terminals in PAG is not always obligatory for their action on PAG neurons) as well as peptide binding sites in the PAG support the idea that these neuropeptides are endogenously affecting PAG neurons. Moreover, estrogen regulation of synthesis of some of these neuropeptides (see below) and defined projections from their synthesis sites to the PAG provides a sufficient basis for their presumptive candidacy as lordosis-facilitating peptides in the PAG. Nevertheless, their electrophysiological actions on PAG neurons are not well defined although accumulated *in vivo* and *in vitro* electrophysiological studies have shown that in some hypothalamic as well as limbic areas, some of these neuropeptides indeed alter neuronal activity (Chan et al., 1983; Kow and Pfaff, 1986; 1988a; Dudley and Moss, 1987; Wong et al., 1990). Thus, in the next section, we will discuss our recent *in vitro* extracellular recordings (Ogawa et al., 1989; 1990) which are aimed at characterizing the electrophysiological action of lordosis-relevant neuropeptides on individual PAG neurons to assess their possible roles in PAG control of lordosis.

Substance P

Substance P, known to be involved in the induction of naloxone-reversible analgesia in the PAG, also participates in the PAG control of lordosis. Infusion of substance P into the caudal half of the dorsal PAG facilitates lordosis as early as 5 min after infusion and for at least 3 h in estrogen-primed rats (Dornan et al., 1987).

We found that substance P has, indeed, an excitatory neurotransmitter-like action on PAG neurons which might be a part of the mechanism by which substance P facilitates lordosis (Fig. 5). Of the tested PAG neurons, 12% (out of 51), 62% (out of 91), or 79% (out of 19) were excited by 1, 10, or 100 nM substance P, respectively. These affected neurons were distributed more or less evenly throughout the PAG and included both spontaneously firing neurons and silent neurons. These electrophysiological findings together with neuroanatomical findings of substance P containing terminals (Ljungdahl et al., 1978; Moss and Basbaum, 1983) and substance P receptors (Mantyh et al., 1984; Wolf et al., 1985) in the PAG imply that endogenous substance P may be influencing the neuronal activity of PAG neurons. Substance P may directly act on lordosis-relevant PAG neurons and excite them since Turcotte and Blaustein (1989) have observed substance P-immunoreactive terminal boutons in close association with estrogen receptor-immunoreactive cell bodies in the PAG of guinea pigs. Substance P may also be involved in the PAG control mechanisms of lordosis by inducing the release of met-enkephalin (Del Rio et al., 1983), the action of which will be discussed below.

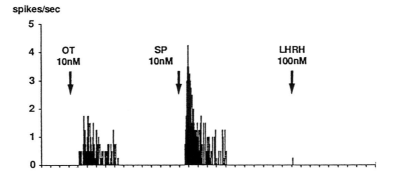

Figure 5: Both oxytocin (OT) and substance P (SP) had excitatory effects on a silent PAG neuron whereas LHRH had no effect.

There is evidence suggesting that the substance P in the PAG may originate from the VMH. Cell bodies containing immunoreactive substance P (Ljungdahl et al., 1978; Panula et al., 1984) and substance P mRNA (Romano et al., 1989) are found in the VMH, most heavily in the ventrolateral VMH. Retrograde tracing from the dorsal PAG combined with substance P immunofluorescence revealed that 17% of substance P immunoreactive cells in the ventrolateral VMH were double labeled (Dornan et al., 1990). In the ventrolateral VMH, it has been shown that 43% of substance P immunoreactive cells accumulate estrogen (Akesson and Micevych, 1988) and estrogen increases the firing rates of VMH cells (Bueno and Pfaff, 1976). However, there are conflicting reports about whether estrogen regulates the synthesis of substance P in this brain region: Brown et al. (1990) claim that estrogen treatment increases substance P mRNA levels in the basomedial hypothalamus whereas Romano et al. (1989) see no effect with either slot blots or *in situ* hybridization.

Oxytocin

Oxytocin has a facilitatory effect on lordosis in the VMH. It has been shown that estrogen plus progesterone regulate oxytocin receptors in the VMH and this may be a part of the mechanisms of oxytocin facilitation of lordosis in this brain region (Schumacher et al., 1989; 1990). An *in vitro* electrophysiological study has shown that oxytocin acts as an excitatory neurotransmitter in the VMH (Kow and Pfaff, 1986) and estrogen treatment which induces oxytocin receptors (Johnson et al., 1989) indeed results in enhancement of neuronal responsiveness to oxytocin in the VMH (Kow et al., 1991).

Interestingly, we found that oxytocin can also act as an excitatory neurotransmitter in the PAG (Fig. 5). Of PAG neurons tested, 11, 34, or 53% were affected by 1, 10, or 100 nM oxytocin, respectively; excitation was the predominant response. Oxytocin-responsive neurons were distributed throughout the PAG although they tended to be more ventral and rostral. Oxytocin actions were mimicked by the more specific agonist, [Thr^4, Gly^7]-oxytocin, in 16 out of 17 tested neurons but not by vasopressin, and the excitatory actions of oxytocin were blocked by an oxytocin antagonist but not by the vasopressin receptor antagonist. In addition, the excitatory effects were not abolished by blockade of synaptic transmission indicating that oxytocin acts directly on PAG neurons through oxytocin receptors to increase their neuronal activity.

Potential behavioral roles of oxytocin in the PAG remain to be determined. Although Caldwell et al. (1989) reported that PAG infusion of oxytocin was ineffective in facilitating lordosis in estrogen-treated females, testing for longer periods of time after infusion, with higher doses and/or in estrogen plus progesterone-primed females may be necessary to obtain a real picture of oxytocin effects on lordosis in this brain region.

LHRH

The PAG is well established as the primary site of LHRH facilitatory effects on lordosis. There is a consistency among different studies that infusion of LHRH into the dorsal PAG (Sakuma and Pfaff, 1980d; 1983; Sirinathsinghji, 1984; 1985) or ventrolateral PAG (Riskind and Moss, 1979; 1983a) facilitates lordosis within a few minutes in estrogen-primed ovariectomized rats. Riskind and Moss (1983a) reported that this facilitatory effect is not due to diffusion of LHRH to the arcuate nucleus-VMH region since serotonin blocked LHRH facilitatory effects in the arcuate nucleus-VMH but not in PAG. In addition, as shown in hypothalamic areas (see Pfaff and Schwartz-Giblin, 1988), facilitatory effects of LHRH on lordosis in PAG may be independent from those on release of gonadotropin since LHRH analogs, inactive in terms of LH release, facilitate lordosis in PAG (Sakuma and Pfaff, 1983; Dudley and Moss, 1988).

PAG immunostaining for LHRH is consistent with a neuromodulatory role for this peptide on lordosis (Shivers et al., 1983a; Buma, 1989). LHRH-containing fibers are found mainly in ventral position in close association with the ependyma throughout the PAG (Liposits and Setalo, 1980; Shivers et al., 1983a; Buma, 1989; Veening et al., this volume). It has been reported that these LHRH-containing terminals make no clear synaptic contacts on dendrites of PAG neurons (Buma, 1989; also see Veening et al., this volume). This suggests that endogenous LHRH released from the terminals in the ependymal thickenings may influence PAG neurons in a nonsynaptic way by diffusion through the intercellular space. Since binding sites for an LHRH agonist are seen throughout the entire PAG (Hsueh and Schaeffer, 1985; Jennes et al., 1988), effects of LHRH may be amplified. This may result in a neuromodulatory action of LHRH on PAG neurons; neuromodulatory as well as neurotransmitter-like actions have been reported for neuropeptides by Kow and Pfaff (1988b). In fact, in our *in vitro* recording, LHRH (10-100 nM) failed to directly and rapidly alter neuronal activity of PAG neurons (Fig. 5) in both estrogen-treated (0 out of 27 tested neurons) and non-treated preparations (1 slightly excited neuron out of 18) although *in vivo*, iontophoretically applied LHRH affected half of PAG neurons (mainly inhibitory effect) in ovariectomized females treated with estrogen plus progesterone (Chan et al., 1985). Since it has been shown in the medial preoptic area slice preparation, that LHRH not only affected baseline neuronal activity but also either potentiated or attenuated responsiveness to norepinephrine or serotonin (Pan et al., 1988), it is plausible to assume that LHRH may also influence in a neuromodulatory manner, the descending and ascending afferents and/or intrinsic neurons in the PAG. This possibility is currently being investigated in this laboratory.

Since infusion of LHRH antiserum into the dorsal PAG completely blocks lordosis (Sakuma and Pfaff, 1980d; 1983; Sirinathsinghji, 1984; 1985), an endogenous supply of LHRH on PAG neurons may be required for facilitation of lordosis. With respect to these well defined behavioral effects and the existence of LHRH-containing terminals in the PAG, the endogenous source of LHRH responsible for the facilitation of lordosis is proposed to be in the preoptic area. Several studies have suggested that the major origin of LHRH-containing fibers found in the subependymal position of the PAG may be in the medial preoptic area (Merchenthaler et al., 1984; Nieuwenhuys, 1985). Abundant cell bodies containing LHRH immunoreactivity (Shivers et al., 1983a; Bergen and Pfaff, 1991) or LHRH mRNA (Shivers et al., 1986; Rothfeld et al., 1989; Bergen and Pfaff, 1991) are found in the medial preoptic area. In this brain region, synthesis of LHRH is facilitated by estrogen (Rothfeld et al., 1989) but this can only be an indirect regulation since estrogen and LHRH do not co-localize in the same neurons (Shivers, et al., 1983b).

Prolactin

Infusion of prolactin into the lateral part of the PAG in estrogen-primed rats facilitates lordosis with a latency of about 40 min (Harlan et al., 1983). Endogenous prolactin may originate in the mediobasal hypothalamus. Thus, cell bodies containing immunoreactive prolactin are found in the arcuate nucleus and adjacent areas ventral to the VMH (Harlan et al., 1983; 1989) where 33% of them are also estrogen-accumulating (Shivers et al., 1989). Densely packed prolactin immunopositive fibers extend from the mediobasal hypothalamus to be found throughout the PAG (Harlan et al., 1983; 1989).

Electrophysiologically, however, prolactin appears to have only a weak direct effect on PAG neurons. Iontophoretically applied prolactin either excited or inhibited only 15-25% of PAG neurons *in vivo* (Chan et al., 1985). Likewise, prolactin (1 µM) affected very few PAG neurons, so far, in our *in vitro* experiments (Ogawa et al., unpublished data). Therefore, behaviorally important aspects of prolactin action in PAG may include non-electrophysiological mechanisms.

Involvement of Opioid Peptides in Control of Lordosis in PAG

Effects of opioid peptides in PAG on lordosis are not yet clearly understood. The main reasons for this are 1) the effects of opioid peptides and their antagonists vary among different brain sites as well as for different application routes; 2) there are multiple opioid receptor subtypes which may contribute to the different effects of opioids, even in opposite directions; and 3) behavioral effects are dependent on endogenous levels of opioids which may be regulated by steroids (Södersten et al., 1989; Wiesner and Moss, 1989). Therefore, no opioid agonist or antagonist with definite unidirectional effects has been reported (Pfaus and Gorzalka, 1987a). For example, while ß-endorphin inhibits lordosis when microinjected into the PAG (Sirinathsinghji, 1984), intracerebroventricular injections of ß-endorphin both inhibited and facilitated lordosis in a dose-dependent fashion (Pfaus and Gorzalka, 1987b).

Among the endogenous opioid peptides, enkephalin (ENK) may be involved in estrogen control of lordosis. Several lines of evidence have been obtained to suggest that ENK may be one of the estrogen-induced polypeptides which mediate the effects of estrogen on lordosis; it is synthesized in the VMH and one relevant feature is its transport to the PAG. Thus, it has been shown in the ventrolateral subdivision of the VMH where estrogen-containing cells are accumulated (Pfaff and Keiner, 1973) and estrogen most effectively facilitates lordosis (Davis et al., 1979), that 1) mRNA for proenkephalin, the main precursor for met-

ENK, is abundant (Harlan et al., 1987); 2) estrogen increases proenkephalin mRNA levels at short latency (Romano et al., 1988; 1989); 3) progesterone prolongs estrogen-induced elevation of proenkephalin mRNA levels after removal of estrogen (Romano et al., 1989); 4) lordosis levels are correlated with proenkephalin mRNA levels (Lauber et al., 1990); 5) estrogen induction of proenkephalin mRNA shows a sex difference (Romano et al., 1991); and 6) estrogen and met-ENK co-localize in 27% of ventrolateral VMH cells (Akesson et al., 1989; 1991) whereas very few neurons in the VMH co-localize estrogen with ß-endorphin or dynorphin (Morrell et al., 1985). In addition, cell bodies containing ENK-like immunoreactivity are abundant in the VMH (Finley et al., 1981) and they actually project to the dorsal PAG (Yamano et al., 1986).

In sexually-receptive females, therefore, PAG neurons may be exposed to increased enkephalinergic influence. This may result in increased activity of those lordosis-relevant neuronal components in the PAG with excitatory output to the medullary centers for motor control of back muscles (Sakuma and Pfaff, 1980a,b; Cottingham and Pfaff, 1987; Cottingham et al., 1987; 1988; Robbins et al., 1990). In our recent *in vitro* extracellular recordings in the PAG, about 8% of the tested neurons (dorsal, 9 out of 111; ventral, 3 out of 38) were found to be excited by met-ENK (Ogawa et al., 1989). They were all located in the caudal two thirds of the PAG. Since the direct action of ENK is inhibitory in the CNS (Nicoll et al., 1977; 1980; Duggan and North, 1983), excitatory responses of PAG neurons to ENK may be through inhibitory interneurons. This will be discussed in the next section.

Integrative Hypotheses of GABAergic and Neuropeptidergic Modulation of Lordosis

Our behavioral data and electrophysiological data together suggest that GABAergic synaptic transmission may have importance in PAG control of lordosis. This GABAergic action is most likely due to interneurons intrinsic to the PAG and adjacent areas since BMI effectively reduced reproductive behavior at sites where many neurons in the slice were found to be tonically inhibited by GABA as long as 5 h after slice preparation. This is supported by anatomical data that GAD-immunoreactive cell bodies and terminals are both abundant in the PAG (Perez de la Mora et al., 1981; Mugnaini and Oertel, 1985) and the report by Reichling (this volume) that only a very small number of GABA-immunoreactive cells (2%) project to the medulla (i.e., raphé magnus). Beart et al. (1988) also reported the retrograde labeling by D-[³H] aspartate but not by [³H] GABA in the VMH following injections into the PAG.

Since muscimol facilitated lordosis and BMI inhibited lordosis in the dorsal PAG and the net output from the PAG to the gigantocellular nucleus and lateral vestibular nucleus of the medulla is excitatory for lordosis and for the back muscles that perform lordosis (Sakuma and Pfaff, 1980a,b; Cottingham and Pfaff, 1987; Cottingham et al., 1987; 1988), we postulate that GABAergic neurons contact other neurons, or their terminals, which exert tonic inhibition on the PAG output neurons (see Fig. 2). β-endorphin is one candidate for the interposed tonic inhibitory synapse since infusion of ß-endorphin into the dorsal PAG inhibited lordosis and infusion of its antiserum facilitated lordosis (Sirinathsinghji, 1984). While the order of opioid and GABA synapses in this hypothetical circuit is different from that in the pain circuit proposed by Moreau and Fields (1986) and Depaulis et al. (1987), such a difference might be predicted since muscimol which increases lordosis causes hyperalgesia in the pain circuit. Furthermore, at our sites, muscimol has a tendency to decrease vocalization (10 fold but not statistically significant). On the other hand, BMI which decreases lordosis in our experiments, causes antinociception in the pain circuit; at our sites, it tends to increase vocalizations (10 fold, but not statistically significant).

Parallel to its action on reproductive behavior, GABAergic neurons may inhibit the PAG neuronal circuit responsible for induction of behaviors antagonistic to lordosis. In the dorsal PAG, close to the injection sites in our behavioral study, there is presumptive tonic GABAergic inhibition of defensive behavior since the $GABA_A$ antagonist, picrotoxin into similar lateral sites increased defensive behaviors such as upright posture, sideway, and retreat locomotion, and decreased offensive behaviors in the residence-intruder paradigm (Depaulis and Vergnes, 1986). The defensive patterns are closely related to rejections by females observed in our behavioral tests (Blanchard and Blanchard, 1977; Adams, 1979) and muscimol actually decreased the Defense Index. However, we do not believe that inhibition of lordosis by BMI is simply due to release from tonic GABAergic inhibition on defensive behavior since BMI did not affect the Defense Index. We assume that the tonic actions of endogenous GABA on a lordosis circuit and on a defense circuit are parallel rather than one being secondary to the other. On the other hand, manipulations of GABAergic transmission at our cannula sites did not alter escape behavior as measured by the time spent in the TEST CHAMBER and the number of exits from the TEST CHAMBER. This is consistent with the observation that escape and lordosis facilitation could not be elicited from the identical PAG sites with electrical stimulation and that at sites where escape was elicited, lordosis reflex scores, obtained by manual stimulation, were not affected by electrical stimulation (see Fig. 1; Sakuma and Pfaff, 1979a). In more caudal sites, PAG infusions of the $GABA_A$ agonist, THIP decreased, while the $GABA_A$ antagonist, BMI or SR95103 increased flight and "explosive motor behavior" which may be an expression of aversive behavior (Di Scala et al., 1984; Schmitt et al., 1984; 1985). Interestingly, the $GABA_B$ agonist, baclofen did not have a similar effect (Audi and Graeff, 1987) in spite of its strong electrophysiological action on PAG neurons (Ogawa et al., 1989; also see "Electrophysiological actions of GABA and GABA-related agents on PAG neurons"). Involvement of $GABA_B$ receptors in GABAergic control of lordosis in the PAG is currently being investigated in this laboratory.

Electrophysiological studies have shown for several CNS regions that ENK, in fact, inhibits GABAergic interneurons and this releases other cells from tonic inhibition (Nicoll et al., 1980; Madison and Nicoll, 1988). In the ventral PAG, it has been proposed that analgesic effects of opiates may be due to disinhibition of output cells (Moreau and Fields, 1986; Depaulis et al., 1987; Jacquet et al., 1987) which are synaptically contacted by GABA cells intrinsic to the PAG (Reichling, this volume). We have electrophysiological evidence for this type of neuronal circuitry in the 8% of PAG neurons cited above (Ogawa et al., 1989); their activity was increased by ENK and also increased by BMI or decreased by GABA. However, since blockade of GABAergic tonic inhibition by BMI, at least in the dorsal and lateral parts of the intermediate third of the PAG, actually inhibits lordosis (McCarthy et al., 1991; also see above), this type of neuronal connection may not be primary for ENK mediation of *estrogen* effects in the PAG. Rather, ENK may release lordosis-relevant output neurons through the inhibition of non-GABAergic inhibitory interneurons such as the β-endorphin inhibitory interneuron postulated above.

Alternatively, increased enkephalinergic influences from the VMH to the PAG may serve tonic inhibitory action on PAG neuronal components responsible for the induction of behaviors antagonistic to lordosis. It has been reported that both GABA agonists and ENK agonists, μ- and δ-, but not κ-receptor agonists, have inhibitory effects on affective defense in the cat when infused into similar PAG regions (Shaikh and Siegel, 1990; Shaikh et al., 1990). In our *in vitro* recordings, 19 out of 21 dorsal PAG neurons inhibited by met-ENK were also inhibited by GABA and 13 neurons out of 19 GABA-responsive neurons were excited by BMI suggesting that they were under tonic inhibition (Ogawa et al., 1989; see Fig. 4). Therefore, both GABA and ENK may exert tonic inhibitory influences on the same neurons in the dorsal PAG to decrease the induction of behavior antagonistic to lordosis such as defense behavior and may assist to increase the occurrence of lordosis whenever the circuit for lordosis is turned on. This type of action of ENK is consistent with the reported ineffectiveness of infusions of ENK as well as anti-ENK to alter lordosis in the dorsal PAG (Sirinathsinghji, 1984).

Finally, it should be mentioned that opioids may exert different behavioral effects through different receptor subtypes in the PAG. One study with lateral ventricle infusion of μ- or δ-receptor specific agonists has shown that the δ-receptor agonist has facilitatory effects on lordosis whereas the μ-receptor agonist has dual effects depending upon whether it binds to high affinity μ1 receptors or low affinity, μ2 receptors (Pfaus and Gorzalka, 1987b). PAG has μ-, δ- and κ-receptors but concentrations of μ- and κ-receptors are much higher than those of δ-receptors (Waksman et al., 1986; Tempel and Zukin, 1987; Blackburn et al., 1988). Therefore, if endogenous ENK is released exclusively at δ-receptors, then these specific effects of ENK cannot be detected by exogenous applications since ENK can act through both μ- and δ-receptors. This possibility is now being tested electrophysiologically in this laboratory using μ- and δ-receptor specific agonists.

References

Adams, D.B., Brain mechanisms for offense, defense, and submission, Behav. Brain Sci., 2 (1979) 201-241.

Akaishi, T. and Sakuma, Y., Projections of estrogen-sensitive neurons from the ventrome-dial hypothalamic nucleus of the female rat, J. Physiol., 372 (1986) 207-220.

Akesson, T.R. and Micevych, P.E., Estrogen concentration by substance P-immunoreactive neurons in the medial basal hypothalamus of the female rat, J. Neurosci. Res., 19 (1988) 412-419.

Akesson, T.R. and Micevych, P.E., Enkephalin-immunoreactive neurons of the hypothala-mic ventromedial nucleus concentrate estrogen in male and female rats, Soc. Neurosci. Abst., 15 (1989) 577.

Akesson, T.R. and Micevych, P.E., Endogenous opioid-immunoreactive neurons of the ventromedial hypothalamic nucleus concentrate estrogen in male and female rats, J. Neurosci. Res., 28 (1991) 359-366.

Audi, E.A. and Graeff, F.G., $GABA_A$ receptors in the midbrain central grey mediate the antiaversive action of GABA, Eur. J. Pharmacol., 135 (1987) 225-229.

Beart, P.M., Nicolopoulos, L.S., West, D.C. and Headley, P.M., An excitatory amino acid projection from ventromedial hypothalamus to periaqueductal gray in the rat: autora-diographic and electrophysiological evidence, Neurosci. Lett., 85 (1988) 205-211.

Behbehani, M.M., Jiang, M., Chandler, S.D. and Ennis, M., The effect of GABA and its antagonists on midbrain periaqueductal gray neurons in the rat, Pain, 40 (1990) 195-204.

Beitz, A. J., The organization of afferent projections to the midbrain periaqueductal gray of the rat, Neuroscience, 7 (1982) 133-159.

Bergen, H. and Pfaff, D.W., Luteinizing hormone-releasing hormone (LHRH) neurons fol-lowing exposure to gamma-aminobutyric acid (GABA) agonists, Exp. Brain Res., 1991, in press.

Blackburn, T.P., Cross, A.J., Hille, C. and Slater, P., Autoradiographic localization of delta opiate receptors in rat and human brain, Neuroscience, 27 (1988) 497-506.

Blanchard, R.J. and Blanchard, D.C., Aggressive behavior in the rat, Behav. Biol., 21 (1977) 197-224.

Bowery, N.G., Hudson, A.L. and Price, G.W., $GABA_A$ and $GABA_B$ receptor site distribu-tion in the rat central nervous system, Neuroscience, 20 (1987) 365-383.

Brown, E.R., Harlan, R.E. and Krause, J.E., Gonadal steroid regulation of substance P (SP) and SP-encoding messenger ribonucleic acids in the rat anterior pituitary and hypo-thalamus, Endocrinology, 126 (1990) 330-340.

Bueno, J. and Pfaff, D.W., Single unit recording in hypothalamus and preoptic area of estrogen-treated and untreated ovariectomized female rats, Brain Res., 101 (1976) 67-78.

Buma, P., Characterization of luteinizing hormone-releasing hormone fibers in the mesen-cephalic central grey substance of the rat, Neuroendocrinology, 49 (1989) 623-630.

Caldwell, J.D., Jirikowski, G.F., Greer, E.R. and Pedersen, C.A., Medial preoptic area oxy-tocin and female sexual receptivity, Behav. Neurosci., 103 (1989) 655-662.

Chan, A., Dudley, C.A. and Moss, R.L., Action of prolactin, dopamine and LHRH on ven-tromedial hypothalamic neurons as a function of ovarian hormones, Neuroendocrinology, 36 (1983) 397-403.

Chan, A., Dudley, C.A. and Moss, R.L., Hormonal modulation of the responsiveness of midbrain central gray neurons to LH-RH, Neuroendocrinology, 41 (1985) 163-168.

Corvaja, N., Mergner, T. and Pompeiano, O., Organization of reticular projections to the vestibular nuclei in the cat, In: Reflex control of posture and movement, Granit R. and Pompeiano O. (Eds.), Elsevier, Amsterdam, 1979, pp. 205-209.

Corodimas, K.P. and Morrell, J., Estradiol-concentrating forebrain and midbrain neurons project directly to the medulla, J. Comp. Neurol., 291 (1990) 609-620.

Cottingham, S.L. and Pfaff, D.W., Electrical stimulation of the midbrain central gray facili-tates lateral vestibulospinal activation of back muscle EMG in the rat, Brain Res., 421 (1987) 397-400.

Cottingham, S.L., Femano, P.A. and Pfaff, D.W., Electrical stimulation of the midbrain central gray facilitates reticulospinal activation of axial muscle EMG, Exp. Neurol., 97 (1987) 704-724.

Cottingham, S.L., Femano, P.A. and Pfaff, D.W., Vestibulospinal and reticulospinal interactions in the activation of back muscle EMG in the rat, Exp. Brain Res., 73 (1988) 198-208.

Davis, P.G., McEwen, B. and Pfaff, D.W., Localized behavioral effects of tritiated estradiol implants in the ventromedial hypothalamus of female rats, Endocrinology, 104 (1979) 893-903.

Del Rio, J., Naranjo, J.R., Yang, H.-Y. and Costa, E., Substance P-induced release of met5-enkephalin from striatal and periaqueductal gray slices, Brain Res., 279 (1983) 121-126.

Depaulis, A. and Vergnes, M., Elicitation of intraspecific defensive behaviors in the rat by microinjection of picrotoxin, a gamma-aminobutyric acid antagonist, into the midbrain periaqueductal gray matter, Brain Res., 367 (1986) 87-95.

Depaulis, A., Morgan, M.M. and Liebeskind, J.C., GABAergic modulation of the analgesic effects of morphine microinjected in the ventral periaqueductal gray matter of the rat, Brain Res., 436 (1987) 223-228.

Di Scala, G., Schmitt, P. and Karli, P., Flight induced by infusion of bicuculline methiodide into periventricular structures, Brain Res., 309 (1984) 199-208.

Dornan, W.P., Akesson, T.R. and Micevych, P.E., A substance P projection from the VMH to the dorsal midbrain central gray: implication for lordosis, Brain Behav. Bull., 25 (1990) 791-796.

Dornan, W.P., Malsbury, C.W. and Penney, R.B., Facilitation of lordosis by injection of substance P into the midbrain central gray, Neuroendocrinology, 45 (1987) 498-506.

Dudley, C.A. and Moss, R.L., Effects of a behaviorally active LHRH fragment and septal area stimulation on the activity of mediobasal hypothalamic neurons, Synapse, 1 (1987) 240-247.

Dudley, C.A. and Moss, R.L., Facilitation of lordosis in female rats by CNS-site specific infusions of an LH-RH fragments, Ac-LH-RH-(5-10), Brain Res., 441 (1988) 161-167.

Duggan, A.W. and North, R.A., Electrophysiology of opioids, Pharmacol. Rev., 35 (1983) 219-282.

Femano, P.A., Schwartz-Giblin, S. and Pfaff, D.W., Brain stem reticular influences on lumbar axial muscle activity. I. Effective sites, Am. J. Physiol., 246 (1984a) R389-R395.

Femano, P.A., Schwartz-Giblin, S. and Pfaff, D.W., Brain stem reticular influences on lumbar axial muscle activity. II. Temporal aspects, Am. J. Physiol., 246 (1984b) R396-R401.

Finley, J.C.W., Maderdrut, J.L. and Petrusz, P., The immunocytochemical localization of enkephalin in the central nervous system of the rat, J. Comp. Neurol., 198 (1981) 541-565.

Floody, O.R. and O'Donohue, T.L., Lesions of the mesencephalic central gray depress ultrasound production and lordosis by female hamsters, Physiol. Behav., 24 (1980) 79-85.

Hamilton, B.L. and Skultety, F.M., Efferent connections of the periaqueductal gray matter in the cat, J. Comp. Neurol., 139 (1970) 105-114.

Harlan, R.E., Shivers, B.D. and Pfaff, D.W., Midbrain microinfusions of prolactin increase the estrogen-dependent behavior, lordosis, Science, 219 (1983) 1451-1453.

Harlan, R.E., Shivers, B.D., Romano, G.J., Howells, R.D. and Pfaff, D.W., Localization of preproenkephalin mRNA in the rat brain and spinal cord by in situ hybridization, J. Comp. Neurol., 258 (1987) 159-184.

Harlan, R.E., Shivers, B.D., Fox, S.R., Kaplove, K.A., Schachter, B.S. and Pfaff, D.W., Distribution and partial characterization of immunoreactive prolactin in the rat brain, Neuroendocrinology, 49 (1989) 7-22.

Hennessey, A.C., Camak, L., Gordon, F. and Edwards, D.A., Connections between the pontine central gray and the ventromedial hypothalamus are essential for lordosis in female rats, Behav. Neurosci., 104 (1990) 477-488.

Hironaka, T., Morita, Y., Hagihira, S., Tateno, E., Kita, H. and Tohyama, M., Localization of $GABA_A$-receptor $\alpha 1$ subunit mRNA-containing neurons in the lower brainstem of the rat, Mol. Brain Res., 7 (1990) 335-345.

Hsueh, A.J.W. and Schaeffer, J.M., Gonadotropin-releasing hormone as a paracrine hormone and neurotransmitter in extra-pituitary sites, J. Steroid Biochem., 23 (1985) 757-764.

Jacquet, Y.F., Saederuo, E. and Squires, R.F., Non-stereospecific excitatory actions of morphine may be due to GABA-A receptor blockade, Eur. J. Pharmacol., 138 (1987) 285-288.

Jennes, L., Dalati, B. and Conn, P.M., Distribution of gonadotropin-releasing hormone agonist binding sites in the rat central nervous system, Brain Res., 452 (1988) 156-164.

Johnson, A.E., Coirini, H., Ball, G.F. and McEwen, B.S., Anatomical localization of the effects of 17ß-estradiol on oxytocin receptor binding in the ventromedial hypothalamic nucleus, Endocrinology, 124 (1989) 207-211.

Kow, L.-M. and Pfaff, D.W., Responses of medullary reticulospinal and other reticular neurons to somatosensory and brainstem stimulation in anesthetized of freely-moving ovariectomized rats with or without estrogen treatment, Exp. Brain Res., 47 (1982) 191-202.

Kow, L.-M. and Pfaff, D.W., Vasopressin excites ventromedial hypothalamic glucose-responsive neurons in vitro, Physiol. Behav., 37 (1986) 153-158.

Kow, L.-M. and Pfaff, D.W., Transmitter and peptide actions on hypothalamic neurons in vitro: Implications for lordosis, Brain Res. Bull., 20 (1988a) 857-861.

Kow, L.-M. and Pfaff, D.W., Neuromodulatory actions of peptides, Ann. Rev. Pharmacol. Toxicol., 28 (1988b) 163-188.

Kow, L.-M.,. Grill, H. and Pfaff, D.W., Elimination of lordosis in decerebrate female rats: observations from acute and chronic preparations, Physiol. Behav., 20 (1978) 171-174.

Kow, L.-M.,. Montgomery, M.O. and Pfaff, D.W., Effects of spinal cord transections on lordosis reflex in female rats, Brain Res., 123 (1977) 75-88.

Kow, L.-M., Johnson, A.E., Ogawa, S. and Pfaff, D.W., Electrophysiological actions of oxytocin on hypothalamic neurons in vitro: neuropharmacological characterization and effects of ovarian steroids, Neuroendocrinology, 1991, in press.

Krieger, M.S., Conrad, L.C.A. and Pfaff, D.W., An autoradiographic study of the efferent connections of the ventromedial nucleus of the hypothalamus, J. Comp. Neurol., 183 (1979) 785-816.

Lauber, A.H., Romano, G.J., Mobbs, C.V., Howells, R.D. and Pfaff, D.W., Estradiol induction of proenkephalin messenger RNA in hypothalamus: dose-response and relation to reproductive behavior in the female rat, Mol. Brain Res., 8 (1990) 47-54.

Liposits, Z. and Setalo, G., Descending luteinizing hormone-releasing (LH-RH) nerve fibers to the midbrain of the rat, Neurosci. Lett., 20 (1980) 1-4.

Ljungdahl, A., Hokfelt, T. and Nilsson, G., Distribution of substance P-like immunoreactivity in the central nervous system of the rat. I. Cell bodies and nerve terminals, Neuroscience, 3 (1978) 861-943.

McCarthy, M.M., Malik, K.F. and Feder, H.H., Increased GABAergic transmission in medial hypothalamus facilitates lordosis but has the opposite effect in preoptic area, Brain Res., 507 (1990) 40-44.

McCarthy, M.M., Pfaff, D.W. and Schwartz-Giblin, S., Midbrain central gray GABA$_A$ receptor activation enhances and blockade reduces sexual behavior in the female rat, Exp. Brain Res., 1991, in press.

Madison, D.V. and Nicoll, R.A., Increases in potassium conductance: Common mechanisms of opiate action in neurons of the central nervous system, In: Neurotransmitters and cortical function: from molecules to mind, Avoli M., Reader T.A., Dykes R.W., and Gloor P. (Eds.), Plenum, New York, 1988, pp. 585-592.

Malsbury, C., Kelley, D.B. and Pfaff, D.W., Responses of single units in the dorsal midbrain to somatosensory stimulation in female rats, In: Progress in Endocrinology, Proceedings of the IV International Congress Endocrinology, Gaul C. (Ed.), Excerpta Medica International Congress Series, 273, Excerpta Medica, Amsterdam, 1972, pp. 205-209.

Manogue, K.R., Kow, L.-M. and Pfaff, D.W., Selective brain stem transections affecting reproductive behavior of female rats: The role of hypothalamic output of the midbrain, Horm. Behav., 14 (1980) 277-302.

Mantyh, P.W., Hunt, S.P. and Maggio, J.E., Substance P receptors: Localization by light microscopic autoradiography in rat brain using [³H]SP as the radioligand, Brain Res., 307 (1984) 147-165.

Menétrey, D., Chaouch, A. and Besson, J.M., Location and properties of dorsal horn neurons at origin of spinoreticular tract in lumbar enlargement of the rat, J. Neurophysiol., 44 (1980) 862-877.

Menétrey, D., Chaouch, A., Binder, D. and Besson, J.M., The origin of the spinomesencephalic tract in the rat: an anatomical study using the retrograde transport of horseradish peroxidase, J. Comp. Neurol., 206 (1982) 193-207.

Merchenthaler, I., Göres, T., Setalo, G., Petrusz, P. and Flerko, B., Gonadotropin-releasing hormone (GnRH) neurons and pathways in the rat brain, Cell Tissue Res., 237 (1984) 15-29.

Mobbs, C.V., Harlan, R.E., Burrous, M.R. and Pfaff, D.W., An estradiol-induced protein synthesized in the ventral medial hypothalamus and transported to the midbrain central gray, J. Neurosci., 8 (1988) 113-118.

Mobbs, C.V., Fink, G. and Pfaff, D.W., HIP-70: A protein induced by estrogen in the brain and LH-RH in the pituitary, Science, 247 (1990) 1477-1479.

Modianos, D.T. and Pfaff, D.W., Brain stem and cerebellar lesions in female rats. II. Lordosis reflex, Brain Res., 106 (1976) 47-56.

Moreau, J.-L. and Fields, H.L., Evidence for GABA involvement in midbrain control of medullary neurons that modulate nociceptive transmission, Brain Res., 397 (1986) 37-46.

Morrell, J.I. and Pfaff, D.W., Characterization of estrogen-concentrating hypothalamic neurons by their axonal projections, Science, 217 (1982) 1273-1276.

Morrell, J.I. and Pfaff, D.W., Retrograde HRP identification of neurons in the rhombencephalon and spinal cord of the rat that project to the dorsal mesencephalon, Am. J. Anat., 167 (1983) 229-240.

Morrell, J.I., Greenberger, L.M. and Pfaff, D.W., Hypothalamic, other diencephalic, and telencephalic neurons that project to the dorsal midbrain, J. Comp. Neurol., 201 (1981) 589-620.

Morrell, J.I., McGinty, J.F. and Pfaff, D.W., A subset of ß-endorphin- or dynorphin-containing neurons in the medial basal hypothalamus accumulates estradiol, Neuroendocrinology, 41 (1985) 417-426.

Moss, M.S. and Basbaum, A.I., The peptidergic organization of the cat periaqueductal gray. II. The distribution of immunoreactive substance P and vasoactive intestinal polypeptide, J. Neurosci., 7 (1983) 1437-1449.

Mugnaini, E. and Oertel, W.H., An atlas of the distribution of GABAergic neurons and terminals in the rat CNS as revealed by GAD immunohistochemistry, In: Handbook of Chemical Neuroanatomy, Vol. 4: GABA and neuropeptides in the CNS, Part I, Björklund A. and Hökfelt T. (Eds.), Elsevier, Amsterdam, 1985, pp. 436-608.

Nicoll, R.A., Siggins, G.R., Ling, N., Bloom, F.E. and Guillemin, R., Neuronal actions of endorphins and enkephalins among brain regions: A comparative microiontophoretic study, Proc. Natl. Acad. Sci., 74 (1977) 2584-2588.

Nicoll, R.A., Alger, B.E. and Jahr, C.E., Enkephalin blocks inhibitory pathways in the vertebrate CNS, Nature, 287 (1980) 22-25.

Nieuwenhuys, R., Chemoarchitecture, Springer-Verlag, Berlin, 1985.

Ogawa, S., Kow, L.-M. and Pfaff, D.W., Neuronal activity of dorsal periaqueductal gray neurons of female rats: Responsiveness to GABA and enkephalin, Soc. Neurosci. Abst., 15 (1989) 146.

Ogawa, S., Kow, L.-M. and Pfaff, D.W., Electrophysiological responses of periaqueductal gray neurons of female rats to LHRH, substance P, oxytocin, and TRH in vitro, Soc. Neurosci. Abst., 16 (1990) 266.

Pan, J.-T., Kow, L.-M. and Pfaff, D.W., Neuromodulatory actions of luteinizing hormone-releasing hormone on electrical activity of preoptic neurons in brain slices, Neuroscience, 27 (1988) 623-628.

Panula, P., Yang, H.-Y.T. and Costa, E., Comparative distribution of bombesin/GRP- and substance P-like immunoreactivities in rat hypothalamus, J. Comp. Neurol., 224 (1984) 606-617.

Perez de la Mora, M., Possani, L.D., Tapia, R., Teran, L., Palacios, R., Fuxe, K., Hökfelt, T. and Ljungdahl, A., Demonstration of central gamma-aminobutyrate-containing nerve terminals by means of antibodies against glutamate decarboxylase, Neuroscience, 6 (1981) 875-895.

Pfaff, D.W., Estrogen and brain function: Neural analysis of a hormone-controlled mammalian reproductive behavior, Springer-Verlag, New York, 1980.

Pfaff, D.W., Patterns of steroid hormone effects on electrical and molecular events in hypothalamic neurons, Mol. Neurobiol., 3 (1989) 135-154.

Pfaff, D.W. and Keiner, M., Atlas of estradiol-concentrating cells in the central nervous system of the female rat, J. Comp. Neurol., 151 (1973) 121-158.

Pfaff, D.W. and Lewis, C., Film analyses of lordosis in female rats, Horm. Behav., 5 (1974) 317-335.

Pfaff, D.W. and Sakuma, Y., Facilitation of the lordosis reflex of female rats from the ventromedial nucleus of the hypothalamus, J. Physiol., 288 (1979a) 189-202.

Pfaff, D.W. and Sakuma, Y., Deficit of the lordosis reflex of female rats caused by lesions in the ventromedial nucleus of the hypothalamus, J. Physiol., 288 (1979b) 203-210.

Pfaff, D.W. and Schwartz-Giblin, S., Cellular mechanisms of female reproductive behaviors, In: The physiology of reproduction, Knobil E, Neill J. et al. (Eds.), Raven Press, New York, 1988, pp. 1487-1568.

Pfaus, J.G. and Gorzalka, B.B., Opioids and sexual behavior, Neurosci. Biobehav. Rev., 11 (1987a) 1-34.

Pfaus, J.G. and Gorzalka, B.B., Selective activation of opioid receptors differentially affects lordosis behavior in female rats, Peptides, 8 (1987b) 309-317.

Riskind, P. and Moss, R.L., Midbrain central gray: LHRH infusion enhances lordotic behavior in estrogen-primed ovariectomized rats, Brain Res. Bull., 4 (1979) 203-205.

Riskind, P. and Moss, R.L., Midbrain LHRH infusions enhance lordotic behavior in ovariectomized estrogen-primed rats independently of a hypothalamic responsiveness to LHRH, Brain Res. Bull., 11 (1983a) 481-485.

Riskind, P. and Moss, R.L., Effects of lesions of putative LHRH-containing pathways and midbrain nuclei on lordotic behavior and luteinizing hormone release in ovariectomized rats, Brain Res. Bull., 11 (1983b) 493-500.

Robbins, A., Schwartz-Giblin, S. and Pfaff, D.W., Ascending and descending projections to medullary reticular formation sites which activate deep lumbar muscles in the rat, Exp. Brain Res., 80 (1990) 463-474.

Robbins, A., Schwartz-Giblin, S. and Pfaff, D.W., Reticulospinal and reticulo-reticular pathways for activating the lumbar back muscles in the rat, submitted.

Romano, G.J., Bonner, T.I. and Pfaff, D.W., Preprotachykinin gene expression in the mediobasal hypothalamus of estrogen-treated and ovariectomized control rats, Exp. Brain Res., 76 (1989) 21-26.

Romano, G.J., Harlan, R.E., Shivers, B.D., Romano, G.J., Howells, R.D. and Pfaff, D.W., Estrogen increases proenkephalin messenger ribonucleic acid levels in the ventromedial hypothalamus of the rat, Mol. Endocrin., 2 (1988) 1320-1328.

Romano, G.J., Mobbs, C.V., Howells, R.D. and Pfaff, D.W., Estrogen regulation of proenkephalin gene expression in the ventromedial hypothalamus of the rat: temporal qualities and synergism with progesterone, Mol. Brain Res., 5 (1989) 51-58.

Romano, G.J., Mobbs, C.V., Lauber, A., Howells, R.D. and Pfaff, D.W., Sex difference in enkephalin gene expression in rat hypothalamus, Brain Res. (1991) in press.

Rose, J.D. and Flynn, F.W., Lordosis can be elicited in chronically-decerebrate rats by combined lumbosacral and vaginocervical stimulation, Soc. Neurosci. Abst., 15 (1989) 110.

Rose, J.D., Functional reconfiguration of midbrain neurons by ovarian steroids in behaving hamsters, Physiol. Behav., 37 (1986) 633-647.

Rothfeld, J., Hejtmancik, J.F., Conn, P.M. and Pfaff, D.W., In situ hybridization for LHRH mRNA following estrogen treatment, Mol. Brain Res., 6 (1989) 121-125.

Sakuma, Y. and Akaishi, T., Cell size, projection path, and localization of estrogen-sensitive neurons in the rat ventromedial hypothalamus, J. Neurophysiol., 57 (1987) 1148-1159.

Sakuma, Y. and Pfaff D.W., Facilitation of female reproductive behavior from mesencephalic central gray in the rat, Am. J. Physiol., 237 (1979a) R278-284.

Sakuma, Y. and Pfaff D.W., Mesencephalic mechanisms for integration of female reproductive behavior in the rat, Am. J. Physiol., 237 (1979b) R285-290.

Sakuma, Y. and Pfaff, D.W., Cells of origin of medullary projections in central gray of rat mesencephalon, J. Neurophysiol., 44 (1980a) 1002-1011.

Sakuma, Y. and Pfaff, D.W., Excitability of female rat central gray cells with medullary projections: Changes produced by hypothalamic stimulation and estrogen treatment, J. Neurophysiol., 44 (1980b) 1012-1023.

Sakuma, Y. and Pfaff, D.W., Convergent effects of lordosis-relevant somatosensory and hypothalamic influences on central gray cells in the rat mesencephalon, Exp. Neurol., 70 (1980c) 269-281.

Sakuma, Y. and Pfaff, D.W., LHRH in the mesencephalic central grey can potentiate lordosis reflex of female rats, Nature, 283 (1980d) 566-567.

Sakuma, Y. and Pfaff, D.W., Modulation of the lordosis reflex of female rats by LHRH, its antiserum and analogs in the mesencephalic central gray, Neuroendocrinology, 36 (1983) 218-224.

Schmitt, P., Di Scala, G., Jenck, F. and Sandner, G., Periventricular structures, elaboration of aversive effects and processing of sensory information, In: Modulation of sensorimotor activity during alternations in behavioral states, Bandler R. (Ed.), Alan R. Liss, New York, 1984, pp. 393-414.

Schmitt, P., Di Scala, G., Brandao, M.L. and Karli, P., Behavioral effects of microinjections of SR 95103, a new GABA-A antagonist, into the medial hypothalamus or the mesencephalic central gray, Eur. J. Pharmacol., 117 (1985) 149-158.

Schumacher, M., Coirini, H., Frankfurt, M. and McEwen, B.S., Localized actions of progesterone in hypothalamus involve oxytocin, Proc. Natl. Acad. Sci., 86 (1989) 6798-6801.

Schumacher, M., Coirini, H., Pfaff, D.W. and McEwen, B.S., Behavioral effects of progesterone associated with rapid modulation of oxytocin receptors, Science, 150 (1990) 691-694.

Schwartz-Giblin, S., Femano, P.A. and Pfaff, D.W., Axial electromyogram and intervertebral length gauge responses during lordosis behavior in rats, Exp. Neurol., 85 (1984) 297-315.

Shaikh, M. and Siegel, A., GABA-mediated regulation of feline aggression elicited from midbrain periaqueductal gray, Brain Res., 507 (1990) 51-56.

Shaikh, M., Dalsass, M. and Siegel, A., Opioidergic mechanisms mediating aggressive behavior in the cat, Agg. Behav., 16 (1990) 191-206.

Shivers, B.D., Harlan, R.E., Morrell, J.I. and Pfaff, D.W., Immunocytochemical localization of luteinizing hormone-releasing hormone in male and female rat brains, Neuroendocrinology, 36 (1983a) 1-12.

Shivers, B.D., Harlan, R.E., Morrell, J.I. and Pfaff, D.W., Absence of oestradiol concentration in cell nuclei of LHRH-immunoreactive neurons, Nature, 304 (1983b) 345-347.

Shivers, B.D., Harlan, R.E., Hejtmancik, J.F., Conn, P.M. and Pfaff, D.W., Localization of cells containing LHRH-like mRNA in rat forebrain using in situ hybridization, Endocrinology, 118 (1986) 883-885.

Shivers, B.D., Harlan, R.E. and Pfaff, D.W., A subset of neurons containing immunoreactive prolactin is a target for estrogen regulation of gene expression in rat hypothalamus, Neuroendocrinology, 49 (1989) 23-27.

Sirinathsinghji, D.J.S., Modulation of lordosis behavior of female rats by naloxone, ß-endorphin and its antiserum in the mesencephalic central gray: Possible mediation via GnRH, Neuroendocrinology, 39 (1984) 222-230.

Sirinathsinghji, D.J.S., Modulation of lordosis behavior of female rats by corticotropin releasing factor, ß-endorphin and gonadotropin releasing hormone in the mesencephalic central gray, Brain Res., 336 (1985) 45-55.

Södersten, P., Forsberg, G., Bednar, I., Eneroth, P. and Wiesenfeld-Hallin, Z., Opioid peptide inhibition of sexual behaviour in female rats, In: Brain opioid systems in reproduction, Dyer R.G. and Bicknell R.J. (Eds.), Oxford Science Publications, Oxford, 1989, pp. 203-215.

Swett, J.E., McMahon, S.B. and Wall, P.D., Long ascending projections to the midbrain from cells of lamina I and nucleus of the dorsolateral funiculus of the rat spinal cord, J. Comp. Neurol., 238 (1985) 401-416.

Tempel, A. and Zukin, R.S., Neuroanatomical patterns of the δ, κ, and gamma opioid receptors of rat brain as determined by quantitative *in vitro* autoradiography, Proc. Natl. Acad. Sci., 84 (1987) 4308-4312.

Turcotte, J.C. and Blaustein, J.D., The distribution of neurons having estrogen receptor-immunoreactivity and substance P innervation overlaps in the midbrain central gray of guinea pigs, Soc. Neurosci. Abst., 15 (1989) 628.

Waksman, G., Hamel, E., Fournie-Zaluski, M.-C. and Roques, B.P., Autoradiographic comparison of the distribution of the neutral endopeptidase "enkephalinase" and of δ and κ opioid receptors in rat brain, Proc. Natl. Acad. Sci., 83 (1986) 1523-1527.

Wiesner, J.B. and Moss, R.L., A psychopharmacological characterization of the opioid suppression of sexual behaviour in the female rat, In: Brain opioid systems in reproduction, Dyer R.G. and Bicknell R.J. (Eds.), Oxford Science Publications, Oxford, 1989, pp. 187-202.

Wolf, S.S., Moody, T.W., Quirion, R. and O'Donohue, T.L., Biochemical characterization and autoradiographic localization of central substance P receptors using [^{125}I]physalaemin, Brain Res., 332 (1985) 299-307.

Wong, M., Eaton, M.J. and Moss, R.L., Electrophysiological actions of luteinizing hormone-releasing hormone: Intracellular studies in the rat hippocampal slice preparation, Synapse, 5 (1990) 65-70.

Yamano, M., Inagaki, S., Kito, S., Matsuzaki, T., Shinohara, Y. and Tohyama, M., Enkephalinergic projection from the ventromedial hypothalamic nucleus to the midbrain central gray matter in the rat: an immunocytochemical analysis, Brain Res., 398 (1986) 337-346.

Zemlan, F.P., Kow, L.-M., Morrell, J.I. and Pfaff, D.W., Descending tracts of the lateral columns of the rat spinal cord: A study using horseradish peroxidase and silver impregnation technique, J. Anat., 128 (1979) 489-512.

PART II
ANATOMICAL AND
NEUROCHEMICAL ASPECTS

Descending Pathways from the Periaqueductal Gray and Adjacent Areas

Gert Holstege

Department of Anatomy and Embryology
Rijksuniversiteit Groningen
The Netherlands

Introduction

The limbic system consists of several structures in the cortex cerebri as well as in the mostly ventral portions of the diencephalon. For a discussion of which structures belong to it, see Holstege (1991). As early as 1958, Nauta pointed out that the limbic system has extremely strong reciprocal connections with mesencephalic structures such as the periaqueductal gray (PAG) and the laterally and ventrally adjoining tegmentum (Nauta's limbic system-midbrain circuit). More recent findings strongly support Nauta's concept and has led Holstege (1990) to consider the mesencephalic periaqueductal gray (PAG) and large parts of the lateral and ventral mesencephalic tegmentum as the caudal pole of the limbic system.

It is important to realize that the definition "periaqueductal gray" is derived from the large portion of gray matter present around the aqueduct of Sylvius. The name "periaqueductal gray" clearly indicates that the function of this "gray" was not known. With the exception of its most rostral and caudal parts, the boundaries of the PAG are formed by two fiberstreams: 1, the tectobulbospinal fibers originating in the intermediate and deep layers of the superior colliculus and 2, the fibers of the mesencephalic trigeminal tract (see Holstege, 1991 and Cowie and Holstege, 1991 for reviews).

Another way to look at this periaqueductal area is to regard it as a large accumulation of neurons penetrated by two fiberstreams, the mesencephalic trigeminal and the tectobulbospinal tract. The caudal one third of this accumulation of neurons is located around the aqueduct between the inferior colliculi and the rostral two thirds are located around the aqueduct between the superficial layers of the superior colliculus, the brachia of the inferior colliculi and/or the

The Midbrain Periaqueductal Gray Matter, Edited by A. Depaulis and
R. Bandler, Plenum Press, New York, 1991

corpora geniculata mediales. This penetration of two fiberstreams resulted in the definition of the PAG as the area medial to the fiber streams. The diffuse area lateral to the fiberstreams received several names, its dorsal part was called "deep layers of the superior colliculus" or "deep tectum", the ventrolateral part "nucleus cuneiformis" and some parts in between were referred to as "adjacent mesencephalic tegmentum". The fact that fiberstreams penetrating certain nuclear masses result in different names for the two parts of these masses is not new. Other examples are the caudate nucleus and putamen, or the central and medial amygdaloid nuclei and the lateral and medial parts of the bed nucleus of the stria terminalis. Both are accumulations of neurons, with almost identical pathways and functions, but split into two cellgroups by the fibers of the internal capsule (see Holstege et al., 1985 for review).

Thus, a concept is presented in which the PAG proper is not a separate entity, but the central portion of a larger accumulation of neurons, forming the caudal pole of the limbic system. It would mean that the boundaries of the PAG proper are totally artificial (see also Fig. 1), which idea is corroborated by several studies on the function of the PAG. These studies indicate that the boundaries of the PAG proper do not correspond with the boundaries of the neuronal cell groups involved in certain functions, such as vocalization and defense behavior.

The PAG and adjacent mesencephalic tegmentum exerts a strong influence on somatic and autonomic motoneurons. Physiological studies have shown that

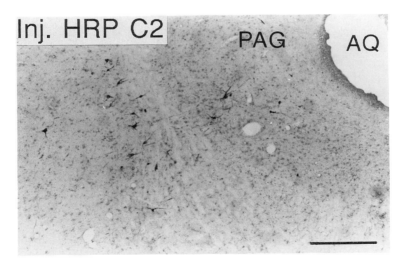

Figure 1. Group of labeled neurons after an HRP injection in the C2 spinal cord. Note that tectobulbospinal fibers split the cell group into two parts, one in the PAG and one lateral to it. Without these fibers the labeled neurons would form one group. Obviously this is not only true for the neurons projecting to the spinal cord, but also for the other neurons in the PAG.

stimulation in the PAG and adjacent areas result in motor activities such as active defense behavior, lordosis, vocalization and locomotion. On the other hand neurons in this area have never been demonstrated to project directly to motoneurons (Holstege, 1988a; 1991). Apparently, "interneurons" play a role in the motor activities obtained after stimulation in this area. Although some specific projections to certain well circumscribed groups of interneurons will be described, the main projection from the PAG and adjacent areas is to the ventral portion of the caudal pontine and medullary medial tegmentum. The fact that the last area plays such an extremely important role in the descending output of the PAG and adjacent areas makes it necessary to first describe the descending projections from the ventral portion of the caudal pontine and medullary medial tegmentum to the spinal cord.

Projections from the Ventral Part of the Caudal Pontine and Medullary Medial Tegmental Field

Retrograde HRP results (Kuypers and Maisky, 1975; Tohyama et al., 1979; Holstege, G. and Kuypers, 1982; Holstege, 1988b) indicate that a great number of neurons in the ventral part of the caudal pontine and medullary medial tegmentum, including the nuclei raphe magnus, pallidus and obscurus project to the spinal cord (Fig. 2). It was also demonstrated by means of retrograde double labeling tracing techniques that many of these neurons project to cervical as well as lumbar levels of the spinal cord and to the caudal spinal trigeminal nucleus (Martin et al.,1981b; Hayes and Rustioni, 1981; Huisman et al., 1982; Lovick and Robinson, 1983).

Basbaum et al. (1978), using the autoradiographic tracing technique, were the first to demonstrate in the cat that the nucleus raphe magnus (NRM) projects to the marginal layer of the caudal spinal trigeminal nucleus and in the spinal cord to laminae I, II, V, VI and VII, and to the thoracolumbar intermediolateral cell column. Similar projections were observed from the tegmentum located next to the NRM, i.e., the ventral part of the medial tegmental field at the level of the facial nuclei, also called the nucleus reticularis magnocellularis. The results of Basbaum et al. (1978) were confirmed in the opossum and rat (Martin et al., 1981a; 1985) and in the cat (Holstege et al., 1979; Holstege, G. and Kuypers, 1982; Fig. 3 left). Moreover, Holstege, G. and Kuypers (1982) demonstrated that the rostral NRM and adjoining reticular formation project to all parts of the dorsal horn and that the caudal NRM and adjoining tegmentum project also to the sacral intermediomedial and intermediolateral cell column. Another very important finding was that the nucleus raphe pallidus (NRP) and its adjoining tegmentum does not project to the dorsal horn of caudal medulla and spinal cord, but to all other parts of the spinal gray matter, i.e., the intermediate zone and the somatic and autonomic motoneuronal cell groups of the spinal cord (Fig. 3 right; Holstege, J.C. and Kuypers, 1982 in the rat; Martin et al., 1981a in the opossum; Holstege et al., 1979, Holstege, G. and Kuypers, 1982 in the cat). The projections to the somatic motoneurons have also been demonstrated at the ultrastructural

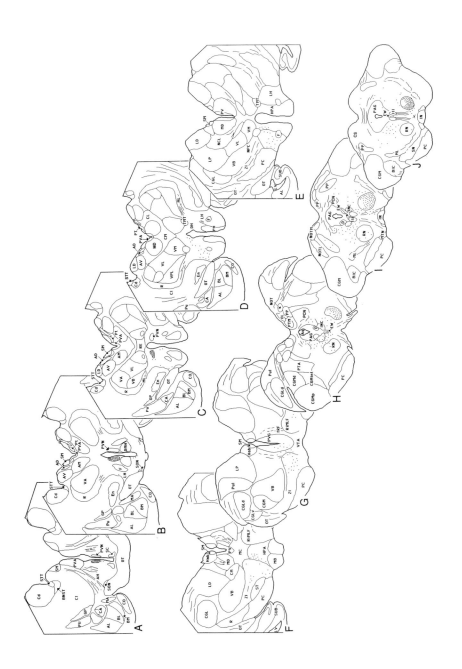

Figure 2. Schematic representation of the distribution of the HRP labeled neurons in brainstem and diencephalon of the cat after hemi-infiltration of HRP in the C2 spinal cord. Reprinted with permission from Holstege (1988) Progress in Brain Research.

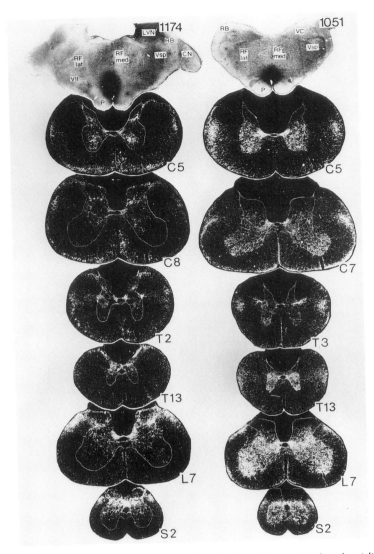

Figure 3: Brightfield photomicrographs of autoradiographs showing tritiated leucine injection sites in the raphe nuclei and darkfield photomicrographs showing the distributions of the labeled fibers in the spinal cord.
On the left an injection is shown in the caudal NRM and adjoining reticular formation. Note that labeled fibers are distributed mainly to the dorsal
horn (laminae I, the upper part of II and V), the intermediate zone and
the autonomic motoneuronal cell groups. On the right the injection is
placed in the NRP and immediately adjoining tegmentum. Note that the
labeled fibers are not distributed to the dorsal horn, but very strongly to
the ventral horn (intermediate zone and autonomic and somatic motoneuronal cell groups). Reprinted with permission from Holstege and
Kuypers (1982) Progress in Brain Research.

level (Holstege J.C and Kuypers, 1982; 1987). Caudal NRM and rostral NRP also project to the thoracolumbar and sacral intermediolateral cell groups (IML), i.e., to autonomic (sympathetic and parasympathetic) preganglionic motoneuronal cell groups (Fig. 3).

Summarizing, NRM, NRP and NRO, with their adjoining reticular formation, send fibers throughout the length of the spinal cord, giving off collaterals to all spinal levels. These descending systems are extremely diffuse and are not topographically organized. Furthermore, a strong heterogeneity exists in these projections, in which 1) the rostral NRM and adjoining reticular formation project to all parts of the dorsal horn; 2) the caudal NRM and adjoining reticular formation project mainly to laminae I and V and the autonomic motoneuronal cell groups and 3) the NRP and adjacent ventromedial medulla project to the intermediate zone and the ventral horn, including the autonomic and somatic motoneuronal cell groups.

Physiological studies are consistent with the anatomy of the descending pathways outlined above. Electrical stimulation in the NRM was found to produce an inhibitory postsynaptic potential (IPSP) in neurons in laminae I and II of the dorsal horn at a latency consistent with a monosynaptic connection (Light et al., 1986). Not only NRM stimulation, but also stimulation in the adjacent ventral part of the caudal pontine and/or upper part of the medullary medial reticular formation produces inhibition of neurons in the dorsal horn (Fields et al., 1977; Akaike et al., 1978).

The diffuse organization of NRP and ventromedial medulla projections to the motoneuronal cell groups suggests that they do not steer specific motor activities such as movements of distal (arm, hand or leg) or axial parts of the body, but have a more global effect on the level of activity of the motoneurons. Stimulation of the raphe nuclei has a facilitory effect on motoneurons (Cardona and Rudomin, 1983). There exist many different neurotransmitter substances in this area, of which serotonin is the best known. Serotonin plays a role in the facilitation of motoneurons, probably directly by acting on the Ca^{2+} conductance or indirectly by reduction of K^+ conductance of the membrane of the motoneuron (McCall and Aghajanian, 1979; White and Neuman, 1980; VanderMaelen and Aghajanian, 1982; Hounsgaard et al., 1986). Thus serotonin enhances the excitability of the motoneurons for inputs from other sources, such as red nucleus or motor cortex (McCall and Aghajanian, 1979). In mammals, there are many serotonergic fibers around the motoneurons (Steinbusch, 1981 and Kojima, 1983b in the rat, Kojima et al., 1982 in the dog, Kojima, 1983a in the monkey). The cell bodies of these serotonergic fibers are located mainly in the NRP, but not in the NRM (Alstermark et al., 1987).

Not only serotonin, but, at least in the rat, several peptides are present also in the spinally projecting neurons in the ventromedial medulla and NRP. Many neurons contain substance P, thyrotropin releasing hormone (TRH), somatostatin, methionine (M-ENK) and leucine-enkephalin (L-ENK), while a relatively small number contains vasoactive intestinal peptide (VIP) and cholecystokinin (CCK). It has been demonstrated that most of these peptides coexist to a variable extent with serotonin in the same neuron (Chan Palay et al., 1978; Hökfelt et al., 1978; Hökfelt et al., 1979; Johansson et al., 1981; Hunt and Lovick, 1982; Bowker et al., 1983; Mantyh and Hunt, 1984; Taber-Pierce et al., 1985; Helke et al., 1986; Léger et al., 1986; Bowker et al., 1988). Johansson et al. (1981) have also demonstrated the coexistence of serotonin, substance P and TRH in one and the same neuron. This coexistence of serotonin with different peptides not only occurs in the neuronal cell bodies, but also in their terminals in the ventral horn (Pelletier et al., 1981; Bowker, 1986; Wessendorf and Elde, 1987). According to Hökfelt et al. (1984), at the ultrastructural level, serotonin, substance P and TRH is stored in the terminals in dense core or granular vesicles, terminals with such vesicles are called G-type terminals (G=granular). Ulfhake et al. (1987) have recently shown that some of the G-type terminals lack synaptic specialization, suggesting that the content of dense core vesicles may be released at non-synaptic sites of the terminal membrane (see also Veening et al., this volume).

It must be emphasized that a major portion of the diffuse descending pathways to the dorsal horn and the motoneuronal cell groups is not derived from serotonergic neurons (Bowker et al., 1982; Johannessen et al., 1984). Other possible neurotransmitters include acetylcholine, since some of the neurons in the ventromedial medulla are ChAT positive (Jones and Beaudet, 1987). Also GABA may play an important role in these non-serotonergic pathways. Holstege, J.C. (1989) in the rat showed that after injection of WGA-HRP in the ventromedial medulla, 40% of the labeled terminals in the L5-L6 lateral motoneuronal cell group were also labeled for GABA. Of the double labeled terminals about 80% contained flattened vesicles, indicating an inhibitory function (Krnjévic and Schwartz, 1966). Holstege, J.C. (1989) also found that about 10% of the labeled terminals containing GABA were of the so-called G-type, which probably contain serotonin and/or peptides such as substance P, TRH or enkephalin-like substances (Pelletier et al., 1981; Holstege, J.C. and Kuypers, 1987). This corresponds with the finding of Belin et al. (1983) and Millhorn et al. (1988), who demonstrated colocalization of serotonin and GABA in neurons in the ventral medulla in the rat. Thus, there exist spinally projecting neurons in the ventromedial medulla that contain serotonin as well as GABA. Nicoll (1988) has found that 5HT1A and GABA$_B$ receptors are coupled to the same ion channel. The functional implication of these findings is that some terminals, taking part in this diffuse descending system, may have inhibitory as well as facilitatory effects on the postsynaptic element (i.e., the motoneuron), although the majority is probably either facilitatory or inhibitory. Spinal motoneurons display a bistable behavior, i.e., they can switch back and forth to a higher excitable level (Hounsgaard et al., 1984; 1986; 1988; Crone et al., 1988). Bistable behavior disappears after spinal transection, but reappears after subsequent intravenous injection of the serotonin precursor 5-hydroxy-tryptophan. Thus, intact descending pathways are essential for this bistable behavior of motoneurons and serotonin is one of the neurotransmitters involved in switching to a higher level of excitation. Possibly, GABA may be involved in switching to a lower level of excitation.

In summary, the diffuse descending pathways originating in the ventromedial medulla, including the nucleus raphe pallidus and obscurus, have very general and diffuse facilitatory or inhibitory effects on motoneurons and probably also on interneurons in the intermediate zone. Although most of the terminals have either a facilitatory or an inhibitory function, recent results suggest that there also exist terminals with both facilitatory and inhibitory functions.

Although similar diffuse projections originate in the area of the locus coeruleus/nucleus subcoeruleus, these will not be described here, also because the emphasis of the PAG projections is not on this area. For a review on these projections see Holstege (1991).

Projections from the Periaqueductal Gray and Adjacent Tegmentum to Caudal Brainstem and Spinal Cord

Descending mesencephalic projections to the ventral part of the caudal pontine and medullary medial tegmentum

Retrograde HRP tracing studies (Abols and Basbaum, 1981; Holstege, 1988a) indicate that an enormous number of HRP labeled neurons in the PAG

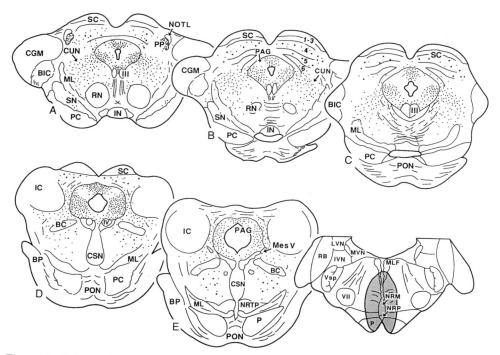

Figure 4. Schematic drawings of HRP-labeled neurons in the mesencephalon after injection of HRP in the NRM/NRP region. Note the dense distribution of labeled neurons in the PAG (except its dorsolateral part) and the tegmentum ventrolateral to it. Reprinted with permission from Holstege (1988) Progress in Brain Research.

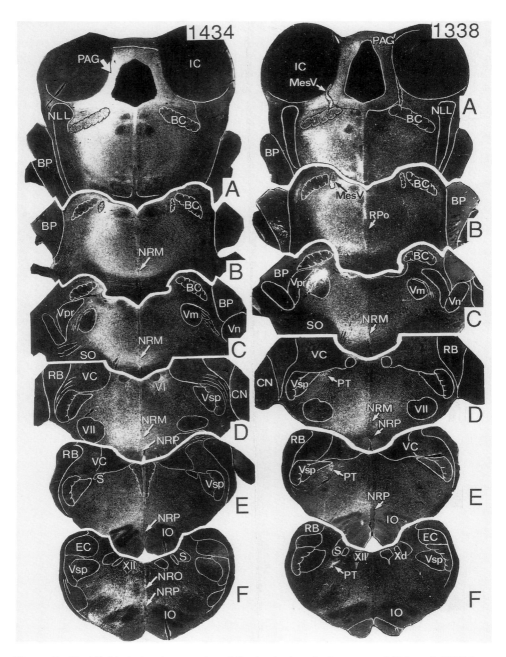

Figure 5. Darkfield photomicrographs of the brainstem in the cases 1434 and 1338 (see
Fig. 6) with injections in respectively the ventrolateral PAG and more rostrally in the lateral
PAG. Note the strong projections to the NRM and the ventral part of the medial
tegmentum of caudal pons and medulla in both cases. Note that in case 1434, but not in
case 1338 labeled fibers were also distributed to the NRP. Reprinted with permission from
Holstege (1988) Progress in Brain Research.

and laterally and ventrolaterally adjoining areas project to NRM, NRP and ventral part of the caudal pontine and medullary medial tegmentum (Fig. 4).

Interestingly, the dorsolateral portion of the PAG and the dorsolaterally adjacent tegmentum are free of labeled neurons (Fig. 4). This is also the case for the PAG-projections to other caudal brainstem areas such as the medullary lateral tegmental field, the nucleus retroambiguus and the spinal cord (Fig. 2). The function of the dorsolateral PAG is not yet known. It receives specific projections from the superior colliculus (Cowie and Holstege, 1991) and a small cellgroup just dorsal to the genu of the facial nerve in the upper medulla (Holstege et al., in preparation). It also differs from other areas of the PAG in respect to the content of certain neurotransmitters (Gundlach, this volume) and it receives different afferent projections from medial and lateral cortical fields (Shipley et al., this volume). The dorsolateral PAG (with its adjacent tegmentum) may serve as a "sensory area", which receives afferents from various distinct areas and sends fibers to other parts of the PAG and the cuneiform nucleus (see also Redgrave et al. 1988; Redgrave and Dean, this volume).

Retrograde and anterograde (autoradiographic) tracing studies (Jürgens and Pratt, 1979; Mantyh, 1983; Holstege, 1988a; Fig. 5) show that different parts of the PAG and adjacent tegmentum, with the exception of the dorsolateral PAG, project in the same basic pattern to the caudal brainstem. The descending fibers pass ipsilaterally through the mesencephalic and pontine lateral tegmental field, but gradually shift ventrally and medially at caudal pontine levels. They terminate mainly ipsilaterally in the ventral part of the caudal pontine and medullary medial tegmentum and in the NRM (Fig. 5). On their way to the medulla they distribute fibers to the area of the locus coeruleus, the nucleus subcoeruleus and the paralemniscal cell group. The latter cell group in turn projects contralaterally to laminae I and V throughout the length of the spinal cord (Holstege, 1988a). It

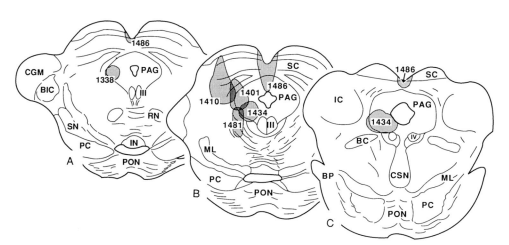

Figure 6. Schematic representation of the ³H-leucine injection sites of 6 cases of which some projections are shown in Figs. 5,7,8 and 10.

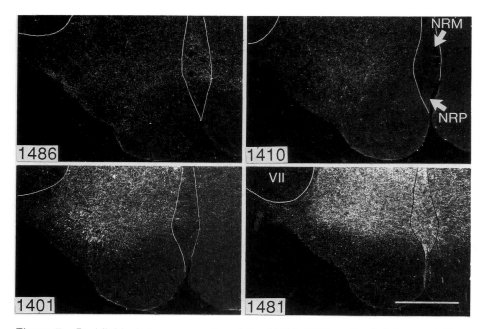

Figure 7. Darkfield photomicrographs of the NRM, NRP and adjoining medullary medial tegmentum in cases 1486,1410,1401 and 1481 (see Fig. 6). Note that the dorsomedial PAG (case 1486) projects mainly to the medial part of the medullary medial tegmentum, i.e., to the NRM, while the laterally adjacent tegmentum (case 1410) projects to the lateral parts of this area, and not to the NRM. Note further that the ventrally adjoining tegmentum (case 1481) projects strongly to the NRP, while such projections do not originate in the lateral PAG (1401). Reprinted with permission from Holstege (1988) Progress in Brain Research. Bar represents 1 mm.

is important to keep in mind that not only PAG neurons project to the ventral part of the caudal pontine and medullary medial tegmentum, but also neurons in the area lateral to it (see case 1410; Figs. 6 B and 7 top right). This corresponds with the view that the boundaries of the PAG do not correspond with the boundaries of the various cell groups giving rise to the descending projections (see also Fig. 1). There exists a mediolateral organization within the ipsilaterally descending pathways from PAG and adjacent areas. The main projection of the neurons in the medial part of the dorsal PAG, is to the medially located NRM and immediately adjacent tegmentum (case 1486, Figs. 6 A-B; 7 top left). On the other hand, neurons in the laterally adjacent tegmentum and the intermediate and deep layers of the superior colliculus project mainly laterally to the ventral part of the caudal pontine and upper medullary medial tegmentum with virtually no projections to the NRM (Holstege, 1988a; Cowie and Holstege, 1991; Fig. 7 top right).

Neurons in the ventrolateral portion of the caudal PAG (case 1434, Figs. 5 left; 6 B,C) and ventrally adjoining tegmentum (case 1481, Figs. 6 B and 7 bottom right), but not in the lateral and dorsal parts of the PAG, send fibers to the NRP (Fig. 7). This group of neurons forms another example of a cell group occupying parts of the PAG as well as parts of the adjacent tegmentum. The functional meaning of these specific NRP projections is not yet known.

Involvement of the descending mesencephalic projections in control of nociception
In animals (see Besson and Chaouch, 1987 and Willis, 1988 for reviews; see also Besson et al.; Lovick; Morgan, this volume) as well as in humans (Hosobuchi, 1988; Meyerson, 1988) the PAG is well known for its involvement in the supraspinal control of nociception. The strong impact on nociception is only partly mediated via its projections to the NRM and adjacent reticular formation, because in cases with reversible blocks of the NRM and adjacent tegmentum, PAG stimulation results in reduced analgesic effects (Gebhart et al., 1983; Sandkühler and Gebhart, 1984). However, the analgesic effects do not completely disappear after blocking the NRM and adjacent tegmentum, which suggests that other brainstem regions also play a role. In this respect the PAG projections to the paralemniscal cell group are of interest, since part of the antinociceptive action of the PAG may be exerted through this pathway (for review see Holstege, 1988a).

Involvement of the descending mesencephalic projections in the lordosis reflex
Stimulation in the PAG facilitates the lordosis reflex, while electrolytic lesions in the dorsal half of the PAG and the adjacent mesencephalic tegmentum produce an immediate decline in performance of the lordosis reflex (Sakuma and Pfaff, 1979a,b; Ogawa et al., this volume). Lordosis, a curvature of the vertebral column with ventral convexity, is an essential element of female copulatory behavior in rodents. The lordosis reflex is facilitated by stimulation of the ventro-medial hypothalamic nucleus (Pfaff and Sakuma, 1979a,b) and the PAG (Sakuma and Pfaff, 1979a,b). Stimulation of the L1 through S1 dermatomes is sufficient for eliciting the lordosis reflex, but several studies suggested that it was oestrogen dependent, i.e., would only occur when copulation can result in fertilization. This led to the concept that the lordosis reflex cannot be produced in the absence of facilitatory forebrain influences. However, it was recently demonstrated that the lordosis reflex can also be elicited in decerebrate rats (Rose and Flynn, 1989). It is also known that descending fibers in the ventrolateral funiculus play a role in the facilitation of the reflex (Kow et al., 1977). It is not possible that these fibers originate from neurons in the ventromedial hypothalamic nucleus or the PAG, because none of the two structures projects directly to the lumbosacral spinal cord (Holstege, 1987; Holstege, 1988a, see also below).

Perhaps, the lordosis reflex should be considered as a spinal reflex, in which the L1-S1 cutaneous input from flank, rump, tailbase and perineum serves as the afferent loop, and the fibers of the back and axial muscle motoneurons form the efferent loop. Both loops are interconnected by spinal interneurons and short and long propriospinal pathways. Neurons in the dorsal two thirds of the pontine and upper medullary medial tegmentum may coordinate the back and axial muscle inter- and motoneuronal activity via their descending pathways through the ventral funiculus of the spinal cord (see section 4a1 of Holstege, 1991). These neurons receive afferents from the PAG, although their number is much lower than the PAG fibers terminating in the ventral parts of the medial tegmentum (Fig. 5). However, lordosis behaviour occurs only when the membrane excitability of the motoneurons is high. This level of excitability is determined by descending pathways, which originate in the ventral part of the medullary medial tegmentum and project diffusely to all inter- and motoneuronal cell groups in the ventral horn throughout the length of the spinal cord (see above). The ventral part of the medial tegmental field receives its afferents from PAG and anteromedial hypothalamus, but not from the ventromedial hypothalamic nucleus (Holstege, 1987).

Thus, a concept is put forward in which the ventromedial hypothalamic nucleus controls the lordosis reflex by means of its projections to the dorsal and ventral parts of the caudal pontine and medullary medial tegmentum, using the anteromedial hypothalamus and the dorsal PAG as relay structures (see also Ogawa et al., this volume). Neurons in the dorsal two thirds of the medial tegmentum coordinate the back and axial muscle motoneuronal activity while the medullary ventromedial tegmentum increases the excitability of the motoneurons to such a level that cutaneous L1-S1 afferent stimulation, which is otherwise ineffective, results in lordosis. Actually, during oestrus the female rat shows highly active behavior, characterized by frequent locomotion and other stress like phenomena (Pfaff, 1980). During various forms of stress, such as aggression, fear or sexual arousal, the motor system needs to be set at a "high" level. In such circumstances spinal reflexes such as the lordosis reflex can easily be elicited. Pfaff (1980) points to the lateral vestibulospinal tract to play an important role in lordosis behavior, although the lateral vestibular nucleus does not receive afferents from the hypothalamus or PAG. Lesions in the lateral vestibular nucleus led to decreases in lordosis (Modianos and Pfaff, 1979). In this respect it should be recalled that the lateral vestibulospinal tract has an important influence on all axial movements, thus including the lordosis movements. The question remains whether the lateral vestibular nucleus is specifically involved in lordosis behavior.

Involvement of the descending pedunculopontine projections in locomotion

Just lateral to the brachium conjunctivum, just ventral to the cuneiform nucleus and just rostral to the parabrachial nuclei is located the so-called pedunculopontine nucleus. This nucleus takes part in the accumulation of neurons that forms the caudal pole of the limbic system (see Introduction). The pedunculopontine nucleus contains many ChAT positive neurons (Jones and Beaudet, 1987). Stimulation in this region induces locomotion in cats (Shik et al., 1966), which is the reason that this area is also termed the mesencephalic locomotor region (MLR). The MLR not only comprises the pedunculopontine nucleus, but extends into the cuneiform nucleus, which is located just dorsal to the pedunculopontine nucleus. Garcia-Rill and Skinner (1988), found that during locomotion neurons in the cuneiform nucleus were related preferentially to rhythmic (bursting) activity, while neurons in the pedunculopontine nucleus are preferentially related to the onset or termination of cyclic episodes (on/off cells).

Anatomical studies (Moon Edley and Graybiel, 1983; Holstege, unpublished results) revealed that the descending projections from this area are organized similar to those from the PAG and adjacent tegmentum. The mainly ipsilateral fiberstream first descends laterally in the mesencephalon and upper pons and then gradually shifts medially to terminate bilaterally, but mainly ipsilaterally in the ventral part of the caudal pontine and medullary medial tegmental field (see also Garcia-Rill and Skinner, 1987b). Only sparse projections exist to the nucleus raphe magnus and almost none to the dorsal portions of the caudal pontine and medullary medial tegmentum. By means of low-amplitude (<70 µA), high frequency (5-60 Hz) stimulation or via injection of cholinergic agonists in this same area, Garcia-Rill and Skinner (1987a) were able to elicit locomotion in the ventral portion of the caudal pontine and medullary medial tegmentum. They also demonstrated that the locomotion in the medioventral medulla could control or override the stepping frequency induced by the mesencephalic locomotor region. Moreover, Garcia-Rill and Skinner (1987b) reported that about 35% of the cells in this area project through the ventrolateral funiculus of the C2 spinal cord and half of these cells received short latency orthodromic input from the mesencephalic locomotor region. Somewhat surprising was that they also found such cells as far rostral as the caudal pontine ventral tegmentum. The latter area, according to the anatomic findings, only projects to the dorsal horn via the dorsolateral funiculus and not to the intermediate zone or ventral horn via the ventrolateral funiculus. Nevertheless, the findings of Garcia-Rill and Skinner (1987a,b) indicate that locomotion, elicited in the mesencephalic locomotor region, is based on the projections from this area to the medial part of the ventral medullary medial tegmentum and on the diffuse projections from the latter area to the rhythm generators in the spinal cord.

The afferent connections of the mesencephalic locomotor area are derived from lateral parts of the limbic system, such as the bed nucleus of the stria terminalis, central nucleus of the amygdala and lateral hypothalamus (Moon Edley and Graybiel, 1983). Strong projections are also derived from the entopeduncular nucleus, subthalamic nucleus and the substantia nigra pars reticulata, but motor cortex projections to the MLR are very scarce (Moon Edley and Graybiel, 1983). These findings indicate that the MLR is influenced by extrapyramidal and lateral limbic structures, and virtually not by somatic motor structures. This corresponds with the fact that the descending projections from the MLR terminate in the ventromedial part of the caudal pontine and medullary tegmental field, which area receives afferents from many other limbic system related areas, but not from the somatic motor structures.

Projections to the ventrolateral medulla; involvement in blood pressure control

It has been shown that neurons in the rostral part of the ventrolateral tegmental field of the medulla (subretrofacial nucleus) are essential for the mainte-

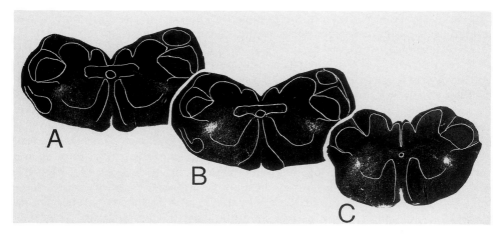

Figure 8. Darkfield photographs of the caudal medulla in a cat (1434, see also Fig. 4 left and 6 B,C) with an injection of ³H-leucine in the ventrolateral part of the caudal PAG. Note the strong bilateral projections to the NRA. Reprinted with permission from Holstege (1989) Journal of Comparative Neurology.

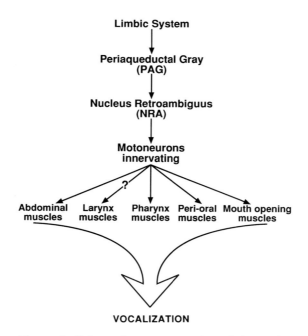

Figure 9. Schematic representation of the pathways for vocalization from the limbic system to the vocalization muscles. Reprinted with permission from Holstege (1989) Journal of Comparative Neurology.

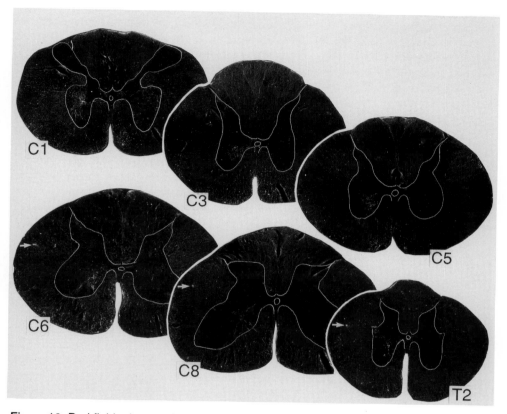

Figure 10. Darkfield micrographs of cervical and upper thoracic sections of the spinal cord after an injection in the lateral PAG, its adjacent tegmentum and the deep collicular layers (case 1401 Fig. 6 B). Note that labeled fibers descend ipsilaterally through the ventral and ventrolateral funiculi to terminate in the medial part of the cervical and upper thoracic intermediate zone. Note also that a small contingent of labeled fibers descend in the ipsilateral dorsolateral funiculus (see arrows). These fibers probably terminate in the upper thoracic intermediolateral cell group (see section T2). Contralaterally a few tectospinal fibers descend in the ventral funiculus of the upper cervical cord to terminate in the lateral part of the upper cervical intermediate zone. Reprinted with permission from Holstege (1988) Progress in Brain Research.

nance of the vasomotor tone and reflex regulation of the systemic arterial blood pressure (see section 3 c of Holstege, 1991). Physiological studies suggest that neurons in the rostral part of the subretrofacial nucleus project specifically to the IML neurons, innervating the kidney and adrenal medulla, while neurons in the caudal part of the subretrofacial nucleus project to IML neurons innervating the hindlimb vasculature (Lovick, 1987; Dampney and McAllen, 1988). From an anatomical study Carrive et al. (1989) have suggested that neurons lateral to the aqueduct in the caudal half of the PAG have an excitatory effect on the neurons in the subretrofacial nucleus (increase of blood pressure; see also Carrive; Lovick, this volume), while neurons lying ventrolaterally in the caudal third of the PAG have an inhibitory effect (decrease of blood pressure). The same authors also have shown that neurons in the caudal third of the PAG project to the rostral part of the subretrofacial nucleus, which in turn projects to IML motoneurons that innervate the kidney and adrenal medulla. On the other hand, neurons in the intermediate third of the PAG project to the caudal subretrofacial nucleus, which projects to IML motoneurons innervating the hindlimb. In conclusion, there exists a precise organization in the mesencephalic control of blood pressure in different parts of the body. All these projections take part in a descending system involved in the elaboration of emotional motor activities. For an extensive review of this control system, see Bandler et al. (1991) and Carrive (this volume).

Projections to the nucleus retroambiguus; involvement in vocalization

In many different species, from leopard frog to chimpanzee (see Holstege, G., 1989 for review), stimulation in the caudal PAG results in vocalization, i.e., the nonverbal production of sound. In humans, laughing and crying are probably examples of vocalization. Holstege, G. (1989) has demonstrated that a specific group of neurons in the lateral and to a limited extent in the dorsomedial part of the caudal two thirds of the PAG send fibers to the nucleus retroambiguus (NRA) in the caudal medulla (Fig. 8). The cell group in the PAG differs from the smaller cells projecting to the raphe nuclei and adjacent tegmentum or the larger cells projecting to the spinal cord. The NRA in turn projects to the somatic motoneurons innervating the pharynx, soft palate, intercostal and abdominal muscles and probably the larynx (Fig. 9). Direct PAG projections to these somatic motoneurons do not exist (Holstege, G., 1989). In all likelihood, the projection from the PAG to the NRA forms, at least in part, the final common pathway for vocalization, because DeRosier et al. (1988) found that during vocalization the NRA neurons were more closely related to the vocalization muscle EMG than the PAG (see also Davis and Zhang, this volume). This finding is important, because it shows that a specific expressive behavior may be similarly based on a distinct descending pathway, suggesting that all the other specific motor activities displayed during expressive behavior are based on separate descending pathways.

PAG projections to the spinal cord

Only limited PAG projections to the spinal cord exist. Some neurons in the lateral part of the caudal one third of the PAG and laterally adjacent tegmentum send fibers through the ipsilateral ventral funiculus of the cervical spinal cord to terminate in laminae VIII and the adjoining part of VII (Martin et al., 1979; Holstege, 1988a,b; Fig. 10). A very few fibers descend ipsilaterally in the lateral funiculus to terminate in the T1-T2 IML (Holstege, 1988a,b; Fig. 10). The projections to the spinal cord may play a role in the defensive behavior evoked by excitatory amino acid microinjections made in the PAG (Bandler and Carrive, 1988; Bandler and Depaulis, this volume). For example, the projection to the medial part of the intermediate zone of the cervical cord may be involved in the contralateral head turning movements as part of defensive behavior, while the projection to the T1-T2 IML may produce the pupil dilation described by Bandler and Carrive (1988).

Epilogue

In this chapter an attempt has been made to give an overview of the many different cell groups in the PAG and adjacent tegmentum, which give rise to descending pathways (Fig. 11). Four main conclusions can be drawn:

1. Anatomically the PAG is an accumulation of neurons with clearcut boundaries. These boundaries are often formed by thick fibers belonging to systems such as the tectobulbospinal tract, the mesencephalic trigeminal tract or the medial longitudinal fasciculus. However, many anatomical and physiological studies clearly indicate that these boundaries are not absolute functional. Cell groups giving rise to descending fibers with specific targets are not limited to the PAG, but very often extend into the adjacent mesencephalic tegmentum. Figure 1 nicely illustrates that the boundaries of the PAG are only artificial and not functional.

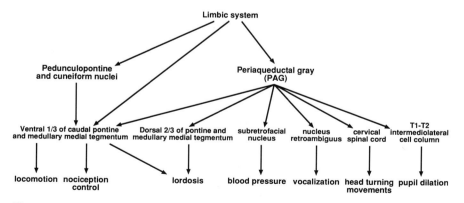

Figure 11. Schematic overview of the descending projections from the PAG and pedunculopontine and cuneiform nuclei to different regions of the caudal brainstem and spinal cord. The functions in which each of the projections might be involved are also indicated. It should be emphasized that these functional interpretations are only tentative.

2. The most impressive projection of the PAG and its adjacent areas is to the ventromedial tegmentum of caudal pons and medulla. These projections are so overwhelming and so much stronger than any other descending projection from PAG and adjacent areas, that a thorough understanding of the anatomy and function of this region in caudal pons and medulla, for example its projections to the spinal cord, is absolutely essential for understanding the motor output of the PAG.

3. Only one area of the PAG and its adjacent tegmentum (its dorsolateral part) seems to behave differently from the other areas. The dorsolateral PAG (with the exception of its most rostral and caudal portion) does not contain neurons giving rise to descending pathways and may serve as the "sensory area" of the PAG. Another area, which has escaped the attention of most PAG investigators is the ventromedial PAG. It is known that this area contains many premotor neurons for the oculomotor system, but its precise function is not well understood.

4. The PAG contains many cell groups with different outputs presumably serving different functions. Many of these cell groups overlap. The same is true for the afferent projections originating in the limbic system, of which large regions, such as hypothalamus, preoptic area, amygdala, bed nucleus of the stria terminalis, orbitofrontal cortex and many other regions, send fibers to the PAG (for a discussion of the specificity of the forebrain projections to the PAG, see Shipley et al., this volume). For example, in respect to vocalization it has been shown that it can be elicited in many regions of the limbic system, but that bilateral lesions in the PAG render all these areas "mute", indicating that the PAG descending projections form the final common pathway for vocalization. If this is so for vocalization it may also be true for the other behaviors indicated in Figure 11. It should be kept in mind that the final emotional behavior is a combination of many components such as changes in blood pressure, heart rate, vocalization, locomotion, neck movements and many other specific behaviors. Different combinations give different behaviors. More rostral portions of the limbic system likely determine, by way of their specific pathways to certain cell groups or longitudinal neuronal columns in the PAG, which combinations of motor activities (i.e., which class of emotional behavior) will be performed (see also Shipley et al., this volume). It is possible that future studies will reveal that certain neurons or neuronal cell groups in more rostral portions of the limbic system project to a certain combination of neurons in the PAG, producing a specific behavior. Finally, it should be stressed that these emotional motor activities form a separate part of the motor system as a whole. According to Holstege (1991; Fig. 12), the motor system consists of 3 "systems", the first system are the premotor interneurons projecting to the motoneurons. This system not only contains the interneurons in the intermediate zone in the spinal cord, but also interneurons such as those giving rise to long descending pathways, projecting to motoneurons innervating respiratory or micturition related musculature. Thus the medullary interneurons projecting to the phrenic and abdominal muscle motoneurons, and the pontine interneurons projecting to the sacral cord moto-

neurons innervating bladder or bladder-sphincter also belong to the first system. The second system contains the neurons involved in the production of voluntary movements. The main example is the corticospinal tract in mammals, but the rubrospinal, pontine reticulospinal, vestibulospinal, tectospinal and interstitiospinal tracts also belong to this second system. The descending projections originating in the PAG belong to the third system. This system contains the pathways originating in the limbic system and related areas such as the caudal raphe nuclei. All the descending projections described in this chapter belong to this "third system".

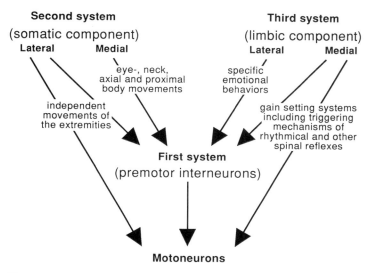

Figure 12. Schematic overview of the three subdivisions of the motor system. Reprinted with permission from Holstege (1991) Progress in Brain Research.

Abbreviations

AD: anterodorsal nucleus of the thalamus; AH: anterior hypothalamic area; AL: lateral amygdaloid nucleus; AM: anteromedial nucleus of the thalamus; Aq: aqueduct of Sylvius; AV: anteroventral nucleus of the thalamus; BC: brachium conjunctivum; BIC: brachium of the inferior colliculus; BL: basolateral amygdaloid nucleus; BM: basomedial amygdaloid nucleus; BNST: bed nucleus of the stria terminalis; BP: brachium pontis; CA: central amygdaloid nucleus ; Cd: caudate nucleus; CGL: lateral geniculate body; CGLd: lateral geniculate body (dorsal part); CGLv: lateral geniculate body (ventral part); CGM: medial geniculate body; CGMd: medial geniculate body, dorsal part; CGMint: medial geniculate body, interior division; CGMp: medial geniculate body, principal part; CI: capsula interna; C L: nucleus centralis lateralis of the thalamus; CM: centromedian thalamic nucleus; CN: cochlear nuclei; CO: cortical amygdaloid nucleus; CR: corpus restiforme; CS: superior colliculus; CSN: nucleus raphe centralis superior; CU: nucleus cuneatus; CUN: cuneiform nucleus; D: nucleus of Darkschewitsch; DH: dorsal hypothalamic area; DMH: dorsomedial hypothalamic nucleus; EC: external cuneate nucleus; ECU: external cuneate nucleus; En: entopeduncular nucleus; E W: nucleus Edinger-Westphal; F: fornix; fRF: fasciculus retroflexus; G: nucleus gracilis; GP: globus pallidus; Hab: habenular nucleus; HPA: posterior hypothalamus area; IC: inferior colliculus; IN: interpeduncular nucleus; I N C: interstitial nucleus of Cajal; IO: inferior olive; IVN: inferior vestibular nucleus; KF: nucleus Kölliker-Fuse; LD: nucleus lateralis dorsalis of the thalamus; LH: lateral hypothalamic area; LL: lateral lemniscus; LP: lateral posterior nucleus of the thalamus ; LRN: lateral reticular nucleus; LV: lateral ventricle; LVN: lateral vestibular nucleus; MA: medial amygdaloid nucleus; MB: mammillary body; MC: nucleus medialis centralis of the thalamus; MD: nucleus medialis dorsalis of the thalamus; MesV: mesencephalic trigeminal tract; ML: medial lemniscus; MLF: medial longitudinal fasciculus; motV: motor trigeminal nucleus; MVN: medial vestibular nucleus; M T N: medial terminal nucleus; NCL: nucleus centralis lateralis; NLL: nucleus of the lateral lemniscus; NOT: nucleus of the optic tract; NOTL: lateral nucleus of the optic tract; NOTM: medial nucleus of the optic tract; NPC: nucleus paracentralis of the thalamus; NRO: nucleus raphe obscurus; NTB: nucleus of the trapezoid body; NRA: nucleus retroambiguus; NRM: nucleus raphe magnus; NRP: nucleus raphe pallidus; NRTP: nucleus reticularis tegmenti pontis; NTS: nucleus tractus solitarius; nVII: facial nerve; OL: olivary pretectal nucleus; OT: optic tract; P: pyramidal tract; PAG: periaqueductal gray; PbL: lateral parabrachial nucleus; PC: pedunculus cerebri; PCN: nucleus of the posterior commissure; P H: periventricular hypothalamic nucleus; PON: pontine nuclei; PP: posterior pretectal nucleus; Pt: parataenial nucleus of the thalamus; PT: Probst tract; P T A: anterior pretectal nucleus; PTM: medial pretectal nucleus; Pu: putamen; Pul: pulvinar nucleus of the thalamus; PV: posterior paraventricular nucleus of the thalamus; PVA: paraventricular nucleus of the thalamus (anterior part); P V G: periventricular gray; PVN: paraventricular hypothalamic nucleus; R: reticular nucleus of the thalamus; RB: restiform body; RB: retractor bulbi motor nucleus; RE: nucleus reuniens of the thalamus; RFmed: medial reticular formation ; RFlat: lateral reticular formation; RiMLF: rostral interstitial nucleus of the MLF; RN: red nucleus; Rpo: nucleus raphe pontis; RST: rubrospinal tract; S: solitary complex; SC: suprachiasmatic nucleus; SC: nucleus subcoeruleus ; SM: stria medullaris; SN: substantia nigra; SO: superior olivary complex; SON: supraoptic nucleus; S T: subthalamic nucleus; STT: stria terminalis; SUB: subiculum; TMT: mammillothalamic tract ; VA: ventroanterior nucleus of the thalamus; VB: ventrobasal complex of the thalamus; VC: vestibular complex; VL: ventrolateral nucleus of the thalamus; VM: ventromedial nucleus of the thalamus; VPL: nucleus ventralis posterolateralis of the thalamus; VTA: ventral tegmental area of Tsai; VTN: ventral tegmental nucleus; ZI: zona incerta; III: oculomotor nucleus; IV: trochlear nucleus; Vm: motor trigeminal nucleus; Vn: trigeminal nerve; Vpr.: principal trigeminal nucleus; Vprinc.: principal trigeminal nucleus; Vsp.: spinal trigeminal complex; Vspin.caud.: spinal trigeminal complex pars caudalis; VI: abducens nucleus; VII: facial nucleus; Xd: dorsal nucleus of the vagal nerve; XII: hypoglossal nucleus.

References

Abols, I. A., and Basbaum, A. I., Afferent connections of the rostral medulla of the cat: a neural substrate for midbrain-medullary interactions in the modulation of pain, J. Comp. Neurol., 201 (1981) 285-297.

Akaike, A., Shibata, T., Satoh, M. and Takagi, H., Analgesia induced by microinjection of morphine into, and electrical stimulation of, the nucleus reticularis paragigantocellularis of rat medulla oblongata, Neuropharmacol. , 17 (1978) 775-778.

Alstermark, B., Kummel, H., Pinter, M. J. and Tantisira, B., Branching and termination of C3-C4 propriospinal neurones in the cervical spinal cord of the cat, Neurosci. Lett., 74 (1987) 291-296.

Bandler, R. and Carrive, P., Integrated defence reaction elicited by excitatory amino acid microinjection in the midbrain periaqueductal grey region of the unrestrained cat, Brain Res., 439 (1988) 95-106.

Bandler, R., Carrive, P. and Zhang, S.P., Integration of somatic and autonomic reactions within the midbrain periaqueductal grey: Viscerotopic, somatotopic and functional organization, Prog. Brain Res., 87 (1991) 269-305.

Basbaum, A. I., Clanton, C. H. and Fields, H. L., Three bulbospinal pathways from the rostral medulla of the cat: an autoradiographic study of pain modulating systems, J. Comp. Neurol., 178 (1978) 209-224.

Belin, M. F., Nanopoulos, D., Didier, M., Aguera, M., Steinbusch, H., Verhofstad, A., Maitre, M. and Pujol, J. -F., Immunohistochemical evidence for the presence of gamma-aminobutyric acid and serotonin in one nerve cell. A study on the raphe nuclei of rat using antibodies to glutamate decarboxylase and serotonin, Brain Res., 275 (1983) 329-339.

Besson, J-M. and Chaouch, A., Peripheral and spinal mechanisms of nociception, Physiological Reviews, 67 (1987) 67-186.

Bowker, R. M., Serotonergic and peptidergic inputs to the primate ventral spinal cord as visualized with multiple chromagens on the same tissue section, Brain Res., 375 (1986) 345-350

Bowker, R. M., Westlund, K. N., Sullivan, M. C. and Coulter, J. D., Organization of serotonergic projections to the spinal cord., Prog. Brain Res., 57 (1982) 239-265.

Bowker, R. M., Westlund, K. N., Sullivan, M. C., Wilber, J. F. and Coulter, J. D., Descending serotonergic, peptidergic and cholinergic pathways from the raphe nuclei: A multiple transmitter complex, Brain Res., 288 (1983) 33-48.

Bowker, R. M., Abbott, L. C. and Dilts, R. P. , Peptidergic neurons in the nucleus raphe magnus and the nucleus gigantocellularis: their distributions, interrelationships, and projections to the spinal cord, Prog. Brain Res., 77 (1988) 95-127.

Cardona, A. and Rudomin, P., Activation of brainstem serotoninergic pathways decreases homosynaptic depression of monosynaptic responses of frog spinal motoneurons, Brain Res., 280 (1983) 373-378.

Carrive, P., Bandler, R. and Dampney, R.A.L., Viscerotopic control of regional vascular beds by discrete groups of neurons within the midbrain periaqueductal gray, Brain Res., 493 (1989) 385-390.

Chan Palay, V., Jonsson, G. and Palay, S. L., Serotonin and substance P coexist in neurons of the rat's central nervous system, Proc. Natl. Acad. Sci., 75 (1978) 1582-1586.

Cowie, R. J. and Holstege, G., Dorsal mesencephalic projections to pons, medulla oblongata and spinal cord in the cat. Limbic and non-limbic components, Neuroscience, 1991, submitted.

Crone, C., Hultborn, H., Kiehn, O., Mazieres, L. and Wigstrom, H., Maintained changes in motoneuronal excitability by short-lasting synaptic inputs in the decerebrate cat, J. Physiol., 405 (1988) 321-343.

Dampney, R. A. L. and McAllen, R. M., Differential control of sympathetic fibres supplying hindlimb skin and muscle by retrofacial neurones in the cat, J. Physiol. (Lond.), 395 (1988) 41-56.

DeRosier, E. A., West, R.A. and Larson, C. R., Comparison of single unit discharge properties in the periaqueductal gray and nucleus retroambiguus during vocalization in monkeys, Soc. Neurosci. Abstr., 14 (1988) 1237.

Fields, H. L., Basbaum, A. I., Clanton, C. H. and Anderson, S. D., Nucleus raphe magnus inhibition of spinal cord dorsal horn neurons, Brain Res., 126 (1977) 441-453.

Garcia-Rill, E. and Skinner, R. D., The mesencephalic locomotor region. I. Activation of a medullary projection site, Brain Res., 411 (1987a) 1-12.

Garcia-Rill, E. and Skinner, R. D., The mesencephalic locomotor region. II. Projections to reticulospinal neurons, Brain Res., 411 (1987b) 13-20.

Garcia-Rill, E. and Skinner, R. D. , Modulation of rhythmic function in the posterior midbrain, Neuroscience, 27 (1988) 639-654.

Gebhart, G. F., Sandkühler, J., Thalhammer, J. G. and Zimmermann, M., Quantitative comparison of inhibition in spinal cord of nociceptive information by stimulation in periaqueductal gray or nucleus raphe magnus of the cat, J. Neurophysiol., 50 (1983) 1433-1445.

Hayes, N. L. and Rustioni, A., Descending projections from brainstem and sensorimotor cortex to spinal enlargements in the cat, Exp. Brain Res., 41 (1981) 89-107.

Helke, C. J., Sayson, S. C., Keeler, J. R. and Charlton, C. G., Thyrotropin-releasing hormone-immunoreactive neurons project from the ventral medulla to the intermediolateral cell column: partial coexistence with serotonin, Brain Res., 381 (1986) 1-7.

Hökfelt, T., Ljungdahl, A., Steinbusch, H., Verhofstad, A., Nilsson, G., Brodin, E., Pernow, B. and Goldstein, M., Immunohistochemical evidence of substance P-like immunoreactivity in some 5-hydroxytryptamine-containing neurons in the rat central nervous system, Neuroscience, 3 (1978) 517-538.

Hökfelt, T., Terenius, T., Kuypers, H. G. J. M. and Dann, O., Evidence for enkephalin immunoreactivity neurons in the medulla oblongata projecting to the spinal cord, Neurosci. Lett., 14 (1979) 55-61.

Hökfelt, T., Johansson, O. and Goldstein, M., Chemical neuroanatomy of the brain, Science, 225 (1984) 1326-1334.

Holstege, G., Some anatomical observations on the projections from the hypothalamus to brainstem and spinal cord: an HRP and autoradiographic tracing study in the cat, J. Comp. Neurol., 260 (1987) 98-126.

Holstege, G., Direct and indirect pathways to lamina I in the medulla oblongata and spinal cord of the cat, Prog. Brain Res., 77 (1988a) 47-94.

Holstege, G., Brainstem-spinal cord projections in the cat, related to control of head and axial movements, In: Neuroanatomy of the oculomotor system, Büttner-Ennever J. (Ed.), Elsevier, Amsterdam, 1988b, pp. 429-468.

Holstege, G., An anatomical study on the final common pathway for vocalization in the cat, J. Comp. Neurol., 284 (1989) 242-252.

Holstege, G., Subcortical limbic system projections to caudal brainstem and spinal cord, In: The human nervous system, Paxinos G. (Ed.), Academic Press, Sydney, 1990, pp. 261-284.

Holstege, G., Descending motor pathways and the spinal motor system. Limbic and non-limbic components, Prog. Brain Res. , 87 (1991) 307-421.

Holstege, G. and Kuypers, H. G. J. M., The anatomy of brain stem pathways to the spinal cord in the cat. A labeled amino acid tracing study, Prog. Brain Res., 57 (1982) 145-175.

Holstege, G., Kuypers, H. G. J. M. and Boer, R. C., Anatomical evidence for direct brain stem projections to the somatic motoneuronal cell groups and autonomic preganglionic cell groups in cat spinal cord, Brain Res., 171 (1979) 329-333.

Holstege, G., Meiners, L. and Tan, K., Projections of the bed nucleus of the stria terminalis to the mesencephalon, pons and medulla oblongata in the cat, Exp. Brain Res., 58 (1985) 379-391.

Holstege, J. C., Ultrastructural evidence for GABA-ergic brainstem projections to spinal motoneurons, Soc. Neurosci. Abstr., 15 (1989) 308.

Holstege, J. C. and Kuypers, H. G. J. M., Brain stem projections to spinal motoneuronal cell groups in rat studied by means of electron microscopy autoradiography, Prog. Brain Res., 57 (1982) 177-183.

Holstege, J. C. and Kuypers, H. G. J. M., Brainstem projections to lumbar motoneurons in rat. I. An ultrastructural study using autoradiography and the combination of autoradiography and horseradish peroxidase histochemistry, Neuroscience, 21 (1987) 345-367.

Hosobuchi, Y., Current issues regarding subcortical electrical stimulation for pain control in humans, Prog. Brain Res., 77 (1988) 189-192.

Hounsgaard, J., Hultborn, H., Jespersen, B. and Kiehn, O., Intrinsic membrane properties causing a bistable behavior of α-motoneurons, Exp. Brain Res., 55 (1984) 391-394.

Hounsgaard, J., Hultborn, H. and Kiehn, O., Transmitter-controlled properties of α-motoneurones causing long-lasting motor discharge to brief excitatory inputs, Prog. Brain Res., 64 (1986) 39-49.

Hounsgaard, J., Hultborn, H., Jespersen, B. and Kiehn, O., Bistability of alpha-motoneurons in the decerebrate cat and in the acute spinal cat after intravenous 5-hydroxytryptophan, J. Physiol., 405 (1988) 345-367.

Huisman, A. M., Kuypers, H. G. J. M. and Verburgh, C. A., Differences in collateralization of the descending spinal pathways from red nucleus and other brain stem cell groups in cat and monkey, Prog. Brain Res., 57 (1982) 185-217.

Hunt, S. P. and Lovick, T. A., The distribution of serotonin, met-enkephalin and β-lipotropin-like immunoreactivity in neuronal perikarya of the cat brainstem, Neurosci. Lett., 30 (1982) 139-145.

Johannessen, J. N., Watkins, L. R. and Mayer, D. J., Non-serotonergic origins of the dorsolateral funiculus in the rat ventral medulla, J. Neurosci., 4 (1984) 757-766.

Johansson, O., Hökfelt, T., Pernow, B., Jeffcoate, S. L., White, N., Steinbusch, H. W. M., Verhofstad, A. A. J., Emson, P. C. and Spindel, E., Immunohistochemical support for three putative transmitters in one neuron: coexistence of 5-hydroxytryptamine, substance P- and thyrotropin releasing hormone-like immunoreactivity in medullary neurons projecting to the spinal cord, Neuroscience, 6 (1981) 1857-1881.

Jones, B. E. and Beaudet, A., Distribution of acetylcholine and catecholamine neurons in the cat brainstem: A choline acetyltransferase and tyrosine hydroxylase immunohistochemical study, J. Comp. Neurol., 261 (1987) 15-32.

Jones, S. L. and Gebhart, G. F., Spinal pathways mediating tonic, coeruleospinal and raphe-spinal descending inhibition in the rat, J. Neurophysiol., 58 (1987) 138-159.

Jürgens, U. and Pratt, R., The cingular vocalization pathway in the squirrel monkey, Exp. Brain Res., 34 (1979) 499-510.

Kojima, M., Takeuchi, Y., Goto, M. and Sano, Y., Immunohistochemical study on the distribution of serotonin fibers in the spinal cord of the dog, Cell Tissue Res., 226 (1982) 477-491.

Kojima, M., Takeuchi, Y., Goto, M. and Sano, Y., Immunohistochemical study on the localization of serotonin fibers and terminals in the spinal cord of the monkey (Macaca fuscata), Cell Tissue Res., 229 (1983a)23-36.

Kojima, M., Takeuchi, Y., Kawata, M. and Sano, Y., Motoneurons innervating the cremaster muscle of the rat are characteristically densely innervated by serotonergic fibers as revealed by combined immunohistochemistry and retrograde fluorescence DAPI-labeling, Anat. Embryol. , 168 (1983b) 41-49.

Kow, L. -M, Montgomery, M. O. and Pfaff, D. W., Effects of spinal cord transections on lordosis reflex in female rats, Brain Res., 123 (1977)75-88.

Krnjević, K. and Schwartz, S., Is gamma-aminobutyric acid an inhibitory transmitter?, Nature , 211 (1966) 1372-1374.

Kuypers, H. G. J. M. and Maisky, V. A., Retrograde axonal transport of horseradish peroxidase from spinal cord to brain stem cell groups in the cat, Neurosci. Lett. , 1 (1975) 9-14.

Léger, L., Charnay, Y., Dubois, P. M. and Jouvet, M., Distribution of enkephalin-immunoreactive cell bodies in relation to serotonin-containing neurons in the raphe nuclei of the cat: immunohistochemical evidence for the coexistence of enkephalins and serotonin in certain cells, Brain Res., 362 (1986) 63-73.

Light, A. R., Casale, E. J. and Menetrey, D. M., The effects of focal stimulation in nucleus raphe magnus and periaqueductal gray on intracellularly recorded neurons in spinal laminae I and II, J. Neurophysiol. , 56 (1986) 555-571.

Lovick, T. A., Differential control of cardiac and vasomotor activity by neurones in nucleus paragigantocellularis lateralis in the cat, J. Physiol. (Lond.), 389 (1987) 23-35.

Lovick, T. A. and Robinson, J. P., Bulbar raphe neurones with projections to the trigeminal nucleus caudalis and the lumbar cord in the rat: A fluorescence double-labelling study, Exp. Brain Res., 50 (1983) 299-309.

Mantyh, P. W., Connections of midbrain periaqueductal gray in the monkey. II. Descending efferent projections, J. Neurophysiol., 49 (1983) 582-595.

Mantyh, P. W. and Hunt, S. P., Evidence for cholecystokinin-like immunoreactive neurons in the rat medulla oblongata which project to the spinal cord, Brain Res., 291 (1984) 49-54.

Martin, G. F., Cabana, T. and Humbertson, A. O. Jr., Evidence for collateral innervation of the cervical and lumbar enlargements of the spinal cord by single reticular and raphe neurons. Studies using fluorescent markers in double-labelling experiments on the North American opossum, Neurosci. Lett., 24 (1981a) 1-6.

Martin, G. F., Cabana, T., Humbertson, A. O. Jr., Laxson, L. C. and Pannetion, W. M., Spinal projections from the medullary reticular formation of the North American Opossum: Evidence for connectional heterogeneity, J. Comp. Neurol., 196 (1981b) 663-682.

Martin, G. F., Humbertson, A. O. Jr., Laxson, L. C., Panneton, W. M. and Tschismadia, I., Spinal projections from the mesencephalic and pontine reticular formation in the north american opossum: A study using axonal transport techniques, J. Comp. Neurol., 187 (1979) 373-401.

Martin, G. F., Vertes, R. P. and Waltzer, R., Spinal projections of the gigantocellular reticular formation in the rat. Evidence for projections from different areas to laminae I and II and lamina IX, Exp. Brain Res., 58 (1985) 154-162.

McCall, R. B. and Aghajanian, G. K., Serotonergic facilitation of facial motoneuron excitation, Brain Res., 169 (1979) 11-29.

Meyerson, B. A., Problems and controversies in PVG and sensory thalamic stimulation as treatment for pain, Prog. Brain Res., 77 (1988) 175-188.

Millhorn, D. E., Hökfelt, T., Seroogy, K. and Verhofstad, A. A. J., Extent of colocalization of serotonin and GABA in neurons of the ventral medulla oblongata in rat, Brain Res., 461 (1988) 169-174.

Modianos, D. and Pfaff, D.W., Medullary reticular formation lesions and lordosis reflex in female rats, Brain Res., 171 (1979) 334-338.

Moon Edley, S. and Graybiel, A. M., The afferent and efferent connections of the feline nucleus tegmenti pedunculopontine, pars compacta, J. Comp. Neurol., 217 (1983) 187-216.

Nauta, W. J. H., Hippocampal projections and related neural pathways to the mid-Brain in the cat, Brain, 80 (1958) 319-341.

Nicoll, R. A., The coupling of neurotransmitter receptors to ion channels in the brain, Science, 241 (1988) 545-551.

Pelletier, G., Steinbusch, H. W. M. and Verhofstad, A. A. J., Immunoreactive substance P and serotonin present in the same dense core vesicles, Nature, 293 (1981) 71-72.

Pfaff, D.W., Estrogens and brain function. Neuronal analysis of a hormone-controlled mammalian reproductive behavior, Springer-Verlag, Berlin, 1980.

Pfaff, D. W. and Sakuma, Y., Deficit in the lordosis reflex of female rats caused by lesions in the ventromedial nucleus of the hypothalamus, J. Physiol., 288 (1979a) 203-211.

Pfaff, D. W. and Sakuma, Y., Facilitation of the lordosis reflex of female rats from the ventromedial nucleus of the hypothalamus, J. Physiol., 288 (1979b) 189-203.

Redgrave, P., Dean, P., Mitchell, I.J., Odekunle, A. and Clark, A., The projection from the superior colliculus to cuneiform area in the rat. I. Anatomical studies, Exp. Brain Res., 72 (1988) 611-625.

Rose, J.D. and Flynn, F.W., Lordosis can be elicited in chronically-decerebrate rats by combined lumbosacral and vagino-cervical stimulation, Soc. Neurosci. Abstr., 15 (1989) 1100.

Sakuma, Y. and Pfaff, D. W., Mesencephalic mechanisms for integration of female reproductive behavior in the rat, Am. J. Physiol., 237 (1979a) R285-R290.

Sakuma, Y. and Pfaff, D. W., Facilitation of female reproductive behavior from mesencephalic central gray in the rat, Am. J. Physiol., 237 (1979b) R278-R284.

Sandkühler, J. and Gebhart, G. F., Characterization of inhibition of a spinal nociceptive reflex by stimulation medially and laterally in the midbrain and medulla in the pentobarbital-anesthetized rat, Brain Res., 305 (1984) 67-76.

Shik, M. L., Severin, F. V. and Orlovski, G. N., Control of walking and running by means of electrical stimulation of the mid-brain, Biophysics, 11 (1966) 756-765.

Steinbusch, H. W. M., Distribution of serotonin-immunoreactivity in the central nervous system of the rat: cell-bodies and terminals, Neuroscience, 6 (1981) 557-618.

Taber-Pierce, E., Lichtenstein, E. and Feldman, S. C., The somatostatin systems of the guinea-pig brainstem., Neuroscience, 15 (1985) 215-235.

Tohyama, M., Sakai, K., Salvert, D., Touret, M. and Jouvet, M., Spinal projections from the lower brain stem in the cat as demonstrated by the horseradish peroxidase technique. I. Origins of the reticulospinal tracts and their funicular trajectories, Brain Res., 173 (1979) 383-405.

Ulfhake, B., Arvidsson, U., Cullheim, S., Hokfelt, T., Brodin, E., Verhofstad, A. and Visser, T., An ultrastructural study of 5-hydroxytryptamine-, thyrotropin-releasing hormone- and substance P-immunoreactive axonal boutons in the motor nucleus of spinal cord segments L7-S1 in the adult cat, Neuroscience, 23 (1987) 917-929.

VanderMaelen, C. P. and Aghajanian, G. K., Serotonin-induced depolarization of rat facial motoneurons in vivo: comparison with amino acid transmitters, Brain Res., 239 (1982) 139-152.

Wessendorf, M. W. and Elde, R., The coexistence of serotonin- and substance P-like immunoreactivity in the spinal cord of the rat as shown by immunofluorescent double labeling, J. Neurosci., 7 (1987) 2352-2363.

White, S. R. and Neuman, R. S., Facilitation of spinal motoneurone excitability by 5-hydroxytryptamine and noradrenaline, Brain Res., 188 (1980) 119-127.

Willis, W. D., Anatomy and physiology of descending control of nociceptive responses of dorsal horn neurons: comprehensive review, Prog. Brain Res., 77 (1988) 1-29.

Induction of the Proto-Oncogene *c-fos* as a Cellular Marker of Brainstem Neurons Activated from the PAG

Jürgen Sandkühler

II. Physiologisches Institut, Universität Heidelberg
Heidelberg, Germany

Introduction

The diverse functions of the midbrain periaqueductal gray (PAG) have been characterized in behavioral, cardiovascular and electrophysiological studies. The widespread ascending and descending efferent connections provide the anatomical substrates for the diversity of output functions of PAG neurons (Mantyh, 1983a; 1983b; Holstege, this volume). However, little is known about which target sites are activated by efferents originating from the PAG or the number of neurons involved, i.e., about the functional, cellular substrate of the output functions of the PAG.

Electrophysiological experiments have provided detailed, real time information about the effects of PAG stimulation on single neurons outside the PAG (e.g., Maciewicz et al., 1984; Moreau and Fields, 1986; Kai et al., 1988; Mason, this volume). But, by the nature of the single cell recording techniques, comprehensive maps of activated neurons cannot be provided. In an early study by Beitz and Buggy (1981) the uptake of [³H]-2-deoxyglucose (2-DG) was used to map brain regions which were activated during electrical stimulation in the PAG. At that time, the technical limits of the 2-DG technique did not permit monitoring neuronal activity at the single cell level. Using electrical stimulation all neuronal elements in the area are inevitably activated including fibers of passage which transmit action potentials both ortho- and antidromically. Further, electrical stimulation also activates efferent and afferent projections of the PAG.

In the present study, the question of which brainstem neurons are activated by PAG efferents has been addressed. We have used the expression of the proto-oncogene *c-fos* as a cellular marker for activated neurons throughout the brainstem (Sagar et al., 1988; Dragunow and Faull, 1989). The gene product, nuclear phosphoprotein FOS, was detected by immunohistochemistry.

The Midbrain Periaqueductal Gray Matter, Edited by A. Depaulis and
R. Bandler, Plenum Press, New York, 1991

For the purpose of the present study PAG stimulation should satisfy the following requirements. It should selectively increase the firing rate of neurons of origin, recruit as many efferent neurons as possible, and last for at least 10 to 20 minutes to allow induction of FOS in most activated neurons. Finally, the method of stimulation should have been proven to be efficacious in behavioral and in electrophysiological experiments.

Gamma-aminobuturic acid (GABA) is a major inhibitory neurotransmitter in the PAG (Sandner et al., 1981; Barbaresi and Manfrini, 1988; Behbehani et al., 1990) and blockade of GABA$_A$ receptors by bicuculline may increase background activity of most neurons in the PAG, both, *in vivo* and *in vitro* (Behbehani et al., 1990; Ogawa et al., this volume). Microinjections of bicuculline at pmol doses have been shown to maximally activate descending antinociceptive systems (Sandkühler et al., 1989; 1991) and induce strong changes in behavior (Jacquet et al., 1987; DiScala et al., 1984).

Here, we report that bicuculline microinjections into dorsal parts of the PAG induce expression of the proto-oncogene *c-fos* in neurons in well defined areas throughout the brainstem of the rat.

Methods

Experiments were performed on male Sprague-Dawley rats weighing 220 to 380 g. Animals were housed in groups of two in one Makrolon type III cage in an air conditioned room (temperature 22 ± 2 °C, water saturation: 55 ± 5 %) with a 12h light-dark cycle, light on at 7:00 a.m.. Food pellets and water were given *ad libitum*. The animals were allowed to adapt to this environment for at least seven days before the experiments were performed. At the day of the experiment anesthesia was initiated between 8:00 and 10:00 a.m. in the animal room. The animals were placed into a transparent box with the air saturated with halothane gas. Induction of anesthesia lasted for about 10 s. Under halothane anesthesia the animals were removed from the box and pentobarbital was injected intraperitoneally at a dose of 60 mg/kg. The animals were then transported to the laboratory while being under deep pentobarbital anesthesia. A surgical level of anesthesia was maintained throughout the course of the experiment by hourly supplemental i.p. bolus injections of 10 mg/kg pentobarbital. Rectal temperature was measured with a small probe (2 mm in diameter) and kept constant at 37.5 ± 0.5 °C by a feedback controlled heating blanket underneath the ventral surface of the animals. The eyes were covered by sterile Vaseline to prevent drying. Three experimental protocols were employed:

1) Fifteen to 30 min after the induction of anesthesia, unoperated control animals were killed by an overdose of pentobarbital i.p. and were then transcardially perfused with phosphate buffered saline followed by ice-cold 4% paraformaldehyde (n=3).

2) Sham treated animals were placed in a stereotaxic frame with the dorsal surface of the skull in a horizontal position. Care was taken not to apply excessive pressure and not to perforate the tympanic membrane. The skin overlying

the temporal bones was shaved and a gel containing the local anesthetic lidocaine was applied to the skin which was then incised sagitally. Lidocaine was then also applied to the wound margins. A craniotomy was performed in the left temporal bone with an electric drill and the dura mater was incised para-sagitally to allow insertion of a fine multibarrel glass probe (tip diameter < 40 μm, see Sandkühler et al., 1991) 5-7.0 mm caudal to bregma (i.e., 4-2 mm anterior to interaural line) , 1.5 mm lateral to the midline, and with a medio-lateral angle of 10°. The atlas of Paxinos and Watson (1982) was used. The tip of the probe was lowered with an electronically controlled microstep motor into the dorsal PAG, 4500 μm below the surface of the cortex. Pressure microinjections of 50 nl of a 0.9% saline solution were used in three sham treated animals.

3) In two PAG stimulated animals, 200 pmoles of bicuculline methiodide (Sigma, Deisenhofen, FRG) (in 50 nl 0.9 % saline, pH adjusted to 2.5 with HCL) was microinjected into the dorsal part of the rostral PAG (Fig. 2C) and the dorsal part of the intermediate PAG (Fig. 3E). Following bicuculline injection, motor effects such as movements of the whiskers, tachypnea, abdominal or facial muscle contractions or movements of the tail were observed. To mark injection sites, 100-200 nl of a saturated fast green dye solution were injected immediately before killing the animals.

The animals were killed by an overdose of pentobarbital given i.p. 90 min after the injections and were then perfused transcardially with phosphate buffered saline followed by ice cold 4% paraformaldehyde. The brains were removed and postfixed overnight in paraformaldehyde and then stored for at least 48h in 30% sucrose for cryoprotection. The brains were cut in a cryostat in 50 μm coronal sections and the sites of fast green dye deposition were located with reference to the atlas of Paxinos and Watson (1982). Free floating sections were incuba-ted with normal goat serum (2% in phosphate buffered saline and 0.2% Triton X-100) for 1 h, followed by the primary antiserum at 1:6,000 for 24 to 48 hr. The polyclonal rabbit antibody was raised against bacterial expressed fusion protein and was generously provided by Dr. R. Bravo, The Squibb Inst. Medical Research, Princeton, U.S.A.. After washing, the sections were incubated in biotinylated goat anti-rabbit antiserum followed by an avidin-peroxidase complex (Vectastain, Vector Laboratories) for 1 h. The sections were then developed in 0.02 % diaminobenzidine with 0.02 % hydrogen peroxide and intensified by addition of 0.02% cobalt chloride and nickel ammonium sulfate.

Results and Discussion

FOS as a cellular marker of activated neurons

To induce FOS immunoreactivity (IR) in neurons, a strong mono- or poly-synaptic activation is effective (Sagar et al., 1988), probably via modification of the open time of voltage-dependent calcium channels (Morgan and Curran, 1986). Hyperpolarizing, i.e., inhibitory agents or conditions have not been reported to induce FOS. In addition to neurons, glial cells, may also express c-fos (Hisanaga et al., 1990), e.g., in response to heat-shock (Dragunow et al., 1989) or following cortical trauma (Dragunow and Robertson, 1988) but not following depolarization (Hisanaga et al., 1990). Here, we have not attempted to prove that all FOS positive cells were neurons. The number and the pattern of stained cells suggest, however, that the vast majority were indeed neurons.

At many brain regions which are activated following stimulation, both FOS-IR and 2-DG uptake are increased. At some sites however, a mismatch was reported which could be due to the fact that 2-DG uptake but not FOS-IR may detect axonal and dendritic activity in addition to the activity of the cell body (Sagar et al., 1988; Dragunow and Faull, 1989). Negative results obtained by the c-fos but not by the 2-DG technique could also indicate that some neurons never express c-fos no matter what stimulus is applied (Dragunow and Faull, 1989) or that stimulation occurred during a 'refractory period' for the induction of c-fos (Morgan et al., 1987). In contrast, FOS immunostaining may also detect activated neurons in some regions of the brain, e.g., in the hypothalamic magnocellular nuclei, which did not show an increased 2-DG uptake upon stimulation (Sagar et al., 1988). Pentobarbital, the general anesthetic used in the present study does not induce FOS by itself but may rather depress its induction (Morgan et al., 1987; Ménétrey et al., 1989). Here, the basal expression of c-fos in unoperated control animals was very low or absent in most brain regions. The functions of the nuclear FOS protein are not known but FOS-associated nuclear proteins bind to DNA (Sassone-Corsi et al., 1988) and it has been suggested that c-fos and its coregulated genes are excellent candidates for coupling excitation of neurons to long term adaptive modifications of transcription (Morgan and Curran, 1986).

Bicuculline microinjections into the PAG

To selectively excite neurons originating from a circumscribed area of the central nervous system, microinjections of excitatory amino acids are a useful tool (Goodchild et al., 1982). At some sites glutamate applied by iontophoresis or by microinjection may, however, fail to excite a significant proportion of neurons (Schneider and Perl, 1985) or may have mixed effects with a short-lasting firing rate increase followed by a long-lasting decrease in firing rate (Lipski et al., 1988). In the PAG, an alternative method of excitation of neurons of origin but not fibers of passage is disinhibition by application of a $GABA_A$ receptor antagonist such as bicuculline (Behbehani et al., 1990). Of course, it is possible that not all neurons in the PAG and not all efferent functions of the PAG are affected by blockade of $GABA_A$-receptors and consequently, bicuculline microinjections might not recruit all efferents at the injection site.

However, it was demonstrated recently that bicuculline largely increases background activity of most neurons in the PAG, both *in vivo* and *in vitro* (Behbehani et al., 1990). Blockade of $GABA_A$ receptors in the PAG was found to strongly modify behavior, e.g., induce analgesia (Moreau and Fields, 1986) elicit defensive or explosive motor behavior (DiScala et al., 1984; 1989; Depaulis and Vergnes, 1986; Jacquet et al., 1987; see also Bandler, 1988 for a recent review). The effects of bicuculline in the PAG are most likely due to specific blockade of $GABA_A$ receptors, as application of other $GABA_A$ receptor antagonists such as picrotoxin have been shown to produce qualitatively identical effects (e.g.,

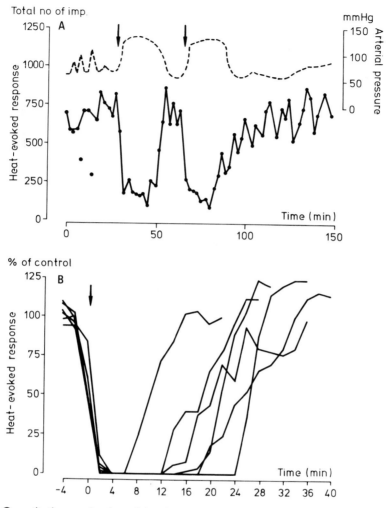

Figure 1. Quantitative evaluation of the time course of two different output functions of the PAG, triggered by the microinjection of 200 pmol bicuculline. A: Responses of one spinal dorsal horn neuron to noxious radiant heating (50°C, 10 s) of the glabrous skin at the ipsilateral hindpaw are expressed as total number of action potentials in 18s. The responses which were evoked in intervals of 2 min are plotted as closed circles connected by a solid line on the ordinate versus time. Bicuculline was microinjected two times into the same site of the PAG, as indicated by the arrows at the top. Heat-evoked responses during electrical stimulation (0.1 ms cathodal pulses given at 100 Hz) at the same site in the dorsolateral PAG at 100, 200 and 400 µA are plotted as closed circles which are not connected. The pressure in one carotid artery was averaged for 18 s during heat-evoked responses and plotted as a broken line at the top (right hand scale). B: The time courses of descending inhibition of six different neurons in six different rats following microinjections of 200 pmol bicuculline into the PAG at time zero are overlayed for comparison. Heat-evoked responses which are expressed as percent of control values were totally depressed and recovered to control values within 16 to 30 min after the injection. Modified from Sandkühler et al., 1991.

Depaulis and Vergnes, 1986; Moreau and Fields, 1986). Further, the effects of bicuculline in the PAG can be antagonized by GABA and vice versa (Behbehani et al., 1990).

We have shown recently that in the pentobarbital anesthetized rat, microinjections at many sites within the PAG of 200 pmoles bicuculline is more effective in producing descending inhibition and changes in arterial blood pressure than microinjection of 10 to 50 nmoles glutamate (Sandkühler et al., 1991). To determine the time course and the efficacy of bicuculline microinjections into the PAG we have quantitatively assessed the effect on two different output functions of the PAG: 1) changes in mean arterial blood pressure and 2) the efficacy of descending inhibition of nociceptive spinal dorsal horn neurons (Sandkühler et al., 1991; Fig. 1A). Following microinjection of 200 pmol bicuculline, noxious heat-evoked responses were totally depressed in six spinal dorsal horn neurons (Fig. 1B). Responses recovered to control values within 16 to 30 min. The time course of inhibition was almost identical to the time course of rise in blood pressure (Fig. 1A) and to the duration of contractions of facial, abdominal or tail muscles. Thus, with respect to the duration of neuronal activation, microinjections of 200 pmol bicuculline are suitable for the purpose of the present study, as neuronal activation for a comparable period of time has been shown to strongly induce c-fos in many neurons of the central nervous system (Morgan and Curran, 1986; Morgan et al., 1987; Sagar et al., 1988).

Expression of c-fos in brainstem neurons

BACKGROUND FOS-IR. The number and the location of cells with FOS-IR in "control" animals largely depends on the immediate past history approximately 6 to 0.5 h prior to the perfusion. Animals which were deeply anesthetized in the animal house as described under Methods and perfused 15-30 min later had no, or only very faint background FOS-IR in some brain regions, especially in the superior colliculus.

In sham injected animals the surgery and especially the trauma of the cerebral cortex at the site of pipette penetration induced additional c-fos expression throughout the ipsilateral cortex, mainly in the primary olfactory cortex and the Islands of Cajal, possibly due to a spreading depression (Dragunow and Robertson, 1988). Strong labeling was also found in the habenular nuclei and moderate FOS-IR was observed bilaterally in the lateral reticular nucleus and the median raphe nucleus. Most nuclei, including the PAG, which contained many labeled cells following PAG stimulation had few or no FOS-IR in sham treated animals. Thus, the quick induction of anesthesia within 10 s with halothane and the surgery under a deep level of pentobarbital anesthesia did not induce FOS in most neurons of the brainstem. We observed that in animals which were exposed to unspecific, eventually stressful stimuli (transporting naive, awake animals to the laboratory, intraperitoneal drug injections), FOS-IR was strongly enhanced in numerous brainstem areas, including the PAG, the superior colliculus, the solitary tract nucleus and the medial hypothalamus. This distribution of FOS-IR overlapped with the pattern of FOS-IR following bicuculline microinjection into the dorsal PAG. Thus, without a detailed description of the immediate past history of the animals, especially the procedure of handling, drug application or control of body temperature (heat-shock), no firm conclusions can be drawn from the background labeling on a basal, constitutive expression of *c-fos*.

c-fos EXPRESSION IN RESPONSE TO PAG STIMULATION. FOS-IR was interpreted as a response to the microinjection of bicuculline into the PAG if labeling was clearly above IR in sham treated animals. All figures of FOS-IR shown here were taken from the same animal (rat A in Table I) which received an injection of bicuculline into the dorsal part of the rostral PAG (Fig. 2A,C). The approximate positions of the photomicrographs are indicated for most figures on standard frontal sections taken from the atlas of Paxinos and Watson (1982). In another rat (rat B) which also received a dorsal injection but approximately 1.5 mm further caudal in the intermediate third of the PAG (Fig. 3E) the pattern of labeling was similar (Table I).

Following microinjection of bicuculline into the dorsal part of the rostral PAG FOS-IR was detected in cells throughout the rostro-caudal extent of the PAG, extending rostrally into the periventricular grey (Fig. 8B). At the level of the rostral PAG (Fig. 2 A,B), labeled neurons were concentrated bilaterally and dorsally, while relatively few labeled cells were found in the lateral and ventrola-

Table I

	Animal A (Rostral PAG)	Animal B (intermediate PAG)
Solitary tract		
Nucleus (comm.)	+++	+++
(rostr.)	+++	+++
Ventrolateral medulla	++	+++
Medial vestibular n.	+	++
Locus coeruleus	+++	+
Raphe pallidus	++	+
Raphe magnus	+	+
NRPGα	++	+
PAG (rostr.)		
dorsal	+++	++
ventral	+	++
PAG (caudal)		
dorsal	+++	++
ventral	+++	++
Superior colliculus	+++	+
Periventricular gray	++	0
Zona incerta	+++	+++
Periventricular hypoth. n.	+	+
Dorsal hypoth.	++	++
Suprachiasmatic n.	++	0
Supraoptic n.	++	++

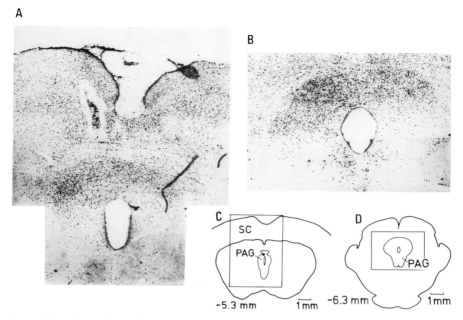

Figure 2. A: Frontal section through the midbrain at the level of the commissure of the superior colliculus, 5.3 mm caudal to bregma (i.e., 3.7 mm from interaural line). The pipette track is visible in the superior colliculus to the left of the midline. FOS-IR is strong bilaterally in the superior colliculus and the dorsal but not ventral PAG. B: FOS positive cells are numerous bilaterally in the dorsal PAG and overlaying SC 6.3 mm caudal to bregma (i.e., 2.7 mm from interaural line). Relatively few cells in the ventral PAG were also labeled. In C and D the injection site and the approximate positions of the photomicrographs are shown on standard frontal section through the midbrain.

teral PAG. At more caudal levels labeled cells were also found within the lateral and ventrolateral divisions of the PAG (Fig. 3A,C), extending into the nucleus cuneiformis (Fig. 3 A) which receives a direct projection from the dorsolateral PAG (Mantyh, 1983a; Redgrave and Dean, this volume). A dorso-ventral gradient of activated cells in the PAG is consistent with reports stressing functional differences induced by dorsal versus ventral stimulation in the PAG (e.g., Fardin et al., 1984; Lovick; Morgan; Bandler and Depaulis; Carrive, this volume) or stimulation of the dorsal raphe (Cannon et al., 1982; Prieto et al., 1983).

If an injection strongly activates output functions of the PAG, e.g., maximally activates descending inhibition or induces large changes in blood pressure, the injection site could be characterized 1) by the presence of highly effective efferents leaving the PAG at that site and/or 2) by the presence of local interneurons with projections within the PAG which could recruit efferents from multiple sites of the PAG. Our finding that activated neurons were identified throughout the entire rostro-caudal extension of the PAG clearly favours the latter hypothe-

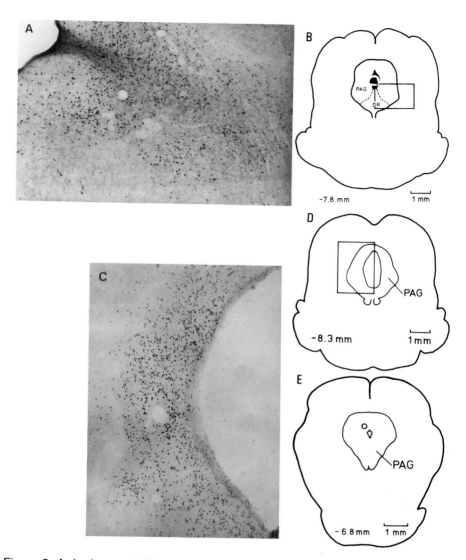

Figure 3. A: In the caudal PAG, approx. 7.8 mm posterior to bregma (i.e., 1.2 mm from interaural line), FOS-IR was found bilaterally in the ventrolateral quadrant. The approximate position of the photomicrographs is indicated on the schematic diagram of a frontal section 7.8 mm caudal to bregma in B. In C a frontal section through the left PAG 8.3 mm caudal to bregma is shown (i.e., 0.7 from interaural line). At this caudal level, labeled cells are found both in dorsal and in ventral aspect of the PAG bilaterally. In E the site of bicuculline injection (open circle) in the other of the two stimulated rats (rat B) is shown.

sis and may explain why brainstem areas for which there is no anatomical evidence of direct projections, could nevertheless be activated. Increased FOS-IR was also found at brainstem sites outside the PAG, as described below in a caudal to rostral order:

The solitary tract nuclei (Sol) contained numerous labeled cells bilaterally (Fig. 4 A and C). Anatomical data have shown that the Sol receive a direct projection from the lateral PAG (Bandler and Tork, 1987) and our results suggest that at least some of the PAG-Sol projections are excitatory. Beitz and Buggy (1981) have found a moderate increase (+14 %) in 2-DG glucose uptake in this area following electrical stimulation in the PAG. The PAG-induced activation of Sol cells may be involved in the modulation of cardiovascular function and nociception (Randich et al., 1988; Carrive, this volume) induced by PAG stimulation.

FOS-IR was found bilaterally in cells of the caudal ventrolateral medulla including the lateral reticular nucleus (LRN) (Fig. 4 B). The caudal ventrolateral medulla receives a direct projection from the PAG (Mantyh, 1983b) and it has been shown that DLH injection in this region evokes a fall in blood pressure

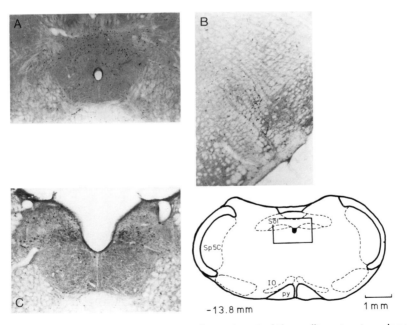

Figure 4. FOS-IR in cells of the commissural part of the solitary tract nucleus (A) and in the ventrolateral medulla contralateral to the injection site (B) on a frontal section 13.8 mm caudal to bregma. Similar IR was found at the ipsilateral site. In C labeled cells are shown in the Sol 13.3 mm caudal to bregma (i.e., 4.3 caudal to interaural line). At further rostral levels no significant staining was found in the Sol.

(Bonham and Jeske, 1988). Stimulation in the LRN may modulate cardiovascular function and produce descending antinociception (Janss et al., 1987). Evidence has been provided in the cat that the LRN might be a relay station for PAG-induced descending inhibition (Morton et al., 1984). Labeling more laterally in the rostral ventrolateral medulla (i.e., in the nucleus paragigantocellularis) was weak and variable both in the stimulated and in control animals.

Labeled cells were consistently found in the **medial vestibular nuclei** (MVe; Fig. 5), both in sham treated animals and following PAG stimulation. The functional role of these projections remains unclear.

Densely stained neurons were identified bilaterally in the **locus coeruleus** (LC) (Fig. 6), a region which might receive direct projection from the PAG (Mantyh, 1983b), see, however, Aston-Jones et al. (1986). Surprisingly, electrical stimulation in the PAG did not significantly increase 2-DG uptake in the LC (Beitz and Buggy, 1981). Noradrenaline containing spinopetal LC neurons are

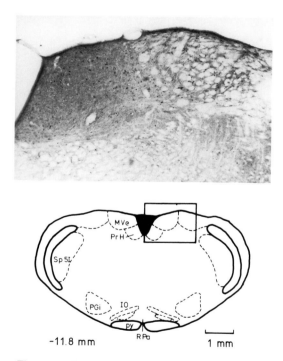

Figure 5. FOS-IR was found bilaterally in the medial vestibular nucleus (MVe). Here, the labeling contralateral to the injection site in the PAG is shown.

excellent relay candidates for descending antinociceptive effects of PAG stimulation (Mokha et al., 1985; Jones and Gebhart, 1986), as alpha-adrenergic receptors have been shown to mediate part of PAG-induced descending inhibition of the nocifensive tail flick reflex (Jensen and Yaksh, 1984; Aimone et al., 1987). However, blockade of spinal monoaminergic receptors failed to affect the descending inhibition of multireceptive spinal dorsal horn neurons (Sandkühler and Zimmermann, 1988). Possibly, noradrenergic LC neurons are also involved in 'stress' responses (Abercrombie and Jacobs, 1987) which may be part of the complex behavioral patterns elicited by stimulation in the lateral PAG (see Bandler and Depaulis; Besson; Fanselow, this volume).

The nucleus raphe magnus (NRM) and the neighboring nucleus reticularis paragigantocellularis pars alpha (NRPGα) have attracted much attention as a relay station for stimulation-produced descending inhibition induced from the PAG (Fields and Basbaum, 1978; see also Mason, this volume), as both areas

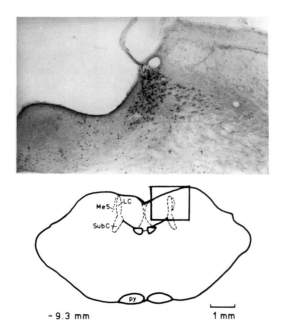

Figure 6. Densely stained cells were found in large numbers bilaterally in the locus coeruleus. Here, the staining contralateral to the injection site is shown.

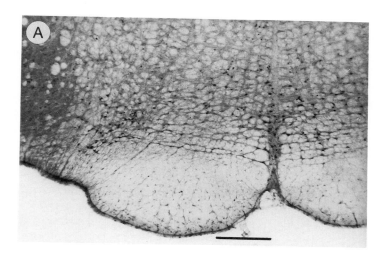

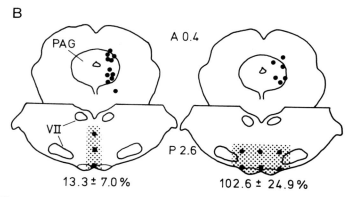

Figure 7. A: In the ventral medulla, FOS-IR was found in cells medially in the nucleus raphe magnus and the nucleus raphe pallidus and bilaterally in the nucleus reticularis paragigantocellularis pars alpha (Scale bar: 500 μm). Some of these activated cells may mediate PAG-induced descending inhibition of the spinally mediated nociceptive tail flick reflex, as illustrated in part B: here lidocaine was microinjected either into medial parts of the ventral medulla (left hand section) or into medial and lateral parts (right hand section) to produce reversible blocks of neuronal activity in the stuppled areas. During medial blocks the threshold of electrical stimulation in the PAG for descending inhibition did not increase significantly (numbers given at the bottom), while blocks medially and bilaterally significantly elevated thresholds for inhibition by more than 100 %.

receive direct projection from the PAG (Mantyh, 1983b; Beitz and Williams, this volume) and project directly to the spinal cord (Basbaum et al., 1978). Simultaneous bilateral local anesthetic blocks of the NRM and the NRPGα increased the threshold for PAG stimulation to inhibit the spinally mediated nocifensive tail flick reflex by more than 100 % (Sandkühler and Gebhart, 1984; Fig. 7 B). These blocks also reduced efficacy of PAG-induced descending inhibition of nociceptive spinal dorsal horn neurons (Gebhart et al., 1983). Here, chemical stimulation in the PAG induced FOS in cells of the NRM and the NRPGα bilaterally (Fig. 7 A), providing evidence that indeed both areas can be simultaneously activated by PAG stimulation and give rise to parallel inhibitory descending pathways.

Compared to the number and the density of labeled cells in some areas of the brainstem, e.g., the LC (Fig. 6), the number of FOS positive cells in the NRM and the NRPGα was relatively low, as was the increase in 2-DG uptake in the NRM (+8%) following electrical stimulation in the PAG (Beitz and Buggy, 1981). A strong increase of 2-DG uptake was, in contrast, measured in the NRPGα (+36%). These results suggest that only relatively few raphe neurons have been activated by stimulation in the PAG. These few NRM neurons may, however, exert a strong inhibitory effect on spinal processing of nociceptive information (see also Hentall et al., 1984).

In the **superior colliculus** (SC) FOS-IR was found in all layers and in high density, similar to the density of staining within the dorsal part of the PAG (Fig. 2A). This suggests that a strong excitatory projection exists from the dorsal PAG to the overlaying SC. A dense projection from the PAG to the SC has indeed been described (Mantyh, 1983b). Excitation of neurons in the SC by microinjections of excitatory amino acids may evoke species specific defense reactions (Dean et al., 1988; Redgrave and Dean, this volume) as well as cardiovascular and respiratory responses (Keay et al., 1990) and some neurons in the SC can be excited by natural noxious stimuli (McHaffie et al., 1989).

Labeled cells were found in an area dorsolateral to the posterior commissure (Fig. 8A) in the dorsal and medial parts of the periventricular gray matter (Fig. 8B) and bilaterally in the zona incerta (Fig. 8C), which receives direct projection from the PAG (Mantyh, 1983a). Following electrical stimulation in the PAG the zona incerta displayed only a moderate increase in 2-DG uptake (Beitz and Buggy, 1981).

Labeled neurons were consistently found in the **medial hypothalamus** (Fig. 9), overlapping with areas which receive direct projection from the dorsal PAG (Mantyh, 1983a) and also overlapping, in part, with areas from which defense and flight behavior (see Siegel and Pott, 1988) and descending inhibition (Carstens, 1986) can be elicited by electrical stimulation. In the lateral hypothalamus FOS-IR was absent or very sparse and not above background labeling in sham treated animals.

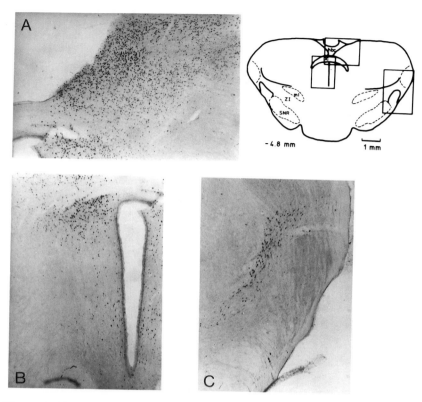

Figure 8. Rostral to the PAG FOS-IR was identified in cells dorsolateral to the posterior commissure (A) and in dorsal and medial parts of the periventricular gray matter (B), as well as in the zona incerta (C).

The supraoptic nuclei (SO) which probably do not receive direct projection from the PAG contained dense FOS-IR in this animal (Fig. 10 B), weaker IR in the other stimulated animal and had no significant basal expression in our study as in another previous report (Sagar et al., 1988). The mechanism of this activation is not known but could involve a disynaptic pathway with a relay in the Sol (Raby and Renaud, 1989). A PAG-SO pathway could be involved in the nonspecific increase in vasopressin release (Shibuki and Yagi, 1986).

Conclusions

The present study has identified FOS-IR in cells in well defined areas throughout the rostro-caudal extent of the brainstem following activation of neu-

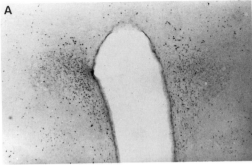

Figure 9. A: The dorsal part of the medial hypothalamus contained numerous cells with FOS-IR. B: labeled cells were also found in the periventricular hypothalamic nucleus and in the median eminence (Scale bar: 600 μm).

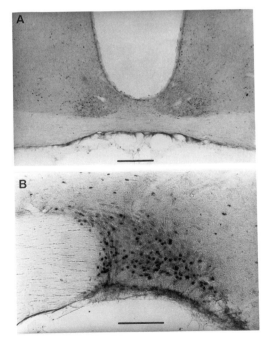

Figure 10. A: FOS-IR in cells of the suprachiasmatic nucleus and in the supraoptic nucleus (Scale bar: 600 μm). B: contralateral to the injection site in the PAG. The staining was similar ipsilaterally (Scale bar: 200 μm).

rons of the PAG. Labeled cells were mainly found in those areas which receive direct or indirect input from some parts of the PAG. Since induction of FOS is generally held to be indicative of neuronal excitation, our results suggest that many of these projections are excitatory.

The clear predominance of labeled cells in areas which are probably only a few synapses away from the injection site in the PAG may be due to the use of pentobarbital as a general anesthetic. Pentobarbital may preferentially depress polysynatic excitatory pathways with little effect on mono- or oligosynaptic excitation. Thus, in awake and drug free animals PAG stimulation could induce FOS in additional areas of the brain. It should be stressed that stimulation at different sites in the PAG, e.g., in the caudal ventrolateral PAG, which induces different behavioral responses (Bandler and Depaulis; Carrive; Lovick; Morgan, this volume) could also induce a different pattern of FOS-IR. Preliminary results obtained in our laboratory support this hypothesis.

Areas which receive direct projection from the dorsal PAG but had no FOS-IR following PAG stimulation such as the anterior lateral hypothalamic area could be characterized by a predominant inhibitory input from the PAG. It has been shown indeed that the vast majority of neurons in this area are inhibited by stimulation in the PAG (Kai et al., 1988).

The pattern of FOS-IR found in the present study suggests a concomitant direct or indirect neuronal excitation in those brainstem areas which are involved in the control of cardiovascular, respiratory, antinociceptive and complex behavioral functions. Thus, our results provide a cellular functional correlate for the diverse output functions of the PAG which can be simultaneously activated by stimulation at selected sites in the PAG.

Acknowledgements

The author wish to express thanks to G. Eilber for technical assistance, A.E. Manisali for graphics, Dr. R. Bravo for generously providing the FOS antibody and Dr. T. Herdegen for valuable help. The work was supported by a grant from the Deutsche Forschungsgemeinschaft (SA 435/3-3).

References

Abercrombie, E.D. and Jacobs, B.L., Single-unit response of noradrenergic neurons in the locus coeruleus of freely moving cats. I. Acutely presented stressful and nonstressful stimuli, J. Neurosci, 7 (1987) 2837-2843.

Aimone, L.D., Jones, S.L. and Gebhart, G.F., Stimulation-produced descending inhibition from the periaqueductal gray and nucleus raphe magnus in the rat: mediation by spinal monoamines but not opioids, Pain, 31 (1987) 123-136.

Aston-Jones, G., Ennis, M., Pieribone, V.A., Nickell, W.T. and Shipley, M.T., The brain nucleus locus coeruleus: restricted afferent control of a broad efferent network, Science, 234 (1986) 734-737.

Bandler, R., Brain mechanisms of aggression as revealed by electrical and chemical stimulation: Suggestion of a central role for the midbrain periaqueductal grey region, In: Progress in Psychobiology and Physiological Psychology, Vol. 13, Epstein A. and

Horrison, A. (Eds.), Academic Press, New York, 1988, pp. 67-154.

Bandler, R. and Tork, I., Midbrain periaqueductal grey region in the cat has afferent and efferent connections with solitary tract nuclei, Neurosci. Lett., 74 (1987) 1-6.

Barbaresi, P. and Manfrini, E., Glutamate decarboxylase- immunoreactive neurons and terminals in the periaqueductal gray of the rat, Neuroscience, 27 (1988) 183-191.

Basbaum, A.I., Clanton, C.H. and Fields, H.L., Three bulbospinal pathways from the rostral medulla of the cat: an autoradiographic study of pain modulating systems, J. Comp. Neurol., 178 (1978) 209-224.

Behbehani, M.M., Jiang, M., Chandler, S.D. and Ennis, M., The effect of GABA and its antagonists on midbrain periaqueductal gray neurons in the rat, Pain, 40 (1990) 195-204.

Beitz, A.J. and Buggy, J., Brain functional activity during PAG stimulation-produced analgesia: A 2-DG study, Brain Res. Bull., 6 (1981) 487-494.

Bonham, A.C. and Jeske, I., Cardiorespiratory effects of DL-homocysteic acid in caudal ventrolateral medulla, Am. J. Physiol., 254 (1988) H686-H692.

Cannon, J.T., Prieto, G.J., Lee, A. and Liebeskind, J.C., Evidence for opioid and non opioid forms of stimulation-produced analgesia in the rat, Brain Res., 243 (1982) 316-321.

Carstens, E., Hypothalamic inhibition of rat dorsal horn neuronal responses to noxious skin heating, Pain, 25 (1986) 95-107.

Dean, P., Mitchell, I. and Redgrave, P., Responses resembling defensive behaviour produced by microinjection of glutamate into superior colliculus of rats, Neuroscience, 24 (1988) 501-510.

Depaulis, A., Morgan, M.M. and Liebeskind, J.C., GABAergic modulation of the analgesic effects of morphine microinjected in the ventral periaqueductal gray matter of the rat, Brain Res., 436 (1987) 223-228.

Depaulis, A. and Vergnes, M., Elicitation of intraspecific defense behaviors in the rat by microinjection of picrotoxin, a gamma-amminobutyric acid antagonist, into the midbrain periaqueductal gray matter, Brain Res., 367 (1986) 87-95.

DiScala, G., Schmitt, P. and Karli, P., Flight induced by infusion of bicuculline methiodide into periventricular structures, Brain Res., 309 (1984) 199-208.

DiScala, G. and Sandner, G., Conditioned place aversion produced by microinjections of semicarbazide into the periaqueductal gray of the rat, Brain Res., 483 (1989) 91-97.

Dragunow, M., Currie, R.W., Robertson, H.A. and Faull, R.L.M., Heat shock induces c-fos protein-like immunoreactivity in glial cells in adult rat brain, Exp. Neurol., 106 (1989) 105-109.

Dragunow, M. and Faull, R., The use of c-fos as a metabolic marker in neuronal pathway tracing, J. Neurosci. Meth., 29 (1989) 261-265.

Dragunow, M. and Robertson, H.A., Brain injury induces c-fos protein(s) in nerve and glial-like cells in adult mammalian brain, Brain Res., 455 (1988) 295-299.

Fardin, V., Oliveras, J.-L. and Besson, J.-M., A reinvestigation of the analgesic effects induced by stimulation of the periaqueductal gray matter in the rat. II. Differential characteristics of the analgesia induced by ventral and dorsal PAG stimulation, Brain Res., 306 (1984) 125-139.

Fields, H.L. and Basbaum, A.I., Brainstem control of spinal pain-transmission neurons, Annu. Rev. Physiol., 40 (1978) 217-248.

Gebhart, G.F., Sandkühler, J., Thalhammer, J.G. and Zimmermann, M., Inhibition of spinal nociceptive information by stimulation in midbrain of the cat is blocked by lidocaine microinjected in nucleus raphe magnus and medullary reticular formation, J. Neurophysiol., 50 (1983) 1446-1459.

Goodchild, A.K., Dampney, R.A.L. and Bandler, R., A method for evoking physiological responses by stimulating cell bodies, but not axons of passage, within localized regions of the central nervous system, J. Neurosci. Meth., 6 (1982) 351-363.

Hentall, I.D., Zorman, G., Kansky, S. and Fields, H.L., An estimate of minimum number of brain stem neurons required for inhibition of a flexion reflex, J. Neurophysiol., 51 (1984) 978-985.

Hisanaga, K., Sagar, S.M., Hicks, K.J., Swanson, R.A. and Sharp, F.R., *c-fos* proto-oncogene expression in astrocytes associated with differentiation or proliferation but not depolarization, Mol. Brain Res., 8 (1990) 69-75.

Jacquet, Y., Saederup, E. and Squirres, R., Non-stereospecific excitatory actions of morphine may be due to GABA-A receptor blockade, Europ. J. Pharmacol., 138 (1987) 285-288.

Janss, A.J., Cox, B.F., Brody, M.J. and Gebhart, G.F., Dissociation of antinociceptive from cardiovascular effects of stimulation in the lateral reticular nucleus in the rat, Brain Res., 405 (1987) 140-149.

Jensen, T.S. and Yaksh, T.L., Spinal monoamine and opiate systems partly mediate the antinociceptive effects produced by glutamate at brain stem sites, Brain Res., 321 (1984) 287-297.

Jones, S.L. and Gebhart, G.F., Quantitative characterization of coeruleospinal inhibition of nociceptive transmission in the rat, J. Neurophysiol., 56 (1986) 1397-1410.

Kai, Y., Oomura, Y. and Shimizu, N., Responses of rat lateral hypothalamic neurons to periaqueductal gray stimulation and nociceptive stimuli, Brain Res., 461 (1988) 107-117.

Keay, K., Dean, P. and Redgrave, P., N-methyl D-aspartate (NMDA) evoked changes in blood pressure and heart rate from the rat superior colliculus, Exp. Brain Res., 80 (1990) 148-156.

Lipski, J., Bellingham, M.C., West, M.J. and Pilowsky, P., Limitations of the technique of pressure microinjection of excitatory amino acids for evoking responses from localized regions of the CNS, J. Neurosci. Meth., 26 (1988) 169-179.

Maciewicz, R., Sandrew, B.B., Phipps, B.S., Poletti, C.E. and Foote, W.E., Pontomedullary raphe neurons: intracellular responses to central and peripheral electrical stimulation, Brain Res., 293 (1984) 17-33.

Mantyh, P.W., Connections of midbrain periaqueductal gray in the monkey. I. Ascending efferent projections, J. Neurophysiol., 49 (1983a) 567-581.

Mantyh, P.W., Connections of midbrain periaqueductal gray in the monkey. II. Descending efferent projections, J. Neurophysiol., 49 (1983b) 582-594.

McHaffie, J.G., Kao, C.-Q. and Stein, B.E., Nociceptive neurons in rat superior colliculus: response properties, topography, and functional implications, J. Neurophysiol., 62 (1989) 510-525.

Menétrey, D., Gannon, A., Levine, J.D. and Basbaum, A.I., Expression of *c-fos* protein in interneurons and projection neurons of the rat spinal cord in response to noxious somatic, articular, and visceral stimulation, J. Comp. Neurol., 285 (1989) 177-195.

Mokha, S.S., McMillan, J.A. and Iggo, A., Descending control of spinal nociceptive transmission. Actions produced on spinal multireceptive coeruleus (LC) and raphe magnus (NRM), Exp. Brain Res., 58 (1985) 213-226.

Moreau, J.-L. and Fields, H.L., Evidence for GABA involvement in midbrain control of medullary neurons that modulate nociceptive transmission, Brain Res., 397 (1986) 37-46.

Morgan, J.I., Cohen, D.R., Hempstead, J.L. and Curran, T., Mapping patterns of *c-fos* expression in the central nervous system after seizure, Science, 237 (1987) 192-197.

Morgan, J.I. and Curran, T., Role of ion flux in the control of *c-fos* expression, Nature, 322 (1986) 552-555.

Morton, C.R., Duggan, A.W. and Zhao, Z.Q., The effects of lesions of medullary midline and lateral reticular areas on inhibition in the dorsal horn produced by periaqueductal grey stimulation in the cat, Brain Res., 301 (1984) 121-130.

Paxinos, G. and Watson, C., The rat brain in stereotaxic coordinates, Academic Press, New York, 1982.

Prieto, G.J., Cannon, J.T. and Liebeskind, J.C., N. Raphe magnus lesions disrupt stimulation-produced analgesia from ventral but not dorsal midbrain areas in the rat, Brain Res., 261 (1983) 53-57.

Raby, W.N. and Renaud, L.P., Dorsomedial medulla stimulation activates rat supraoptic oxytocin and vasopressin neurons through different pathways, J. Physiol., 417 (1989) 279-294.

Randich, A., Roose, M.G. and Gebhart, G.F., Characterization of antinociception produced by glutamate microinjection in the nucleus tractus solitarius and the nucleus reticularis ventralis, J. Neurosci., 8 (1988) 4675-4684.

Sagar, S.M., Sharp, F.R. and Curran, T., Expression of c-fos protein in brain: metabolic mapping at the cellular level, Science, 240 (1988) 1328-1331.

Sandkühler, J. and Gebhart, G.F., Relative contributions of the nucleus raphe magnus and adjacent medullary reticular formation to the inhibition by stimulation in the periaqueductal gray of a spinal nociceptive reflex in the pentobarbital-anesthetized rat, Brain Res., 305 (1984) 77-87.

Sandkühler, J., Willmann, E. and Fu, Q.-G., Blockade of $GABA_A$ receptors in the midbrain periaqueductal gray abolishes nociceptive spinal dorsal horn neuronal activity, Eur. J. Pharmacol., 160 (1989) 163-166.

Sandkühler, J., Willmann, E. and Fu, Q.-G., Characteristics of midbrain control of spinal nociceptive neurons and non- somatosensory parameters in the pentobarbital anaesthetized rat, J. Neurophysiol., 65 (1991) 33-48.

Sandkühler, J. and Zimmermann, M., Neuronal effects of controlled superfusion of the spinal cord with monoaminergic receptor antagonists in the cat, Prog. Brain Res., 77 (1988) 321-327.

Sandner, G., Dessort, D., Schmitt, P. and Karli, P., Distribution of GABA in the periaqueductal gray matter. Effects of medial hypothalamic lesions, Brain Res., 224 (1981) 279-290.

Sassone-Corti, P., Lamph, W.W., Kamps, M. and Verma, I.M., Fos- associated Cellular P39 is related to nuclear transcription factor AP-1, Cell, 54 (1988) 553-560.

Schneider, P.S. and Perl, E.R., Selective excitation of neurons in the mammalian spinal dorsal horn by aspartate and glutamate in vitro: correlation with location and excitatory input, Brain Res., 360 (1985) 339-343.

Shibuki, K. and Yagi, K., Synergistic activation of rat supraoptic neurosecretory neurons by noxious and hypovolemic stimuli, Exp. Brain Res., 62 (1986) 572-578.

Siegel, A. and Pott, C.B., Neural substrates of aggression and flight in the cat, Prog. Neurobiol., 31 (1988) 261-283.

The Nociceptive Modulatory Effects
of Periaqueductal Gray Activation are Mediated
by Two Neuronal Classes
in the Rostral Ventromedial Medulla

Peggy Mason

Department of Neurology
UCSF, San Francisco, U.S.A.

Introduction

The periaqueductal gray (PAG) has long been implicated in the central modulation of nociceptive transmission (for reviews see Basbaum and Fields, 1984; Besson and Chaouch, 1987). PAG activation suppresses nociceptive reflexes and inhibits the responses of nociceptive cells in the spinal and medullary dorsal horns. The antinociceptive effects of PAG stimulation are mediated, at least in part, by a synaptic relay in the rostral ventromedial medulla (RVM), a region that contains many neurons that project directly to the dorsal horn and spinal trigeminal nucleus. This chapter will review our current knowledge on the role of the PAG-RVM-dorsal horn pathway in the modulation of nociceptive transmission.

The PAG - RVM - dorsal horn pathway

A direct connection from PAG to the dorsal horn is unlikely to underlie the nociceptive modulatory effects of PAG activation as only a few ventrolateral PAG cells project directly to the spinal cord or to the laminar trigeminal nucleus (Kuypers and Maisky, 1975; Castiglioni et al., 1978; Holstege, this volume). In contrast, there is a very large projection from PAG to the medullary nucleus raphe magnus and the adjacent medial reticular formation, the region that I refer to as the RVM (Holstege, this volume; Abols and Basbaum, 1981; Carlton et al., 1983). RVM neurons in turn, project, via the dorsolateral fasciculus, to the medullary and spinal dorsal horns (Basbaum et al., 1978; Basbaum and Fields, 1979; Holstege and Kuypers, 1982; Basbaum et al., 1986).

A number of lesion studies support the conclusion that PAG activation by either electrical stimulation or chemical microinjection (of opiates or excitatory amino acids) inhibits nociceptive transmission through a relay in the RVM (Behbehani and Fields, 1979; Gebhart et al., 1983; Prieto et al., 1983; Morton et al., 1984; Sandkühler and Gebhart, 1984; Chung et al., 1987). Nociceptive reflex suppression produced by PAG stimulation is attenuated but rarely abolished by chemical, electrical or knife lesions in the RVM (see Fig. 1). Lesions of the entire RVM, including raphe magnus and the adjacent medullary reticular formation bilaterally, are necessary to completely block tail flick suppression evoked by ventral PAG stimulation (see Figure 1E; Sandkühler and Gebhart, 1984).

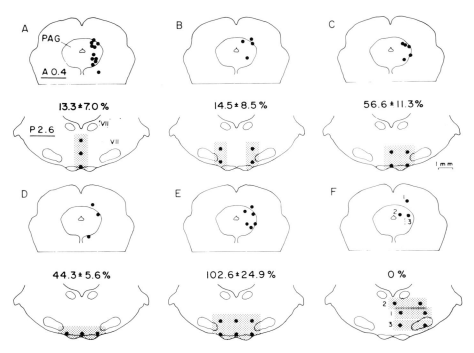

Figure 1. The effect of lidocaine microinjections (stippled areas) on the threshold required for tail flick reflex suppression by electrical stimulation in the PAG. The PAG stimulation sites are shown on midbrain sections. The average percent increase in the stimulation threshold is indicated for each microinjection. Reprinted with permission from Sandkühler and Gebhart (1984) Brain Research.

There are reports that the antinociception produced by dorsal PAG stimulation is not affected by lesions of the RVM and is only attenuated by lesions of the ventrolateral medulla (Prieto et al., 1983; Lovick, 1985). However, in a large series of animals, Sandkühler and Gebhart (1984) studied the effects of RVM inactivation on the antinociception produced by stimulation of either dorsal or ventral PAG sites, and found no difference. Furthermore, lesions of the ventrolateral medulla may be expected to produce similar effects as complete RVM lesions since RVM axons that descend toward the spinal cord travel within the ventrolateral medulla (Mason and Fields, 1989). Although the antinociception produced by both dorsal and ventral PAG sites relays through the RVM, other differences may well exist between the nociceptive modulatory effects of dorsal and ventral PAG activation (Morgan, this volume).

If the nociceptive modulatory effects of PAG are mediated by an excitatory relay in the RVM, then activation of the RVM should largely mimic the nociceptive modulation evoked by PAG stimulation. This is the case as RVM stimulation selectively blocks the behavioral reactions elicited by noxious pinch or tooth pulp stimulation in awake animals (Oliveras et al. 1975; Besson and Oliveras, 1980). In anesthetized animals, electrical stimulation or opioid microinjection into the RVM suppresses nociceptive reflexes (Oliveras et al., 1975; Jensen and Yaksh, 1989) and inhibits the responses of sensory trigeminal or spinal dorsal horn cells to noxious stimuli (Fields et al., 1977; Willis et al., 1977; Sessle and Hu,

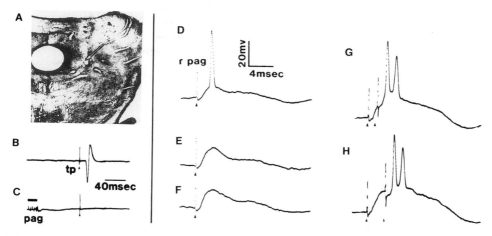

Figure 2. Stimulation in the PAG (A) blocks the tooth pulp evoked jaw-opening reflex. The jaw-opening reflex is measured by recording the digastric muscle EMG (B,C) before (B) and after (C) conditioning stimulation of the PAG. D-H: Antinociceptive PAG stimulation also evokes an EPSP in most RVM neurons. The intracellularly recorded responses of a RVM neuron to single shock stimulation of the PAG are shown during injection of increasing amounts of hyperpolarizing current (0.0, 5.0, 6.0nA in D-F). G-H: The depolarization of this RVM neuron in response to paired PAG stimulations summates temporally. Reprinted with permission from Mason et al. (1985) Brain Research.

1981; Sessle et al., 1981). Further support for the idea that RVM neurons relay, at least in part, the effects of PAG stimulation, is the observation that RVM stimulation produces an inhibitory postsynaptic potential (IPSP) in the same superficial dorsal horn neurons that receive an IPSP from PAG stimulation (Light et al., 1986).

Since PAG activation suppresses nociceptive transmission via a relay in the RVM, RVM neurons must receive an excitatory input from PAG stimulation that is antinociceptive. When stimulating electrodes are placed in the PAG such that they are effective in suppressing a trigeminal nociceptive reflex, electrical stimulation at these sites excites about 90% of cat RVM neurons (Mason et al., 1985) (see Fig. 2). Antinociceptive PAG stimulation evokes a constant latency excitatory postsynaptic potential (EPSP) in RVM neurons; the EPSP occurs at a short latency (average 2.4 ms) consistent with a monosynaptic connection. Anatomical studies confirm that PAG neurons synapse directly on RVM cells, including neurons that project to the spinal cord (Lakos and Basbaum, 1988).

Nociceptive Facilitation

Recently, research on central pain modulation, which has heretofore focussed on analgesia and pain suppression, has been challenged to explain the phenomenon of pain enhancement. One model for this phenomenon is the enhancement of nociceptive reflexes that occurs during the naloxone precipitated withdrawal from opioid administration. Figure 3 shows that the jaw-opening reflex evoked by noxious heat applied to the lip is enhanced during naloxone precipitated withdrawal from a single dose of systemic morphine. The enhancement of this reflex is evident both in an increase in the magnitude of the EMG response and in a decrease in the reflex latency.

The role of the PAG in enhancing nociceptive transmission is not clear. There are several reports of dorsal horn units that are excited by PAG stimulation (Yezierski, this volume). However, without a correlation between the neuronal firing pattern and the behavioral state of the animal (with respect to nociception), one cannot know whether the firing rate of these neurons is directly or inversely related to nociception. Thus, the PAG's excitation of dorsal horn cells is consistent with a role for the PAG in nociceptive facilitation but it is not sufficient evidence to establish such a role.

In a behavioral study, Wilcox et al. (1979) reported that following chronic, systemic opioid administration, naloxone microinjections into the PAG decrease the latency of the nociceptive "flinch-jump" reflex in mice. The latency decrease observed in animals receiving naloxone microinjections into the PAG was similar to that observed in animals given systemic naloxone. These results demonstrate that the PAG contributes to the increase in nociceptive transmission that occurs during opioid withdrawal.

A role for the RVM in the phenomenon of nociceptive enhancement is somewhat better established. Kaplan and Fields (1991) recently demonstrated that the latency decrease in the nociceptive tail flick reflex observed during sys-

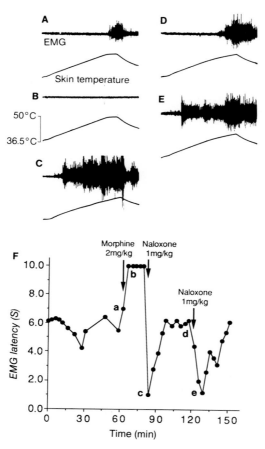

Figure 3. Opioid modulation of the jaw-opening reflex evoked by noxious radiant heat applied to the upper lip. A-E: Heating of the lip (bottom trace) evokes an EMG response in the digastric muscle (upper trace). This reflex is suppressed by systemic morphine (2mg/kg, i.v.) (B) and enhanced after systemic naloxone (1mg/kg, i.v.) (C). The reflex recovers from the short acting naloxone (D) and can be enhanced again by subsequent naloxone (E). F: The latency of the nociceptive reflex is plotted for the duration of the experiment; the timing of the morphine and naloxone injections is shown.

temic opioid withdrawal is also observed after acute withdrawal from opiates microinjected into the RVM. These results are evidence that RVM neurons contribute to the nociceptive enhancement seen during opioid withdrawal. Electrical stimulation in the RVM can also produce a decrease in the latency of the nociceptive tail flick reflex (Zhuo and Gebhart, 1990). The same sites that support tail flick enhancement produce nociceptive reflex suppression when higher stimulation intensities are used, evidence that neural units with opposing effects on nociceptive transmission are intermingled within the RVM. Glutamate microinjection into these RVM sites suppresses the tail flick reflex but never produces a decrease in the tail flick latency. One interpretation of these results is that nociceptive inhibition is mediated by RVM somata while axons of passage mediate nociceptive facilitation. Alternatively, it may be difficult to activate sufficient numbers of somata that support nociceptive facilitation without activating more somata that support nociceptive suppression. Evidence for the latter possibility is presented below.

Any enhancement of nociceptive transmission could be due to an active facilitation or to a decrease in tonic inhibition. To address this question, Kaplan and Fields (1991) administered naloxone (i.v.) to precipitate a withdrawal syndrome from systemic opiates and then blocked RVM activity with a local anesthetic microinjection. The local anesthetic block of the RVM reversed the enhancement of the tail flick reflex, normally seen during the withdrawal syndrome. These results imply that enhanced nociceptive transmission results from the activation of at least a subset of RVM neurons that in turn directly facilitates nociceptive transmission at the level of the dorsal horn. Local anesthetic block of the RVM in the absence of opioid withdrawal slightly decreases tail flick latencies, evidence for a tonic inhibitory output from the RVM. Thus, the RVM contains neural units that, when activated, either facilitate or inhibit nociceptive transmission. The following sections explore the mechanisms through which the activation of different RVM neuronal cell classes can result in either facilitation or inhibition of nociceptive transmission.

RVM On- and Off-Cells

Physiological studies have demonstrated that there are two types of RVM neurons that respond to noxious stimuli. In lightly anesthetized animals, RVM off-cells decrease and on-cells increase their discharge rate just prior to and during withdrawal reflexes evoked by noxious heat or pinch (see Fig. 4; Fields et al., 1983a; for review, see Fields et al., 1991). Both on- and off-cells respond with a consistent excitation or inhibition, respectively, to noxious stimuli applied anywhere over the body surface.

About 30% of RVM on- and off-cells project to the spinal cord (Vanegas et al., 1984). In order to determine the role that on- and off-cells play in the descending modulation of spinal and trigeminal nociceptive transmission, Fields and coworkers have recorded their firing patterns during conditions of either inhibited or facilitated nociceptive responsiveness. By determining how the activity of on-

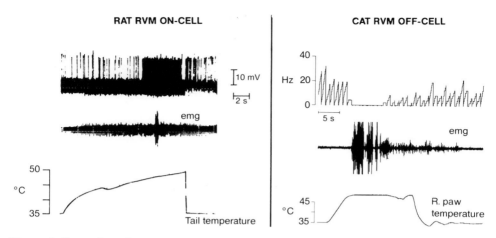

Figure 4. Recordings from an on- and off-cell during a nociceptive withdrawal reflex evoked by noxious heat. On the left, noxious heat applied to the tail (bottom trace) excites a rat on-cell (intracellular recording is shown in the upper trace) and evokes a reflex (EMG from the paraspinous muscles is shown in the middle trace). On the right, noxious heat applied to the right hind paw (bottom trace) inhibits a cat off-cell (firing rate shown in the upper trace) and evokes a reflex (middle trace).

and off-cells is correlated with the latency of a nociceptive reflex evoked by a standard stimulus, a functional role for each of these cell types has been proposed.

Off-cells are excited by morphine, administered either systemically or by microinjection into the PAG (Fields et al. 1983b; Barbaro et al., 1986; Cheng et al., 1986). In both cases, the off-cell is excited at a time when nociceptive reflexes are inhibited. The microstimulation and lesion data cited above are evidence that the activation of a neuronal class is required to inhibit spinal nociceptive transmission. Since off-cells are the only RVM cell class that is excited by opioid administration, they are thought to have a net inhibitory effect on nociceptive transmission at the level of the dorsal horn.

On-cells, in contrast to off-cells, are inhibited by morphine and are activated during periods of increased responsiveness to noxious stimuli (Barbaro et al., 1986; Cheng et al., 1986). For instance, when nociceptive reflexes are enhanced during naloxone-precipitated opioid withdrawal, on-cell but not off-cell activity is correlated with the degree of reflex facilitation (Bederson et al., 1990). Since activation of at least a subset of neurons is required to enhance nociceptive reflexes during opioid withdrawal (see above), the on-cell is hypothesized to facilitate nociception.

Since on-cells facilitate and off-cells inhibit nociceptive transmission, their spontaneous activity patterns are important in determining an animal's receptivity to any given somatic stimulus. The spontaneous discharge of both on- and off-cells alternates between periods of a high discharge rate and periods of low activity (Barbaro et al., 1989). When the spontaneous activity of two RVM cells is recorded simultaneously, neurons of the same class are active together whereas cells of opposite classes have reciprocal firing patterns. Consistent with the idea that on- and off-cells modulate nociceptive responsiveness in an on-going fashion, the latency of the nociceptive tail flick reflex is shorter during periods of on-cell activity than during periods of off-cell activity (Heinricher et al., 1989). Furthermore, nociceptive reflexes are enhanced immediately following a noxious stimulus, during the time when off-cells are still silent and on-cells actively firing due to the conditioning stimulus (Ramirez and Vanegas, 1989). These results are consistent with the idea that the "gain" in nociceptive transmission is constantly changing, increasing during periods of on-cell activity and decreasing when off-cells are active.

Since the on-going activity of RVM on- and off-cells contributes to setting the gain of nociceptive transmission, it is important to determine what influences the spontaneous firing of these neurons. By pairing two noxious stimuli, it has previously been shown that nociceptive input modulates both RVM neural activity and nociceptive responsiveness (see above). It is likely that complex limbic and telencephalic inputs also modulate on- and off-cell activity, and thereby nociceptive transmission. However, in order to know which afferents contact RVM neurons directly and may therefore be important in coordinating the correlated spontaneous activity of RVM neurons, it is important to determine the size and shape of on- and off-cell dendritic arbors. Thus, Mason et al. (1990) studied the somatodendritic morphology of physiologically characterized RVM neurons.

The responses of RVM neurons during nociceptive reflexes evoked by noxious pinch or heat were recorded intracellularly (Mason et al., 1990). Physiologically characterized on- and off-cells were then labeled with intracellular injections of horseradish peroxidase or neurobiotin in the rat and cat. There are no differences between the dendritic arbors of RVM on- and off-cells in either rat or cat. On- and off-cell dendrites extend beyond their nucleus of origin and extend throughout the RVM, including both the raphe magnus and the nuclei of the adjacent ventromedial reticular formation. The most notable feature of the dendritic arbors of on- and off cells is their marked extension in the mediolateral direction (see Fig. 5). On- and off-cell dendritic arbors are always bilateral and extend to the lateral edge of the pyramids or trapezoid body. The extensive mediolateral range of on- and off-cell dendrites is consistent with the finding that the electrical threshold or opiate dose required to suppress nociceptive reflexes does not change significantly as the stimulation or injection site is moved from the nucleus raphe magnus laterally to the adjacent ventromedial reticular formation (Satoh et al., 1980; Zorman et al., 1981; Dostrovsky et al., 1982; Sandkühler and Gebhart, 1984; Barbaro et al., 1985; Jensen and Yaksh, 1986). In contrast, on- and off-cell dendritic arbors do not enter the dorsal nuclei of the pontomedullary reticular formation, a region where the electrical threshold or opioid dose required for nociceptive reflex suppression is much greater than in the RVM.

Although the dendritic domains of individual on- or off-cells do not rami-

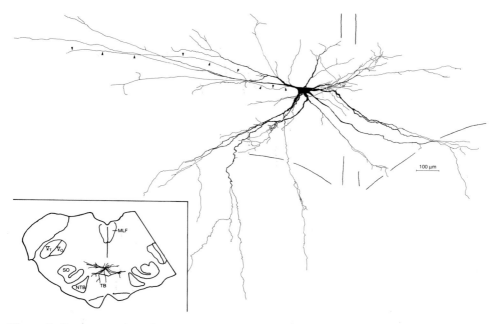

Figure 5. Reconstruction of the somatodendritic arbor of a cat off-cell. Reprinted with permission from Mason et al. (1990) Journal of Comparative Neurology.

fy throughout the RVM, the composite dendritic domain of the population of on- and off-cells includes the entirety of the RVM (Mason et al., 1990). This same area is where other on- and off-cells are located. The extensive dendritic fields of individual RVM neurons also provide a substrate for integration of many neuronal inputs from extrinsic nuclei (Leontovich and Zhukova, 1963; Ramon-Moliner and Nauta, 1966). In fact, the RVM receives large projections from neurons in the PAG, the midbrain nucleus cuneiformis and the parabrachial nuclei (Holstege, this volume; Gallagher and Pert, 1978; Abols and Basbaum, 1981; Carlton et al., 1983; Lakos and Basbaum, 1988) as well as from the limbic forebrain, including the medial and anterior hypothalamic regions (Hosoya, 1985; Holstege, 1987; Luppi et al., 1988). Other afferent inputs to the RVM include projections from the solitary tract nucleus (Beitz, 1982) and the subcoeruleus nucleus (Sakai, 1980).

Although somatosensory input affects on- and off-cell activity, there is little afferent input from spinal and trigeminal sensory neurons directly to the RVM (Kevetter and Willis, 1983). Instead, somatosensory input likely reaches RVM cells indirectly, possibly through a relay in the PAG which receives a large projection from the spinal cord and spinal trigeminal nucleus (Yezierski; Blomqvist and Craig, this volume).

One further feature of the dendritic arbors of labeled on- and off-cells is their restriction in the sagittal plane, with most of the dendrites located within the coronal plane that contains the soma. The planar organization of RVM on- and off-cell dendritic domains is unlikely to relate to a somatotopic organization since no somatotopy has been demonstrated within RVM using either anatomical or physiological techniques. However, it is possible that the afferents to or efferents from the RVM may be organized along the sagittal plane. The distribution of neurotransmitters or receptors may similarly be distributed to either restricted or widespread sagittal regions of the RVM.

The planar organization of RVM on- and off-cells is also interesting in light of the previous results concerning the tightly correlated spontaneous activity patterns of on- and off-cells. In that experiment, recordings were obtained from units that were separated by at least 1mm rostrocaudally (Barbaro et al., 1989). Since on- and off-cell dendritic arbors are restricted sagittally, any overlap between the dendritic arbors of recorded units is unlikely. This implies that the afferents responsible for coordinating RVM activity patterns must collateralize at a number of different rostrocaudal levels.

In a previous study of on- and off-cell axonal projection patterns, it was observed that the off-cell axon collateralizes within the RVM at several rostrocaudal levels (Mason and Fields, 1989). This arrangement makes it possible for off-cells to contact multiple on- and/or off-cells throughout the length of the RVM, enabling off-cell activity to have a widespread and direct effect on the activity of other RVM neurons (see Fig. 6). It is possible that afferents from extrinsic nuclei also contribute to the generation of coordinated on- and off-cell activity. In contrast to off-cell projections, the intra-RVM terminals of on-cell axons are restricted within the coronal plane of the parent soma. On-cells therefore cannot play a major role in the coordination of RVM on- and off-cell activity.

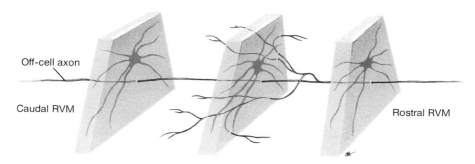

Figure 6. Diagram that shows proposed circuitry within the RVM. The dendritic arbors of single on- or off-cells occupy restricted coronal planes within the RVM. Off-cell axons terminate throughout the rostrocaudal length of the RVM, thereby contacting a number of on- and/or off-cells. Reprinted with permission from Mason et al. (1990) Journal of Comparative Neurology.

Excitatory connections between off-cells but not between on-cells would explain the finding that although on- and off-cells have opposing effects on nociceptive transmission, RVM stimulation consistently results in antinociception. According to this hypothesis, electrical stimulation of, or excitatory amino acid microinjection into the RVM directly excites a small number of off-cells that, in turn, indirectly activate many other off-cells in RVM. In contrast, the direct activation of RVM on-cells would not lead to the excitation of other on-cells throughout RVM. These proposed connections imply that the activation of large numbers of on-cells is rare compared with the activation of many off-cells, consistent with the fact that RVM stimulation rarely produces nociceptive facilitation (Zhuo and Gebhart, 1990).

Under some physiological conditions, there may be pharmacological methods that will selectively activate on-cells and not off-cells. As discussed above, on-cells but not off-cells are excited during opioid withdrawal. In addition, iontophoretic norepinephrine excites on-cells, through an action at the α1 adrenergic receptor, and has no effect on off-cells (Heinricher et al., 1988). In awake (but not anesthetized) animals, α1 adrenergic agonists decrease the tail flick latency, consistent with the increase in on-cell activity (Sagen and Proudfit, 1985; Haws et al., 1991).

On-cell axons, in contrast to off-cell axons, have few terminations within RVM but do project strongly and specifically to the ventrolateral medulla (VLM) at rostral and caudal levels (Mason and Fields, 1989). Figure 7 shows an on-cell axon that terminates in the lateral portion of the nucleus reticularis paragigantocellularis lateralis (NRPGL) and in the region just ventral to the facial nucleus in the cat. Several labeled rat on-cells project to this region as well (unpublished observations). On-cell axonal terminations are also seen within more caudal portions of the VLM, at the level of the inferior olivary complex, and within the raphe pallidus.

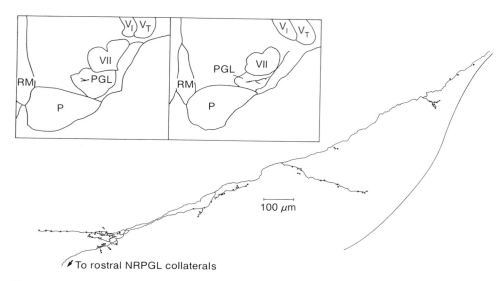

Figure 7. Reconstruction of the axonal projection of a cat on-cell to caudal NRPGL. The insets show, at low magnification, where each axonal collateral is located. Reprinted with permission from Mason et al. (1989) Journal of Comparative Neurology.

The significance of the on-cell projection to the ventrolateral medulla is unclear; several possibilities exist. First, neurons in the VLM play an important role in the control of autonomic function through their projection to the thoracic intermediolateral cell column where they contact preganglionic sympathetic neurons (Guertzenstein and Silver, 1974; Amendt et al., 1979; Ross et al., 1984). Lovick (1988) has demonstrated that many VLM cells receive convergent input from the nucleus of the tractus solitarius, the RVM, the parabrachial nuclei and the dorsal PAG. This convergence of input onto VLM neurons could be important in the integration of somatic and autonomic responses to noxious stimulation (Carrive; Lovick, this volume). RVM on-cells may either modulate or activate this system through their projections to the VLM where they may contact either interneurons or cells that project spinally.

An intriguing possibility is that on-cells contribute to the mediation of the withdrawal syndrome from opiates via their projection to the VLM. On-cells project directly to the NRPGL, one of the nuclei of the VLM, which provides the major input to the locus coeruleus (Guyenet and Young, 1987; Ennis and Aston-Jones, 1988). A role for the locus coeruleus has been implicated in several of the behavioral consequences of opioid withdrawal (Esposito et al., 1987; Grant et al., 1988; Taylor et al., 1988).

The present state of knowledge leaves abundant opportunities for future research. The afferents that control both the spontaneous and evoked activity of RVM on- and off-cells are unclear; further research in this area is needed. As stated above, the somatosensory input to RVM may be relayed through a synapse in the PAG and/or the neighboring parabrachial nuclei. Both the PAG and the parabrachial nuclei receive a major ascending projection from spinal and sensory trigeminal neurons to the PAG and, in turn, project directly to the RVM. The spontaneous activity of RVM on- and off-cells is also likely to be heavily influenced by PAG afferents. The spontaneous activity of these neurons appears related to the behavioral state of the animal as well as to incoming somatosensory information. Information on behavioral state may well be relayed, at least in part, from the limbic forebrain and hypothalamus to the RVM via the PAG. However, whether PAG input to the RVM is distributed widely or is amplified by intrinsic RVM circuitry is unknown. To answer this question, it will be important to determine the termination patterns of individual PAG afferents to the RVM. Knowledge of this type of single cell anatomy will likely provide clues as to the role of PAG input in RVM on- and off-cell circuitry.

Several issues raised in this review, concerning the role of the PAG in nociceptive modulation, bear further investigation. The role of the PAG-RVM pathway in nociceptive facilitation is unclear. Although PAG activation is reported to excite some dorsal horn neurons (see above; Yezierski, this volume), there is little information on how PAG may facilitate nociceptive reflexes and behaviors. It seems likely, however, that the PAG is involved in the pathway that leads to the on-cell facilitation of dorsal horn neurons and spinal reflexes. Finally, if the PAG is involved in nociceptive facilitation, then how does this function interact with the other behavioral and autonomic behaviors that PAG activation produces?

As discussed in several other chapters, stimulation of the PAG, in addition to modulating nociceptive reflexes, produces behaviors that resemble the freezing and the running-escape components of the defense reaction (Bandler and Depaulis; Fanselow, this volume). It is possible that the neuronal circuitry that underlies nociceptive modulation and that which subserves defense reactions may not be activated together under physiological conditions. An alternative possibility is that the nociceptive modulatory aspects associated with PAG stimulation are an essential part of the defense reaction behavior that is evoked by each site. There may be one group of defense and cardiovascular behaviors that occur concurrently with enhanced nociceptive responsiveness and a second set of correlates that occur during inhibition of nociceptive transmission. One can imagine, for instance, that during escape, an analgesic state would allow the animal to retreat undistracted by any injuries or painful inputs. On the other hand, during recovery in a safe environment, hyperalgesia may serve to highlight the injured regions and therefore keep them inactive, thereby avoiding any further injury. Further behavioral research, measuring both defense reactions and nociceptive responsiveness, is needed to clarify these issues.

References

Abols, I.A. and Basbaum, A.I., Afferent connections of the rostral medulla of the cat: a neural substrate for midbrain-medullary interactions in the modulation of pain, J. Comp. Neurol., 201 (1981) 285-297.

Amendt, K., Czachurski, J., Dembowsky, K. and Seller, H., Bulbospinal projections to the intermediolateral cell column: a neuroanatomical study, J. Auton. Nerv. Syst., 1 (1979) 103-117.

Barbaro, N.M., Fields, H.L. and Heinricher, M.M., Putative nociceptive modulatory neurons in the rostroventromedial medulla of the rat display highly correlated firing patterns, Somatosens. Res., 6 (1989) 413-425.

Barbaro, N.M., Hammond, D.L. and Fields, H.L., Effects of intrathecally administered methysergide and yohimbine on microstimulation-produced antinociception in the rat, Brain Res., 343 (1985) 223-229.

Barbaro, N.M., Heinricher, M.M. and Fields, H.L., Putative pain modulating neurons in the rostral ventral medulla: reflex-related activity predicts effects of morphine, Brain Res., 366 (1986) 203-210.

Basbaum, A.I., Clanton, C.H. and Fields, H.L., Three bulbospinal pathways from the rostral medulla of the cat: an autoradiographic study of pain modulating systems, J. Comp. Neurol., 178 (1978) 209-224.

Basbaum, A.I. and Fields, H.L., The origin of descending pathways in the dorsolateral funiculus of the spinal cord of the cat and rat: further studies on the anatomy of pain modulation, J. Comp. Neurol., 187 (1979) 513-531.

Basbaum, A.I. and Fields, H.L., Endogenous pain control systems: Brainstem spinal pathways and endorphin circuitry, Annu. Rev. Neurosci., 7 (1984) 309-338.

Basbaum, A.I., Ralston, D.D. and Ralston, H.J. III, Bulbospinal projections in the primate: A light and electron microscopic study of a pain modulating system, J. Comp. Neurol., 187 (1986) 513-531.

Bederson, J.B., Fields, H.L. and Barbaro, N.M., Hyperalgesia during naloxone-precipitated withdrawal from morphine is correlated with increased on-cell activity in the rostral ventromedial medulla, Somatosens. Mot. Res., 7 (1990) 185-203.

Behbehani, M.M. and Fields, H.L., Evidence that an excitatory connection between the periaqueductal gray and nucleus raphe magnus mediates stimulation produced analgesia, Brain Res., 170 (1979) 85-93.

Beitz, A.J., The nuclei of origin of brainstem enkephalin and substance-P projections to the rodent nucleus raphe magnus, J. Neurosci., 7 (1982) 2753-2768.

Besson, J.M. and Oliveras, J.L., Analgesia induced by electrical stimulation of the brainstem in animals: involvement of serotonergic mechanisms, Acta Neurochir. Suppl., 30 (1980) 201-217.

Besson, J.M. and Chaouch, A., Peripheral and spinal mechanisms of nociception, Physiol. Rev., 67 (1987) 67-186.

Carlton, S.M., Leichnetz, G.R., Young, E.G. and Mayer, D.J., Supramedullary afferents of the nucleus raphe magnus in the rat: a study using the transcannula HRP gel and autoradiographic technique, J. Comp. Neurol., 214 (1983) 43-58.

Castiglioni, A.J., Gallaway, M.C. and Coulter, J.D., Spinal projections from the midbrain in monkey, J. Comp. Neurol., 178 (1978) 329-346.

Cheng, Z.F., Fields, H.L. and Heinricher, M.M., Morphine microinjected into the periaqueductal gray has differential effects on three classes of medullary neurons, Brain Res., 375 (1986) 57-65.

Chung, R.Y., Mason, P., Strassman, A. and Maciewicz, R., Suppression of the jaw-opening reflex by periaqueductal gray stimulation is decreased by paramedian brainstem lesions, Brain Res., 403 (1987) 172-176.

Dostrovsky, J.O., Shah, Y. and Hu, J.W., Stimulation sites in periaqueductal gray, nucleus raphe magnus and adjacent regions effective in suppressing oral-facial reflexes, Brain Res., 252 (1982) 287-297.

Ennis, M. and Aston-Jones, G., Activation of locus coeruleus from nucleus paragigantocellularis: a new excitatory amino acid pathway in brain, J. Neurosci., 8 (1988) 3644-3657.

Esposito, E., Kruszewska, A., Ossowska, G., and Saminin, R., Noradrenergic and behavioral effects of naloxone injected in the locus coeruleus of morphine-dependent rats and their control by clonidine, Psychopharmacology, 93 (1987) 393-396.

Fields, H.L., Basbaum, A.I., Clanton, C.H., and Anderson, S.D., Nucleus raphe magnus inhibition of spinal cord dorsal horn neurons, Brain Res., 126 (1977) 441-453.

Fields, H.L., Bry, J., Hentall, I.D., and Zorman, G., The activity of neurons in the rostral medulla of the rat during withdrawal from noxious heat, J. Neurosci., 3 (1983a) 2545-2552.

Fields, H.L., Heinricher, M.M., and Mason, P., Neurotransmitters in nociceptive modulatory circuits, Annu. Rev. Neurosci., 14 (1991) 219-245.

Fields, H.L., Vanegas, H., Hentall, I.D. and Zorman, G., Evidence that disinhibition of brainstem neurones contributes to morphine analgesia, Nature, 306 (1983b) 684-686.

Gallagher, D.W. and Pert, A., Afferents to brainstem nuclei (brainstem raphe, nucleus reticularis pontis caudalis and nucleus gigantocellularis) in the rat as demonstrated by microiontophoretically applied horseradish peroxidase, Brain Res., 144 (1978) 257-275.

Gebhart, G.F., Sandkühler, J., Thalhammer, J.G. and Zimmerman, M., Inhibition of spinal nociceptive information by stimulation in midbrain of the cat is blocked by lidocaine microinjected in nucleus raphe magnus and medullary reticular formation, J. Neurophysiol., 50 (1983) 1446-1459.

Grant, S.J., Huang, Y.H. and Redmond, D.E., Behavior of monkeys during opiate withdrawal and locus coeruleus stimulation, Pharmacol. Biochem. Behav., 30 (1988) 13-19.

Guertzenstein, P.G. and Silver, A., Fall in blood pressure produced from discrete regions of the ventral surface of the medulla by glycine and lesions, J. Physiol., 242 (1974) 489-503.

Guyenet, P.G. and Young, B.S., Projections of nucleus paragigantocellularis lateralis to locus coeruleus and other structures in rat, Brain Res., 406 (1987) 171-184.

Haws, C.M., Heinricher, M.M., and Fields, H.L., Microinjection of α-adrenergic receptor agonists, but not antagonists, into the rostral ventromedial medulla alters tail flick latency in the lightly anesthetized rat, Brain Res., 1991, in press.

Heinricher, M.M., Barbaro, N.M., and Fields, H.L., Putative nociceptive modulating neurons in the rostral ventromedial medulla of the rat: firing of on- and off-cells is related to nociceptive responsiveness, Somatosens. Mot. Res., 6 (1989) 427-439.

Heinricher, M.M., Haws, C.M. and Fields, H.L., Opposing actions of norepinephrine and clonidine on single pain-modulating neurons in the rostral ventromedial medulla, In: Proc. 5th World Congr. on Pain, Dubner R., Gebhart G.F. and Bond M.R. (Eds.), Elsevier, Amsterdam, 1988, pp. 590-594.

Holstege, G., Some anatomical observations on the projections from the hypothalamus to brainstem and spinal cord: An HRP and autoradiographic tracing study in the cat, J. Comp. Neurol., 260 (1987) 98-126.

Holstege, G. and Kuypers, H.G.J.M., The anatomy of brainstem pathways to the spinal cord in cat. A labeled amino acid tracing study, Prog. Brain Res., 57 (1982) 145-175.

Hosoya, Y., Hypothalamic projections to the ventral medulla oblongata in the rat, with special reference to the nucleus raphe pallidus: a study using autoradiographic and HRP techniques, Brain Res., 344 (1985) 338-350.

Jensen, T.S. and Yaksh, T.L., I. Comparison of antinociceptive action of morphine in the periaqueductal gray, medial and paramedial medulla in rat, Brain Res., 363 (1986) 99-113.

Jensen, T.S. and Yaksh, T.L., Comparison of the antinociceptive effect of morphine and glutamate at coincidental sites in the periaqueductal gray and medial medulla in rats, Brain Res., 476 (1989) 1-9.

Kaplan, H.J. and Fields, H.L., Hyperalgesia during acute opioid abstinence: evidence for a nociceptive facilitative function of the rostral ventromedial medulla, J. Neurosci., 11 (1991) 1433-1439.

Kevetter, G.A. and Willis, W.D., Collaterals of spinothalamic cells in the rat, J. Comp. Neurol., 215 (1983) 453-464.

Kuypers, H.G.J.M. and Maisky, V.A., Retrograde axonal transport of horseradish peroxidase from spinal cord to brainstem cell groups in the cat, Neurosci. Lett., 1 (1975) 9-14.

Lakos, S. and Basbaum, A.I., An ultrastructural study of the projections from the midbrain periaqueductal gray to spinally-projecting, serotonin-immunoreactive neurons of the medullary nucleus raphe magnus, Brain Res., 443 (1988) 383-388.

Leontovich, T.A. and Zhukova, G.P., The specificity of the neuronal structure of the reticular formation in the brain and spinal cord of carnivora, J. Comp. Neurol., 121 (1963) 347-379.

Light, A.R., Casale, E.J., Menetrey, D.M., The effects of focal stimulation in nucleus raphe magnus and periaqueductal gray on intracellularly recorded neurons in spinal laminae I and II, J. Neurophysiol., 56 (1986) 555-571.

Lovick, T.A., Ventrolateral medullary lesions block the antinociceptive and cardiovascular responses elicited by stimulating the dorsal periaqueductal gray matter in rats, Pain, 21 (1985) 241-252.

Lovick, T.A., Convergent afferent inputs to neurones in nucleus paragigantocellularis lateralis in the cat, Brain Res., 456 (1988) 183-187.

Luppi, P.H., Sakai, K., Fort, P., Salvert, D. and Jouvet, M., The nuclei of origin of monoaminergic, peptidergic, and cholinergic afferents to the cat nucleus reticularis magnocellularis: a double-labeling study with cholera toxin as a retrograde tracer, J. Comp. Neurol., 277 (1988) 1-20.

Mason, P. and Fields, H.L., Axonal trajectories and terminations of on- and off-cells in the cat lower brainstem, J. Comp. Neurol., 288 (1989) 185-207.

Mason, P., Floeter, M.K. and Fields, H.L., Somatodendritic morphology of on- and off-cells in the rostral ventromedial medulla, J. Comp. Neurol., 301 (1990) 23-43.

Mason, P., Strassman, A. and Maciewicz, R., Pontomedullary raphe neurons: monosynaptic excitation from midbrain sites that suppress the jaw opening reflex, Brain Res., 329 (1985) 384-389.

Morton, C.R., Duggan, A.W. and Zhao, Z.Q., The effects of lesions of medullary midline and lateral reticular areas on inhibition in the dorsal horn produced by periaqueductal grey stimulation in the cat, Brain Res., 301 (1984) 121-130.

Oliveras, J.L., Redjemi, F., Guilbaud, G. and Besson, J.M., Analgesia induced by electrical stimulation of the inferior centralis nucleus of the raphe in the cat, Pain, 1 (1975) 139-145.

Prieto, G.J., Cannon, J.T. and Liebeskind, J.C., Nucleus raphe magnus lesions disrupt stimulation-produced analgesia from ventral but not dorsal midbrain areas in the rat, Brain Res., 261 (1983) 53-57.

Ramirez, F. and Vanegas, H., Tooth pulp stimulation advances both medullary off-cell pause and tail flick, Neurosci. Lett., 100 (1989) 153-156.

Ramon-Moliner, E. and Nauta, W.J.H., The isodendritic core of the brain stem, J. Comp. Neurol., 126 (1966) 311-336.

Ross, C.A., Ruggiero, D.A., Joh, T.H., Park, D.H. and Reis, D.J., Rostral ventrolateral medulla: Selective projections to the thoracic autonomic cell column from the region containing C1 adrenaline neurons, J. Comp. Neurol., 228 (1984) 168-185.

Sagen, J. and Proudfit, H.K., Evidence for pain modulation by pre- and postsynaptic noradrenergic receptors in the medulla oblongata, Brain Res., 331 (1985) 285-293.

Sakai, K., Some anatomical and physiological properties of pontomesencephalic tegmental neurons with special reference to the PGO waves and postural atonia during paradoxical sleep in the cat, In: The Reticular Formation Revisited, Hobson J.A. and Brazier M.A.B. (Eds.), Raven Press, New York, 1980, pp. 427-448.

Sandkühler, J. and Gebhart, G.F., Relative contributions of the nucleus raphe magnus and adjacent medullary reticular formation to the inhibition by stimulation in the periaqueductal gray of a spinal nociceptive reflex in the pentobarbital-anesthetized rat, Brain Res., 305 (1984) 77-87.

Satoh, M., Akaike, A., Nakazawa, T. and Takagi, H., Evidence for involvement of separate mechanisms in the production of analgesia by electrical stimulation of the nucleus reticularis paragigantocellularis and nucleus raphe magnus in the rat, Brain Res., 194 (1980) 525-529.

Sessle, B.J. and Hu, J.W., Raphe-induced suppression of the jaw-opening reflex and single neurons in trigeminal subnucleus oralis, and influence of naloxone and subnucleus caudalis, Pain, 10 (1981) 19-36.

Sessle, B.J., Hu, J.W., Dubner, R. and Lucier, G.E., Functional properties of neurons in cat trigeminal subnucleus caudalis (medullary dorsal horn). II Modulation of response to noxious and non-noxious stimulation by periaqueductal gray, raphe magnus, cerebellar cortex, and afferent influences, and effect of naloxone, J. Neurophysiol., 45 (1981) 193-207.

Taylor, J.R., Elsworth, J.D., Garcia, E.J., Grant, S.J., Roth, R.H. and Redmond, D.E., Clonidine infusion into the locus coeruleus attenuates behavioral and neurochemical changes associated with naloxone-precipitated withdrawal, Psychopharmacology, 96 (1988) 121-134.

Vanegas, H., Barbaro, N.M. and Fields, H.L., Tail-flick related activity in medullospinal neurons, Brain Res., 321 (1984) 135-141.

Wilcox, R.E., Mikula, J.A. and Levitt, R.A., Periaqueductal gray naloxone microinjection in morphine-dependent rats: Hyperalgesia without "classical" withdrawal, Neuropharmacology, 18 (1979) 639-641.

Willis, W.D., Haber, L.H. and Martin, R.F., Inhibition of spinothalamic tract cells and interneurons by brainstem stimulation in the monkey, J. Neurophysiol., 40 (1977) 968-981.

Zhuo, M. and Gebhart, G.F., Descending facilitation (and inhibition) from the nuclei reticularis gigantocellularis (NRGC) and gigantocellularis pars alpha (NRGCα) in the rat, Soc. Neurosci. Abst., 15 (1990) 152.

Zorman, G., Hentall, I.D., Adams, J.E. and Fields, H.L., Naloxone-reversible analgesia produced by microstimulation in the rat medulla, Brain Res., 219 (1981) 137-148.

Localization of Putative Amino Acid Transmitters in the PAG and their Relationship to the PAG-Raphe Magnus Pathway

Alvin J. Beitz and Frank G. Williams

Department of Veterinary Biology
University of Minnesota
U.S.A.

Introduction

The vast majority of synapses in the central nervous system (CNS) appear to use excitatory amino acids as their neurotransmitters (Cotman et al., 1987; Monaghan et al., 1989; Watkins et al., 1990). Although Curtis et al. (1959; 1960) first provided conclusive evidence over 3 decades ago that glutamate and aspartate exert a powerful excitatory action on neurons, it was only during the past 15 years that glutamate and aspartate have been seriously considered as excitatory neurotransmitters in the CNS. Several other amino acids and dipeptides have been suggested as putative excitatory amino acid transmitter candidates in the CNS, however, the majority of evidence to date favors glutamate and aspartate as the most likely candidates for neurotransmitters in the brain. Thus glutamate and aspartate have largely been shown to fulfill the criteria for a neurotransmitter, e.g., Ca^{2+}-dependent release upon stimulation, high affinity uptake into nerve terminals, presence of the amino acids and synthetic enzymes in nerve terminals, blockade of synaptic transmission by excitatory amino acid antagonists, and identity of action (Cotman et al., 1987; Fonnum, 1984; Nicholls, 1989; Watkins and Evans, 1981).

From a historical perspective the first indication that glutamate may play a role in the periaqueductal gray matter (PAG) stems from a report by Sherman and Gebhart (1975) which indicated that pain significantly reduced glutamate levels in the PAG, while systemic morphine significantly elevated glutamate levels in this region. Although it was not clear whether these changes reflected alterations in the metabolic or neurotransmitter pool of glutamate, this study did suggest that neurochemical changes in the PAG may underlie its role in morphine analgesia. Behbehani and Fields (1979) provided the first indication that glu-

The Midbrain Periaqueductal Gray Matter, Edited by A. Depaulis and
R. Bandler, Plenum Press, New York, 1991

305

tamate may serve a transmitter role in the PAG based on their demonstration that microinjections of glutamate into the rodent PAG produced a potent analgesia presumably by acting on excitatory amino acid receptors in this region. It was subsequently shown that microinjections of glutamate or aspartate into the midbrain PAG also produced an immediate defense reaction in the cat and rat (Bandler, 1982; Bandler et al., 1985; Bandler and Depaulis, this volume) and that injections of glutamate into the monkey PAG induced vocalization (Jürgens and Richter, 1986). Moreover, *in vitro* receptor binding studies have demonstrated high levels of excitatory amino acid receptors in the PAG (Cotman et al., 1987; Greenamyer et al., 1984; Halpain et al., 1984). Together these studies suggest that glutamate and aspartate act on excitatory amino acid receptors within the PAG to activate components of the endogenous analgesia system, to initiate defensive behavior or to induce vocalization. Further evidence to support a transmitter role for endogenous glutamate and aspartate in the PAG comes from immunocytochemical studies demonstrating glutamate and aspartate immunostaining in neurons, axons and axon terminals in this region (Clements et al., 1987) and from electrophysiological and microdialysis studies (see below).

This chapter will present new data concerning the ultrastructural localization of these two amino acids in the PAG and will summarize the association of excitatory amino acids with certain PAG afferent systems. In addition we will discuss the distribution of excitatory amino acid receptors in the PAG and will review data related to the electrophysiological responses of excitatory amino acids in this region. Finally, recent data will be presented that strongly implicates glutamate and aspartate as neurotransmitters of the PAG-raphe magnus pathway.

Immunohistochemical Localization of Glutamate and Aspartate in the PAG

The first demonstration of glutamate immunoreactivity in the PAG was provided by Ottersen and Storm-Mathisen (1984a,b). These investigators described high proportions of glutamate immunostained cells in the peripheral PAG and low proportions of immunoreactive neurons near the mesencephalic aqueduct. Subsequently our laboratory mapped the distribution of glutamate, its putative synthesizing enzyme, glutaminase, aspartate and aspartate aminotransferase (AATase, an enzyme which interconverts glutamate and aspartate) within the PAG (Clements et al., 1987). Immunoreactivity for all four of these substances was found throughout the rostrocaudal extent of the PAG with the heaviest concentration of stained neurons occurring in the periphery. These data confirm Ottersen and Storm-Mathisen's description of a peripheral concentration of glutamate immunoreactivity in this region and further indicate that aspartate has a similar distribution.

We have recently repeated our immunohistochemical analysis of excitatory amino acids in the PAG using a new monoclonal glutamate antibody that has been extensively characterized (McDonald et al., 1989) as well as polyclonal antibodies against glutamate and aspartate (obtained from Drs. A. Rustioni and P. Petrusz) that have also been well characterized (Hepler et al., 1988). The results of this recent analysis confirm our previous findings and in addition indicate that equal numbers of aspartate- and glutamate-like immunoreactive neurons are present in the PAG. Glutamate and aspartate immunoreactive neurons display a similar distribution throughout the PAG and, as indicated in Figure 1, these neurons are most concentrated in the outer portion of this structure. Glutamate and aspartate immunoreactivities are present in all three of the major cell types (triangular, fusiform and multipolar) within this region. Typically, aspartate immunoreactive neurons display higher intensities of staining throughout the PAG than do glutamate immunoreactive cells.

Since the distribution of aspartate and glutamate immunoreactive neurons was very similar, double-labeling experiments were performed to determine if

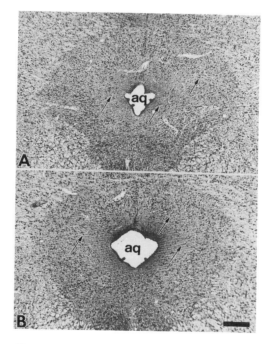

Figure 1: A. Low magnification photomicrograph of glutamate-like immunoreactivity in the PAG at the level of the oculomotor nucleus. Glutamate immunoreactive neurons appear as small black dots (arrows) and are concentrated in a peripheral position. aq = mesencephalic aqueduct. B. Low magnification photomicrograph of aspartate immunoreactive neurons (arrows) at the level of the oculomotor nucleus. Bar = 250 μm.

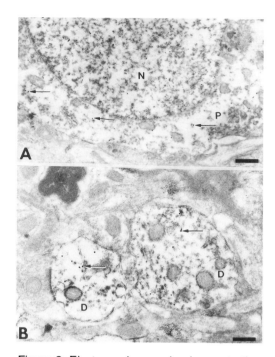

Figure 2: Electron micrographs demonstrating the colocalization of glutamate and aspartate in PAG neurons. A. Neuronal perikarya (P) displaying both glutamate immunoreactivity (indicated by the dark homogeneous DAB reaction product which is evident throughout the cytoplasm and the nucleus) and aspartate immunoreactivity (10 nm gold particles, arrows). The nucleus (N) is also evident. Bar = 0.5 μm. B. Two double labeled dendrites (D) are illustrated. Note the aspartate immunolabeling indicated by the colloidal gold particles (arrows) and the homogeneous DAB reaction product (which can be seen surrounding mitochondria and microtubules) indicating glutamate immunoreactivity. Bar = 0.5 μm.

glutamate and aspartate immunostaining was colocalized in PAG neurons (Beitz, 1990b). The results of these experiments indicated that 95.2% of neurons containing glutamate-like immunoreactivity also contained aspartate immunostaining. Double-labeled perikarya were found throughout the rostrocaudal extent of the PAG and within all PAG subdivisions. These light microscopic studies indicate that neurons throughout the PAG that contain glutamate also co-contain asparate. Ultrastructural examination of the PAG also revealed double labeled perikarya (Fig. 2A) and dendrites (Fig. 2B) confirming that these two amino acids are co-contained in many PAG neurons. Colocalization of glutamate and aspartate in

PAG neurons is not surprising since, as indicated above, the neurons in this region also exhibit immunostaining for the enzyme, aspartate aminotransferase, which converts aspartate to glutamate in a reversible manner.

One caveat in the interpretation of these immunocytochemical data is that the presence of a neuroactive compound in a neuron does not necessarily indicate that it is releasable as a transmitter. This is especially true when dealing with excitatory amino acids since both glutamate and aspartate also serve metabolic roles. In fact, the metabolic pools of glutamate and aspartate may be equal to or exceed the sizes of the respective transmitter pools in terms of total brain content (Fonnum, 1984). This may be particularly relevant with respect to glutamate at the level of the neuronal perikaryon, which is distant from the site of transmitter

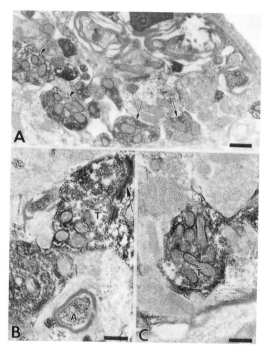

Figure 3: Electron micrographs illustrating glutamate or aspartate immunoreactivity in axon terminals in the PAG. A. Glutamate-like immunoreactivity is present in four terminals (arrows) each containing clear rounded vesicles and several mitochondria. Bar = 0.75 μm. B & C. Aspartate-like immunoreactivity in axonal terminals (T) in the ventrolateral PAG. The terminal in B makes synaptic contact (arrow) with an aspartate immunoreactive dendrite. A lightly myelinated axon (A) is also present and displays immunoreactivity. for aspartate. Bar in B & C = 0.35 μm.

release and where the amount of glutamate immunostaining may be more close-ly related to the size of the metabolic pool than to the transmitter pool (Meeker et al., 1989; Yingcharoen et al., 1989). The intensity of glutamate and aspartate immunostaining varies from neuron to neuron in the PAG, which probably reflects different concentrations of these two amino acids among PAG neurons. Whether the most intensely immunostained PAG cells are neurons that utilize glutamate or aspartate as transmitter substances or simply are neurons that utili-ze a large amount of these two amino acids for metabolic purposes is a difficult question to answer based on our present state of knowledge.

Of greater relevance to possible neurotransmitter roles of glutamate and aspartate in the PAG is data demonstrating that these amino acids are localized in synaptic terminals in this region. As indicated in Figure 3 both glutamate- and aspartate-like immunoreactivities can be identified in PAG axonal terminals sug-gesting that these amino acids may be released from synaptic boutons and affect postsynaptic excitatory receptors. Consistent with such an interpretation are data presented by Ottersen and Storm-Mathisen (1984b) which demonstrate that within the mesencephalic tegmentum the highest uptake of D-[³H]aspartate (a metabolically inert substrate for the acidic amino acid uptake systems) occurs in the PAG, indicating the presence of a relatively large number of excitatory amino acid containing nerve terminals within this region. Glutamate and aspartate immunoreactive terminals were dome-shaped or elongated, contained round, clear synaptic vesicles and formed asymmetrical synapses (Fig. 3). In many ins-tances the active zone was quite elongated (Fig. 3C). Axodendritic contacts were the most common type formed by both glutamate and aspartate immunoreactive terminals. These data together with preliminary microdialysis data (see below) provide strong evidence favoring a neurotransmitter role for glutamate and aspartate in the PAG.

Origin of Excitatory Amino Acid Projections to the PAG

Since the PAG contains a substantial number of glutamate and aspartate immunoreactive terminals it is important to define the origin of these putative excitatory amino acid containing axonal inputs to this region. Data addressing this issue stems primarily from studies employing the retrograde transport of D-[³H]aspartate or the retrograde transport of wheat germ agglutinin-HRP (WGA-HRP) in combination with immunocytochemistry for glutamate. Kalén et al. (1985) provided some insight into the origin of excitatory amino acid projections to the PAG in their study of excitatory amino acid afferents to the nucleus raphe dorsalis. Based on the results of injections of D-[³H]aspartate that spread outside the raphe dorsalis into the surrounding PAG, these investigators suggested that the PAG may receive excitatory amino acid projections from the central amygda-loid nucleus, the mammillary area and possibly the inferior colliculus. Christie and coworkers (1986) subsequently demonstrated an excitatory amino acid pro-jection from the prefrontal cortex to the PAG using lesioning and neurochemical procedures. An excitatory amino acid projection from the ventromedial hypothalamus to the PAG has also been shown using retrograde tracing of D-[³H]aspartate and electrophysiological procedures (Beart et al., 1988). Most recently Jiang and Behbehani (1990) have demonstrated an excitatory amino acid projection from the insular cortex to the PAG using electrophysiological techniques.

The first comprehensive studies of possible excitatory amino acid inputs to the PAG were performed by Beart et al. (1990) using the retrograde D-[³H]aspartate procedure and by our laboratory using retrograde tracing of WGA-HRP in combination with glutamate immunocytochemistry (Beitz, 1989). The origins of the major excitatory amino acid inputs to the PAG based on this work are summarized in Table I.

Table I: Major sources of excitatory amino acid afferents to the PAG.

Source	D-[³H]aspartate Labeling	WGA-HRP-glutamate
Zona Incerta	+++*	+++
Ventromedial Hypothalamus	+++	+
Posterior Hypothal. Area	+++	0
Cingulate Cortex	++	+++
Nucleus Cuneiformis	++	+++
Perirhinal Cortex	++	+
Inferior Colliculus	++	0
Frontal Cortex	+	++
Dorsomedial Hypoth. Nucleus	+	++
Anterior Pretectal Nucleus	+	++
Amygdala	+	+
Contralateral PAG	+	+

* The relative density of labeled perikarya in each CNS area is indicated by the number of plus signs (+). Absence of labeling is indicated by "0".

These data indicate that the zona incerta, hypothalamus and the cerebral cortex provide substantial excitatory amino acid projections to the PAG, while minor contributions arise from several other CNS areas. The lack of glutamate-like immunoreactive retrogradely-labeled neurons in the posterior hypothalamic area and the inferior colliculus and the small amount of such labeling in the ventromedial hypothalamic nucleus and the perirhinal cortex may imply that these projections utilize aspartate rather than glutamate as a neurotransmitter. However, definite conclusions regarding the exact transmitter of these pathways cannot be drawn based solely on immunocytochemical localization of amino acids in their perikarya of origin because of the problem of distinguishing the transmitter pool from the metabolic pool as discussed above. It is interesting to note that the contralateral PAG was labeled in both the D-[³H]aspartate retrograde tracing study and the WGA-HRP retrograde labeling-immunocytochemical study suggesting that some PAG interneurons may utilize excitatory amino acids as transmitters.

Excitatory Amino Acid Binding Sites in the PAG

Traditionally excitatory amino acid receptors have been classified based on their prototype agonists into three major types: N-methyl-D-aspartate (NMDA), alpha-amino-3-hydroxy-5-methyl-4-isoxazolepropionic acid (AMPA, originally termed the quisqualate receptor) and kainate, all of which act by control of ion channels (Collingridge and Lester, 1989;Watkins et al., 1990; Young and Fagg, 1990). In addition to these 3 receptor subtypes, a receptor activated by quisqualate, ibotenate and trans-ACPD and linked to inositol phospholipid metabolism (the so-called metabotropic receptor) has been described (Watkins et al., 1990; Young and Fagg, 1990). Although autoradiographic studies have demonstrated the existence of excitatory amino acid binding sites in the PAG (Cotman et al., 1987; Greenamyre et al., 1984; Halpain et al., 1984; Rainbow et al., 1984) as indicated above, it is only recently that specific receptor subtypes have been analyzed and compared among PAG subdivisions (Albin et al., 1990). Albin and coworkers studied the localization of NMDA, AMPA, kainate and quisqualate-metabotropic binding sites within the PAG and found that they exhibited a heterogeneous distribution. Although it is important to keep in mind that the differences between the greatest and lowest density of binding sites in the study by Albin et al. (1990) was less than 2x, the fact that these investigators identified the greatest density of all receptor subtypes in the dorsolateral PAG and the lowest density in the ventrolateral PAG is relevant to the concept of PAG subdivisions. Unfortunately these investigators limited their analysis to one rostrocaudal level of the PAG (located approximately 6.0 mm caudal to bregma) and thus information concerning the distribution of these receptors in the caudal PAG is lacking. However, Gundlach (see Gundlach, this volume) has recently examined the distribution of kainate binding sites throughout the rostrocaudal extent of the PAG. Gundlach's data confirms the high density of kainate binding sites in the rostral, dorsolateral PAG as reported by Albin et al., but in addition, his data indicate that in the caudal PAG kainate binding is also high in the ventrolateral and ventromedial areas. Whether the other excitatory amino acid subtypes are also present in high levels in the ventrolateral and ventromedial areas of the caudal PAG remains to be determined.

With respect to excitatory amino acid subtypes Albin et al. (1990) found that relative to regions of the CNS with high densities of excitatory amino acid receptors, the quisqualate-metabotropic receptor subtype had the highest relative density in the PAG, while NMDA receptors were least dense. Based on this data one would predict that quisqualate should have a greater effect on neurons in this region than NMDA. Recent electrophysiological data from Dr. M. Behbehani's laboratory (see below) indicates that both quisqualate and glutamate are highly effective in activating PAG neurons, which is consistent with the pharmacological profile of a high density of quisqualate sensitive receptors in the PAG.

Since aspartate appears to be a selective NMDA receptor agonist (Patneau and Mayer, 1990), it might be reasonable, based on the binding data, to assume that this amino acid plays a less important role in PAG function than glutamate. Recent microdialysis data from our laboratory indicate that KCl depolarization of the PAG results in similar percent increases in the amount of aspartate and glutamate in dialysates obtained from this region. Thus following perfusion through a dialysis probe implanted into the PAG in four rats with a solution containing 200 mM KCl, there was a 260.5% (±18% SEM) increase in the amount of aspartate in the dialysate compared to baseline levels and a 241% (±13% SEM) increase in glutamate (see below for details of the microdialysis procedure). Interestingly, when veratridine was perfused through the microdialysis system, aspartate increased approximately 1,300% and glutamate approximately 865%. These results indicate that both amino acids are released in the PAG following depolarization of this region. However, the mean baseline concentration of aspartate (2.18 picomoles) in PAG dialysates is approximately one tenth the mean baseline concentration of glutamate (18.42 picomoles). Thus although both amino acids show similar percent increases following potassium depolarization and aspartate shows a larger percent increase upon depolarization by veratridine, there is actually a larger amount of glutamate released. This would support the hypothesis that glutamate may be the more important of the two putative excitatory amino acid transmitters in the PAG. In addition the lower levels of releasable aspartate in this region parallels the low density of NMDA receptors.

Some caution must be used in interpreting the above receptor binding data since it has recently been suggested that the pharmacologically defined kainate binding site and the physiologically defined kainate receptor are separate entities (Johnson and Koerner, 1988; Wenthold, et al., 1990). In fact the electrophysiological effects of kainate appear to be mediated by AMPA receptors (Keinänen et al., 1990; Zorumski and Yang, 1988) indicating that the kainate receptor and AMPA receptor are one and the same. Finally, Keinänen and coworkers (1990) have demonstrated the existence of multiple glutamate receptors that display characteristic AMPA pharmacology and which are abundantly and differentially expressed in the brain. It appears that a reclassification of excitatory amino acid receptors on a molecular basis is warranted.

Electrophysiological Effects of Excitatory Amino Acids in the PAG

There have been very few electrophysiological studies that have examined the effects of excitatory amino acid agonists and antagonists on PAG neurons. The initial report by Behbehani and Fields (1979) demonstated that glutamate microinjected into the PAG increased the threshold of a flexion reflex elicited by thermal stimuli to the hindpaw of a rat and also caused excitation of neurons in the nucleus raphe magnus via the PAG-raphe magnus pathway. These data suggested that glutamate has an excitatory effect on neurons in the PAG presumably by acting on excitatory amino acid receptors. More recently Beart and coworkers (1988) examined the effects of iontophoretically applied excitatory amino acid agonists and antagonists on PAG neurons in their studies of ventromedial hypothalamic excitatory input to the PAG. These investigators found that the nonspecific excitatory amino acid antagonist, kynurenate reduced the evoked responses of 8 out of 18 PAG neurons to below 25% of control while the specific NMDA antagonist, APV reduced the responses of 7 out of 16 neurons. Kynurenate was also found to reduce the responses of PAG neurons to NMDA, quisqualate and kainate equally, while APV was selective for NMDA. These data indicate that neurons in the PAG are responsive to all three of the traditional receptor subtype agonists. There are some limitations to these studies that should be considered in the interpretion of this data. Perhaps the most important relates to the interpretation of NMDA effects. Although the weak effects of NMDA agonists might indicate a minor contribution of NMDA receptors to actions in the PAG, these studies suffer from the weakness that high magnesium concentrations are present. Thus these weak effects may have been due to blockade of NMDA receptors by magnesium (e.g., see Collingridge and Lester, 1989) and NMDA may in fact play a greater role than these studies would suggest.

In addition Beart and colleagues (1988) demonstrated two types of cell sensitivity to excitatory amino acid antagonists. One cell type was sensitive to both kynurenate and APV, while the other cell type was selective for kynurenate. This finding is consistent with the binding data mentioned above that indicates a lower density of NMDA than nonNMDA receptors in the PAG. More recently Behbehani (personal communication) has shown that neurons in this region are

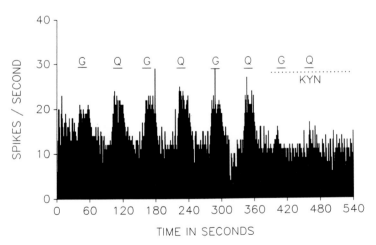

Figure 4: The effect of iontophoretically applied quisqualate (Q) and glutamate (G) on a PAG cell recorded in vivo. The concentration of glutamate was 100 mM and quisqualate was 5 mM (pH 8.5). Both drugs were ejected with a current of -40 nA and were retained with +10 nA. Kynurenate (KYN) blocked the effect of both glutamate and quisqualate.

sensitive to both glutamate and quisqualate. Both compounds caused an increase firing rate of PAG neurons upon iontophoretic application (see Fig. 4) which was blocked by administration of kynurenate. This finding further supports a transmitter role for glutamate in the PAG. The effects of aspartate or the selective excitatory amino acid agonist, AMPA, remain to be determined.

Excitatory Amino Acids and the PAG-Raphe Magnus Projection

The PAG was originally shown by Reynolds (1969) to be an effective site for stimulation produced analgesia and this was subsequently confirmed by many investigators (see reviews by Basbaum and Fields, 1984; Fields and Besson, 1988; Besson et al., this volume). There are several descending pathways from the PAG that participate in pain modulation, but the projection from the PAG to the ventrolateral medulla terminating in the nucleus raphe magnus (NRM) and adjacent reticular nuclei appears to be a key component of this endogenous pain modulation system (Beitz, 1990a; Fields and Besson, 1988; Mason, this volume). Although previous studies have shown that this pathway is predominantly excitatory (Behbehani and Fields, 1979) and contains serotonin and several neuropeptides (Beitz et al., 1983; Luppi et al., 1988), recent evidence suggests that excitatory amino acids may also be important neurotransmitters in this projection (Aimone and Gebhart, 1986; Wiklund et al., 1988). In order to determine if glutamate and/or aspartate are possible transmitters of this projection, we have recently employed a combined retrograde D-[³H]aspartate tracing-immunocytochemical procedure (Mullett et al., 1989) as well as *in vivo* microdialysis combined with HPLC (Beitz, 1990b). The results of these studies will be summarized below.

D-[³H]aspartate retrograde labeling in combination with immunocytochemistry

Several lines of evidence now suggest that the high affinity uptake and retrograde transport of D[³H]aspartate represents a powerful technique for localizing pathways that utilize excitatory amino acids as transmitters (Beart et al., 1988; 1990; Wiklund et al., 1988). Unfortunately since glutamate and aspartate appear to be taken up by the same high affinity reuptake mechanism associated with nerve terminals, this procedure does not distinguish between D-aspartate, L-aspartate and L-glutamate. In collaboration with Drs. L. Wiklund and S. Araneda (C.N.R.S., Gif-sur-Yvette, France) we have developed a procedure which combines retrograde D-[³H]aspartate labeling with immunocytochemical detection of L-glutamate and L-aspartate (Araneda et al., 1991; Mullett et al., 1989). Fifteen to 19 hr following an injection of 50 nl of radiolabeled D-aspartate (10^2 M) into the NRM, rats were anaesthetized and fixed by transcardiac perfusion with 3.5% glutaraldehyde. Brains were sectioned at 30 μm thickness on a freezing microtome and alternate brainstem sections were processed for immunocytochemical detection of L-Asp or L-Glu, respectively. Following immunocytochemical processing the sections were mounted on subbed slides and processed for autoradiographic localization of D-[³H]aspartate using Kodak NTB-2 emulsion. The autoradiograms were developed in Dektol and the sections examined by both brightfield and darkfield microscopy (see Araneda et al., 1991 for further details of the methodology). It should be pointed out that there are several limitations associated with the retrograde D-[³H]aspartate tracing procedure and these have been reviewed by Cuénod and Streit (1983). These include the appearance of 'false-negative' and 'false positive' results. It is unlikely that these limitations play a significant role in the interpretation of the data presented

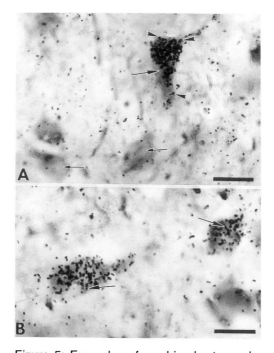

Figure 5: Examples of combined retrograde D-[³H]aspartate-immunohistochemical labeling in the ventrolateral PAG. The D-[³H]aspartate labeling is indicated by the presence of black silver grains in the overlying photographic emulsion (arrowheads in A, arrows in B) while the immunohistochemical labeling appears as a homogeneous DAB reaction product. A. A neuron (large arrow) containing both the radiolabeled D-aspartate (arrowheads) and glutamate immunoreactivity. Several other single labeled glutamate immunoreactive neurons are present in the field (small arrows). B. Two double labeled neurons that display both aspartate immunoreactivity and retrogradely transported D-[³H]aspartate (arrows). Bar in C & D = 10 μm.

below because there is significant physiological, pharmacological and behavioral evidence for the presence of an excitatory amino acid projection from the PAG to the raphe magnus as indicated above.

Figure 5 illustrates an example of the resulting double labeling of neuronal perikarya in the ventrolateral midbrain PAG. Approximately 98% of D-[³H]aspartate retrogradely labeled cells were found to be immunoreactive for L-Glu, while on alternate sections 97% of D-[³H]aspartate retrogradely labeled neurons were immunoreactive for L-Asp. This data in combination with the immunocytochemical studies described above showing that glutamate and aspartate are colocalized in PAG neurons raises the possibility that PAG-NRM projection neurons corelease glutamate and aspartate. This concept is further supported by recent microdialysis data demonstrating that both glutamate and aspartate are released in the NRM following PAG stimulation (see below).

Although this is an attractive hypothesis, it should be pointed out that there are several other possible interpretations from this work. Since the presence of glutamate or aspartate immunoreactivity in neuronal perikarya may also reflect metabolic pools of these two amino acids, it is also possible that the immunostaining in PAG perikarya represents the nontransmitter, metabolic pool of one or both amino acids. On the other hand, the presence of AATase in PAG perikarya implies that one of these two amino acids may serve as a precursor for the other (see Beitz, 1990b for a further discussion of this issue). The exact relationship between the perikaryal contents of glutamate and aspartate and the releasable neurotransmitter pool associated with the axonal terminals derived from these perikarya obviously requires further clarification.

Analysis of amino acid release into the NRM following PAG stimulation

In vivo microdialysis represents an important tool for monitoring extracellular levels of amino acids or other neurochemicals within specific brain regions (Benveniste, 1989). Since this procedure can be used in awake, freely moving animals, it allows analysis of extracellular levels of amino acids before, during and after stimulation of CNS pathways without the complicating effects of anesthetics. This technique has recently been employed in our laboratory to investigate amino acids associated with the NRM and the PAG-NRM pathway (Beitz, 1990b). Rats are first anesthetized and implanted with chronic guide cannulae into both the PAG and NRM (Beitz, 1990b). The PAG cannula is used for the microinjection of drugs, while a microdialysis probe is inserted into the NRM cannula 24 hr prior to the start of a microdialysis experiment. Our microdialysis probes are constructed from 25 gauge stainless steel tubing into which a dialysis fiber (200 μm diameter; 5,000 molecular weight cutoff) is inserted approximately 1-2 mm distance and glued into place with epoxy as previously described (Beitz, 1990b). The opposite end of the stainless steel tubing is attached to a fluid swivel via a piece of microline tubing. Silica capillary tubing (145 μm o.d.) is threaded into the dialysis fiber via a small slit in the microline tubing and serves as the outflow route from the dialysis probe. The dialysis system is attached to a peristaltic pump and 60 µl samples are collected via a Gilson microfraction collector. The microdialysis probe is perfused with Ringers solution at a flow rate of 3-5 µl/min and samples are collected at 12 min intervals in polypropylene tubes and maintained at 5°C until analyzed for amino acids by HPLC.

The mean basal concentration of amino acids in the dialysate samples obtained from the NRM in eight rats was 15.83 ± 4.55 picomoles of glutamate, 1.95 ± 0.45 picomoles of aspartate, 67.80 ± 10.78 picomoles of glycine and 11.83 ± 1.12 picomoles of taurine. Glycine and glutamate were present in the highest amounts in the dialysis samples, whereas aspartate was present in the lowest. The basal concentration of glutamate was 8-10 times that of aspartate. In more recent experiments we have also measured the concentration of GABA in our dialysate samples. The mean basal concentration of GABA in the dialysate was 0.83 ± 0.28 picomoles (n=4) and thus it is present in lower amounts than aspar-

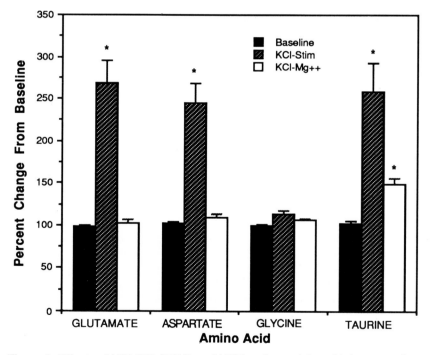

Figure 6: Effects of KCl (KC-STIM) and KCl in a low calcium, high magnesium solution (KC-Mg++) on basal levels of amino acids in the dialysate obtained from the NRM. The mean baseline concentration of each amino acid in the dialysate is set at 100% and is based on the mean concentration of 5 baseline samples per rat obtained from 8 rats. An asterisk indicates p<0.025 by Student's paired t-test.

tate. In order to analyze the releasable pool of aspartate and glutamate in the NRM, the NRM was depolarized with 200 mM KCl perfused through the dialysis probe for 10 min. Perfusion with KCL increased the concentration of glutamate in the dialysate by 168%, of aspartate by 145% and of taurine by 163% as shown in Figure 6.

The concentration of glycine was not significantly increased following treatment with KCl suggesting that this amino acid may not play a significant transmitter role in the NRM. This is consistent with the low levels of glycine binding sites observed in this region. On the other hand, when the $CaCl_2$ in our KCl solution was replaced with $MgCl_2$, this prevented the increase in glutamate and aspartate in the dialysate (Fig. 6), suggesting that the release of these amino acids is calcium dependent. Since the potassium evoked efflux of these two transmitter amino acids was calcium dependent it was considered to reflect release from the transmitter pool (Paulsen and Fonnum, 1989). In contrast the addition of $MgCl_2$ only partially reduced the increase in taurine evoked by KCl stimulation, an observation consistent with the numerous proposed nontransmitter roles for this amino acid (Huxtable, 1989). In addition to this demonstration of a calcium sensitive, potassium evoked release of glutamate and aspartate, we have recently perfused 200 µM veratridine (a voltage sensitive, sodium channel activator) through the dialysis system in three animals. Veratridine caused a mean increase of 329.8% in the amount of aspartate in our dialysate and a mean increase of 231% in the amount of glutamate in the dialysate. These data are also consistent with release of glutamate and aspartate from a transmitter pool in the NRM. Interestingly, 200 µM veratridine also caused a mean increase of 902% in the baseline amount of GABA in the dialysate, suggesting that a relatively large releasable pool of GABA is present in the NRM.

In order to determine if stimulation of the PAG causes release of excitatory amino acids from the PAG-NRM pathway, microinjections of the excitatory amino acid agonist, D-L-homocysteic acid (DLH) were made into the PAG and microdialysis was performed in the NRM. All rats used for these experiments were behaviorally tested with the tail flick test to assure that PAG injections of DLH caused analgesia. The tail flick reflex was evoked by focused radiant heat applied to the underside of the tail of unanesthetised, unrestrained rats using a tail flick analgesia meter obtained from IITC Life Sciences U.S.A. A cutoff latency of 7 seconds was used in all experiments. Only animals that demonstrated analgesia in the tail flick test following DLH microinjections into the PAG were used for these studies. Following an injection of 100 mM DLH into the caudal ventrolateral PAG, the amount of aspartate in the dialysate was found to increase by 253.8% compared to the injection of vehicle in the same animal (see Fig. 7).

The mean concentration of taurine in the dialysate samples increased 158% and the mean concentration of glutamate increased 56% following DLH injection into the PAG. The fact that aspartate increased 253.8% following PAG stimulation compared to a 56% increase in glutamate would imply that aspartate may be a more important transmitter in the PAG-NRM pathway. However, it is important to keep in mind that the baseline concentration of glutamate is

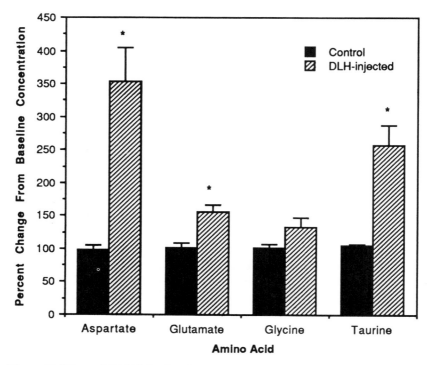

Figure 7: Effect of DLH injection into the PAG on the concentration of amino acids in the NRM dialysate compared to injections of vehicle into the PAG. The mean baseline concentration of each amino acid in the dialysate is set at 100% and is based on the mean concentration of 5 baseline samples per rat obtained from 6 rats. The effect of injections of DLH or vehicle (control) into the PAG on the concentration of each amino acid is represented as a percent change from baseline. Values represent the mean ± S.E.M. of 3 dialysate samples from each of 6 rats obtained following injection of vehicle or DLH into the PAG. An asterisk indicates $p<0.05$ by Student's paired t-test.

approximately 10 times higher than aspartate and thus a 56% increase in glutamate is greater in terms of total amino acid concentration than a 253% increase in the concentration of aspartate.

Although we have demonstrated that both glutamate and aspartate are released upon activation of the PAG-NRM pathway, it is important to determine if this release is from PAG axonal terminals in the NRM rather than release from a nontransmitter pool activated by PAG-NRM stimulation. In order to determine if the release is from a transmitter pool, DLH was injected into the PAG, while the NRM was perfused with a solution of Ω-conotoxin (which blocks N-type calcium channels associated with presynaptic terminals). Data from three animals indicates that the release of glutamate and aspartate induced by DLH injection into the PAG was significantly reduced by administration of Ω-conotoxin (aspartate only increased 80% and glutamate increased 21%), while taurine release induced by PAG stimulation was not blocked. Although one possible explanation of this data is that some glutamate and aspartate are released upon PAG stimulation from a nontransmitter pool, another likely possibility is that the Ω-conotoxin, because of its higher molecular weight, does not diffuse outward from the dialysis probe far enough to block release from more distant terminals. The glutamate and aspartate released from these distant terminals would still diffuse to the region of the dialysis probe because of their small size compared to the Ω-conotoxin. Despite the fact that these two possibilities need to be resolved, the fact that Ω-conotoxin blocked a major portion of the PAG-stimulated release provides strong support for the hypothesis that glutamate and aspartate play a transmitter role in the PAG-NRM pathway.

In several experiments DLH was microinjected into the dorsolateral or dorsomedial PAG or into the overlying colliculus at the same AP level as previously (A 1.0 from interaural line). When DLH was injected into either the dorsomedial PAG or superior colliculus no release of excitatory amino acids was observed in the NRM dialysates and no analgesia was observed in the tail flick test. With dorsolateral PAG injections small increases in glutamate and aspartate were observed in 2 of 3 rats and there was some increase in the tail flick latency in the tail flick test. In addition injections of DLH into the dorsolateral or dorsomedial PAG often resulted in behavioral reactions in the rats. Animals that received DLH microinjections into these regions often appeared agitated with rotations, jumps around the cage and sonic vocalizations. Rats with injections in the lateral edge of the PAG often exhibited circling behavior. Injections into the ventrolateral PAG of some animals also caused a defensive reaction that typically lasted from 1-2 minutes.

Effects of morphine on the release of excitatory amino acids in the NRM

Numerous studies have demonstrated the ability of opiates to preferentially inhibit pain transmission. This ability of opiates to elicit analgesia depends to a large degree on supraspinal activation of an endogenous inhibitory system capable of modulating noxious input at the level of the spinal cord (Fields and Besson, 1988; Yaksh and Rudy, 1978; Young et al., 1984). One of the key components of this endogenous modulatory system is the midbrain PAG (Yaksh and Rudy, 1978). Both systemic opiate injection, as well as direct injection of opiates into the PAG, appear to activate a descending projection from the PAG to the ventromedial medulla terminating in part in the NRM (Beitz, 1991; Fields and Besson, 1988; Young et al., 1984). Recent evidence indicates that opiates act by inhibiting an inhibitory interneuron in the PAG which in effect disinhibits the PAG output neuron to the ventromedial medulla (Beitz, 1991; Fields and Besson, 1988; Mason; Morgan, this volume). Since opiates appear to act predominantly in the caudal ventrolateral PAG (Beitz, 1991; Yaksh and Rudy, 1978) in areas where glutamate and aspartate immunoreactive neurons are quite dense and if in fact excitatory amino acids are neurotransmitters in the PAG-NRM projection as the evidence presented above suggest, then one would predict that they should be released in the NRM following activation of this pathway by opiates. To test this hypothesis we administered morphine systemically (5-8 mg/kg i.p.) in 5 rats or directly into the PAG (7.5 µg in 0.5µl) in 5 rats and examined possible release into the NRM with our microdialysis system as described above. All animals involved in these experiments were further tested following the microdialysis experiments to assure that systemic or PAG-injected opiates produced analgesia as measured by the tail-flick test. As indicated in Figure 8, neither systemic nor direct microinjections of morphine caused any significant changes in the spontaneously released excitatory amino acids present in the dialysate samples compared to vehicle injected controls.

Since the PAG-NRM pathway plays a key role in the ability of morphine to produce analgesia following injection into the PAG (Young et al., 1984; Fields and Besson, 1988), and since activation of this pathway causes release of glutamate and aspartate, one would predict that both amino acids should be released upon morphine administration into the PAG. It was thus surprising that morphine administered either systemically or directly into the PAG failed to cause an increase in amino acids in the NRM dialysate. These results lead to the obvious conclusion that morphine activates a nonexcitatory amino acid component of the PAG-NRM pathway, whereas nonselective stimulation of the PAG (using DLH) at the same sites activates an excitatory amino acid component of this pathway. There are several other possible interpretations of the above data. For instance, DLH injection might also have activated indirect excitatory pathways to the NRM. The PAG projects to the parabrachial nucleus, the nucleus paragigantocellularis and several hypothalamic nuclei, which in turn project to the NRM (Beitz, 1991). Several of these indirect projections to the NRM may utilize excitatory amino acid transmitters (Beitz, 1991) and they may have been indirectly activated by DLH microinjection into the PAG, but not by microinjection of morphine into this region. A second possibility is that DLH microinjection into the PAG provides such profound stimulation of the NRM that it activates intrinsic excitatory amino acid systems in the NRM, that are not activated by morphine. Perhaps a more likely explanation, however, is based on electrophysiological work from Behbehani's laboratory (Behbehani and Pomeroy, 1978) which indicates that morphine microinjected into the PAG does not have a very drastic effect on NRM neurons. This data coupled with evidence showing that lesions of the ventromedial medulla adjacent to the NRM block the analgesia produced by morphine microinjected into the PAG but not that produced by electrical stimulation of the PAG (Mohrland et al., 1982; Mason, this volume) suggests that morphine injections into the PAG might cause release of excitatory amino acid transmitters in the region adjacent to the NRM without causing release in the NRM. We are currently examining this latter possibility. Whichever interpretation turns out to be correct, the present data does indicate a chemical distinction (involving excitatory amino acids) between opiate-induced and PAG stimulation produced analgesia at the level of the NRM.

In conclusion this chapter has attempted to provide data indicating that

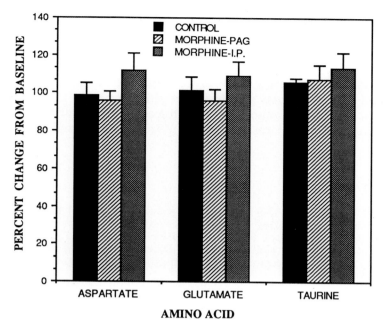

Figure 8: Effects of systemic (morphine-i.p; n=3) versus direct PAG (Morphine-PAG; n=4) injections of morphine on the concentration of amino acids in the NRM dialysate compared to control injections of vehicle into the PAG. The mean baseline concentration of each amino acid in the dialysate is set at 100% and is based on the mean concentration of 5 baseline samples per rat obtained from the 7 rats used in this study. The effect of systemic morphine or direct PAG injection of morphine or vehicle on the concentration of each amino acid is represented as the percent change from baseline. No significant differences were observed.

the excitatory amino acid neurotransmitters, glutamate and aspartate play an important role in the midbrain PAG. Not only are these substances localized in axonal terminals in this region but they are released into the PAG in a calcium dependent fashion and glutamate, at least, has been shown to directly activate PAG neurons. This data together with the results of studies demonstrating excitatory amino acid binding sites in the PAG strongly support a role for excitatory amino acids in this region. Finally, our data demonstrating that both glutamate and aspartate are released into the NRM in a calcium dependent manner following excitatory amino acid stimulation of the PAG suggests that both amino acids serve transmitter roles in the PAG-NRM pathway.

Acknowledgements

The authors would like to thank Mary Mullett and Joan Hautman for their excellent technical assistance. We would also like to thank Drs. Peter Kalivas, Alice Larson, Stephen Skilling and David Smullin for their help and advice in establishing a microdialysis system in our laboratory and Drs. Wiklund and Araneda for their collaboration on the retrograde D[³H]aspartate studies. This work was supported by grants DA06687, NS19208, DE06682 and NS28016.

References

Aimone, L.D. and Gebhart, G.F., Stimulation-produced spinal inhibition from the midbrain in the rat is mediated by an excitatory amino acid neurotransmitter in the medial medulla, J. Neurosci., 6 (1986) 1803-1813.

Albin, R.L., Makowiec, R.L., Hollingsworth, Z., Dure, L.S., Penney, J.B. and Young, A.B., Excitatory amino acid receptors in the periaqueductal gray of the rat, Neurosci. Lett., 118 (1990) 112-115.

Araneda, S., Ghilini, G., Mullett, M., Beitz, A.J. and Wiklund, L., Combination of D-[³H]aspartate retrograde labeling and immunocytochemical detection of L-GLU and L-ASP in neurons of the periaqueductal gray (PAG) projecting to the raphe magnus of the rat, J. Chem. Neuroanat., submitted.

Bandler, R.J., Induction of 'rage' following microinjection of glutamate into midbrain but not hypothalamus of cats, Neurosci. Lett., 30 (1982) 183-188.

Bandler, R.J., Depaulis, A. and Vergnes, M., Identification of midbrain neurons mediating defensive behavior in the rat by microinjection of excitatory amino acids, Behav. Brain Res., 15 (1985) 107-119.

Basbaum, A.I. and Fields, H.L., Endogenous pain control systems: brainstem spinal pathways and endorphin circuitry, Ann. Rev. Neurosci., 7 (1984) 309-338.

Beart, P.M., Summers, R.J., Stephenson, J.A., Cook, C.J. and Christie, M.J., Excitatory amino acid projections to the periaqueductal gray in the rat: a retrograde transport study utilizing D[³H]aspartate and [³H]GABA, Neuroscience, 34 (1990) 163-176.

Beart, P.M., Nicolopoulos, L.S., West, D.C. and Headley, P.M., An excitatory amino acid projection from the ventromedial hypothalamus to periaqueductal gray in the rat: autoradiographic and electrophysiological evidence, Neurosci. Lett., 85 (1988) 205-211.

Behbehani, M.M. and Fields, H.L., Evidence that an excitatory connection between periaqueductal gray and nucleus raphe magnus mediates stimulation produced analgesia, Brain Res., 170 (1979) 85-93.

Behbehani, M.M. and Pomeroy, S.L., Effect of morphine injected in periaqueductal gray on the activity of single units in nucleus raphe magnus of the rat, Brain Res., 149 (1978) 266-271.

Beitz, A.J., Possible origin of glutamatergic projections to the midbrain periaqueductal gray and deep layer of the superior colliculus of the rat, Brain Res. Bull., 23 (1989) 25-35.

Beitz, A.J., Central Gray, In: The Human Nervous System, Paxinos G., (Ed.), Academic Press, San Diego, 1990a, pp. 307-320.

Beitz, A.J., The relationship of glutamate and aspartate to the periaqueductal gray-raphe magnus projection: analysis using immunocytochemistry and microdialysis, J. Histochem. Cytochem, 38 (1990b) 1755-1765.

Beitz, A.J., The Anatomical and Chemical Organization of Descending Pain Modulation Systems, In: Animal Pain and Its Control, Short C.E., (Ed.), Churchill Livingstone, Inc., New York, 1991, in press.

Beitz, A.J., Shepard, R.D. and Wells, W.L., The periaqueductal gray-raphe magnus projection contains somatostatin, neurotensin, and serotonin but not cholecystokinin, Brain Res., 261 (1983) 132-137.

Benveniste, H., Brain microdialysis, J. Neurochem., 52 (1989) 1667-1674.

Christie, M.J., James, L.B. and Beart, P.M., An excitatory amino acid projection from rat prefrontal cortex to periaqueductal gray, Brain Res. Bull., 16 (1986) 127-129.

Clements, J.R., Madl, J.E., Johnson, R.L., Larson, A.A. and Beitz, A.A., Localization of glutamate, glutaminase, aspartate and aspartate aminotransferase in the rat midbrain periaqueductal gray, Exp. Brain Res., 67 (1987) 594-602.

Collingridge, G.L. and Lester, R.A.J., Excitatory amino acid receptors in the vertebrate central nervous system, Pharmacol. Rev., 40 (1989) 143-210.

Cotman, C.W., Monaghan, D.T., Ottersen, O.P. and Storm-Mathisen, J., Anatomical organization of excitatory amino acid receptors and their pathways, Trends Neurosci., 10 (1987) 273-280.

Cuénod, M. and Streit, P., Neuronal tracing using retrograde migration of labeled transmitter-related compounds, In: Methods in Chemical Neuroanatomy, Vol. 1, Björklund A. and Hökfelt T. (Eds.), Elsevier, Amsterdam, 1983, pp. 365-397.

Curtis, D.R., Phillis, J.W. and Watkins, J.C., Chemical excitation of spinal neurones, Nature, 183 (1959) 611-612.

Curtis, D.R., Phillis, J.W. and Watkins, J.C., The chemical excitation of spinal neurones by certain acidic amino acids, J. Physiol. (Lond), 150 (1960) 656-682.

Fields, H.L. and Besson, J.M., Pain Modulation, Prog. Brain Res.,Vol. 77, Elsevier, Amsterdam, 1988.

Fonnum, F., Glutamate: a neurotransmitter in mammalian brain, J. Neurochem., 42 (1984) 1-11.

Greenamyre, J.T., Young, A.B. and Penney, J.B., Quantitative autoradiographic distribution of L-[^{3}H] glutamate-binding sites in rat central nervous system, J. Neurosci., 4 (1984) 2133-2144.

Halpain, S., Wieczorek, C.M. and Rainbow, T.C., Localization of l-glutamate receptors in rat brain by quantitative autoradiography, J. Neurosci., 4 (1984) 2247-2258.

Hepler, J.R., Toomim, C.S., McCarthy, K.D., Conti, F., Battaglia, G., Rustioni, A. and Petrusz, P., Characterization of antisera to glutamate and aspartate, J. Histochem. Cytochem., 36 (1988) 13-22.

Huxtable, R.J., Taurine in the central nervous system and the mammalian actions of taurine, Prog. Brain Res., 32 (1989) 471-533.

Jiang, M. and Behbehani, M.M., Interaction between the insular cortex (IC) and the periaqueductal gray (PAG), Soc. Neurosci. Abstr., 16 (1990) 563.

Johnson, R.L. and Koerner, J.F., Excitatory amino acid neurotransmission, J. Med. Chem., 31 (1988) 2057-2066.

Jürgens, U. and Richter, K., Glutamate-induced vocalization in the squirrel monkey, Brain Res., 373 (1986) 349-358.

Kalén, P., Karlson, M. and Wiklund, L., Possible excitatory amino acid afferents to nucleus raphe dorsalis of the rat investigated with retrograde wheat germ agglutinin and D-[^{3}H]aspartate tracing, Brain Res., 360 (1985) 285-297.

Keinänen, K., Wisden, W., Sommer, B., Werner, P., Herb, A., Verdoorn, T.A., Sakmann, B. and Seeburg, P.H., A family of AMPA-selective glutamate receptors, Science, 249 (1990) 556-560.

Luppi, P.-H., Kazuya, S., Fort, P., Salvert, D. and Jouvet, M., The nuclei of origin of monoaminergic, peptidergic and cholinergic afferents to the cat nucleus reticularis magnocellularis: a double labeling study with cholera toxin as a retrograde tracer, J. Comp. Neurol., 277 (1988) 1-20.

McDonald, A.J., Beitz, A.J., Larson, A.A., Kuriyama, R., Sellitto, C. and Madl, J.E., Colocalization of glutamate and tubulin in putative excitatory neurons of the hippocampus: An immunohistochemical study using monoclonal antibodies, Neuroscience, 30 (1989) 405-421.

Meeker, R.B., Swanson, D.J. and Hayward, J.N., Light and electron microscopic localization of glutamate immunoreactivity in the supraoptic nucleus of the rat hypothalamus, Neuroscience, 16 (1989) 157-167.

Monaghan, D.T., Bridges, R.J. and Cotman, C.W., The excitatory amino acid receptors: Their classes, pharmacology, and distinct properties in the function of the central nervous system, Annu. Rev. Pharmacol. Toxicol., 29 (1989) 365-402.

Mohrland, J.S., McManus, D.Q. and Gebhart, G.F., Lesions in the nucleus reticularis gigantocellularis: Effects on the antinociception produced by microinjection of morphine and focal electrical stimulation in the periaqueductal gray matter, Brain Res., 231 (1982) 143-152.

Mullett, M.A., Araneda, S., Ghilini, G., Wiklund, L. and Beitz, A.J., Combination of D-[^{3}H]Asp retrograde labelling and immunocytochemical detection of L-Glu and L-Asp in the PAG projection to raphe magnus, Soc. Neurosci. Abstr., 15 (1989) 941.

Nicholls, D.G., Release of glutamate, aspartate and gamma-aminobutyric acid from isolated nerve terminals, J. Neurochem., 52 (1989) 331-341.

Ottersen, O.P. and Storm-Mathisen, J., Glutamate- and GABA-containing neurons in the mouse and rat brain, as demonstrated with a new immunocytochemical technique, J. Comp. Neurol., 229 (1984a) 374-392.

Ottersen, O.P. and Storm-Mathisen, J., Neurons containing or accumulating transmitter amino acids, In: Handbook of Chemical Neuroanatomy, Vol. 3: Classical transmitters and transmitter receptors in the CNS, Part II, Björklund A., Hökfelt T. and Kuhar M.J. (Eds.), Elsevier, Amsterdam, 1984b, pp. 141-246.

Patneau, D.K. and Mayer, M.L., Structure-activity relationships for amino acid transmitter candidates acting at N-methyl-D-aspartate and Quisqualate receptors, J. Neurosci., 10 (1990) 2385-2399.

Paulsen, R.E. and Fonnum, F., Role of glial cells for the basal and Ca^{2+}-dependent K$^+$-evoked release of transmitter amino acids investigated by microdialysis, J. Neurochem., 52 (1989) 1823-1827.

Rainbow, T.C., Wieczorek, C.M. and Halpain, S., Quantitative autoradiography of binding sites for [^{3}H]AMPA, a structural analogue of glutamic acid, Brain Res., 309 (1984) 173-177.

Reynolds, D.V., Surgery in the rat during electrical analgesia induced by focal brain stimulation, Science, 164 (1969) 444-445.

Sherman, A.D. and Gebhart, G.F., Pain-induced alteration of glutamate in periaqueductal central gray and its reversal by morphine, Life Sci., 15 (1975) 1781-1789.

Watkins, J.C. and Evans, R.H., Excitatory amino acid transmitters, Annu. Rev. Pharmacol. Toxicol., 21 (1981) 165-204.

Watkins, J.C., Krogsgaard-Larsen, P. and Honoré, T., Structure-activity relationships in the development of excitatory amino acid receptor agonists and competitive antagonists, Trends Pharmacol. Sci., 11 (1990) 25-33.

Wenthold, R.J., Dechesne, C.J. and Wada, K., Isolation, localization and cloning of a kainic acid binding protein from frog brain, J. Histochem. Cytochem., 38 (1990) 1717-1723.

Wiklund, L., Behzadi, G., Kalen, P., Headley, P.M., Nicolopoulos, L.S., Parsons, C.G. and West, D.C., Autoradiographic and electrophysiological evidence for excitatory amino acid transmission in the periaqueductal gray projection to nucleus raphe magnus in the rat, Neurosci. Lett., 93 (1988) 158-163.

Yaksh, T.L. and Rudy, T.A., Narcotic analgetics: CNS sites and mechanisms of action as revealed by intracerebral injection techniques, Pain, 4 (1978) 299-360

Yingcharoen, K., Rinvik, E., Storm-Mathisen, J. and Ottersen, O.P., GABA, glycine, glutamate, aspartate and taurine in the perihypoglossal nuclei: an immunocytochemical investigation in the cat with particular reference to the issue of amino acid colocalization, Exp. Brain Res., 78 (1989) 345-357.

Young, A.B. and Fagg, G.E., Excitatory amino acid receptors in the brain: membrane binding and receptor autoradiographic approaches, Trends Pharmacol. Sci., 11 (1990) 126-133.

Young, E.G., Watkins, L.R. and Mayer, D.J., Comparison of the effects of ventral medulla-
    ry lesions on systemic and microinjection morphine analgesia, Brain Res., 290 (1984)
    119-129.
Zorumski, C.F. and Yang, J., AMPA, kainate and quisqualate activate a common receptor-
    channel complex on embryonic chick motoneurons, J. Neurosci., 8 (1988) 4277-4286.

GABAergic Neuronal Circuitry
in the Periaqueductal Gray Matter

David B. Reichling

Department of Physiology and Cellular Biophysics
Columbia University, College of Physicians and Surgeons
New York, USA

Introduction

The midbrain periaqueductal gray matter (PAG) has been implicated in a wide range of possible functions including antinociception, reproductive behavior, and components of the defense reaction (Besson; Ogawa et al.; Bandler and Depaulis, this volume). Stimulation of the PAG, using electrodes or excitatory amino acids, has been very useful for characterizing the effects that can be elicited from the PAG, by essentially treating the PAG as a "black box," bypassing intrinsic circuitry to ultimately activate efferent axons. Thus, our knowledge is limited concerning the neuronal circuitry internal to the PAG that regulates individually or collectively, its putative functions. This chapter describes results of anatomical, pharmacological, and electrophysiological studies suggesting that GABAergic neuronal elements play a prominent role in the intrinsic neuronal circuitry of the PAG. Based on this information, it seems likely that GABAergic neurons in the PAG are local circuit interneurons, and that GABAergic elements exert potent tonic inhibitory control over a variety of putative PAG functions including antinociception. To begin characterizing the anatomical basis for such GABAergic controls in the PAG, our work has combined the methods of immunocytochemistry with retrograde tracing at the electron microscopic level. These studies demonstrated that GABA-immunoreactive axon terminals synapse directly on neurons that give rise to the major projection from the PAG to the medullary nucleus raphe magnus (NRM). Furthermore, since this projection sends collateral branches to regions of the medial diencephalon, this finding is probably relevant to other putative functions of the PAG in addition to antinociception.

The Midbrain Periaqueductal Gray Matter, Edited by A. Depaulis and
R. Bandler, Plenum Press, New York, 1991

Anatomical Mapping of GABAergic Circuitry in the PAG

The distribution of GABAergic elements

GABA and its metabolites appear to be unevenly distributed throughout the PAG. For example, Sandner et al. (1981) have detected a gradient in GABA concentration along the dorsoventral axis of the PAG, with the highest levels occurring at depths corresponding to the midbrain aqueduct. The levels of GABA in the PAG are increased by noxious stimuli (Sherman and Gebhart, 1976), implying that some GABAergic neuronal circuitry in the PAG is involved in nociception-related functions. It is difficult to draw any more detailed conclusions concerning changes in activity in neuronal circuits based on such information about changes in the tissue content of a transmitter. In the future, valuable information about the functional regulation of GABAergic transmission in the PAG may emerge as temporal (and perhaps spatial) resolution is increased with the use of microdialysis probes to measure changes in the extracellular GABA concentration *in vivo*.

A variety of methods have been developed to specifically label individual GABAergic neuronal elements. These methods each access different stages in the life cycle of GABA as a neurotransmitter, from synthesis to storage, reuptake, and degradation. As described below, while all methods confirm the presence of GABAergic circuitry in the PAG, the results do not always agree in detail. The earliest method of labeling putative GABAergic neurons exploited the specific reuptake of GABA by neurons, using [³H]GABA autoradiography (Hökfelt and Ljungdahl,1970). Some neurons accumulate exogenously applied [³H]GABA more readily than do other cells, and the former are assumed to be GABAergic. With this technique, Belin et al. (1979) observed labeled cells only in the ventromedial part of the PAG and in the dorsal raphe nucleus. These results must be interpreted cautiously since it is not known if all GABAergic neurons take up [³H]GABA equally well, and more importantly, non-GABAergic neurons and even glial cells can accumulate [³H]GABA (Csillag et al., 1987; Kisvárday et al., 1986; Zucker et al., 1984). Another cytochemical method for the indirect detection of GABAergic neurons is the detection of the GABA degradative enzyme, GABA transaminase. Cell bodies stained histochemically for GABA-T are found only in the ventromedial PAG (Nagai et al., 1983). This distribution of cell bodies is very similar to the pattern of cells that accumulate [³H]GABA.

A possibly more reliable technique for the detection of GABAergic neuronal elements focuses on the biosynthesis of the GABA molecule, i.e., the decarboxylation of glutamate by glutamic acid decarboxylase (GAD; Saito et al., 1974). Thus, immunocytochemical detection of GAD selectively labels only a subpopulation of neurons (and their axon terminals) in the CNS, that appear to synthesize GABA for use as a transmitter. This method reveals labeled cell bodies throughout the rostrocaudal extent of the rat, rabbit, and opossum PAG (Penny et al., 1984; Mugnaini and Oertel, 1985; Barbaresi and Manfrini, 1988). These cells are most common in the dorsal, ventrolateral, and ventral PAG (and dorsal raphe nucleus) of the rat where they comprise up to 50% of the total population

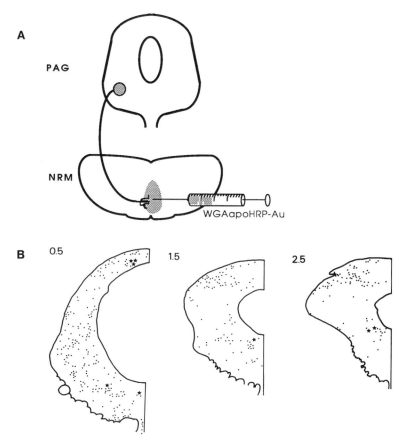

Figure 1. A: Schematic illustration of an experiment designed to detect GABAergic neurons in the PAG that project to the NRM. As illustrated, the retrograde tracer, WGAapoHRP-Au, was microinjected in the NRM. The rat was later perfused and the PAG was sectioned. Single sections were silver-enhanced to disclose the tracer, and immunostained with an antiserum directed against the glutaraldehyde-fixed GABA molecule. B: Camera lucida drawings illustrating the distribution of GABA-immunoreactive neurons in the PAG (black dots). The positions of eight GABA-immunoreactive neurons that were retrogradely labeled from the NRM are indicated by black stars. Numbers indicate the distance (in mm) rostral to interaural zero. Reprinted with permission from Reichling and Basbaum (1990) Journal of Comparative Neurology.

of neurons (Mugnaini and Oertel, 1985); lateral parts of the PAG are relatively poor in GAD-immunoreactive cell bodies. Labeled fibers, observed throughout the PAG, are especially dense around the aqueduct. A possible limitation of this method is that the concentration of GAD usually must be artificially enhanced in neuronal cell bodies by pretreating the animal with the drug colchicine.

An alternative approach to labeling GABAergic neuronal elements became available with the development of antibodies directed against the GABA molecule itself (Storm-Mathisen et al., 1983). The immunogen used to produce these antisera is the GABA molecule cross-linked to a protein by aldehyde fixation. Although GABA- and GAD-immunocytochemistry can produce markedly different patterns of staining in some regions of the brain (Newton and Maley, 1987; Spreafico et al., 1988; Hokoç et al., 1990), results from our laboratory using GABA immunocytochemistry in the PAG agree well with those of Mugnaini and Oertel (1985) who used GAD immunocytochemistry. As shown in Figure 1, without colchicine pretreatment, GABA immunoreactive cell bodies are widespread in the PAG (Reichling et al., 1988; Reichling and Basbaum, 1990a). At the level of the inferior colliculus, GABA-immunoreactive cell bodies account for approximately 15% of the total population of PAG neurons, and are most dense in the ventrolateral PAG and in a wedge-shaped region in the dorsolateral PAG. At more rostral levels, the dorsolateral group of cells is most prominent. This dorsolateral group of GABA-immunoreactive cells was also observed by Ottersen and Storm-Mathisen (1984) and by Clements et al. (1985). GABA-immunoreactive fibers are common throughout the PAG, but are most dense near the aqueduct, and at the level of the inferior colliculus, this region of dense staining extends outwards into the ventrolateral PAG. A potential pitfall of this method is that a certain amount of GABA is produced by cell metabolism; for example, putrescine is metabolized to GABA in mammalian brain (Seiler and Al-Therib, 1974). However, it is unlikely that our immunocytochemical method is detecting this pool of GABA since, using electron microscopy, GABA immunoreactivity at levels above background, is observed only in a subpopulation of neuronal cell bodies and vesicle-containing axon terminals (Reichling and Basbaum, 1990b).

The distribution of GABA receptors

The methods described above mapped the distribution of putative GABAergic neuronal elements, which presumably represent the presynaptic, transmitter-containing elements in GABAergic synapses. However, the distribution of a transmitter in a region is not always well-correlated with the distribution of its receptor (Herkenham, 1987). Thus, to map the location of postsynaptic elements in GABAergic synapses, a variety of methods have been used to detect GABA receptors in the PAG. Two subtypes of GABA receptor have been defined (for review, see Sivilotti and Nistri, 1991). Briefly, the $GABA_A$ receptor inhibits neurons by opening chloride channels, and it is associated with benzodiazepine binding. On the other hand, the $GABA_B$ receptor subtype may regulate K^+ and Ca^{2+} channels as well as adenylate cyclase activity.

Putative receptor sites are commonly mapped using autoradiographic detection of bound, labeled ligands. Binding of [³H]bicuculline or [³H]GABA in the presence of baclofen (to block binding to GABA$_B$ receptors), reveals a high density of GABA$_A$ binding sites throughout the PAG (McCabe and Wamsley, 1986; Bowery et al., 1987). Similarly, [^{3}H]benzodiazepine binding, used to map the GABA$_A$-associated benzodiazepine receptor, is also dense in the PAG (Young and Kuhar, 1980; McCabe and Wamsley, 1986; Gundlach, this volume). These studies indicate that GABA$_A$ receptors are especially dense in the dorsolateral part of the rostral PAG, and more caudally in the dorsal and dorsolateral PAG and in the midline dorsal raphe nucleus. On the other hand, only low levels of GABA$_B$ binding sites are detected in the PAG, except for a dorsolateral patch of moderate density binding in the rostral PAG (McCabe and Wamsley, 1986; Bowery, et al., 1987). Patterns of autoradiographic staining can be affected by the quenching of emitted beta particles by white matter (for further discussion see Gundlach, this volume). In this regard it may be significant that there is a trend toward increased myelinization in the PAG from the aqueduct outwards, and the dorsolateral part of the rostral PAG is especially myelin-poor (Mantyh, 1982).

Autoradiographic techniques provide relatively low spatial resolution, and consequently it may be difficult to associate emulsion grains with individual neurons or their processes. Much better spatial resolution can be afforded by immunocytochemical techniques using monoclonal antibodies raised against the GABA$_A$/benzodiazepine receptor complex. Moderate levels of GABA$_A$ receptor immunoreactivity in the PAG were shown in a survey of the rat brain, but finer details were not resolved (Schoch et al., 1985). Such antibodies should be very useful for associating GABA receptors with individual neurons and processes in the PAG at both the light and electron microscopic levels of resolution (Williams and Beitz, 1990).

Recently, five subunits of the GABA$_A$ receptor have been cloned (Levitan et al., 1988), and the distribution of messenger RNA for the alpha1 subunit has been mapped using *in situ* hybridization histochemistry with radiolabeled oligo-nucleotide probes (Hironaka et al., 1990). Neurons that express messenger RNA for the alpha1 subunit are numerous throughout the PAG of the rat. Since the great majority of GABA-immunoreactive synaptic terminals in the PAG contact dendrites (Moss et al., 1983; Reichling and Basbaum, 1990b), staining of cell bodies by *in situ* hybridization probably does not accurately reflect the distribution of GABAergic synapses. However, the abundance of stained cell bodies throughout the PAG suggests that many GABAergic synapses in the PAG may occur on dendrites that originate from cells within the PAG itself. This method will probably be most useful in the future for the detection of changes in the regulation of GABA receptors in response to various stimuli and drugs.

Possible Sources and Organization of GABAergic Elements in the PAG

GABAergic interneurons in the PAG

Since GABAergic neurons are common in the PAG, it seems likely that many of the GABAergic terminals in that region must originate locally from these neurons. However, in recent years evidence has accumulated that some GABAergic neurons can project axons over long distances, and so, we must consider the possibility that GABAergic axon terminals found in the PAG are derived from extrinsic sources as well as from local interneurons. Arguing against this possibility is the observation that lesions placed at the sites of origin of major inputs to the PAG do not decrease GAD activity in the PAG (Belin et al., 1979), and cause only a small decrease in GABA levels in the PAG (Sandner et al., 1981). Furthermore, [³H]GABA microinjected in the PAG, retrogradely labels cell bodies within the PAG itself (Beart et al., 1990), but labels only a small number of cell bodies outside the PAG, in the dorsal raphe and parabrachial nuclei. Thus, it seems that many GABA terminals in the PAG are likely to arise from local interneurons.

Conversely, we can ask if most GABAergic cell bodies in the PAG are local interneurons or projection neurons. Arguing against the latter possibility is the observation that while GAD- or GABA-immunoreactive neurons in the PAG tend to be small, neurons projecting from the PAG tend to be larger in size (Barbaresi and Manfrini, 1988; Reichling and Basbaum, 1990b; Holstege, this volume). To address this issue more directly, we combined retrograde tracing techniques with GABA immunocytochemistry to search for GABAergic components in a major efferent pathway from the PAG. As illustrated in Figure 1, although single injections of retrograde tracer in the nucleus raphe magnus in the rat labeled approximately 20% of all neurons in the PAG, and 14% of PAG neurons were GABA-immunoreactive, only about 2% of either population of neurons was double labeled (Reichling and Basbaum, 1990a). Similar results were obtained when GABA immunocytochemistry was combined with retrograde labeling of PAG neurons after tracer injections in the pontine nuclei and reticular tegmental nucleus of the cat (Aas and Brodal, 1990). Although not conclusive, the lack of positive evidence for the existence of GABAergic projection neurons in the PAG suggests that many GABAergic PAG neurons are likely to be local interneurons.

Although local circuit axons have been observed in Golgi-stained PAG tissue (Liu and Hamilton, 1980; Tredici et al., 1983), there is little information available concerning the organization of local inhibitory circuits within the PAG. Interestingly, Sandner et al. (1986) found that when electrically stimulating and recording at nearby sites in the PAG, local circuit inhibitory effects were most often recorded when the stimulation electrode was in the same coronal plane as the recording electrode. As described below, electrophysiological studies of neurons in the PAG further support the notion of dense local circuit GABAergic connections in the PAG.

GABA-sensitivity of PAG neurons
In view of the anatomical evidence for abundant GABAergic elements in the PAG, it is not surprising that GABA inhibits spontaneous activity in the majority of neurons in rat PAG *in vivo* and in the PAG slice (Ogawa et al.,1989; this volume; Behbehani et al., 1990). Furthermore, approximately half of PAG

neurons are excited by GABA antagonists (Ogawa et al., 1989; this volume, Behbehani et al., 1990). This is an important observation since it implies that much of the GABAergic inhibitory control in the PAG is tonically active. In addition, since the tonically active inhibitory synaptic connections survive in the slice preparation, they must arise from nearby interneurons. Complementary evidence for the existence of tonically active GABAergic controls in the PAG is the observation that microinjection of the GABA antagonist bicuculline, increases the expression of the immediate gene, c-fos, in some PAG neurons (Sandkühler, this volume).

Possible functional role of GABA in PAG circuitry

Antinociception has been the singlemost intensively studied putative function of the PAG, and numerous studies have examined the effects of GABA receptor ligands on PAG antinociceptive circuitry. Furthermore, since some of the relevant afferent and efferent pathways for this phenomenon are known, it is possible to propose detailed, testable hypotheses concerning the postsynaptic targets of GABAergic terminals and the inputs to GABAergic cells in this circuitry.

Since electrical stimulation of the PAG produces analgesia (Mayer and Liebeskind, 1974; Besson et al., this volume) and lesions of the PAG do not cause analgesia (Dostrovsky and Deakin, 1977), stimulation-produced analgesia must involve activation of neurons projecting from the PAG. On the other hand, antinociceptive circuitry is also potently activated by narcotics microinjected in the PAG (Yaksh et al., 1976). However, the postsynaptic actions of opiates are thought to be exclusively inhibitory (however see Crain and Shen, 1990), and so it has been proposed that opiates activate the projection from the PAG to the NRM indirectly, by inhibiting tonically active inhibitory inputs to the projection neurons (Yaksh et al., 1976). More specifically, it was proposed, by analogy with the hippocampus (Zieglgänsberger et al., 1979), that opiates could activate PAG projection neurons by inhibiting tonically active GABAergic interneurons in the PAG (see Fig. 2; Basbaum and Fields, 1984).

Pharmacological evidence supports this proposal. First, if GABAergic inhibition tonically depresses activity in PAG projection neurons that mediate antinociception, then blocking this inhibition should produce analgesia. Indeed, GABA antagonists microinjected in the PAG can produce antinociception (Moreau and Fields, 1986; Sandkühler, this volume), and at low doses can potentiate the analgesic effect of morphine injected at the same site (Depaulis et al., 1987). Obversely, inhibiting the projection neurons by activating the postsynaptic GABA receptors should antagonize opiate-induced antinociception. Thus, the antinociceptive effect of systemically applied morphine is reduced by GABA agonists administered systemically (Ho et al., 1976), or microinjected in the PAG (Zambotti et al., 1982; Romandini and Samanin, 1984). Furthermore, the GABA/opiate interaction likely occurs in the PAG itself since GABA antagonizes morphine antinociception when the two agents are microinjected at the same sites in the PAG (Depaulis et al., 1987; Moreau and Fields, 1986).

The pharmacological studies described above used manipulations that specifically affect GABA$_A$ receptors or nonspecifically increase GABAergic transmission. Together, with the reported scarcity of GABA$_B$ type binding sites in the PAG, these data suggest that the proposed GABAergic control of antinoci-

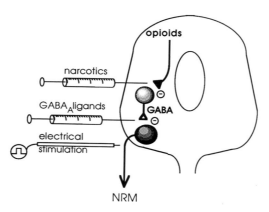

Figure 2. A schematic illustration of the proposed tonically active interneuron that controls activity in the projection from the PAG to the NRM. Narcotics activate the antinociceptive projection to the NRM, in part, indirectly by reducing the tonic GABAergic inhibition. The principal sites at which various manipulations influence this circuitry are indicated.

ception in the PAG must be mediated by tonically activated receptors of the GABA$_A$ subtype. This would be consistent with anatomical evidence that presynaptic inhibition (typically associated with GABA$_B$ receptors) is extremely rare in the PAG (Moss et al., 1983; Reichling and Basbaum, 1990b). This would also provide a site of action for the analgesic effects of benzodiazepine antagonists (Morgan et al., 1987).

Anatomical Studies of GABAergic Synaptic Relationships in the PAG

Although it is simplest to propose that the axon terminals of tonically active GABAergic interneurons are directly presynaptic to the PAG projection neuron, based on the pharmacological data, we cannot rule out the possibility that one or more excitatory interneurons are interposed between the GABAergic terminal and projection neuron. Therefore, we performed experiments to directly test the hypothesis that projections from the PAG that mediate antinociception are directly contacted by GABAergic synaptic terminals (Reichling and Basbaum, 1990b). These experiments focused on the projection from the PAG to the medullary NRM since it is the best characterized antinociceptive efferent pathway from the PAG (Beitz and Williams; Mason, this volume). Furthermore, observations by Moreau and Fields (1986) show that this pathway specifically, is subject to tonic

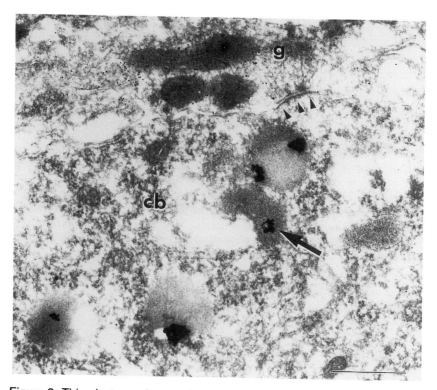

Figure 3. This electron micrograph shows the ultrastructural appearance of a synaptic contact (arrowheads) between a GABA-immunoreactive axon terminal (g) labeled by colloidal gold particles, and a cell body (cb) retrogradely labeled from the NRM with silver-enhanced and gold-toned particles of WGAapoHRP-Au (arrow). Scale bar = 0.5 µm. Reprinted with permission from Reichling and Basbaum (1990) Journal of Comparative Neurology.

GABAergic control. Our subsequent anatomical studies combined methods of GABA immunocytochemistry and retrograde tracing at the electron microscopic level. As illustrated in Figure 1A, neurons in the PAG were retrogradely labeled by an injection of the tracer, WGAapoHRP-Au (Basbaum and Menetrey, 1987), into the NRM. Plastic-embedded ultrathin sections of the caudal ventrolateral PAG were then immunostained with an anti-GABA antiserum.

GABA-immunoreactive neurons were found to be very common, constituting approximately 40% of all synaptic terminals in the ventrolateral PAG. These terminals contain small, agranular vesicles that are round or occasionally pleomorphic. Approximately half of the GABA-immunoreactive terminals also contain a few dense-cored vesicles, suggesting that peptide transmitters might coexist with GABA in these terminals (see also Veening et al., this volume). The majority of immunoreactive terminals form axo-dendritic synapses, with only 5% synapsing upon cell bodies. Although putative GABAergic synapses in some brain regions may be "asymmetrical" (Ribak and Roberts, 1990), synaptic contacts in the PAG exhibit "symmetrical" morphology common among inhibitory synapses. GABA-immunoreactive axo-axonic synapses were not observed, suggesting that GABAergic presynaptic inhibition is rare or absent in this region.

Numerous cell bodies and proximal dendrites were retrogradely labeled by tracer injected into the NRM. Most of these retrogradely labeled cell bodies received only a few synaptic inputs in a single plane of section, but some had more than 30% of their perimeters occupied by synaptic contacts. Overall, approximately half of the profiles in synaptic contact with retrogradely labeled cell bodies were GABA-immunoreactive (Fig. 3). Such proximally located inhibitory inputs might be in a very favorable position to prevent excitatory inputs on dendrites from successfully activating the axon hillock. In this way, this GABAergic control could gate activity in axons that make up the projection to the NRM. In a recent study using pre-embedding GABA immunocytochemistry, Williams and Beitz (1990) similarly demonstrated GABA immunoreactive contacts onto PAG neurons projecting to the NRM, and these authors detected immunoreactivity for the $GABA_A$ receptor subtype on projection neurons.

As illustrated in Figure 2, we have proposed that tonically active GABAergic interneurons receive inhibitory opioidergic inputs. This connection is suggested only indirectly by pharmacological data, however, electrophysiological observations lend additional support to this idea. While recording in the PAG slice, Pfaff and colleagues found that some neurons that were inhibited by GABA were excited during enkephalin application (Ogawa et al., this volume). In the future, electrophysiological and anatomical studies should be designed to verify and characterize this opioid/GABA connection in the PAG.

GABAergic Control over other Putative Functions of the PAG

The ultrastructural double-labeling study described above focused on the presumed antinociceptive function of projections from the PAG to the NRM. However, it is not clear that the PAG-NRM projections are purely antinociceptive (e.g., the NRM also projects to the intermediolateral cell column in the thoraco-lumbar spinal cord, Holstege, 1988), and therefore the results might reflect a mechanism that more generally modulates activity in PAG efferent pathways. This point is illustrated by a double-label retrograde tracing study of collaterals of the projection from the PAG to the NRM (Reichling and Basbaum, 1991). To examine the relationships between the descending pathway to the NRM and other, ascending, efferent pathways from the PAG, pairs of injections of two retrograde tracers were placed in the NRM and in a site rostral to the PAG (Fig. 4). Most injections of tracer in the diencephalon or forebrain labeled neurons in the dorsal raphe nucleus but not in the PAG itself. On the other hand, injections in the ventromedial hypothalamus and the medial thalamus labeled large numbers of PAG neurons, and between 13 and 20% of those neurons were double-labeled by injections in the NRM.

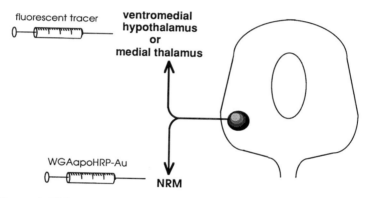

Figure 4. This schematic illustrates a double-label retrograde tracer experiment (described in text) demonstrating that neurons in the PAG that project to the NRM can also send an axon collateral to the medial thalamus (nucleus parafascicularis) or the ventromedial hypothalamus (arcuate nucleus).

These results demonstrate that the projection to the NRM is not a wholly exclusive one, but rather is a component of a more promiscuous system of collateral projections. As a result, activity in the PAG-NRM pathway, stimulated by whatever means, must be accompanied by activity in a number of rostrally-directed collaterals.

The wide range of effects that can be elicited by benzodiazepines in the PAG further suggests that tonic $GABA_A$ controls are not unique to antinociception. The anxiolytic effects of benzodiazepines are thought to be largely due to inhibition of the widely divergent projections from serotonergic neurons in the DRN (Stein et al., 1975). Direct synaptic contacts by GABA-accumulating or GABA immunoreactive terminals onto serotonin-immunoreactive neurons of the

DRN have been demonstrated (Harandi et al., 1987; Reichling et al., in preparation).

Also consistent with the idea that a variety of PAG functions are subject to tonically active GABAergic inhibitory controls, behavioral studies demonstrate that the microinjection of GABA antagonists in the PAG mimics many of the effects of electrical stimulation in the PAG. For example, the defensive reactions and related cardiovascular changes that are evoked by excitatory amino acid activation of cell bodies in the lateral PAG (Bandler and Depaulis; Carrive, this volume) can be reproduced by microinjection of a GABA antagonist in the same areas (Bandler, 1988; Depaulis and Vergnes 1986; Keay et al., 1988; Schenberg et al., 1983). Similarly, electrical stimulation of the dorsal PAG can be strongly aversive, and microinjection of GABA antagonists can mimic this aversive effect (Imperato and DiChiara, 1981; Brandão et al., 1982; DiScala et al., 1984; Jenck et al., 1988). In related studies (Jürgens, this volume), vocalizations were elicited from the monkey by electrical stimulation in the forebrain. Injection of GABA antagonists in the PAG facilitated this vocalization, indicating that there are tonically active GABAergic synapses within the PAG that modulate this behavior.

So, it seems likely that tonic intrinsic GABAergic inhibitory control of activity in efferent pathways is a common feature in the integration of a wide range of PAG functions. It would, however, be overly simplistic to suppose that tonic inhibition of efferents is the sole contribution of GABAergic elements to neuronal circuitry in the PAG. Thus, a large body of data shows that GABA receptor agonists can produce potent analgesia (e.g., DeFeudis, 1983), apparently in contradiction to the proposal of PAG GABAergic circuitry outlined above. This conflict is partially resolved by considering that analgesic actions of GABA occur at sites outside the PAG. For instance, baclofen, a $GABA_B$ receptor agonist elicits analgesia when applied to the spinal cord (Wilson and Yaksh, 1978; Hammond and Drower, 1984), lower brainstem (Levy and Proudfit, 1979), or hypothalamus (Lim et al., 1985), and $GABA_A$ analgesic effects can be elicited from the ventral medulla (Drower and Hammond, 1988). Nevertheless, some analgesic effects, both $GABA_A$ (Proudfit and Levy, 1978) and $GABA_B$ (Retz and Holaday, 1986), ostensibly occur within the PAG itself (see also Mason, this volume, for discussion of nociceptive facilitation in the PAG), and further investigations will be required to characterize the GABAergic circuitry in the PAG which mediates these effects that cannot be accounted for by the present model.

References

Aas, J.E. and Brodal, P. , GABA and glycine as putative transmitters in subcortical pathways to the pontine nuclei, combined immunocytochemical and retrograde tracing study in the cat with some observations in the rat, Neurosci., 34 (1990) 149-162.

Bandler, R., Brain mechanisms of aggression as revealed by electrical and chemical stimulation: Suggestion of a central role for the midbrain periaqueductal grey region, In: Prog. Psychobiol. Physiol. Psychol., Vol. 13., Epstein A. and Morrison A. (Eds.), Academic Press, New York, 1988, pp. 67-154.

Barbaresi, P. and Manfrini, E., Glutamate decarboxylase-immunoreactive neurons and terminals in the periaqueductal gray of the rat, Neuroscience, 27 (1988) 183-191.

Basbaum, A.I. and Fields, H.L., Endogenous pain control systems: brainstem spinal pathways and endorphin circuitry, Ann. Rev. Neurosci., 7 (1984) 309-338.

Basbaum, A.I. and Menetrey, D., Wheat germ agglutinin-apoHRP gold: a new retrograde tracer for light- and electron-microscopic single- and double-label studies, J. Comp. Neurol., 261 (1987) 306-318.

Beart, P.M., Summers, R.J., Stephenson, J.A., Cook, C.J. and Christie, M.J., Excitatory amino acid projections to the periaqueductal gray in the rat: a retrograde transport study utilizing D[³H]aspartate and [³H]GABA, Neuroscience, 34 (1990) 163-176.

Behbehani, M.M., Jiang, M., Chandler, S.D. and Ennis, M., The effect of GABA and its antagonists on midbrain periaqueductal gray neurons in the rat, Pain, 40 (1990) 195-204.

Belin, M.F., Aguera, M., Tapaz, A., McRae-DeQueurce, A., Bobillier, P. and Pujol, J.F., GABA-accumulating neurons in the nucleus raphe dorsalis and periaqueductal gray matter in the rat: a biochemical and radioautographic study, Brain Res., 170 (1979) 279-297.

Bowery, N.G., Hudson, A.L. and Price, G.W., GABA$_A$ and GABA$_B$ receptor site distribution in the rat central nervous system, Neuroscience, 20 (1987) 365-383.

Brandão, M.L., DeAguiar, J.C. and Graeff, F.G., Mediation of the anti-aversive action of minor tranquilizers, Pharmacol. Biochem. Behav., 16 (1982) 397-402.

Clements, J.R., Beitz, A.J., Larson, A.A. and Madl, J.E., The ultrastructural localization of polyclonal GABA and monoclonal glutamate immunoreactivity in the rat midbrain periaqueductal gray, Soc. Neurosci. Abst., 11 (1985) 126.

Crain, S.M. and Shen, K.F., Opioids can evoke direct receptor-mediated excitatory effects on sensory neurons, Trends Pharmacol. Sci., 11 (1990) 77-81.

Csillag, A., Stewart, M.G. and Curtis, E.M., GABAergic structures in the chick telencephalon: GABA immunocytochemistry combined with light and electron microscope autoradiography, and Golgi impregnation, Brain Res., 437 (1987) 283-297.

DeFeudis, F.V., Central GABAergic systems and analgesia, Drug Dev. Res., 3 (1983) 1-15.

Depaulis, A. and Vergnes, M., Elicitation of intraspecific defensive behaviors in the rat by microinjections of picrotoxin, a GABA antagonist, into the midbrain periaqueductal gray matter, Brain Res., 367 (1986) 87-95.

Depaulis, A., Morgan, M.M. and Liebeskind, J.C., GABAergic modulation of the analgesic effects of morphine microinjected in the periaqueductal gray matter of the rat, Brain Res., 436 (1987) 223-228.

DiScala, G., Schmitt, P. and Karli, P., Flight induced by infusion of bicuculline methiodide into periventricular structures, Brain Res., 309 (1984) 199-208.

Dostrovsky, J.O. and Deakin, J.F.W., Periaqueductal gray lesions reduce morphine analgesia in the rat, Neurosci. Lett., 4 (1977) 99-103.

Drower, E.J. and Hammond, D.L., GABAergic modulation of nociceptive threshold: effects of THIP and bicuculline microinjected in the ventral medulla of the rat, Brain Res., 450 (1988) 316-324.

Hammond, D.L. and Drower, E.J., Effects of intrathecally administered THIP, baclofen, and muscimol on nociceptive threshold, Eur. J. Pharmacol., 103 (1984) 121-125.

Harandi, M., Aguera, M., Gamrani, H., Didier, M., Maitre, M., Calas, A. and Belin, M.F., Gamma-aminobutyric acid and 5-hydroxytryptamine interrelationship in the rat nucleus raphe dorsalis: combination of radioautographic and immunocytochemical techniques at light and electron microscopy levels, Neuroscience, 21 (1987) 237-251.

Herkenham, M., Mismatches between neurotransmitter and receptor localizations in the brain: observation and implications, Neuroscience, 23 (1987) 1-38.

Hironaka, T., Morita, Y., Hagihira, S., Tateno, E., Kita, H. and Tohyama, M., Localization of GABA$_A$-receptor alpha$_1$ subunit mRNA-containing neurons in the lower brainstem of the rat, Molec. Brain Res., 7 (1990) 335-345.

Hokoç, J.N., Ventura, A.L.M., Gardino, P.F. and DeMello, F.G., Developmental immuno-reactivity for GABA and GAD in the avian retina: possible alternative pathway for GABA synthesis, Brain Res., 532 (1990) 197-202.

Hökfelt, T. and Ljundahl, Å., Cellular localization of labeled gamma-aminobutyric acid (³H-GABA) in rat cerebellar cortex: an autoradiographic study, Brain Res., 22 (1970) 391-396.

Holstege, G., Direct and indirect pathways to lamina I in the medulla oblongata and spinal cord of the cat, Prog. Brain Res., 77 (1988) 47-94.

Imperato, A. and DiChiara, G., Behavioural effects of GABA-agonists and antagonists infused in the mesencephalic reticular formation-deep layers of superior colliculus, Brain Res., 224 (1981) 185-194.

Jenck, F., Moreau, J.L. and Karli, P., Modulation by morphine of aversive-like behavior induced by GABAergic blockade in periaqueductal gray or medial hypothalamus, Pharmacol. Biochem. Behav., 31 (1988) 193-200.

Keay, K.A., Redgrave, P. and Dean, P., Cardiovascular and respiratory changes elicited by stimulation of rat superior colliculus, Brain Res. Bull., 20 (1988) 13-26.

Kisvárday, Z.F., Cowey, A., Hodgson, A.J. and Somogyi, P., The relationship between GABA immunoreactivity and labelling by local uptake of [³H]GABA in the striate cortex of monkey, Exp. Brain Res., 62 (1986) 89-98.

Levitan, E.S., Schofield, P.R., Burt, D.R., Rhee, L.M., Wisden, W., Khöler, M., Fujita, N., Rodriguez, H., Stephenson, F.A., Darlison, M.G., Barnard, E.A. and Seeburg, P.H., Structural and functional basis for GABA$_A$ receptor heterogeneity, Nature, 335 (1988) 76-79.

Levy, R.A. and Proudfit, H.K., Analgesia produced by microinjection of baclofen and morphine at brain stem sites, Eur. J. Pharmacol., 57 (1979) 43-55.

Lim, C.R., Garant, D.S. and Gale, K., GABA agonist induced analgesia from the lateral preoptic area in the rat, Eur. J. Pharmacol., 107 (1985) 91-94.

Liu, R.P.C. and Hamilton, B.L., Neurons of the periaqueductal gray matter as revealed by Golgi study, J. Comp. Neurol., 189 (1980) 403-418.

Mantyh, P.W., The midbrain periaqueductal gray in the rat, cat, and monkey: a Nissl, Weil, and Golgi analysis, J. Comp. Neurol., 204 (1982) 349-363.

Mayer, D.J. and Liebeskind, J.C., Pain reduction by focal electrical stimulation of the brain: an anatomical and behavioral analysis, Brain Res., 68 (1974) 73-93.

McCabe, R.T. and Wamsley, J.K., Autoradiographic localization of subcomponents of the macromolecular GABA receptor complex, Life Sci., 39 (1986) 1937-1945.

Moreau, J.L. and Fields, H.L., Evidence for GABA involvement in midbrain control of medullary neurons that modulate nociceptive transmission, Brain Res., 397 (1986) 37-46.

Morgan, M.M., Levin, E.D. and Liebeskind, J.C., Characterization of the analgesic effects of the benzodiazepine antagonist, Ro 15-1788, Brain Res., 415 (1987) 367-370.

Moss, M.S., Glazer, E.J. and Basbaum, A.I., The peptidergic organization of the cat periaqueductal gray: I. the distribution of enkephalin-containing neurons and terminals, J. Neurosci., 3 (1983) 603-616.

Mugnaini, E. and Oertel, W.H., An atlas of the distribution of GABAergic neurons and terminals in the rat CNS as revealed by GAD immunohistochemistry, In: Handbook of Chemical Neuroanatomy. Vol. 4: GABA and Neuropeptides in the CNS, Part 1, Björklund A. and Hökfelt T. (Eds.), Elsevier, Amsterdam, 1985, pp. 436-608.

Nagai, T., McGeer, T.L. and McGeer, E.G., Distribution of GABA-T intensive neurons in the rat forebrain and midbrain, J. Comp. Neurol., 218 (1983) 220-238.

Newton, B.W. and Maley, B.E., A comparison of GABA- and GAD-like immunoreactivity within the area postrema of the rat and cat, J. Comp. Neurol., 255 (1987) 208-216.

Ogawa, S., Kow, L.M. and Pfaff, D.W., Neuronal activity of dorsal periaqueductal gray neurons of female rats: responsiveness to GABA and enkephalin, Soc. Neurosci. Abst., 15 (1989) 146.

Ottersen, O.P. and Storm-Mathisen, J., Glutamate- and GABA-containing neurons in the mouse and rat brain, as demonstrated with a new immunocytochemical technique, J. Comp. Neurol., 229 (1984) 374-392.

Penny, G.R., Conley, M., Diamond, I.T. and Schmechel, D.E., The distribution of glutamic acid decarboxylase immunoreactivity in the diencephalon of the opossum and rabbit, J. Comp. Neurol., 228 (1984) 38-57.

Proudfit, H.K. and Levy, R.A., Delimitation of neuronal substrates necessary for the analgesic action of baclofen and morphine, Eur. J. Pharmacol., 47 (1978) 159-166.

Reichling, D.B. and Basbaum, A.I., The contribution of brainstem GABAergic circuitry to descending antinociceptive controls. I. GABA-immunoreactive projection neurons in the periaqueductal gray and nucleus raphe magnus, J. Comp. Neurol., 302 (1990a) 370-377.

Reichling, D.B. and Basbaum, A.I., The contribution of brainstem GABAergic circuitry to descending antinociceptive controls. II. Electron microscopic immunocytochemical evidence of GABAergic control over the projection from the periaqueductal gray matter to the nucleus raphe magnus, J. Comp. Neurol., 302 (1990b) 378-393.

Reichling, D.B. and Basbaum, A.I., Collateralization of periaqueductal gray neurons to forebrain or diencephalon and nucleus raphe magnus, Neuroscience, 42 (1991) 183-200.

Reichling, D.B., Chazal, G. and Basbaum, A.I., Electron microscopic immunocytochemical evidence of GABAergic synaptic contacts onto serotonergic neurons in the dorsal raphe nucleus of rat, in preparation.

Reichling, D.R., Kwait, G.C. and Basbaum, A.I., Anatomy, physiology, and pharmacology of the periaqueductal gray contribution to antinociceptive controls, Prog. Brain Res., 77 (1988) 31-46.

Retz, K.C. and Holaday, L.M., Analgesia and motor activity following administration of THIP into the periaqueductal gray and lateral ventricle, Drug. Dev. Res., 9 (1986) 133-142.

Ribak, C.E. and Roberts, R.C., GABAergic synapses in the brain identified with antisera to GABA and its degradative enzyme, glutamate decarboxylase, J. Elec. Microsc. Tech., 15 (1990) 34-48.

Romandini, S. and Samanin, R., Muscimol injections in the nucleus raphe dorsalis block the antinociceptive effects of morphine in rats: apparent lack of 5-hydroxytryptamine involvement in muscimol's effect, Br. J. Pharmacol., 81 (1984) 25-29.

Saito, K., Barber, R., Wu, J.Y., Matsuda, T., Roberts, E. and Vaughn, J.E., Immunohistochemical localization of glutamic acid decarboxylase in rat cerebellum, Proc. Nat. Acad. Sci., 71 (1974) 269-273.

Sandner, G., Dessort, D., Schmitt, P. and Karli, P., Distribution of GABA in the periaqueductal gray matter. Effects of medial hypothalamic lesions, Brain Res., 224 (1981) 279-290.

Sandner, G., Schmitt, P. and Karli, P., Unit activity alterations induced in the mesencephalic periaqueductal gray by local electrical stimulation, Brain Res., 386 (1986) 53-63.

Schenberg, L.C., DeAguiar, J.C. and Graeff, F.G., GABA modulation of the defence reaction induced by brain electrical stimulation, Physiol. Behav., 31 (1983) 429-437.

Schoch, P., Richards, J.G., Häring, P., Takacs, B., Stähli, C., Staehelin, T., Haefely, W. and Möhler, H., Co-localization of $GABA_A$ receptors and benzodiazepine receptors in the brain shown by monoclonal antibodies, Nature, 314 (1985) 168-170.

Seiler, N. and Al-Therib, M.J., Putrescine catabolism in mammalian brain, Biochem. J., 144 (1974) 29-35.

Sherman, A.D. and Gebhart, G.F., Morphine and pain: effects on aspartate GABA and glutamate in four discrete areas of mouse brain, Brain Res., 110 (1976) 273-281.

Sivilotti. L. and Nistri, A., GABA receptor mechanisms in the central nervous system, Prog. Neurobiol., 36 (1991) 35-92.

Spreafico, R., DeBiase, S., Frassoni, C. and Battaglia, G., A comparison of GAD- and GABA-immunoreactive neurons in the first somatosensory area (SI) of the rat cortex, Brain Res., 474 (1988) 192-196.

Stein, L., Wise, C.D. and Beluzzi, J.D., Effects of benzodiazepines on central serotonergic mechanisms, Adv. Biochem. Psychopharmacol., 14 (1975) 29-44.

Storm-Mathisen, J., Leknes, A.K., Bore, A.T., Vaaland, J.L., Edminson, P., Haug, F.M.S. and Ottersen, O.P., First visualization of glutamate and GABA in neurones by immunocytochemistry, Nature, 301 (1983) 517-520.

Tredici, G., Bianchi, R. and Gioia, M., Short intrinsic circuit in the periaqueductal gray matter of the cat, Neurosci. Lett., 39 (1983) 131-136.

Williams, F.G. and Beitz, A.L., Ultrastructural morphometric analysis of GABA-immunoreactive terminals in the ventrocaudal periaqueductal grey: analysis of the relationship of GABA terminals and the $GABA_A$ receptor to periaqueductal grey-raphe magnus projection neurons, J. Neurocytol., 19 (1990) 686-696.

Wilson, P.R. and Yaksh, T.L., Baclofen is antinociceptive in the intrathecal space of animals, Eur. J. Pharmacol., 51 (1978) 323-330.

Yaksh, T.L., Yeung, Y.C. and Rudy, T.A., Systematic examination in the rat of brainstem sites sensitive to the direct application of morphine: observation of differential effects within the periaqueductal gray, Brain Res., 114 (1976) 83-103.

Young, W.S. and Kuhar, M.J., Radiohistochemical localization of benzodiazepine receptors in rat brain, J. Pharmacol. Exp. Ther., 212 (1980) 337-346.

Zambotti, F., Zonta, N., Parenti, M., Tommasi, R., Vicentini, L., Conci, F. and Montegazza, P., Periaqueductal gray matter involvement in the muscimol induced decrease of morphine antinociception, Naunyn-Scheideberg's Arch. Pharmacol., 318 (1982) 368-369.

Zieglgänsberger, W., French, E.D., Siggins, G.R. and Bloom, F.E., Opioid peptides may excite hippocampal pyramidal neurons by inhibiting adjacent inhibitory interneurons, Science, 205 (1979) 415.

Zucker, C., Yazulla, S. and Wu, J.Y., Non-correspondence of [³H]GABA uptake and GAD localization in goldfish amacrine cells, Brain Res., 298 (1984) 154-158.

Organization of Spinal and Trigeminal Input to the PAG

A. Blomqvist and A.D. Craig*

Department of Cell Biology, Faculty of Health Sciences
University of Linköping, Sweden
and *Divisions of Neurobiology and Neurosurgery
Barrow Neurological Institute, Phoenix, AR, U.S.A.

Introduction

Many of the functions in which the PAG plays an important role, such as defensive and reproductive behavior, antinociception and cardiovascular regulation (see chapters by, e.g., Bandler and Depaulis; Ogawa et al.; Besson et al.; Lovick; Carrive, this volume), are carried out as a response to somatosensory stimulation. In accordance with the behavioral data, electrophysiological experiments have identified neurons in the PAG that are activated by somatic sensory stimuli (Eickhoff et al., 1978; Rose, 1975; 1979). Yet, little is known about the mechanisms underlying the somatosensory functions of the PAG. A prerequisite for the understanding of these mechanisms is knowledge of how sensory information reaches the PAG. Such knowledge involves both anatomical and electrophysiological characterization (see Yezierski, this volume) of the somatosensory pathways that transmit information to the PAG. With regard to anatomical data, there are sparse reports in the classical literature on the termination in the PAG of the ascending sensory pathways (Mehler, 1969; Nauta and Kuypers, 1958), and only with the advent of modern tract tracing techniques has it been possible to analyse these projections in detail. In the present chapter we report the results of light and electron microscopic studies on the origin and termination of somatosensory projections to the PAG. These studies involve anterograde and retrograde tracing with horseradish peroxidase (HRP) and HRP conjugated to wheat

germ agglutinin (WGA-HRP), as well as anterograde tracing with the plant lectin *Phaseolus vulgaris* leucoagglutinin (PHA-L). Parts of the findings have been
reported elsewhere (Flink et al., 1983; Wiberg and Blomqvist, 1984a,b; Wiberg et
al., 1986; 1987; Craig, 1988).

Materials and Methods

Experiments were carried out in adult cats and monkeys (*Macaca fascicularis*). The animals were anesthetized with pentobarbital. Access to the brain stem
for stereotaxic injections was provided by a craniectomy. Access to the lower
medulla was obtained by removing the occipital bone and the atlanto-occipital
membrane and pushing the cerebellum aside. The spinal cord was exposed by
dorsal laminectomies over appropriate segments. Single or multiple pressure
injections of WGA-HRP or free HRP were made by using a glass micropipette
(outer diameter 10-15 μm) or with a Hamilton syringe. Iontophoretic injections
of PHA-L were made through a micropipette (15 μm) with 5-8 microamperes
pulsed positive current for 8-15 min. Following survival periods varying between 1 and 5 days for the HRP and WGA-HRP experiments, and between 12
and 74 days for the PHA-L experiments, the animals were killed by transcardial
perfusion. Animals injected with HRP or WGA-HRP were fixed with a 2% gluta-
raldehyde-1% paraformaldehyde phosphate-buffered solution. Animals injected
with PHA-L lectin were fixed according to a pH-shift method with 4% parafor-
maldehyde-0.2% picric acid in acetate buffer, followed by 4% paraformaldehyde-
0.05% glutaraldehyde in phosphate buffer. Sections were cut at 40-60 μm on a
freezing microtome or a vibratome. Peroxidase activity was demonstrated with
tetramethyl benzidine. PHA-L was visualized immunohistochemically using
either the peroxidase-antiperoxidase (PAP) method or a biotin-avidin-Texas Red
immunofluorescence technique. For electron microscopic analysis of PHA-L
labeling, vibratome sections were processed with the PAP-method, osmicated
and flat-embedded in Epon between acetate foils (see Blomqvist et al., 1985).

Results

Termination pattern of ascending somatosensory projections
The termination in the cat's mesencephalon of fibers from the spinal cord
and the laminar spinal trigeminal nucleus (nucleus caudalis), as demonstrated
by anterograde transport of WGA-HRP, is shown in Figure 1. Although there are
projections to several structures such as the intercollicular nucleus (Inc), the posterior and anterior pretectal nuclei (PTP, PTA) and the nucleus of
Darkschewitsch (D), the PAG is the major target of the projection from the spinal
cord (see also Björkeland and Boivie, 1984b), and it is almost the sole target of
the projection from the laminar spinal trigeminal nucleus. Note, however, that
with the exception of the projection to the nucleus of Darkschewitsch, spinal or
trigeminal fibers do not terminate in the rostral, pretectal, portion of the PAG.
The projection to the PAG is predominantly crossed, but there is also a signifi-

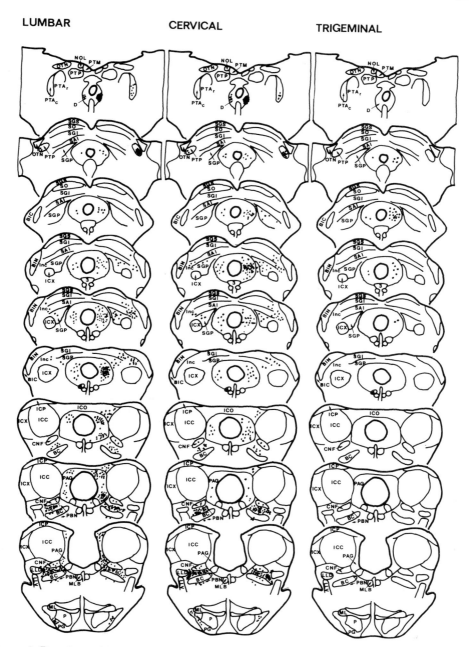

Figure 1. Drawings of frontal sections through the midbrain of the cat, showing the distribution of anterogradely transported WGA-HRP after tracer injection into the lumbar spinal cord, the cervical spinal cord and the laminar spinal trigeminal nucleus. All injections were made on the left side. The terminal labeling is indicated by dots. The sections are ordered, beginning at the bottom, from caudal to rostral. Note that the projection to the PAG is somatotopically organized. Note also that, with the exception of the projection to nucleus of Darkschewitsch, there is no spinal or trigeminal input to the most rostral part of the PAG.

cant ipsilateral component. Most of the fibers terminate in the lateral part of the PAG; however, in the very caudal part of the PAG, the terminal field is divided into two areas, separated by a terminal-sparse zone. Comparison of the projection patterns from the lumbar spinal segments, the cervical segments and the laminar spinal trigeminal nucleus (cf. left, middle and right row of sections in Fig. 1) suggests that the terminal field within the lateral part of the PAG is somatotopically organized in the rostrocaudal direction. The PAG projection from the lumbar enlargement is focussed at the level of the inferior colliculus and the intercollicular region, that from the cervical spinal cord is localized at the middle level of the superior colliculus, and that from the laminar spinal trigeminal nucleus is at the level of the rostral part of the superior colliculus.

Studies of the spinal and trigeminal projections to the PAG in the cynomolgus monkey (Fig. 2) reveal a termination pattern almost identical to that seen in the cat. The major target of the spinal and trigeminal fibers is the lateral part of the PAG, again with the exception of the most caudal region where the terminal field is split into two areas. As in the cat, the projections are predominantly crossed, and they do not extend into the most rostral portion of the PAG. The same topography as in the cat is also present.

The distribution of spinal fibers in the PAG of the rat has been investigated in detail only for the projection from the lumbosacral segments (Yezierski, 1988), but appears to be very similar to that in the cat and monkey. Also, electrophysiological studies in rats have shown a partial somatotopic organization in the PAG, with the face and anterior limbs best represented rostrally and the tail and posterior limbs best represented in the middle and caudal parts (Liebeskind and Mayer, 1971). Studies in a variety of other mammalian species (Mehler, 1969; Jane and Schroeder, 1971; Hazlett et al., 1972; Schroeder and Jane, 1976) have reported projections from the spinal cord to the lateral portion of the PAG. Thus, taken together, the available data seem to indicate that the spinal projection to the PAG maintains a phylogenetic constancy across mammalian species.

There seems to be little, if any, projection to the PAG in cats, monkeys and rats from other somatosensory relay nuclei such as the lateral cervical nucleus (Flink et al., 1983; Björkeland and Boivie, 1985; Wiberg et al., 1987), the dorsal column nuclei (Hand and VanWinkle, 1977; Berkley and Hand, 1978; Björkeland and Boivie, 1984a; Wiberg and Blomqvist, 1984a; Wiberg et al., 1986; 1987; Weinberg and Rustioni, 1989), the alaminar spinal trigeminal nucleus and the principal trigeminal nucleus (Wiberg et al., 1986; 1987). However, with the exception of the latter structure, which in the cat and monkey seems to project exclusively to the thalamus (for refs. see Wiberg et al., 1986; 1987), these somatosensory nuclei project heavily onto several other regions of the dorsal midbrain (Fig. 3). Of particular interest are the projections to the intercollicular nucleus and the deep gray layers of the superior colliculus. The somatotopic organization of these projection (see Fig. 3) is consistent with the findings in electrophysiological experiments, that a somatotopic representation of the body similar to that in the PAG, also exists in the intercollicular nucleus and in the superior colliculus (Stein et al., 1976; Danielsson and Norrsell, 1986; Blomqvist et al., 1990).

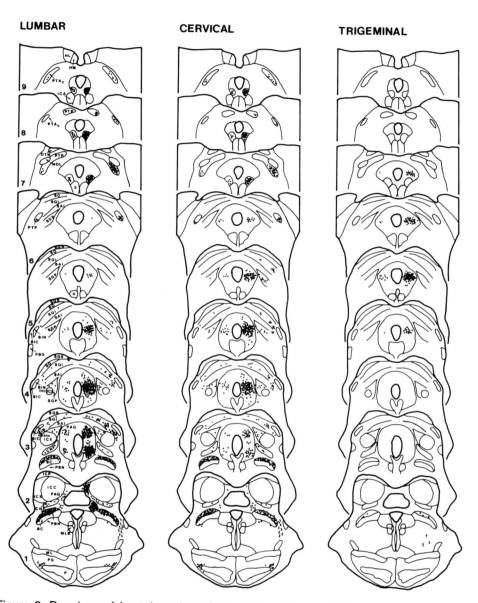

Figure 2. Drawings of frontal sections through the midbrain of the cynomolgus monkey, showing the distribution of anterogradely transported WGA-HRP after tracer injection in the lumbar spinal cord, the cervical spinal cord and the laminar spinal trigeminal nucleus. All injections were made on the left side. Modified from Wiberg et al. (1987).

Cells of origin of spinal and trigeminal projections to PAG

The origin of the spinal and trigeminal projections to the PAG has been investigated with retrograde tracing methods. Figure 4 shows the distribution pattern in the spinal cord of the cynomolgus monkey of neurons that were retrogradely labeled by an injection of WGA-HRP into the dorsal part of the midbrain, including the PAG. The majority of the labeled neurons were situated in lamina I with a smaller fraction located in lamina V. Between 70 and 80% of the labeled neurons were found on the side contralateral to the injection site. In lamina I most of the retrogradely labeled neurons were concentrated in its dorsolateral part at the apex of the dorsal horn. The retrogradely labeled neurons in lamina V were preferentially located in its lateral, reticulated part. Particularly in the upper cervical segments, but to some extent also in other parts of the spinal cord, a third group of labeled neurons was seen in the dorsolateral part of lamina VII. In the nucleus caudalis of the spinal trigeminal nucleus, the retrograde labeling pattern was similar to that in the dorsal horn of the spinal cord. The majority of the labeled neurons were situated in the marginal zone (lamina I), but some were also found along the border between the magnocellular layer (lamina IV) and the reticular formation (lamina V; Gobel et al., 1977).

Since the injection site in the experiment shown in Figure 4 involved parts of the superior and inferior colliculi in addition to the PAG, labeling of spinal neurons projecting to these structures may have influenced the distribution pattern of the retrogradely labeled neurons. However, the spinal projection to the colliculi is sparse in comparison to that to the PAG (Fig. 2). Furthermore, the findings are consistent with those of other reports on the spinal input to the PAG in the monkey (Trevino, 1976; Mantyh, 1982; Zhang et al., 1990), although it has been suggested that the medial part of the PAG receives input primarily from deep laminae (V and VII) (Mantyh, 1982). Of particular interest is the recent work by Zhang et al. (1990). These investigators reported retrograde labeling of neurons in laminae I, V, and in the upper cervical segments also in lamina VII, after small injections of fluorescent microspheres into the PAG. This labeling pattern is almost identical to that obtained by Wiberg et al. (1987) after larger injection with WGA-HRP (Fig. 4). In addition, Zhang et al. (1990) found that a larger percentage of the spino-PAG neurons were labeled at lower spinal levels when the tracer was injected into the caudal PAG, supporting the somatotopic organization demonstrated by anterograde tracing studies (Wiberg et al., 1987).

Neurons in the spinal cord of the cat that project to the dorsal part of the midbrain are distributed in a pattern similar to that described above for the monkey, but there are a few differences. In the cat there are proportionately more neurons in the upper cervical segments. Furthermore, in the cervical, but not in the lumbar segments, retrogradely labeled neurons are found not only in lamina I and V, but also in lamina IV (Wiberg and Blomqvist, 1984). As in the monkey (Fig. 4), scattered neurons in the deep laminae project to the dorsal mesencephalon but are few in number (Wiberg and Blomqvist, 1984). Proportionately more spinomesencephalic neurons are labeled in deep laminae after injections aimed at the parabrachial nucleus (Hylden et al., 1986); however, in those cases there

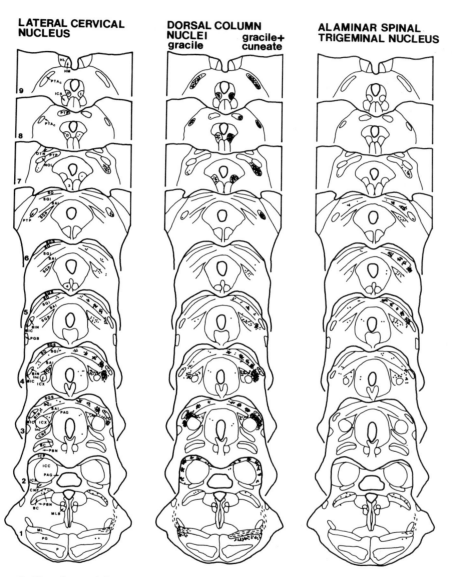

Figure 3. Drawings of frontal sections through the midbrain of the cynomolgus monkey, showing the distribution of anterogradely transported WGA-HRP after tracer injection in the lateral cervical nucleus, the dorsal column nuclei and the alaminar spinal trigeminal nucleus. All injections were made on the left side. Note the somatotopic termination pattern in the intercollicular nucleus and in the deep gray layers of the superior colliculus. Modified from Wiberg et al. (1987).

was little or no involvement of the PAG. In the nucleus caudalis of the spinal trigeminal nucleus (Wiberg et al., 1986), the retrogradely labeled neurons are located in the marginal layer, the lateral part of the magnocellular layer, and in the
part of the reticular formation that has been suggested to correspond to lamina V
of the spinal gray matter (Gobel et al., 1977).

This appraisal of the laminar distribution of neurons that project to the
PAG in monkey and cat is also generally consistent with available data in the rat
(e.g., Menétrey et al., 1982; Liu, 1983; Harmann et al., 1988). However, some studies in the rat (e.g., Liu, 1983; Lima and Coimbra, 1989) have reported a larger
proportion of PAG-projecting neurons in the deep laminae than seems to be present in the cat and monkey (Wiberg and Blomqvist, 1984; Wiberg et al., 1987;
Zhang et al., 1990).

The lamina I projection to the PAG

The details of the lamina I projection to the PAG from the lumbar, cervical
and trigeminal dorsal horns were investigated in the cat with the PHA-L

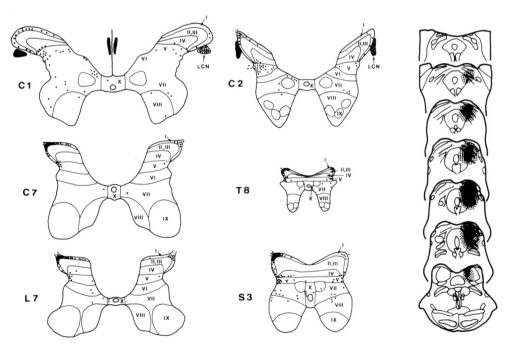

Figure 4. Drawings of spinal cord segments showing the location of neurons (dots) that
were retrogradely labeled after injection of WGA-HRP into the dorsal midbrain, including
the PAG, of the cynomolgus monkey. The injected area is shown in the drawings to the
right. Most of the retrogradely labeled neurons were located in laminae I and V. Reprinted
with permission from Wiberg et al. (1987) Journal of Comparative Neurology.

method. An example of an injection site is shown in Figure 5. The ascending lamina I fibers to the PAG traveled with the spinothalamic tract in the ventrolateral aspect of the brain stem. Most of the fibers coursed dorsally along the lateral edge of the brain stem at the level of the isthmus, passed through the ventral spinocerebellar tract and entered the PAG through the cuneiform nucleus and the ventral aspect of the inferior colliculus in a lateral to medial trajectory (Fig. 6). Other fibers ascended at a slightly more rostral level through the lateral lemniscus. The PHA-L labeling that was observed in the PAG was distributed in a pattern similar to that found in the WGA-HRP tracing experiments. Thus, the predominant projection to the lateral portion of the PAG (Fig. 6) and the two separated PAG regions more caudally were confirmed, as was the rostrocaudal topography. The labeled lamina I fibers, which arborized through the PAG in the lateromedial direction (Fig. 6), seemed to be restricted to a rather narrow rostrocaudal plane; accordingly, many fibers could be followed for considerable distances in a transverse section (Fig. 7). Along their course, the fibers displayed boutons

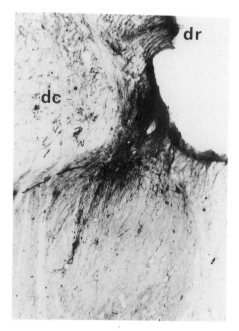

Figure 5. Microphotograph showing the distribution of reaction product in the dorsal horn of the C8 segment after an iontophoretic injection of *Phaseolus vulgaris* leucoagglutinin (PHA-L) into lamina I and peroxidase-antiperoxidase immunohistochemistry. Dorsal to the top, lateral to the right. dc: dorsal columns. dr: dorsal rootlets. x 75.

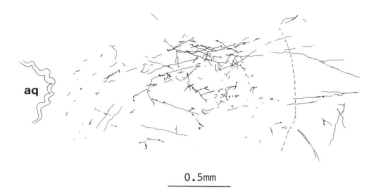

0.5mm

Figure 6. Camera lucida drawing of a transverse sections through
the lateral part of the PAG at the level of the superior colliculus
showing the distribution of labeled fibers after a PHA-L injection
into the lamina I of the cervical spinal cord on the contralateral
side. Medial to the left, dorsal upwards. aq: cerebral aqueduct.
Dashed line represents the lateral border of the PAG. Note that the
fibers enter the PAG in a lateral to medial trajectory. Also note that
the densest termination is found in the outer region of the PAG and
that only scattered fibers and terminals are seen in area adjacent
to the cerebral aqueduct.

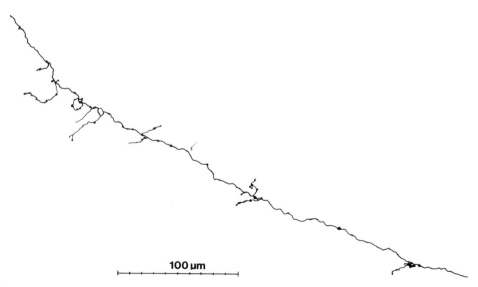

100 μm

Figure 7. Camera lucida drawing of a PHA-L labeled fiber which traversed the lateral
PAG. Note the boutons of passage and the thin collaterals.

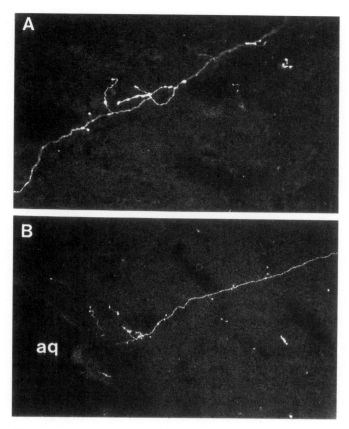

Figure 8. Fibers in the lateral PAG, labeled after injection of PHA-L into lamina I of the cervical spinal cord. The PHA-L was demonstrated with a biotin-avidin-Texas Red immuno-fluorescent technique. Note that the labeled fibers display boutons of passage and thin collaterals with terminal boutons. aq: cerebral aqueduct. A: x 190; B: x 95.

of passage and isolated short axon collaterals with terminal boutons (Figs. 7, 8). As judged from the density of the labeling, most of the fibers appeared to end within the outer region of the PAG (Fig. 6), but some fibers were traced to the immediate vicinity of the cerebral aqueduct (Fig. 8B).

Preliminary observations in the electron microscope revealed that the PHA-L labeled boutons were filled with rounded synaptic vesicles (Figs. 9, 10). The boutons terminated on dendritic profiles of varying sizes. Synaptic speciali-zations were seen, of the asymmetric type (Fig. 9). When followed in serial sec-tions, the labeled boutons were found to originate from thin unmyelinated fibers (Fig. 10).

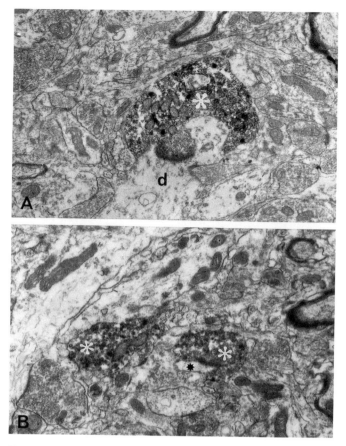

Figure 9. A. Electron micrograph showing a PHA-L labeled
bouton (asterisk) in synaptic contact with a dendrite (d).
x 14,000. B. Two PHA-L labeled boutons (asterisks). One of
the boutons is in synaptic contact with a small dendritic profile (star). The labeled boutons contain round synaptic
vesicles. x 14,000.

Conclusion

The results of the anatomical tracing studies demonstrate the presence of a
topographically organized, predominantly crossed projection from the spinal
cord and the laminar spinal trigeminal nucleus to a restricted region of the PAG.
In addition, the studies demonstrate that the other somatosensory pathways, i.e.
the ascending projections from the lateral cervical nucleus, the dorsal column
nuclei and the alaminar spinal trigeminal nucleus, do not significantly contribute
to a direct somatosensory input to the PAG.

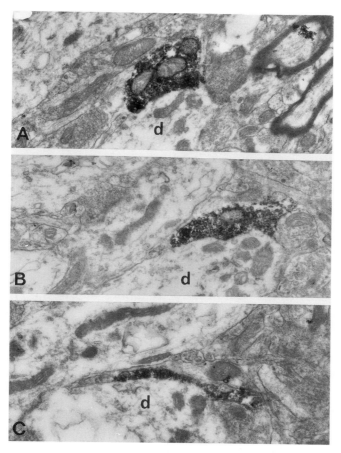

Figure 10. A: PHA-L labeled bouton in contact with a dendrite (d). When this bouton was followed in serial sections (B
and C), it was found to originate from a thin, longitudinally
running fiber (C). The bouton contains round synaptic
vesicles (best seen in B). x 17,500.

A major part of the spinal and trigeminal projections to the PAG seems to
originate in lamina I neurons. The lamina I fibers run mediolaterally in the transverse plane of the PAG and make synaptic contacts through boutons of passage
and terminal boutons on thin collaterals with dendrites in the PAG. The mediolateral distribution in the PAG of the spinal and trigeminal fibers suggests that
these projections have a restricted rostrocaudal extent, which is in line with a
somatotopic organization of these inputs.

Since different types of information are carried by the different spinal cord
and trigeminal pathways (see e.g. Willis and Coggeshall, 1978), the somatosenso-

ry information that is transmitted to the PAG is likely to be different from that transmitted to other regions of the dorsal midbrain. The finding that the spinal and trigeminal projections to the PAG originates from regions in which small myelinated and unmyelinated fibers terminate (Light and Perl, 1979; Craig and Mense, 1983; Cervero and Connell, 1984; Craig et al., 1988; Mense and Craig, 1988; Sugiura et al., 1986; Sugiura et al., 1989), suggests that the somatosensory input to the PAG conveys nociceptive, thermoreceptive and visceroceptive information from all organs of the body. The results of electrophysiological studies support this suggestion, since nociceptive spinal neurons that project to the dorsal midbrain have been identified (Yezierski and Schwartz, 1986; Yezierski et al., 1987). Of particular interest is the large lamina I projection to the PAG. The lamina I cells that project to the thalamus are specifically activated by nociceptive or thermoreceptive stimuli (Craig and Kniffki, 1985) and lamina I cells with ascending axons that respond to visceral stimuli have been recorded (Cervero and Tattersall, 1987). Whether this is also the case for lamina I cells that project to the PAG remains to be clarified, since so far the response characteristics of only few such cells have been investigated. However, there is indirect evidence that the lamina I projection to the PAG transmits nociceptive-specific information, since lamina I cells that project both to the thalamus and to the PAG have been demonstrated in the primate (Zhang et al., 1990). Furthermore, most lamina I spinomesencephalic cells that project to the parabrachial nucleus are nociceptive specific (Hylden et al., 1986; Light et al., 1987).

Thus, the available data would seem to indicate that the spinal and trigeminal projections to the PAG carry nociceptive, thermoreceptive and visceroceptive information, and that this information is relayed onto a restricted (lateral) region of the contralateral PAG in a somatotopic fashion. It is of particular interest that the same (lateral) region of the PAG controls defensive behavior, and that defense reactions elicited by microinjections of excitatory amino acid into the PAG in conjunction with tactile stimulation show both a clear asymmetry and a somatotopic gradient: tactile stimulation of the contralateral head and forelimb evoked strong defense reactions when excitatory amino acids had been injected into the pretentorial PAG, whereas tactile stimulation of the ipsilateral side as well as the contralateral hindlimb evoked weak reactions or no reaction (Bandler and Carrive, 1988; Depaulis et al., 1989; Bandler and Depaulis, this volume). This behavioral pattern is consistent with the somatotopic organization of the afferent spinal and trigeminal input. It is also of interest to note that different types of defense reactions are evoked from the pre- and subtentorial PAG. Thus, excitation of neurons in the pretentorial PAG (which receives sensory input from the face and forelimbs) evokes threat display, associated with faciovocal activity (Bandler and Carrive, 1988; Carrive et al., 1987), whereas excitation of neurons in the subtentorial PAG (which receives input from the hindlimbs) results in a flight reaction which is associated with limb movements (Carrive et al., 1989a; Zhang et al., 1990). These findings seem to indicate that information about the location of a sensory stimulus may be of importance for the organization in the PAG of an appropriate motor response (Bandler et al., 1991), and reinforces the idea of somatotopic specificity within the PAG. Also, the results of studies of the endogenous pain suppressive mechanisms that can be activated by peripheral noxious stimulation (Fleischmann and Urca, 1988) fit in with the idea that somatotopically specific afferent input results in a highly localized PAG excitation (Bandler et al., 1991) by demonstrating that different behavioral responses are elicited by nociceptive stimulation of different body regions. Thus, noxious stimulation of the nape of the neck, which is likely to signal a severe predatory situation with the threat of lethal outcome (Fleischmann and Urca, 1988), can suppress pain responsiveness both at the site of stimulation and at sites remote from the stimulated area while noxious stimulation of the tail produces analgesia only at sites remote from the stimulated area (Fleischmann and Urca, 1988). These different analgesia systems are associated with the motor responses (immobility and fight or flight, respectively) which appear to be most adaptive given the location of the nociceptive information (Fleischmann and Urca, 1988).

With respect to the topographic organization of the sensory input to the PAG it is also of interest to note that the autonomic reactions (i.e., blood flow through particular vascular beds) that can be elicited by stimulation of the lateral part of the PAG seem to be topographically organized in an appropriate manner to support the somatic responses (Carrive et al., 1989 a,b; Zhang et al., 1990; Carrive, this volume). This observation suggests a coordination in the PAG of somatic and autonomic reactions. An association between the antinociception that can be elicited from the PAG and autonomic regulation may also exist, since electrical stimulation of the PAG that suppresses spinal transmission of impulses from C fiber primary afferents has been shown to result in an increase in muscle blood flow at the expense of cutaneous flow (Duggan and Morton, 1983; however, see Depaulis et al., 1988). Taken together, these observations suggest that the spinal and trigeminal projections to the PAG may constitute an ascending loop in a neuronal circuitry which is involved in the organization of behavioral and homeostatic responses to nociceptive and thermoreceptive information that is vital for the survival of the organism as a whole.

Abbreviations

3: oculomotor nucleus; 4: trochlear nucleus; BC: brachium conjunctivum; BIC: brachium of the inferior colliculus; BIN: nucleus of the brachium of the inferior colliculus; CNF: cuneiform nucleus; D: nucleus of Darkschewitsch; HL: lateral habenular nucleus; HM: medial habenular nucleus; ICA: interstitial nucleus of Cajal; ICC: central nucleus of the inferior colliculus; ICO: commissure of the inferior colliculus; ICP: pericentral nucleus of the inferior colliculus; ICX: external nucleus of the inferior colliculus; Inc: intercollicular nucleus; LCN: lateral cervical nucleus; LLD: dorsal nucleus of the lateral lemniscus; LLV: ventral nucleus of the lateral lemniscus; ML: medial lemniscus; MLB: medial longitudinal bundle; NOL: olivary pretectal nucleus; OTN: nucleus of the optic tract; P: pyramidal tract; PAG: periaqueductal gray matter; PBG: parabigeminal nucleus;

PBN: parabrachial nucleus; PG: pontine gray; PTAc: anterior pretectal nucleus, pars compacta; PTAr: anterior pretectal nucleus, pars reticularis; PTM: medial pretectal nucleus; PTP: posterior pretectal nucleus; SAI: stratum album intermedium of the superior colliculus; SGI: stratum griseum intermedium of the superior colliculus; SGP: stratum griseum profundum of the superior colliculus; SGS: stratum griseum superficiale of the superior colliculus; SO: stratum opticum of the superior colliculus

Acknowledgements

We thank Ms. Tracey Fleming and Ms. Ludmila Mackerlova for skilful technical assistance. This work was supported by the Swedish Medical Research Council (project # 7879), the Barrow Neurological Foundation and NIH grant NS 25616.

References

Bandler, R. and Carrive, P., Integrated defence reaction elicited by excitatory amino acid microinjection in the midbrain periaqueductal grey region of the unrestrained cat, Brain Res., 439 (1988) 95-106.

Bandler, R., Carrive, P. and Zhang, S.P., Integration of somatic and autonomic reactions within the midbrain periaqueductal grey: viscerotopic, somatotopic and functional organization, Prog. Brain Res., 87 (1991) 269-305.

Berkley, K.J. and Hand, P.J., Efferent projections of the gracile nucleus in the cat, Brain Res., 153 (1978) 263-283.

Björkeland, M. and Boivie, J., An anatomical study of the projections from the dorsal column nuclei to the midbrain in the cat, Anat. Embryol., 170 (1984a) 29-43.

Björkeland, M. and Boivie, J., The termination of spinomesencephalic fibers in the cat. An experimental anatomical study, Anat. Embryol., 170 (1984b) 265-277.

Björkeland, M. and Boivie, J., Anatomy of the midbrain projections from the lateral cervical nucleus in the cat, In: Development, organization and processing in somatosensory pathways, Rowe M. and Willis W.D. (Eds.), Alan R. Liss, New York, 1985, pp. 203-214.

Blomqvist, A., Flink, R., Westman, J. and Wiberg, M., Synaptic terminals in the ventroposterolateral nucleus of the thalamus from neurons in the dorsal column and lateral cervical nuclei: an electron microscopic study in the cat, J. Neurocytol., 14 (1985) 869-886.

Blomqvist, A., Danielsson, I. and Norrsell, U., The somatosensory intercollicular nucleus of the cat´s mesencephalon, J. Physiol. (Lond.), 429 (1990) 191-203.

Carrive, P., Dampney, R.A.L. and Bandler, R., Excitation of neurones in a restricted portion of the midbrain periaqueductal gray elicits both the behavioral and cardiovascular components of the defense reaction in the unanaesthetized cat, Neurosci. Lett., 81 (1987) 272-278.

Carrive, P., Bandler, R. and Dampney, R.A.L., Somatic and autonomic integration in the midbrain of the unanesthetized decerebrate cat: a distinctive pattern evoked by excitation of neurons in the subtentorial portion of the midbrain periaqueductal grey, Brain Res., 483 (1989a) 251-258.

Carrive, P., Bandler, R. and Dampney, R.A.L., Viscerotopic control of regional vascular beds by discrete groups of neurons within the midbrain periaqueductal gray, Brain Res., 493 (1989b) 385-390.

Cervero, F. and Connell, L.A., Distribution of somatic and visceral primary afferent fibres within the thoracic spinal cord of the cat, J. Comp. Neurol., 230 (1984) 88-98.

Cervero, F. and Tattersall, J.E.H., Somatic and visceral inputs to the thoracic spinal cord of the cat: marginal zone (lamina I) of the dorsal horn, J. Physiol. (Lond.), 383 (1987) 383-395.

Craig, A.D. Jr., Cervical lamina I spinothalamic projections in the cat, Soc. Neurosci. Abstr., 14 (1988) 120.

Craig, A.D. and Kniffki, K.-D., Spinothalamic lumbosacral lamina I cells responsive to skin and muscle stimulation in the cat, J. Physiol. (Lond.), 365 (1985) 197-221.

Craig, A.D. and Mense, S., The distribution of afferent fibers from the gastrocnemius-soleus muscle in the dorsal horn of the cat, as revealed by the transport of horseradish peroxidase, Neurosci. Lett., 41 (1983) 233-238.

Craig, A.D., Heppelmann, A.D. and Schaible, H.-G., The projection of the medial and posterior articular nerves of the cat's knee to the spinal cord, J. Comp. Neurol., 276 (1988) 279-288.

Danielsson, I. and Norrsell, U., Somatosensory units in the cat's intercollicular region, Acta Phys. Scand., 128 (1986) 579-586.

Depaulis, A., Pechnick, R.N. and Liebeskind, J.C., Relationship between analgesia and cardiovascular changes induced by electrical stimulation of the mesencephalic periaqueductal gray matter in the rat, Brain Res., 451 (1988) 326-332.

Depaulis, A., Bandler, R. and Vergnes, M., Characterization of pretentorial periaqueductal gray neurons mediating intraspecific defensive behavior in the rat by microinjections of kainic acid, Brain Res., 486 (1989) 121-132.

Duggan, A.W. and Morton, C.R., Periaqueductal grey stimulation: an association between selective inhibition of dorsal horn neurones and changes in peripheral circulation, Pain, 15 (1983) 237-248.

Eickhoff, R., Handwerker, H.O., McQueen, D.S. and Schick, E., Noxious and tactile input to medial structures of midbrain and pons in the rat, Pain, 5 (1978) 99-113.

Fleischmann, A. and Urca, G., Different endogenous analgesia systems are activated by noxious stimulation of different body regions, Brain Res., 455 (1988) 49-57.

Flink, R., Wiberg, M. and Blomqvist, A., The termination in the mesencephalon of fibres from the lateral cervical nucleus. An anatomical study in the cat, Brain Res., 259 (1983) 11-20.

Gobel, S., Falls, W.M. and Hockfield, S., The division of the dorsal and ventral horns of mammalian caudal medulla into eight layers using anatomical criteria, In: Pain in the trigeminal region, Anderson D.J. and Matthews B. (Eds.), Elsevier, Amsterdam, 1977, pp. 443-453.

Hand, P.J. and Van Winkle, T., The efferent connections of the feline nucleus cuneatus, J. Comp. Neurol., 171 (1977) 83-110.

Harmann, P.A., Carlton, S.M. and Willis, W.D., Collaterals of spinothalamic tract cells to the periaqueductal gray: a fluorescent double-labeling study in the rat, Brain Res., 441 (1988) 87-97.

Hazlett, J.C., Dom, R. and Martin, G.F., Spino-bulbar, spino-thalamic and medial lemniscal connections in the American opossum, Didelphis marsupialis virginiana, J. Comp. Neurol., 146 (1972) 95-118.

Hylden, J.L.K., Hayashi, H. and Bennett, G.J., Lamina I spinomesencephalic neurons in the cat ascend via the dorsolateral funiculi, Somatosens. Res., 4 (1986) 31-41.

Hylden, J.L.K., Hayashi, H., Dubner, R. and Bennett, G.J., Physiology and morphology of the lamina I spinomesencephalic projection, J. Comp. Neurol., 247 (1986) 505-515.

Jane, J.A. and Schroeder, D.M., A comparison of dorsal column nuclei and spinal afferents in the European hedgehog (Erinaceus europeaus), Exp. Neurol., 30 (1971) 1-17.

Liebeskind, J.C. and Mayer, D.J., Somatosensory evoked responses in the mesencephalic central gray matter of the rat, Brain Res., 27 (1971) 133-188.

Light, A.R. and Perl, E.R., Spinal termination of functionally identified primary afferent neurons with slowly conduction myelinated fibers, J. Comp. Neurol., 186 (1979) 133-150.

Light, A.R., Casale, E. and Sedivec, M., The physiology and anatomy of spinal laminae I and II neurons antidromically activated by stimulation in the parabrachial region of the midbrain and pons, In: Fine afferent nerve fibers and pain, Schmidt R.F., Schaible G.-H. and Vahle-Hinz C. (Eds.), VCH, Weinheim, 1987, pp. 347-356.

Lima, D. and Coimbra, A., Morphological types of spinomesencephalic neurons in the marginal zone (lamina I) of the rat spinal cord, as shown after retrograde labeling with cholera toxin subunit B, J. Comp. Neurol., 279 (1989) 327-339.

Liu, R.P.C., Laminar origin of spinal projection neurons to the periaqueductal gray of the rat, Brain Res., 264 (1983) 118-122.

Mantyh, P.W., The ascending input to the midbrain periaqueductal gray of the primate, J. Comp. Neurol., 211 (1982) 50-64.

Mehler, W.R., Some neurological species differences - a posteriori, Ann. N. Y. Acad. Sci., 167 (1969) 424-468.

Menétrey, D., Chaouch, A., Binder, D. and Besson, J.-M., The origin of the spinomesencephalic tract in the rat: an anatomical study using retrograde transport of horseradish peroxidase, J. Comp. Neurol., 206 (1982) 193-207.

Mense, S. and Craig, A.D. Jr., Spinal and supraspinal terminations of primary afferent fibers from the gastrocnemius-soleus muscle in the cat, Neurosci., 26 (1988) 1023-1035.

Nauta, W.J.H. and Kuypers, H.G.J.M., Some ascending pathways in the brain stem reticular formation, In: Reticular formation of the brain. Henry Ford Hospital International Symposium, Jasper H.H., Proctor L.D., Knighton R.S., Noshay W.C. and Costello R.T. (Eds.), Little, Brown and Company, Boston, 1958, pp. 3-30.

Rose, J.D., Responses of midbrain neurons to genital and somatosensory stimulation in estrous and anestrous cats, Exp. Neurol., 49 (1975) 639-652.

Rose, J.D., Anatomical distribution and sensory properties of brain stem and posterior diencephalic neurons responding to genital, somatosensory and nociceptive stimulation in the squirrel monkey, Exp. Neurol., 66 (1979) 169-185.

Schroeder, D.M. and Jane, J.A., The intercollicular area of the inferior colliculus, Brain Behav. Evol., 13 (1976) 125-141.

Sugiura, Y., Lee, C.L. and Perl, E.R., Central projections of identified, unmyelinated (C) afferent fibers innervating mammalian skin, Science, 234 (1986) 358-361.

Sugiura, Y., Terui, N. and Hosoya, Y., Difference in distribution of central terminals between visceral and somatic unmyelinated (C) primary afferent fibers, J. Neurophysiol., 62 (1989) 834-840.

Stein, B.E., Magalhães-Castro, B. and Kruger, L., Relationship between visual and tactile representations in cat superior colliculus, J. Neurophysiol., 39 (1976) 401-419.

Trevino, D.L., The origin and projections of a spinal nociceptive and thermoreceptive pathway, In: Sensory functions of the skin in primates, with special reference to man, Zotterman Y. (Ed.), Pergamon Press, New York, 1976, pp. 367-376.

Weinberg, R.J. and Rustioni, A., Brainstem projections to the rat cuneate nucleus, J. Comp. Neurol., 282 (1989) 142-156.

Wiberg, M. and Blomqvist, A., The projection to the mesencephalon from the dorsal column nuclei. An anatomical study in the cat, Brain Res., 311 (1984a) 225-244.

Wiberg, M. and Blomqvist, A., The spinomesencephalic tract in the cat: its cells of origin and termination pattern as demonstrated by the intraaxonal transport method, Brain Res., 291 (1984b) 1-18.

Wiberg, M., Westman, J. and Blomqvist, A., The projection to the mesencephalon from the sensory trigeminal nuclei. An anatomical study in the cat, Brain Res., 399 (1986) 51-68.

Wiberg, M., Westman, J. and Blomqvist, A., Somatosensory projection to the mesencephalon: an anatomical study in the monkey, J. Comp. Neurol., 264 (1987) 92-117.

Willis, W.D. and Coggeshall, R.E., Sensory mechanisms of the spinal cord, Plenum Press, New York, 1978.

Yezierski, R.P., Spinomesencephalic tract: projections from lumbosacral spinal cord of the rat, cat and monkey, J. Comp. Neurol., 267 (1988) 131-146.

Yezierski, R.P. and Schwartz, R.H., Response and receptive-field properties of spinomesencephalic tract cells in the cat, J. Neurophysiol., 55 (1986) 76-96.

Yezierski, R.P., Sorkin, L.S. and Willis, W.D., Response properties of spinal neurons projecting to the midbrain or midbrain-thalamus in the monkey, Brain Res., 437 (1987) 165-170.

Zhang, D., Carlton, S. M., Sorkin, L.S. and Willis, W.D., Collaterals of primate spinothalamic tract neurons to the periaqueductal gray, J. Comp. Neurol., 296 (1990) 277-290.

Zhang, S.P., Bandler, R. and Carrive, P., Flight and immobility evoked by excitatory amino acid microinjection within distinct parts of the subtentorial midbrain periaqueductal gray of the cat, Brain Res., 520 (1990) 73-82.

Somatosensory Input to the Periaqueductal Gray: A Spinal Relay to a Descending Control Center

Robert P. Yezierski

Department of Neurological Surgery
University of Miami, School of Medicine
Miami, Florida, U.S.A.

Introduction

The spinomesencephalic tract (SMT) with its varied origins (Mantyh, 1982; Menétrey et al., 1982; Swett et al., 1985; Wiberg and Blomqvist, 1984; Wiberg et al., 1987; Yezierski and Mendez, 1991; Zhang et al., 1990), spinal trajectories (Hylden et al., 1986b; Kerr, 1975; McMahon and Wall, 1985; Yezierski and Schwartz, 1986; Zemlan et al., 1978), and sites of termination (Anderson and Berry, 1959; McMahon and Wall 1985; Mehler, 1969; Morin, 1953; Björkeland and Boivie, 1984; Blomqvist and Craig, this volume; Yezierski, 1988) is often described as having a role in nociception (Bowsher, 1976; Mehler, 1969; Willis, 1985; Willis and Coggeshall, 1978; Yezierski, 1988). Consistent with this hypothesis are the responses of SMT cells to noxious mechanical and thermal stimuli (Hylden et al., 1986a; 1989; Menétrey et al., 1980; Yezierski and Schwartz, 1986; Yezierski et al., 1985). Furthermore, recent studies have shown SMT cells in the upper cervical and lumbosacral spinal cord respond to inputs from cutaneous and/or deep structures, including joints, muscles, and viscera (Yezierski and Broton, 1991; Yezierski and Schwartz, 1986; Yezierski et al., 1987; Yezierski, 1990). These observations as well as the varied functions associated with SMT projection targets supports a role of the SMT in sensory, motor and visceral functions.

As described by Blomqvist and Craig (this volume), one of the principal targets of the spinomesencephalic projection is the periaqueductal gray (PAG). Originating from cells in all regions of the gray matter and at all levels of the spinal cord, the spino-PAG component of the SMT is well suited to relay a wide spectrum of afferent information to projection targets throughout the midbrain. Although the functional significance of this input is unknown, evidence from different laboratories has shown the PAG to be involved in the control of blood pressure (Carrive; Lovick, this volume), defence reactions (Bandler and Depaulis, this volume), respiration and vocalization (Harper; Jürgens; Larson, this volume), reproductive behavior (Ogawa et al., this volume), stomach and bladder motility (Skultety, 1959), grooming (Spruijt et al., 1986), and analgesic mechanisms (Besson et al.; Mason, this volume). The SMT may, therefore, constitute an important component of afferent systems contributing input to neuronal substrates directly or indirectly involved in carrying out these functions. In the sections below the response and receptive field (RF) properties of SMT cells in the lumbosacral and upper cervical spinal cord are discussed in relation to their relevance to functions associated with the PAG.

Response Properties of SMT Cells

In recent years the response properties of SMT cells have been studied in the rat, cat and monkey (Hylden et al., 1989; Menétrey et al., 1980; Yezierski and Broton, 1991; Yezierski and Schwartz, 1986; Yezierski et al., 1987). The present discussion will focus on the results of studies in the cat and monkey which have been obtained from animals anesthetized with alpha chloralose and sodium pentobarbital.

The experimental set-up used in these studies has been described in detail (Yezierski, 1990; Yezierski and Schwartz, 1986; Yezierski et al., 1987). Briefly, following a laminectomy and craniectomy animals are placed in a stereotaxic device and spinal unit. Animals are artificially ventilated and paralyzed. Expired CO_2 and core temperature are maintained within physiological limits. An array of stimulating electrodes is then positioned at different rostrocaudal levels of the midbrain. Electrodes in this array are used to deliver antidromic search stimuli of 100 μsec pulses (10/sec) at stimulus intensities ranging from 350-500 μA. Recordings are made from cells identified by antidromic activation at recording sites in the upper cervical or lumbosacral spinal cord. As shown in Figure 2 recording sites are generally located in nucleus proprius, the intermediate gray, ventral horn and in the region around the central canal. This preferential distribution of recording sites is due to the recording characteristics of the carbon filament electrodes used in these studies and to the difficulty of maintaining stable recordings from small cells in the superficial dorsal horn.

The response profiles of SMT cells in the lumbosacral and upper cervical spinal cord can be divided into three major categories (Yezierski and Broton, 1991; Yezierski and Schwartz, 1986). Based on responses to varying intensities of mechanical stimuli and input from deep structures (muscles and joints) these categories include: (a) wide dynamic range (WDR); (b) high threshold (HT); and (c) deep/tap (D). Wide dynamic range or multi-convergent neurons respond in a graded fashion to light tactile stimuli (brush and light pressure), but respond best to noxious stimuli (pinch and squeeze). High-threshold or nociceptive specific cells respond exclusively to pinch and/or squeeze, whereas deep cells receive input from subcutaneous structures including muscles and joints. Responses to these various stimulus conditions include both excitatory and inhibitory effects

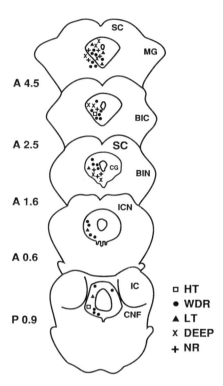

Figure 1. Distribution of antidromic stimulation sites used to backfire cells in the
upper cervical and lumbosacral spinal
cord of the cat. Cells in different functional
classes backfired from the central gray or
adjacent reticular formation are represented by different symbols at the stimulation
site of lowest threshold for each cell.
Coordinates for different midbrain levels
were taken from the atlas of Berman
(1968). Abbreviations: SC, superior colliculus; MG, medial geniculate; BIC, brachium of the inferior colliculus; BIN,
nucleus of the brachium of the inferior colliculus; CG, central gray; ICN, intercollicular nucleus; IC, inferior colliculus; CNF,
nucleus cuneiformis; HT, high threshold;
WDR, wide dynamic range; LT, low threshold; NR, nonresponsive; A, anterior; P,
posterior.

(Yezierski, 1990; Yezierski and Broton, 1991; Yezierski and Schwartz, 1986; Yezierski et al., 1987). A fourth class of SMT cell encountered at both cervical and lumbosacral cord levels are those unresponsive to any form of natural stimuli (Yezierski and Schwartz, 1986; Yezierski and Broton, 1991). Although the functional properties of SMT cells in different laminae have been most frequently studied in the cat (Hylden et al., 1986a; Yezierski, 1990; Yezierski and Schwartz, 1986; Yezierski and Broton, 1991), similar cell types have been described in the rat (Hylden et al., 1990; Menétrey et al., 1980) and monkey (Yezierski et al., 1987; Zhang et al., 1990).

Antidromic stimulation sites used to identify cells belonging to the SMT include all regions of its terminal domain (Hylden et al., 1986a; McMahon and Wall, 1985; Yezierski, 1990; Yezierski and Broton, 1991; Yezierski and Schwartz, 1986; Yezierski et al., 1987; Zhang et al., 1990). In the present discussion only cells antidromically identified from the PAG will be considered. A summary of the distribution of stimulation sites used to backfire spino-PAG cells in the cat is shown in Figure 1 (Yezierski, 1990; Yezierski and Broton, 1991; Yezierski and Schwartz, 1986; Yezierski et al., 1985). From this distribution it can be seen that cells in 2-4 functional classes can be antidromically activated from each of the

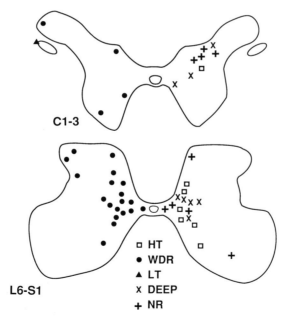

Figure 2. Distribution of recording sites in the upper cervical (C1-3) and lumbosacral (L6-S1) spinal cord of the cat for cells identified from stimulation sites in Fig. 1. Recording sites are indicated by symbols representing the functional class of each cell. Abbreviations are the same as in Fig. 1.

five midbrain levels represented. Thus, the full complement of sensory information represented in the spino-PAG projection is distributed throughout the entire rostrocaudal extent of the cat midbrain. Spinal recording sites for cells backfired from the PAG are located in the dorsal and ventral horns and in the region surrounding the central canal (Fig. 2). Wide dynamic range cells have the most wide spread distribution, whereas cells in other classes are concentrated in the neck of the dorsal horn, intermediate gray and ventral horn. Mean conduction velocities (CV) for cells projecting to the PAG reflect a heterogeneous population of cells with those in laminae IV-VIII and X having faster CVs (37.5m/sec) than cells in laminae I-III (12.2m/sec).

The heterogeneity of the SMT cell population is further exemplified by the response properties and receptive field (RF) organizations of cells belonging to this pathway. In both the upper cervical and lumbosacral cord, cells with small excitatory RFs confined to a single limb are most commonly observed. Cells with complex RFs that often include the hindlimbs, forelimbs and/or oral-facial structures have also been described (Ménétrey et al., 1980; Yezierski, 1990; Yezierski and Broton, 1991; Yezierski and Schwartz, 1986).

The largest class of cells backfired from the PAG are multi-convergent or wide dynamic range neurons (52%). High-threshold or nociceptive specific cells represent 4% of the cells with low threshold and deep/tap cells representing 12% and 16%, respectively. Cells in the non-responsive category account for 16% of the population. From this breakdown it is evident that the majority of responsive spino-PAG cells are excited best or exclusively by nociceptive stimuli.

An example of the mechanical and thermal responses indicative of WDR spino-PAG cells is shown in Figure 3. This cell, recorded in the monkey, was antidromically activated from the contralateral ventral posterolateral nucleus of the thalamus and the contralateral PAG at the level of the inferior colliculus. The recording site in the lumbosacral spinal cord was located in lamina V. Consistent with a WDR response profile, the cell responded in a graded fashion to increasing intensities of mechanical stimuli applied to the ipsilateral foot (Fig. 3A). Similarly, a series of 30sec heat pulses also delivered to the ipsilateral foot produced graded responses to stimuli of increasing intensities (Fig. 3B-E).

Although the responses of SMT cells are most commonly evaluated with mechanical and thermal stimuli, chemical agents including intramuscular injection of hypertonic saline and intravenous serotonin (i.v. 5HT) have also been used to study SMT cell responses (Yezierski and Broton, 1991). An example of the responses of a WDR cell in the upper cervical cord of the cat to i.v. 5HT is shown in Figure 4. This cell had a cutaneous RF on the ipsilateral forepaw; no deep input was detected. The cell was backfired from the contralateral PAG at the intercollicular level of the midbrain and the recording site was located in lamina I. Following the administration of 60 µg of 5HT into the cephalic vein contralateral to the excitatory receptive field a decrease in activity followed by a modest increase in the response of the cell was observed (Fig. 4A). Subsequent to this initial dose of 5HT a series of increasing doses, ranging from 2.5-150 µg, were administered. Doses ranging from 2.5 µg-30 µg produced a dose dependent

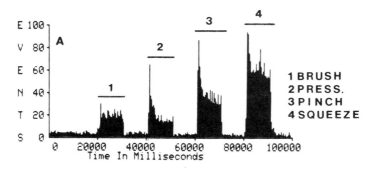

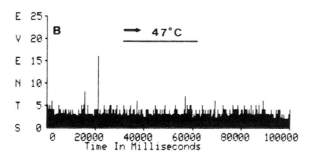

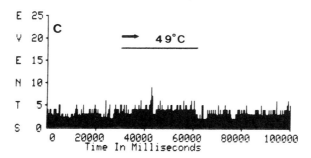

Figure 3. Responses of a wide dynamic range SMT cell to mechanical and thermal stimuli. The cell was recorded in the lumbosacral spinal cord (lamina V) of the monkey. (A) Responses to the application of a series of graded mechanical stimuli (1, brush; 2, light pressure; 3, pinch; 4, squeeze) delivered to the glabrous skin on the ipsilateral foot. (B-E) Responses to 30sec heat pulses delivered from an adapted temperature of 35 °C to 47 °C (B), 49 °C (C), 51 °C (D), and 53 °C (E). Heat pulses were delivered to the same area of skin as in (A). (F) Sensitized responses to mechanical stimuli following delivery of 53 °C heat pulse. Duration of mechanical and thermal stimuli are represented by the stimulus bars in (A-F). Time in milliseconds is represented on the X-axis and the number of spike discharges on the Y-axis (EVENTS). Bin width in A-F equals 400 msec.

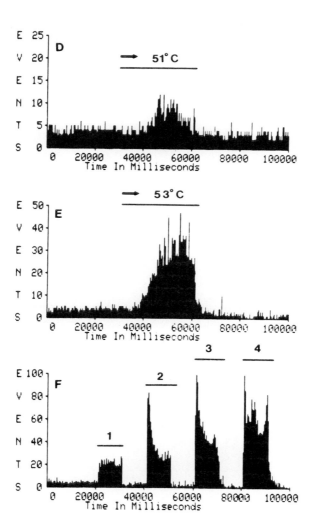

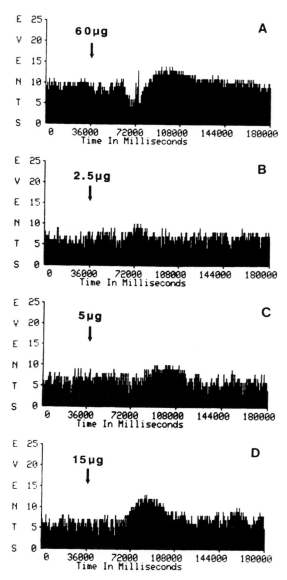

Figure 4. Responses of a wide dynamic range SMT cell to seven different doses of intravenous serotonin creatine phosphate (i.v. 5HT). Time of drug administration in the cephalic vein contralateral to the excitatory receptive field is represented by the arrows. This cell was backfired from the contralateral periaqueductal gray at the intercollicular level of the midbrain and the recording site was located in lamina I of the upper cervical cord (C2) in the cat. (A) Response of cell to initial dose of i.v. 5HT (60 μg); note initial depression of spontaneous activity preceding the excitatory response of the cell. (B-H) Dose-response relationship to a series of 5HT administrations that included 2.5 μg (B), 5 μg (C), 15 μg (D), 30 μg (E), 60 μg (F) 100 μg (G), and 150 μg (H). (I) Effect of a control injection of isotonic saline (2ml) that was used as the vehicle to deliver 5HT in A-H. Time in milliseconds is represented on the X-axis and number of spike discharges on the Y-axis (EVENTS). Bin width in A-I equals 720 msec.

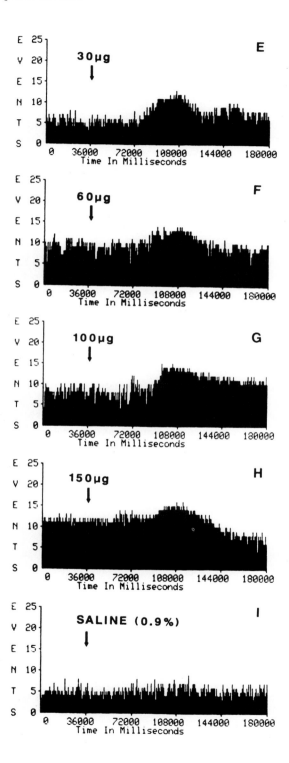

increase in activity (Fig. 4B-E). Doses of 5HT between 60 µg and 150 µg also produced excitatory responses, but these responses never exceeded that obtained with 30 µg (Fig. 4F-H). It should be noted that the initial inhibition of background activity following the first 60 µg dose of 5HT was not observed following the second 60 µg dose. Control injections of isotonic saline had no effect on the response of the cell (Fig. 4I).

To date, only WDR cells backfired from the PAG have been responsive to i.v. 5HT; spino-PAG cells classified as deep or nonresponsive have not been affected. Although the site of action of i.v. 5HT is unknown, algesic chemicals are known to excite cutaneous and group IV muscle afferent fibers (Handwerker et al., 1990; Kniffki et al., 1978; Mense and Schmidt, 1974). Recently, it has also been proposed that i.v. 5HT may represent a noxious stimulus that activates vagal afferents (Meller et al., 1990a,b). Additional studies will be needed to identify the afferent population(s) and central circuitry responsible for the excitatory and inhibitory effects of i.v. 5HT on SMT cells.

The excitatory receptive fields of the cells described in Figures 3 and 4 were relatively small, i.e. confined to a single limb. Although spino-PAG cells with small RFs are commonly observed, another class of cells belonging to this projection are those with large, complex RFs (Yezierski, 1990; Yezierski and Broton, 1991; Yezierski and Schwartz, 1986; Yezierski et al., 1987). An example of the complex RF characteristics of a cell projecting to the PAG in the cat is shown in Figure 5. This cell was antidromically activated from the contralateral, ventrolateral (vl) PAG in the caudal midbrain. The recording site was located in lamina VII. The cell had a WDR response profile by virtue of its responses to nonnoxious and noxious mechanical stimuli applied to the ipsilateral hindlimb. These responses, however, had an inhibitory effect on background activity (Fig. 5F). By contrast, excitatory responses were observed with a noxious squeeze applied to other parts of the body. Squeezing the skin on the contralateral face, for example, produced a transient excitation followed by a burst of activity lasting several minutes (Fig. 5A). A similar response was observed following stimulation of the contralateral forepaw

Figure 5 (right page). Responses of a complex wide dynamic range cell to stimuli applied to different parts of the body. The cell was antidromically activated from the contralateral periaqueductal gray in the caudal midbrain, and the recording site was located in lamina VII (inset). The cell had an inhibitory receptive field on the ipsilateral hindlimb (diagonal lines) and an excitatory field (stippled area) encompassing the remainder of the body (see figurine). (A,B): excitatory responses to noxious squeeze applied to the contralateral face (A) or contralateral forepaw (B). Note excitatory responses during times indicated by stimulus bars and delayed afterdischarges. (C): responses evoked by squeezing the tail with serrated forceps. Following removal of the stimulus, the cell maintained a discharge above background level until a second stimulus was applied. (D,E): responses to noxious squeeze applied to the ipsilateral face (D) or ipsilateral forepaw (E). Note long afterdischarges following cessation of each stimulus. (F): inhibitory effects of brush (1), light pressure (2), pinch (3), and squeeze (4) applied to the ipsilateral hindpaw (bin width for all histograms, 720 msec). Reprinted with permission from Yezierski and Schwartz (1986) Journal of Neurophysiology.

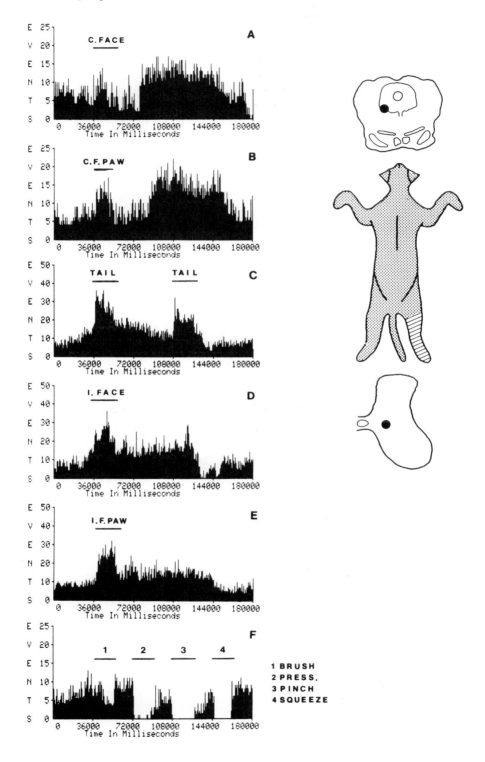

(Fig. 5B). Squeezing the ipsilateral face or forepaw produced an initial burst of activity followed by a long afterdischarge (Fig. 5D,E). A similar response was observed with squeezing the tail, at least for the first stimulus application. When the stimulus was applied a second time, the prolonged afterdischarge was not present (Fig. 5C). Spino-PAG cells with complex RFs are generally, although not exclusively, found in the deeper laminae of the spinal gray, including lamina X.

The functional properties described above are representative of the adequate stimuli and RF organizations of spinal neurons projecting to the PAG and provide insight into the potential functional significance of this pathway. Responses to noxious mechanical, thermal, and chemical stimuli are consistent with this projection having a role in the relay of nociceptive information. Responses to light tactile stimuli, however, suggest that the function of this pathway should not be considered limited to pain mechanisms. It is also important to point out that in order to gain a thorough appreciation of the potential importance of this pathway, one must acknowledge the varied functions associated with different regions of the PAG.

For example, if one considers the autonomic and somatomotor responses elicited by noxious stimuli (Hardy et al., 1952) and couples this with the control of blood pressure (Carrive, this volume), respiration and vocalization (Harper; Jürgens; Larson, this volume), and defence reactions (Bandler et al., 1991; Bandler and Depaulis, this volume) associated with PAG regions overlapping the SMT projection, what emerges is a functional relationship between SMT input to the PAG and behavioral responses to noxious stimuli. Similarly, the relationship between responses of SMT cells to stimulation of the genitals and perineal region and the role of PAG in reproductive behavior (Ogawa et al., this volume) suggests yet another functional role of SMT input to this region. Although the relationships described above undoubtedly depend upon a variety of conditions, including the behavioral state of the animal, they are proposed as a working hypothesis for future discussions related to the functional importance of somatosensory input to the PAG.

Spinal Input to a Descending Control Center

As one considers the functional significance of SMT input to the PAG, another possibility that must be included is the afferent limb of a descending control system. To this end, it has been shown that the spinomesencephalic projection overlaps with sites producing analgesia and the inhibition of nociceptive spinal neurons (Duggan and Griersmith, 1979; Gebhart, 1986; Gerhart et al., 1984; Gray and Dostrovsky, 1983; Hammond, 1986; Lewis and Gebhart, 1977; Liebeskind et al., 1973; Oliveras et al., 1974; 1979; Sandkühler and Gebhart, 1984; Willis, 1988; Yaksh et al., 1976). The distribution of midbrain neurons projecting to medullary raphespinal and reticulospinal neurons also overlaps with the terminal domain of the SMT (Abols and Basbaum, 1981; Chung et al., 1983; Gallager and Pert, 1978). Consistent with a feedback loop between SMT projection targets and neurons belonging to this pathway are reports that stimulation

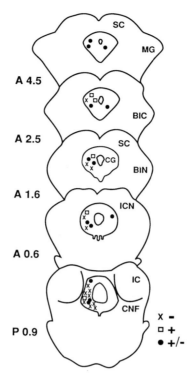

Figure 6. Distribution of stimulation
sites at different rostrocaudal
levels of the midbrain used to produce effects on SMT cells in the
lumbosacral spinal cord of the cat.
Each site and the corresponding
effect obtained at that site is represented by the symbols: X (inhibition only); square (excitation only);
and closed circle (mixed effects).
Abbreviations same as in Fig. 1.

at midbrain sites used to antidromically activate SMT cells also inhibits the evoked responses of these cells (Yezierski and Schwartz, 1986; Yezierski, 1990).

The distribution of PAG stimulation sites used in a recent study to produce effects on SMT cells in the lumbosacral spinal cord is shown in Figure 6
(Yezierski, 1990). It should be noted that the effects of PAG stimulation are not
limited to inhibition. Given the distribution of PAG sites producing analgesia
and aversive reactions (Besson et al.; Mason; Morgan, this volume), it is not surprising that excitatory and mixed effects can be obtained with PAG stimulation.

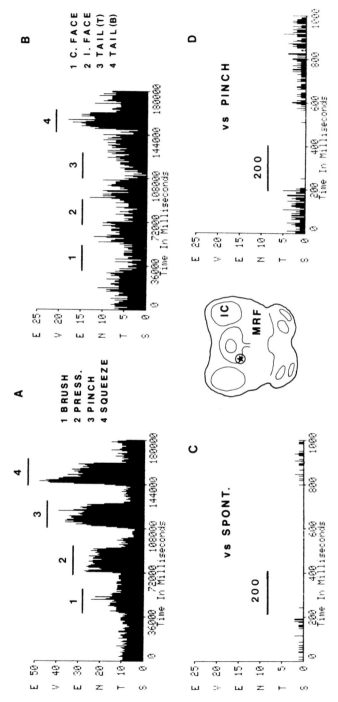

Figure 7. Response properties and inhibitory effects of periaqueductal gray stimulation on a wide dynamic range SMT cell in the cat. (A): excitatory responses to graded intensities of mechanical stimuli (1, brush; 2, pressure; 3, pinch; 4, squeeze) applied to the ipsilateral hindpaw. (B): Inhibitory responses obtained with the application of a noxious squeeze to the contralateral face (C.FACE), ipsilateral face (I.FACE), and tip of tail (TAIL, T). Excitatory response was obtained by squeezing the base of the tail (TAIL, B). (C,D): Inhibitory effects obtained during the delivery of a 200msec duration train (333Hz) at the site of antidromic activation (circled star) in the ventrolateral PAG. Stimulation effects were evaluated against the cells background discharge (C) and against a pinch evoked response (D). Note the duration of the inhibitory effect varied as a function of the cells discharge. Time in milliseconds is represented on the X-axis and number of spike discharges on the Y-axis (EVENTS). Bin width in A,B equals 720msec and 4msec in C,D.

There are, however, sites in the PAG that produce only inhibitory effects on SMT cells. For cells in the lumbosacral spinal cord these are distributed preferentially in the caudal midbrain (Fig. 6). Additional studies will be needed to determine if this distribution shifts rostrally for cells in the cervical spinal cord. Such a shift would parallel the proposed somatotopic organization of spinal input to the PAG (Blomqvist and Craig, this volume; Yezierski, 1988).

A typical example of a stimulation site in the caudal midbrain used to antidromically activate and produce inhibitory effects on a cat SMT cell is shown in Figure 7. This cell, located in lamina VII, was backfired at a threshold of 350 μA from the contralateral vlPAG (inset, Fig. 7). The cell had an excitatory RF on the ipsilateral hindpaw and an inhibitory RF on the tail and ipsi/contralateral face (Fig. 7B). The graded responses of this cell to increasing intensities of mechanical stimuli applied to the ipsilateral hindpaw are shown in Figure 7A. At the site of antidromic activation, lowering the stimulus strength from 350 μA (antidromic threshold) to 200 μA and delivering 200 msec trains once every two seconds produced a long duration inhibition of the cells spontaneous discharge (Fig. 7C). When the same stimulus parameters were used to evaluate effects against responses evoked by pinching the hindpaw a similar, but shorter duration, inhibitory effect was observed (Fig. 7D).

In addition to the inhibitory effects described above, excitatory and mixed effects are also obtained with stimulation in the PAG. An example of mixed effects are shown in Figure 8. This cell responded to light pressure, pinch and squeeze of the ipsilateral hindpaw (Fig. 8A). The recording site for this cell was located in lamina VII. Delivery of a 200 msec train to the contralateral vlPAG at stimulus strengths ranging from 100-300 μA produced a variable duration inhibition of a squeeze evoked response (Fig. 8B-D). The squeeze stimulus was applied continuously throughout the histogram while the 200 msec conditioning train was delivered once every 2 sec (10 epochs) at a delay of 200 msec after the start of each sweep. At 200 μA and 300 μA the inhibitory effects were preceded by a short latency excitation unrelated to the antidromic activation of the cell. The squeeze evoked responses of the same cell were also inhibited by stimulation in nucleus raphe magnus in the rostral medulla (Fig. 8E-G).

Functional Considerations

The involvement of the SMT in the relay of sensory input from cutaneous, muscle, joint, and visceral structures is consistent with the extensive laminar and segmental distribution of cells belonging to this pathway. At present one of the limitations to understanding the functional significance of this projection is the need for a more through evaluation of afferent inputs, e.g., visceral and muscle, to these cells and of responses resulting from the interaction of different stimulus conditions. In spite of this, however, given the response properties of SMT cells and the varied functions associated with different subdivisions of the PAG, it is possible to discuss the probable functional impact of SMT input to the PAG.

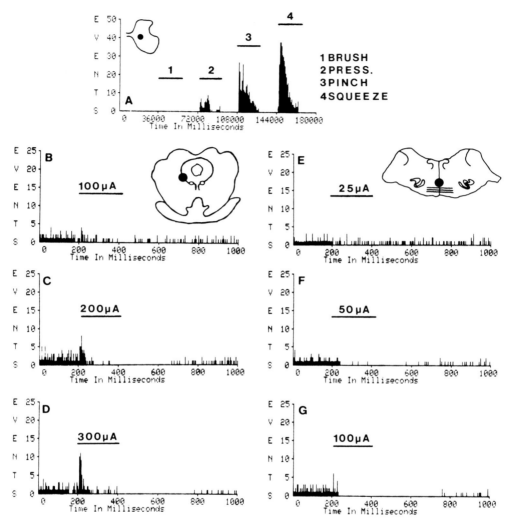

Figure 8. Excitatory and inhibitory effects of midbrain and medullary stimulation produced by different stimulus intensities. (A): responses of cell to graded intensities of mechanical stimuli (1-4) applied to the ipsilateral hindpaw. Recording site for this cell was located in lamina VII (inset). (B-G): effects of 200msec trains (100 μsec pulses, 333Hz) delivered once every two seconds at varying stimulus intensities to sites in the contralateral midbrain (B-D) and rostral medulla (E-G). Stimulation in the midbrain at 100 μA (B) resulted in the inhibition of a squeeze-evoked response. Increasing the stimulus intensity to 200 μA (C) or 300 μA (D) produced a short-latency excitation followed by inhibition of the squeeze response. Stimulation in the medulla (nucleus raphe magnus) with the same conditioning trains at intensities of 25 μA (E), 50 μA (F) and 100 μA (G) produced a progressive increase in inhibition of a squeeze-evoked response. Histograms in B-G are summed responses during 10 stimulus trials (bin width 4 msec). Reprinted with permission from Yezierski (1990) Journal of Neurophysiology.

One of the distinctive characteristics of the PAG is its complex afferent and efferent connections. Because of this it is not surprising that this region, based on its afferent input alone, is equally diverse in its functional organization. The importance of somatosensory input to the PAG is underscored by the overlap between the terminal distribution of the spino-PAG projection (Blomqvist and Craig, this volume) and the functional subdivisions described for different PAG regions.

In recent years, the careful delineation of spinal input to different midbrain regions has led to speculation concerning the functional significance of the SMT projection (McMahon and Wall, 1989; Willis, 1989; Yezierski, 1988; 1990). Much of this speculation has focused on an involvement in the activation of a descending control system that inhibits nociceptive transmission in the spinal cord (McMahon and Wall, 1989; Willis, 1989; Yezierski, 1990). Although this may be one of the functions of this pathway, this view neglects the functional diversity of the PAG and the wide range of adequate stimuli and response properties described for SMT cells. The proposed involvement of somatosensory input, especially nociceptive input, to the PAG in the control of respiratory and cardiovascular parameters, defensive behaviors, and vocalization is consistent with the sensory, motor, and autonomic sequelae to noxious stimuli (Hardy et al., 1952) and may represent additional functions of the spino-PAG projection.

As discussed elsewhere (Bandler et al., 1990) and in this volume (Bandler and Depaulis) the PAG has been linked with the control of defensive behaviors including the elaboration of different defensive strategies. These reactions can be triggered by a variety of stimuli, including noxious, and involve motor and autonomic reactions, e.g., changes in blood pressure, heart rate, and resistance in peripheral vascular beds (Bandler et al., 1991; Carrive; Lovick, this volume). Since the PAG is the region where various components of the defense reaction can be elicited, it is reasonable to assume that the spino-PAG projection has some role in evoking defensive behaviors. Consistent with this are reactions commonly associated with the delivery of noxious stimuli, including changes in motor behavior, respiration and alterations in cardiovascular function (Hardy et al., 1952). Although a straightforward relationship exists between noxious stimuli and defense reactions, in the proper behavioral setting low threshold mechanical stimuli can also elicit different components of this reaction.

Another possible relation between PAG function and the SMT involves cells with complex response and RF characteristics. These cells, regardless of segmental location, receive input from large portions of the body. Because of this extensive input, it is unlikely that these cells are important in stimulus localization. Since the complex RFs of these cells are due, in part, to descending influences from supraspinal structures (Yezierski, 1990), these cells may constitute a component of the SMT involved in a positive feedback loop responsible for autonomic and motor reflexes in the spinal cord. Alternatively, these cells may provide input compatible with the known functions of the reticular activating system or arousal mechanisms (Bowsher, 1976; Magoun et al., 1936; Melzack and Casey, 1968; Spiegel et al., 1954). Again, the reactions elicited by different environmental stimuli could be viewed as essential components of defensive or aggressive behavior.

An important aspect of PAG organization is the existence of functional columns involved in the control of cardiovascular function (Carrive, this volume). Since somatosensory stimuli, both noxious and non-noxious, can influence blood pressure and heart rate (Hardy et al., 1952; Zamir and Maixner, 1986), the relationship between the spino-PAG projection and the behavioral correlates of stimuli known to influence SMT cell responses provides evidence consistent with the SMT influencing selected parameters of cardiac function. Unfortunately, the approaches used in the antidromic activation of SMT cells do not permit detailed comments concerning differences in the response properties of cells projecting to columnar regions associated with hypertension, hypotension or changes in resistance of peripheral vascular beds. One might predict that cells providing input to hypotensive columns would have different functional properties from those projecting to hypertensive areas. Additional studies will be needed to determine if such differences exist.

With regard to somatosensory function one of the most significant findings of the last two decades has been the discovery of supraspinal regions that inhibit nociceptive transmission in the spinal cord and produce behavioral signs of analgesia. One of these regions, the PAG, is the recipient of a major projection from the spinal cord (Blomqvist and Craig, this volume; Yezierski, 1988). The information relayed by this pathway is consistent with somatosensory input to this region being involved in activating this supraspinal control center.

It is important to point out, however, that the effects elicited by PAG stimulation are not limited to inhibition and probably have a functional significance much broader than the analgesic effects that have been described. For example, the excitatory effects of PAG stimulation may facilitate segmental reflex activity. It is also possible that there are different functional compartments within the PAG that exert varied effects on different populations of tract cells and interneurons in different spinal laminae. As a result of this organization a noxious stimulus delivered to the hindpaw may result in the excitation of ventral horn cells involved in withdrawal reflexes while dorsal horn cells signalling stimulus intensity may be inhibited as part of a central damping mechanism. Similarly, descending excitatory, inhibitory or mixed effects may be exerted on cells at different levels of the cord for reflex, e.g., autonomic and motor, control or for purposes of stimulus localization.

Although the spinal effects of PAG stimulation are often emphasized, one should not overlook the ascending connections between PAG and diencephalic structures (Barbaresi et al., 1982; Hamilton, 1974; Mantyh, 1983) or the effects of PAG stimulation on the control of neuronal activity in the thalamus (Andersen, 1986; Barone et al., 1981; Emmers, 1979; Kayser et al., 1983). These results are important to consider not only in discussing the impact of SMT input to the PAG, but also with relationship to the functional significance of PAG efferent projections in somatosensory function.

The above discussion has focused on the role of spinal somatosensory input to the PAG, but one should not ignore the importance of other afferent systems or underestimate the role of integration of multisensory inputs that may be necessary to produce responses appropriate for different environmental conditions. As described elsewhere in this volume (Redgrave and Dean, this volume), input reaching the PAG via the superior colliculus is important in the orchestration of physiological and behavioral changes resulting from visual, auditory and somatosensory stimuli.

Conclusions

The results of anatomical investigations (anterograde and retrograde studies) have provided definitive proof for the existence of a spinal projection to different regions of the PAG. The spinal distribution of cells responsible for this projection, the distribution of primary afferent fibers, and the response properties of spino-PAG cells are all consistent with an involvement of this pathway in conveying information related to the modalities of touch, pressure, pain and temperature to supraspinal targets. Although these results provide a basis to speculate about the functional significance of the spino-PAG projection, any assessment of these functions cannot be made without acknowledging the functional organization of the PAG. To this end, discussions in this volume related to cardiovascular and reproductive function, respiration, vocalization and analgesic mechanisms must be considered. Somatosensory input to different regions of the PAG may not alone trigger any of these functions, but in concert with other afferent systems it may be involved in producing a repertoire of behaviorally appropriate responses to different stimulus conditions.

Aknowledgments

The author would like to express appreciation to Theresa Whittingham for her expert secretarial assistance, Nancy Olson and Carol Mendez for their technical expertise, and Drs. J. Broton, H. Hirata, L. Sorkin, and W. Willis for their collaboration. This work was supported by NS19509 (NIH) and by funds from the Miami Project to Cure Paralysis.

References

Abols, I.A. and Basbaum, A.I., Afferent connections of the rostral medulla of the cat: a neural substrate for midbrain-medullary interactions in the modulation of pain, J. Comp. Neurol., 201 (1981) 285-297.
Andersen, E., Periaqueductal gray and cerebral cortex modulate responses of medial thalamic neurons to noxious stimulation, Brain Res., 375 (1986) 30-36.
Anderson, F.D. and Berry, C.M., Degeneration studies of long ascending fiber systems in the cat brain stem, J. Comp. Neurol., 111 (1959) 195-229.
Bandler, R., Carrive, P., Zhang, S.P., Integration of somatic and autonomic reactions within

the midbrain periaqueductal grey: viscerotopic, somatotopic and functional organization, Prog. Brain Res., 87 (1991) 269-305.

Barbaresi, P., Conti, F. and Manzoni, T., Periaqueductal gray projection to the ventrobasal complex in the cat: an HRP study, Neurosci. Lett., 30 (1982) 205-209.

Barone, F.C., Wagner, M.J. and Tsai, W.H., Effects of periaqueductal gray stimulation on diencephalic neural activity, Brain Res. Bull., 7 (1981) 195-207.

Berman, A.L., The Brain Stem of the Cat, a Stereotaxic Atlas with Stereotaxic Coordinates, Univ. of Wisconsin Press, Madison, 1968.

Björkeland, M. and Boivie, J., The termination of spinomesencephalic fibers in the cat, an experimental anatomical study, Anat. Embryol. (Berl.), 170 (1984) 265-277.

Bowsher, D., Termination of the central pain pathway in man: the conscious appreciation of pain., Brain, 80 (1957) 606-622.

Bowsher, D., Role of the reticular formation in responses to noxious stimulation, Pain, 2 (1976) 361-378.

Chung, J.M., Kevetter, G.A., Yezierski, R.P., Haber, L.H., Martin, R.F. and Willis W.D., Midbrain nuclei projecting to the medial medulla oblongata in the monkey, J. Comp. Neurol., 214 (1983) 93-102.

Duggan, A.W. and Griersmith, B.T., Inhibition of the spinal transmission of nociceptive information by supraspinal stimulation in the cat, Pain, 6 (1979) 149-161.

Emmers, R., Dual alterations of thalamic nociceptive activity by stimulation of the periaqueductal gray matter, Exp. Neurol., 65 (1979) 186-201.

Gallager, D.W. and Pert, A., Afferents to brain stem nuclei in the rat as demonstrated by microiontophoretically applied horseradish peroxidase, Brain Res., 144 (1978) 257-275.

Gebhart, G.F., Modulatory effects of descending systems on spinal dorsal horn neurons, In: Spinal Afferent Processing, Yaksh T. (Ed.), Plenum, New York, 1986, pp. 391-416.

Gerhart, K.D., Yezierski, R.P., Wilcox, T.K. and Willis, W.D., Inhibition of primate spinothalamic tract neurons by stimulation in periaqueductal gray or adjacent midbrain reticular formation, J. Neurophysiol., 51 (1984) 450-466.

Gray, B.G. and Dostrovsky, J.O., Descending inhibitory influences from periaqueductal gray, nucleus raphe magnus, and adjacent reticular formation. I. Effects on lumbar spinal cord nociceptive and nonnociceptive neurons, J. Neurophysiol., 49 (1983) 932-947.

Hamilton, B.L., Projections of the nuclei of the periaqueductal gray matter in the cat., J. Comp. Neurol., 152 (1974) 45-58.

Hammond, D.L., Control systems for nociceptive afferent processing: The descending inhibitory pathways, In: Spinal Afferent Processing, Yaksh T. (Ed.), Plenum, New York, 1986, pp. 363-390.

Handwerker, H.O., Reeh, P.W. and Steen, K.H. Effects of 5HT on nociceptors, In: Serotonin and Pain, Besson J.-M. (Ed.), Elsevier, Amsterdam, 1990, pp. 1-15.

Hardy, J.D., Wolff, H.G, and Goodell, H., Pain Sensations and Reactions, Williams and Wilkins, Baltimore, 1952.

Hylden, J., Hayashi, H. and Bennett, G., Physiology and morphology of the lamina I spinomesencephalic projection, J. Comp. Neurol., 247 (1986a) 505-515.

Hylden, J., Hayashi, H. and Bennett, G., Lamina I spinomesencephalic neurons in the cat ascend via the dorsolateral funiculi, Somatosen. Res., 4 (1986b) 31-41.

Hylden, J.L.K., Nahin, R.L., Anton, F. and Dubner, R., Characterization of lamina I projection neurons: physiology and anatomy, In: Processing of Sensory Information in the Superficial Dorsal Horn of the Spinal Cord, Cervero F., Bennett G.J., Headley P.M. (Eds.), Plenum, New York, 1989, pp. 113-128.

Kayser, V., Benoist, J.-M. and Guilbaud, G., Low doses of morphine microinjected in the ventral periaqueductal gray matter of the rat depresses responses of nociceptive ventrobasal thalamic neurons, Neurosci. Lett., 37 (1983) 193-198.

Kerr, F., The ventral spinothalamic tract and other ascending systems of the ventral funiculus of the spinal cord, J. Comp. Neurol., 159 (1975) 335-356.

Kniffki, K.-D., Mense, S. and Schmidt, R.F., Response of group IV afferent units from skeletal muscle to stretch, contraction and chemical stimulation, Exp. Brain Res., 31 (1978) 511-522.

Lewis, V.A. and Gebhart, G.F., Evaluation of the periaqueductal central gray (PAG) as a morphine-specific locus of action and examination of morphine-induced and stimulation-produced analgesia as a coincident periaqueductal gray loci, Brain Res., 124 (1977) 283-303.

Liebeskind, J.C., Guilbaud, G., Besson, J.M. and Olivéras, J.L., Analgesia from electrical stimulation of the periaqueductal gray matter in the cat: behavioral observations and inhibitory effects on spinal cord interneurons, Brain Res., 50 (1973) 441-446.

Magoun, H.W., Atlas, D., Ingersoll, E.H. and Ranson, S.W., Associated facial, vocal, and respiratory components of emotional expression. An experimental study, J. Neurol. Psychopath., 17 (1936) 241-255.

Mantyh, P.W., The ascending input to the midbrain periaqueductal gray of the primate, J. Comp. Neurol., 211 (1982) 50-64.

Mantyh, P.W., Connections of midbrain periaqueductal gray in the monkey. I. Ascending efferent projections, J. Neurophysiol., 49 (1983) 567-581.

McMahon, S.B. and Wall, P.D., Electrophysiological mapping of brainstem projections of spinal cord lamina I cells in the rat, Brain Res., 333 (1985) 19-26.

McMahon, S.B and Wall, P.D., The significance of plastic changes in lamina I systems, In: Processing of Sensory Information in the Superficial Dorsal Horn of the Spinal Cord, Cervero F., Bennett G.J., Headley P.M. (Eds.) Plenum, New York, 1989, pp. 249-271.

Mehler, W.R., Some neurological species differences-a posteriori, Ann. NY Acad. Sci., 167 (1969) 424-468.

Meller, S.T., Lewis, S.J., Ness, T.J., Brody, M.J. and Gebhart, G.F., Vagal afferent-mediated inhibition of a nociceptive reflex by intravenous serotonin in the rat. I. Characterization, Brain Res., 524 (1990a) 90-100.

Meller, S.T., Lewis, S.J., Brody, M.J. and Gebhart, G.F., Is intravenous serotonin noxious?, Pain, 5 (1990b) S408.

Melzack, R. and Casey, K.L., Sensory, motivational and central control determinants of pain. A new conceptual model, In: The Skin Senses, Kenshalo D.R. (Ed.), Thomas, Springfield, 1968, pp. 423-443.

Menétrey, D., Chaouch, A. and Besson, J.M., Location and properties of dorsal horn neurons at origin of spinoreticular tract in lumbar enlargement of the rat, J. Neurophysiol., 44 (1980) 862-877.

Menétrey, D., Chaouch, A., Binder, D. and Besson, J.M., The origin of the spinomesencephalic tract in the rat: an anatomical study using the retrograde transport of horseradish peroxidase, J. Comp. Neurol., 206 (1982) 193-207.

Mense, S. and Schmidt, R.F., Activation of group IV afferent units from muscle by algesic agents, Brain Res., 72 (1974) 305-310.

Morin, F., Afferent projections to the midbrain tegmentum and their spinal course, Am. J. Physiol., 172 (1953) 483-496.

Olivéras, J.-L., Besson, J.-M., Guilbaud, G. and Liebeskind, J.C., Behavioral and electrophysiological evidence of pain inhibition from midbrain stimulation in the cat, Exp. Brain Res., 20 (1974) 32-44.

Olivéras, J.-L., Guilbaud, G. and Besson, J.M., A map of serotoninergic structures involved in stimulation producing analgesia in unrestrained freely moving cats, Brain Res., 164 (1979) 317-322.

Sandkühler, J. and Gebhart, G.F., Relative contributions of the nucleus raphe magnus and adjacent medullary reticular formation to the inhibition by stimulation in the periaqueductal gray of a spinal nociceptive reflex in phenobarbital anesthetized rat, Brain Res., 305 (1984) 77-87.

Skultety, F., Relation of periaqueductal gray matter to stomach and bladder motility, Neurology, 9 (1959) 190-197.

Spiegel, E.A., Kletzkin, M. and Szekely, E.G., Pain reactions upon stimulation of the tectum mesencephali, J. Neuropathol. Exp. Neurol., 13 (1954) 212-220.

Spruijt, B.M., Cools, A.R. and Gispen, W.H., The periaqueductal gray: a prerequisite for ACTH-induced excessive grooming, Behav. Brain Res., 20 (1986) 19-25.

Swett, J.E., McMahon, S.B. and Wall, P.D., Long ascending projections to the midbrain from cells of lamina I and nucleus of the dorsolateral funiculus of the rat spinal cord, J. Comp. Neurol., 238 (1985) 401-416.

Wiberg, M. and Blomqvist, A., The spinomesencephalic tract in the cat: its cells of origin and termination pattern as demonstrated by the intraxonal transport method, Brain Res., 291 (1984) 1-18.

Wiberg, M., Westman, J. and Blomqvist, A., Somatosensory projections to the mesencephalon: an anatomical study in the monkey, J. Comp. Neurol., 264 (1987) 92-117.

Willis, W.D., The pain system, In: Pain and Headache, Vol. 8, Karger, New York, 1985.

Willis, W.D., Anatomy and physiology of descending control of nociceptive responses of dorsal horn neurons: comprehensive review, Prog. Brain Res., 77 (1988) 1-29.

Willis, W.D., Projections of the superficial dorsal horn to the midbrain and thalamus, In: Processing of Sensory Information in the Superficial Dorsal Horn of the Spinal Cord, Cervero F., Bennett G.J., Headley P.M. (Eds.), Plenum, New York, 1989, pp. 217-237.

Willis, W.D. and Coggeshall, R.E., Sensory mechanisms of the spinal cord, Plenum, New York, 1978.

Yaksh, T.L., Yeung, J.C. and Rudy, T.A., Systematic examination in the rat of brain sites sensitive to the direct application of morphine: observations of differential effects within the periaqueductal gray, Brain Res., 114 (1976) 83-104.

Yezierski, R.P., The spinomesencephalic tract: projections from the lumbosacral spinal cord of the rat, cat and monkey, J. Comp. Neurol., 267 (1988) 131-146.

Yezierski, R.P., The effects of midbrain and medullary stimulation on spinomesencephalic tract cells in the cat, J. Neurophysiol., 63 (1990) 240-255.

Yezierski, R.P. and Broton, J.G., Functional properties of spinomesencephalic tract (SMT) cells in the upper cervical spinal cord of the cat, Pain, 45 (1991) 187-196.

Yezierski, R.P., Hirata, H. and Olson, N.A., Responses of spinomesencephalic tract (SMT) cells to thermal stimuli, Soc. Neurosci. Abstr., 11 (1985) 172.

Yezierski, R.P. and Schwartz, R.H., Response and receptive field properties of spinomesencephalic tract cells in the cat, J. Neurophysiol., 55 (1986) 76-96.

Yezierski, R.P., Sorkin, L.S. and Willis, W.D., Response properties of spinal neurons projecting to midbrain or midbrain and thalamus in the monkey, Brain Res., 437 (1987) 165-170.

Yezierski, R.P. and Mendez C.M., Spinal distribution and collateral projections of rat spinomesencephalic tract cells, Neurosci., 1991, in press.

Zamir, N. and Maixner, W., The relationship between cardiovascular and pain regulatory systems, Annals NY Acad. Sci., 467 (1986) 371-384.

Zemlan, F.P., Leonard, C.M., Kow, L. and Pfaff, D.W., Ascending tracts of the lateral columns of the rat spinal cord: a study using the silver impregnation and horseradish peroxidase techniques, Exp. Neurol., 62 (1978) 298-334.

Zhang, D., Carlton, S.M., Sorkin, L.S. and Willis, W.D., Collaterals of primate spinothalamic tract neurons to the periaqueductal gray, J. Comp. Neurol., 296 (1990) 277-290.

Hypothalamic Projections to the PAG in the Rat: Topographical, Immuno-Electronmicroscopical and Functional Aspects

J. Veening*, P. Buma**, G.J. Ter Horst***,
T.A.P. Roeling*, P.G.M. Luiten**** and R. Nieuwenhuys*

*Department of Anatomy and Embryology
Faculty of Medicine, University of Nijmegen, Nijmegen
**Department of Orthopedics, Nijmegen
***Department of Neurobiology & Oral Physiology, Groningen
****Department of Animal Physiology, Groningen
The Netherlands

Introduction

Projections descending from the hypothalamus to the PAG are in the rat characterized by the fact that they have in common with other ascending and descending projections the principle of termination within restricted parts of the PAG (see Blomqvist and Craig; Shipley et al., this volume). For that reason, we have to pay first some attention to the cytoarchitecture and the possible subdivisions of the PAG. On that basis, the descending projections will be described and compared with data obtained from the literature.

In the second part of this chapter, electronmicroscopical features of the PAG will be described, with much emphasis upon the immuno-electronmicroscopical characteristics of a number of peptidergic varicosities and terminals (Dynorphin, CRH, Met-Enk, LHRH and ACTH). These data strongly suggest that non-synaptic communication may play an important role in the PAG. The possible functional significance of these findings will be discussed at the end of this chapter.

Subdivisions of the PAG in the Rat

The presence of subdivisions in the PAG is a controversial issue since the reports of Mantyh (1982a: rat, cat and monkey) and Gioia et al. (1984: cat), who concluded that anatomic subdivisions cannot be discerned in this structure.

The Midbrain Periaqueductal Gray Matter, Edited by A. Depaulis and
R. Bandler, Plenum Press, New York, 1991

A number of connectivity studies seemed to support this lack of anatomical differentiation (Mantyh, 1982a,b; Marchand and Hagino, 1983).

However, most anatomical studies agree on the presence of more or less distinct subdivisions in the PAG (Taber, 1961; Hamilton, 1973: cat; König and Klippel, 1963; Paxinos and Watson, 1986; Beitz, 1985: rat). This anatomical differentiation is supported by differences in the occurrence of various cell types as shown in Golgi-studies (Liu and Hamilton, 1980: cat; Beitz and Shepard, 1985: rat), by obvious differences in afferent and efferent relationships (Beitz, 1985; Beitz and Williams, this volume) as well as by differences in cytochrome oxidase staining (Conti et al., 1988).

Since Ramón y Cajal (1911) most authors agree on some kind of concentric organization of the PAG, with an internal or medial zone surrounding the cerebral aqueduct, and an external or peripheral zone containing a number of cell condensations. However, there has been an obvious lack of agreement about the details of the width of these concentric zones relative to each other, as well as about the number and the borders of the subdivisions in the peripheral zone. Several different parts have been distinguished in the peripheral zone by different authors (Hamilton, 1973; König and Klippel, 1963; Paxinos and Watson, 1986; Beitz, 1985; Reichling et al., 1988; Conti et al., 1988). This lack of agreement has formed a serious handicap for the correlation between anatomical and functional subdivisions of the PAG. Even if it is to be expected that anatomical boundaries show at the most a limited correspondence with functionally or biochemically visualized borders, they still seem to be valuable by forming the structural basis upon which the presence or lack of correspondence can be described properly.

In order to avoid confusion, we decided to make a limited number of new drawings in the frontal plane (coordinates: 3.2, 2.2 and 1.2 anterior to interaural line, according to Paxinos and Watson, 1986) to be able to describe accurately the descending hypothalamic projections to this part of the PAG, although they extended into more rostral and caudal parts of the PAG as well. This small series of drawings (Fig. 1), that we hope to improve and extend in the future is based upon our series of Nissl-, Klüver-Barrera- and Bodian-stained sections of the rat brain (Geeraedts et al., 1990a,b). Our preliminary cytoarchitectonic parcellation results in a blend of the subdivisions proposed by the above mentioned authors, and consists basically of three zones: a periventricular, a central and a peripheral zone. Without going into details, two aspects have to be mentioned specifically. The internal or medial zone has been subdivided in a small 'periventricular zone', containing almost no neurons but only neuropil and exclusively formed

Figure 1 (right page). The 3 frontal planes of the PAG, used in the present study. Coordinates according to Paxinos and Watson Ant 3.2 (A), 2.2 (B) and 1.2 (C). At the left: Klüver-Barrera stained sections. At the right: Bodian stained sections. In the middle: subdivisions of the PAG, as used in the present study. Noteworthy are the cell condensations in the dorsal and lateral parts of the peripheral zone (left), as well as the lack of some fiber contingents in the dorsolateral part (right).

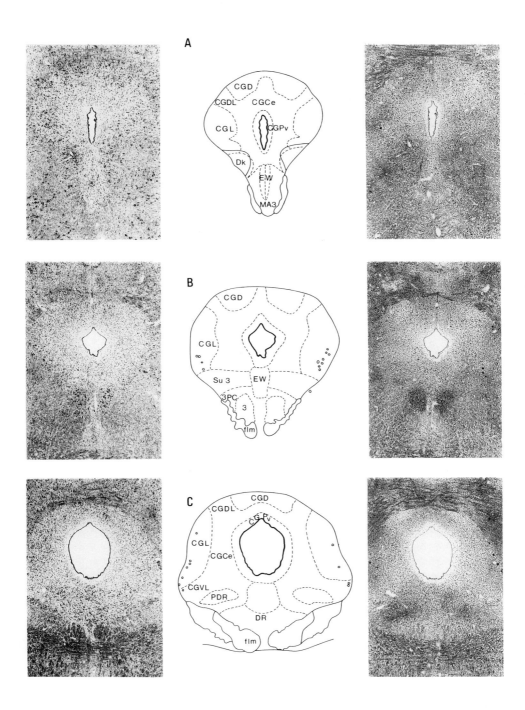

by dendrites and very small unmyelinated axons. It is completely surrounded by the 'central zone', containing moderate numbers and various types of cell bodies, as well as axons of variable diameter, many of them myelinated. Sometimes, the borders between the central and the peripheral zones are not clear. However, in the peripheral zone the dorsomedial aggregation of darkly staining neurons seems to be bordered by dorsolateral areas containing lightly staining neurons and lacking some fiber contingents, as shown in the Bodian-stained sections (Fig. 1). In the lateral part of the PAG, other condensations of cell bodies can be observed which seem to be disrupted in the ventrolateral parts by the massive fiber streams entering or leaving the PAG from or to the adjoining parts of the mesencephalic tegmentum. This part is bordered ventrally by the dorsal raphe nucleus. More caudal parts of the PAG have not been investigated so far.

The subdivisions mentioned in the peripheral zone (dorsomedial, dorsolateral, lateral and ventrolateral) are in agreement with the subdivisions agreed upon at the workshop (see Bandler and Depaulis, introduction of this volume). The adjoining parts of the central zone do not, in our opinion, show a similar clear cytoarchitectonic differentiation in subdivisions. In how far this may reflect functional differences between the central and peripheral zones, is not at all clear. It is remarkable, however, that some projections entering the PAG show a clear preference for, e.g., the central zone, while fibers from other origins are distributed much more equally throughout the PAG (see below: anterograde tracing studies and Shipley et al., this volume).

Retrograde tracing studies

The interpretation of results obtained by the use of retrograde tracers in an area like the PAG, is complicated by the presence of numerous fibers traversing the PAG longitudinally, that are unavoidably damaged during injection. Especially, the issue of a possible topographical differentiation in the distribution of PAG-afferents is hard to settle with the use of retrograde tracers alone. Some studies concluded that there is no such differentiation (Mantyh, 1982b; Marchand and Hagino, 1983). Other studies, however, clearly showed that the numbers of labeled neurons observed in specific hypothalamic nuclei were very much dependent on the exact location of the injection site within the PAG (Beitz, 1982: rat; Meller and Dennis, 1986: rabbit; personal observations). Figure 2 shows how the ventromedial hypothalamic nucleus, but not the lateral hypothalamus, contained retrogradely labeled neurons after a small rostral dorsolateral PAG injection. Many lateral hypothalamic neurons however became labeled after a caudal, ventrolateral PAG injection.

Summarizing the available data from the literature (Beitz, 1982; Mantyh, 1982b; Marchand and Hagino, 1983; Ter Horst et al., 1984; Meller and Dennis, 1986; Veening et al., 1987; 1990), the hypothalamus seems to provide quantitatively the largest input to the PAG. Virtually all hypothalamic nuclei contribute to these descending projections. Especially prominent projections, however, arise from areas like the median and medial preoptic nuclei, the anterior, ventrome-dial, lateral and dorsal premammillary nuclei. Moderate projections originate, e.g., in the paraventricular, perifornical, dorsal, dorsomedial, para-arcuate, tube-ral, supramammillary and posterior hypothalamic nuclei.

Some topographical differentiation has been described (Beitz, 1982; Meller and Dennis, 1986) and the main differences were observed between the dorsal PAG, receiving mainly projections from medial hypothalamic regions and the more caudal and ventral parts of the PAG, receiving most of their afferents from the lateral hypothalamus. The situation in the cat does not seem to be very diffe-rent from the rat (Mantyh, 1982b; Bandler and McCulloch, 1984; Holstege, 1987; Siegel and Pott, 1988).

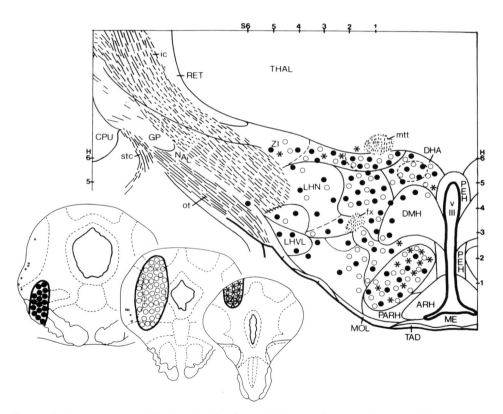

Figure 2 shows at a specific level of the hypothalamus of the rat how the distribution of retrogradely labeled neurons varies depending on the specific location of the PAG injec-tion site. Abbreviations: ARH, arcuate hypothalamic nucleus; DHA, dorsal hypothalamic area; DMH, dorsomedial hypothalamic nucleus; fx, fornix; LHN, lateral hypothalamic nucleus; LHVL, ventrolateral part of the lateral hypothalamic area; ZI, zona incerta. For further abbreviations: see Geeraedts et al., 1990a,b.

Apart from the hypothalamus, a very impressive projection arises also from the zona incerta. In how far this projection is also topographically organized, remains to be elucidated.

In a small number of studies, the retrograde transport of a neuronal tracer has been combined with immunocytochemistry. It has been observed that the galanin-innervation of the PAG partially originates in the median and medial preoptic nuclei (Gray and Magnuson, 1987). The POMC derived peptides (ACTH, β-endorphin, α-MSH) as present in PAG-fibers appear to originate from the rostral three-fifths of the arcuate and para-arcuate nuclei (Yoshida and Taniguchi, 1988). The glutamate containing projections to the PAG arise from many different parts of the brain, including the PAG itself, among them the zona incerta and the dorsomedial hypothalamic nucleus (Beitz, 1989).

Anterograde Tracing Studies

Many autoradiographic studies on hypothalamic nuclei have shown descending projections entering the PAG (rat: Swanson, 1976; Saper et al., 1976a,b;1979; Conrad and Pfaff, 1976a,b; Krieger et al., 1979; Berk and Finkelstein, 1982; cat: Holstege, 1987). Some of these projections have been described in more detail in studies using Phaseolus-L as an anterograde tracer (Luiten et al., 1985; Ter Horst and Luiten, 1986; Behbehani et al., 1988; Simerly and Swanson, 1988). The descending projections from the zona incerta have been described in an autoradiographic study as well (Watanabe and Kawana, 1982), while a [^{14}C] deoxyglucose mapping study showed how specific parts of the PAG became activated after lateral hypothalamic stimulation (Roberts, 1980).

On close inspection, the figures in all of these publications show that hypothalamic projections are never evenly distributed over the PAG, but always tend to concentrate in specific parts of it. This can be observed most clearly after PHA-L experiments as shown in our Figure 3A,B,C,D. Before discussing these figures, it has to be stated that all projections from the hypothalamus into the PAG tend to descend ipsilaterally, with only a few fibers descending into the identical contralateral parts of the PAG. In addition, fibers "en passage" have not been indicated separately since they were not common inside the PAG and tended to be interspersed in between the terminating fibers, for which the presence of varicosities was taken as an indication (Gerfen and Sawchenko, 1985). Figure 3A shows the fibers descending from a lateral hypothalamic injection. Apart from the dorsal raphe nucleus, the fibers tend to be concentrated in the central zone of the ventrolateral PAG.

The projections from the ventromedial hypothalamus (caudal ventrolateral part: plate B, rostral dorsomedial part: plate D) appear to target the dorsolateral PAG at the rostral two thirds of the PAG. The ventrolateral VMH projections appear to be more restricted to central zone of the PAG, and tend to concentrate caudally on the ventrolateral PAG. The dorsomedial VMH projections, however, targets the dorsolateral PAG also at its caudal third. The dorsomedial hypothalamic nucleus (plate C) shows an intermediate pattern of PAG innervation, since it targets mainly the central zone of the lateral PAG at all AP-levels.

Many hypothalamic injection sites were available for the present study (Fig. 4). Except for a ventral premammillary injection, that did not result in any significant PAG labeling, all injections resulted in longitudinal zones of PAG labeling.

These zones are overlapping extensively, and can be wider or more restricted. These zones show a clear and sometimes complete preference for the central part of the PAG, in agreement with the PHA-L and autoradiographic studies, mentioned above. This pattern seems to be different from the single zona incerta injection that we have been able to study so far, that resulted in a much more equal distribution of labeled fibers over both the central and peripheral parts of the PAG. In our caudal sections, however, in the ventral parts of the PAG, along the lateral borders of the dorsal raphe nucleus, this preference may be less obvious since many fibers enter or leave the PAG in that area. Finally, the paraventricular hypothalamic projections seem to cover the PAG more extensively than the other hypothalamic nuclei studied so far, suggesting an extensive autonomic control of the PAG.

Summarizing our findings, it can be stated that every part of the PAG receives a unique combination of descending hypothalamic projections. In figure 4 this has been worked out for the rostral and dorsal PAG, compared with the lateral and with the caudal and ventral PAG. In agreement with the available data from the literature, the ventromedial hypothalamic nucleus (with the rostrally and caudally adjoining parts of the medial hypothalamus), projects mainly to the rostral and intermediate dorsal PAG. The lateral PAG appears to be mainly innervated by the dorsomedial and intermediate parts of the hypothalamus, while the lateral hypothalamus projects most extensively upon the intermediate and caudal ventrolateral PAG.

This conclusion of a global and overlapping topographic organization in the descending hypothalamic projections may be helpful in unraveling the functional relationships between hypothalamus and PAG. Agonistic behavior has been elicited by electrical stimulation (Kruk et al., 1983; Lammers et al., 1988),

Figure 3 (next two pages) shows the injection sites of the anterograde tracer PHA-L into the lateral hypothalamus (3A), into the caudal ventrolateral part of the ventromedial hypothalamic nucleus (3B), into the dorsomedial hypothalamic nucleus (3C) and into the rostral dorsomedial part of the ventromedial hypothalamic nucleus (3D). N.B.: All hypothalamic projections descend ipsilaterally, with only small numbers of fibers to the identical contralateral parts of the PAG. Note the preference of hypothalamic fibers for the central zone.

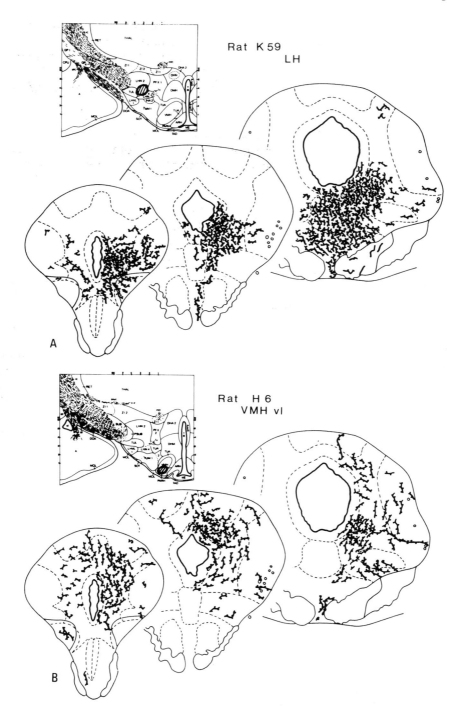

Rat K 59
LH

A

Rat H 6
VMH vl

B

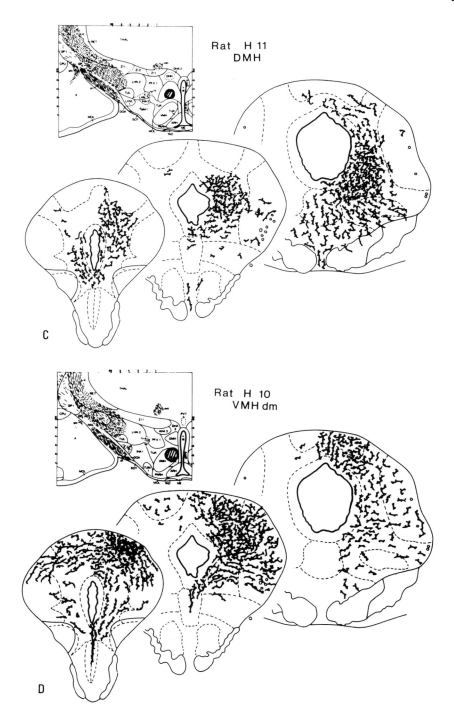

Rat H 11
DMH

C

Rat H 10
VMH dm

D

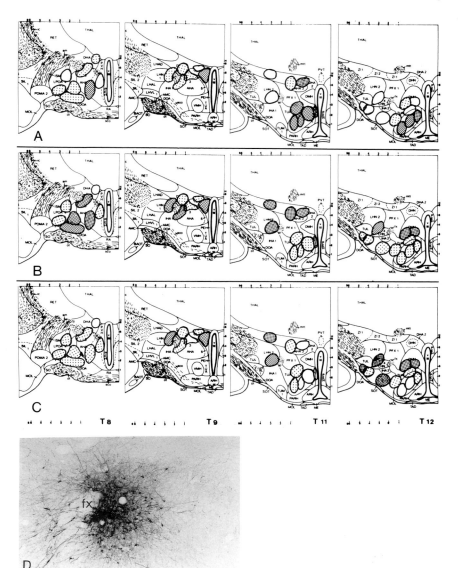

Figure 4 shows a series of 4 frontal sections of the hypothalamus of the rat, from the atlas of Geeraedts et al. (1990a,b). On these sections, a number of PHA-L injection sites have been indicated. The injections sites have been labeled when they projected strongly (dark), moderately (light) or weakly/not at all (empty) to specific parts of the PAG: rostral-dorsal PAG (A), lateral PAG (B) or caudal-ventrolateral PAG (C). One of the injections of PHA-L, bordering the fornix (fx) is shown in fig.4D. Abbreviations: AHA, anterior hypothalamic area; DMH, dorsomedial hypothalamic nucleus; IHA, intermediate hypothalamic area; LHN, lateral hypothalamic nucleus; PFX, perifornical nucleus; VMH, ventromedial hypothalamic nucleus.

and grooming behavior by electrical and chemical stimulation (ACTH, excitatory amino acids) in another part of the hypothalamus (Lammers et al., 1987; Roeling et al., 1990, 1991; Van Erp et al., 1991). The available neuroanatomical data suggest that different parts of the PAG may be involved with certain aspects of these different kinds of behavior.

Female reproductive behavior is controlled by the PAG as a relay station between the hypothalamus and the lower brainstem (see Ogawa et al., this volume). Afferent projections from the ventromedial hypothalamic nucleus to the rostral and intermediate dorsal and lateral PAG, as well as efferent descending projections from the caudal ventrolateral PAG apparently form important parts of the neuronal circuit involved with lordosis (see Ogawa et al., this volume). In agreement with these studies, we did observe that the descending fibers from the VMH target mainly the rostral and intermediate, dorsal and lateral PAG, with an extension towards the caudal ventrolateral PAG especially from the ventrolateral VMH.

The lateral part of the PAG is involved in the control of defensive behavior in the cat and in the rat (see Bandler and Depaulis, this volume). Active forms of defensive behavior can be elicited especially in the intermediate and caudal, lateral PAG, while inactive, immobility-patterns of behavior can be observed after excitatory aminoacid injections into the caudal ventrolateral PAG. The main hypothalamic areas projecting to the lateral PAG seem to be the dorsomedial hypothalamic nucleus and the intermediate hypothalamic area, ventral to the fornix. From these parts of the hypothalamus, components of defensive reactions have been elicited in rat and cat (Yardley and Hilton, 1986; Bandler, 1988; Barrett et al., 1990). This hypothalamic area is bordered by the ventromedial hypothalamic nucleus on the one, and by the lateral hypothalamic areas on the other side. Taking into consideration that these areas also project into the lateral PAG, with a dorsal tendency for the ventromedial hypothalamus and a ventral tendency for the lateral hypothalamus, it would be interesting to see in how far differences in defensive behavior from the different parts of the PAG are reflected by similar differences in the behavior elicited in these adjoining parts of the hypothalamus.

The lateral and ventrolateral parts of the PAG are also different in their cardiovascular control function (see Carrive, this volume) with hypertensive effects from the lateral and hypotensive effects from the ventrolateral PAG. The correlation of these data with functional and anatomical findings in the hypothalamus is at first sight rather complicated. Figure 5A shows the ventrolateral hypothalamus in the rat brain, adjoining the subthalamus, where an interesting convergence occurs of descending fibers from the central amygdaloid nucleus, ascending fibers from the parabrachial region and numerous lateral hypothalamic neurons projecting to the parabrachial and 'dorsal vagal complex' regions of the brainstem (Veening et al., 1987).

This convergence of fibers and neurons related to several brain areas involved with cardiovascular regulation makes this part of the hypothalamus especially interesting. Co-occurrence of cardiovascular changes with defensive reactions by electrical stimulation in the hypothalamus is well documented (Yardley and Hilton, 1986; Hilton and Redfern, 1986; Bandler, 1988). Excitatory

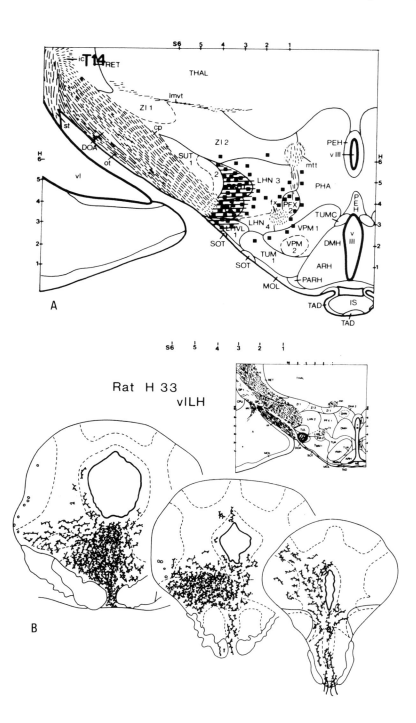

A

Rat H 33
vILH

B

aminoacids injections, however, into the lateral hypothalamus of the rat, including the area indicated in Figure 5A induces hypotensive reactions (Hinrichsen, 1988; Gelsema et al., 1989; Spencer et al., 1989). This part of the lateral hypothalamus mainly projects to the ventrolateral PAG (Fig. 5B, Fig. 3A). On the other hand, the intermediate hypothalamus, below the fornix, and the dorsomedial hypothalamus, related to defensive behavior, project mainly to the lateral PAG (Figs. 3C, 4). In these areas, micro-injections of DL-homocysteate also induced hypotensive reactions (Hilton and Redfern, 1986; Gelsema et al., 1989) with only a limited number of sites from where small pressor responses could be obtained (Gelsema et al., 1989). These limited hypertensive effects were observed at sites in the dorsomedial parts of the hypothalamus dispersed between sites from where hypotensive effects were obtained. Waldrop et al. (1988) injected GABA-antagonists into the hypothalamus of the cat and they obtained hypertensive reactions, especially in the dorsal hypothalamic area. In the rat, large decreases in blood pressure have been observed after glutamate injections into the posterior periventricular area (Spencer et al., 1990). From these anatomical and functional studies, it can be concluded that the correlation between hypotensive parts of the PAG, and the descending lateral hypothalamic projections is rather good, but the functional correlation between hypertensive parts of the PAG and descending intermediate hypothalamic projections is not clear since no clearcut hypertensive part of the hypothalamus has been described so far.

General Aspects of PAG Ultrastructure

Very little is known about the ultrastructural organization of the rat PAG. Gioia et al. (1983) studied the ultrastructural organization of the neuropil of the PAG of the cat at the electronmicroscopical level. With respect to the rat, electron microscopical observations have been done on immunostained fibers containing serotonin (Clements et al., 1985), amino acids (Clements et al., 1987), neurotensin (Shipley et al., 1987) and enkephalin (Moss and Basbaum, 1983). In the light of the extensive literature on functions of the PAG, it is surprising that the general ultrastructure of the rat PAG has not been studied more in detail so far. The present study was designed in order to study the PAG of the rat at the ultrastructural level. Attention is focussed on the ultrastructure of synaptic terminals.

All levels of the PAG studied showed some common characteristics. The

Figure 5A (left page) shows, on a frontal section through the posterior hypothalamus (PHA) of the rat brain, the position of the descending fibers from the central amygdaloid nucleus and the ascending fibers from the parabrachial region, indicated by the striped region bordering the subthalamic nucleus (SUT) and the cerebral peduncle (cp). The dots indicate the position of retrogradely labeled neurons projecting to the parabrachial region of the brainstem. See Veening et al., 1987. Figure 5B shows how fibers from a slightly more rostral ventrolateral hypothalamic injection site descend into the caudal and ventrolateral part of the PAG.

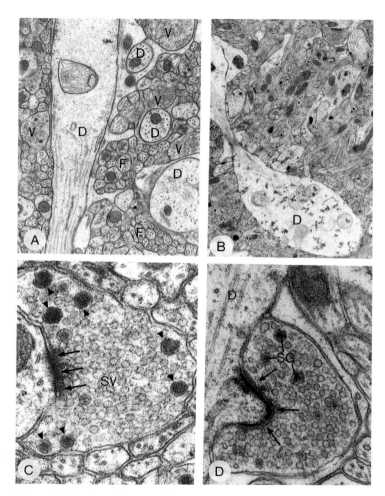

Figure 6. Routine electronmicrographs of typical varicose dendrites
(D) of PAG cells of the rat. A Part of rostral PAG showing cross-
sectioned dendrites, numerous thin unmyelinated fibers (F), and
axonal varicosities (V) filled with electron lucent synaptic vesicles.
x 25000. B Low power electron micrograph. Note varicose dendrite
(D). x 10000. C, D Routine electronmicrographs of synapses in
PAG. C Note large clusters of synaptic vesicles (SV) located in the
direct vicinity of the synaptic active site *(arrows)*, and fewer electron-dense secretory granules *(arrow heads)* located at some distance from the active site. x 50000. D Synapse *(arrows)* on spine
of varicose dendrite (D). Note many synaptic vesicles, and few
dense-core secretory granules (SG). x55000. Reprinted with permission from Buma (1989a) Neuroendocrinology.

periventricular part was always very cell-poor. The cells in this part were small, spindle-shaped, and located parallel to the ependymal lining. In the outer parts of the PAG the cells did not show a clear orientation to the aqueduct. The cells seemed to be located in small clusters, in many cases in apparent direct contact with each other. In some cases the membranes at these locations have a trilaminar appearance, suggesting the presence of gap junctions. In general a large heterogeneity was found with respect to the size of the cells. The cells were not intensely innervated by axo-somatic synapses. In general, only few (1-3) synapses were present on a single profile of a cell. The synapses were mainly of the symmetrical type, with many electron lucent synaptic vesicles.

With respect to the fiber distribution, HRP-studies showed that most fibers within the PAG showed a rostro-caudal orientation, parallel to the aqueduct. Differences were found between the number and distribution of myelinated and very thin unmyelinated fibers. Myelinated fibers could be clearly visualized in the paraphenylene diamino-stained semithin sections. In general, the myelinated fibers were very thin as compared to fibers in compact bundles like the posterior commissure, or the commissure of the superior colliculus. The periventricular cell-poor part of the PAG was almost devoid of myelinated fibers, except for its rostral part of the PAG, immediately ventral to the posterior commissure. The myelinated fibers were rather homogeneously dispersed over the PAG. In general the fibers showed a round profile in the sections indicating that they run in a rostrocaudal direction. In the lateral PAG cross-sections of laterally running fibers were found, indicating that myelinated fibers connect the PAG with the pretectal region. In the last section studied the dorsal fibers seemed to run preferentially in a ventro-lateral direction, to the cuneiform nucleus. Ventrally, an accumulation of cross-sectioned fibers was present.

The distribution of very thin unmyelinated fibers was completely different. Irrespective of the section studied, most thin fibers were found in the periventricular part of the neuropil. In the intermediate and outer parts of the PAG the numbers of fibers were lower.

The dendrites of the PAG neurons were varicose (Fig. 6A,B). Dendritic spines were extremely scarce (1-2 per section of the PAG: Fig. 6B). When present they were always located on the thin interconnecting shafts. Most synaptic contacts on dendrites are directly located on the varicose part. The axo-dendritic synaptic contacts were extremely diverse with respect to the length of the synaptic contact site and the nature of the synaptic vesicles. Many synapses in the PAG contained clusters of clear vesicles located in the direct vicinity of the synaptic contact zone (Fig. 6C,D). Most synapses contained in addition to these clear vesicles, a few dense-core secretory granules of variable diameter (Fig. 6C). In most synapses these dense-core secretory granules were located at some distance from the active synaptic site (Fig. 6C), but occasionally dense-core secretory granules were located close to the synaptic specialization. Most synaptic varicosities contained round synaptic vesicles, but occasionally varicosities with ellipsoid lucent vesicles were found. Irrespective of the area investigated in the PAG, a second type of axonal profile could be distinguished on the basis of the vesicular content. This type of varicosity always contained many large, round dense-core

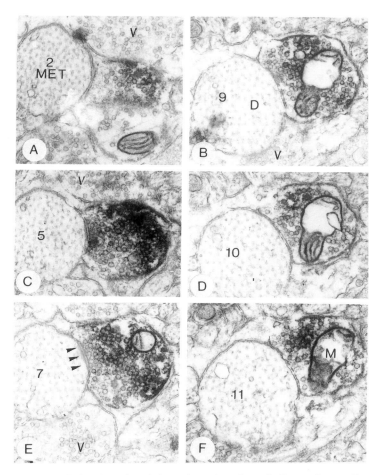

Figure 7. Selection of serial sections (numbers indicate the section numbers) of met-enkephalin stained varicosity. Immunoreactivity is located at some distance from the active site *(arrow heads)*. Unstained varicosities (V) are found in the direct vicinity of the met-enkephalin varicosity. x 40000. D, dendrite; M, mitochondrium. Reprinted with permission from Buma (1989b) Acta Morphol. Neerl.-Scand.

secretory granules; electron lucent synaptic vesicles were scarce (Fig. 8J). The mean diameter of the secretory granules varied between 110 and 160 nm (data not shown). Analysis of serial sections revealed that synaptic specializations were not present on this type of varicosity.

Immunocytochemical Characterization of Peptidergic PAG Varicosities

In the PAG many (peptidergic) transmitters are present (Nieuwenhuys, 1985; Buma, 1989b). We have studied the localization of several peptidergic neuroactive substances in the PAG. The aim of this study was to describe the different types of varicosity in the PAG, with special attention to the possibility that transmitters may be released from unspecialized, nonsynaptic varicosities. The light microscopical immunocytochemical staining procedure was performed on 50 μm vibratome sections of paraformaldehyde fixed brains; Somogii-fixation without glutaraldehyde. For details of the PAP-staining procedure, specificity and sources of the antibodies see Buma (1989a,b) and Buma et al. (1989). In general, the background staining was very low or absent. Controls were all completely negative, indicating that the staining with the antibodies was specific for the different neuronal elements. At the light microscopical level, the transmitters were all located in varicose fibers (Fig. 9B). To further characterize different types of fibers at the ultrastructural level, and the chemical nature of the dense-core secretory granules, the pre-embedding PAP-method was used (electron microscopy was performed on vibratome sections of glutaraldehyde fixed brains; see Buma et al., 1989 for details of staining procedure). Fibers containing different neuroactive substances in the PAG were serially sectioned. In general, immunoreactivity was present in dense-core secretory granules but also present in the cytoplasm near the dense-core secretory granules, and, although to a lower degree, near other cytoplasmic organelles, as a subpopulation of the electron lucent synaptic vesicles, microtubules and mitochondria (Fig. 7A-F).

The immunocytochemical staining procedure showed that the peptides investigated were not preferentially located in the direct vicinity of the active synaptic site. For instance, all varicosities immunoreactive for met-enkephalin made synaptic contacts with dendrites of PAG neurons (Fig. 7). All immunopositive varicosities contained many clear synaptic vesicles. Strikingly, the immunocytochemical reaction product was mainly located at some distance from the synaptic cleft. The dense-core granules were always located in parts of the terminals that were heavily stained. However, in many terminals no dense-core secretory granules could be found, probably because they were obscured by the intense immunocytochemical reaction product (Fig. 7C). In some varicosities the reaction product was only located in a very small part of the terminal. In none of the serially sectioned met-enkephalin positive synapses were high levels of immunoreactivity located in the direct vicinity of the synaptic contact zone. The same results were found for the dynorphin- and CRF-immunoreactivity (data not shown). To answer the question whether varicosities always make synaptic contacts, more transmitters were examined in the same way. Firstly, the LHRH fibers in the PAG were studied in some detail.

LHRH-Immunoreactivity

Extrahypothalamic LHRH pathways are believed to have a function in the regulation of sexual behavior (for references see Buma, 1989a). Particularly well documented is the role of LHRH fibers that project to the PAG. Descending LHRH fibers have been found light microscopically in a subependymal position in the PAG (Liposits and Sétáló, 1980; Shivers et al., 1983). Extracts from the PAG show considerable LHRH activity (Endröczi and Hilliard, 1965; Samson, et al., 1980). LHRH infused into the PAG facilitates the induction of the lordotic reflex (Riskind and Moss, 1979; Sakuma and Pfaff, 1980; Sirinathsinghji, 1985). Passive immunization against endogenous LHRH by anti-LHRH-globulins diminishes the reflex (Sakuma and Pfaff, 1980; Sirinathsinghji et al., 1983). Although contro- versy exists about the effects of iontophoretically applied LHRH on the excitabi- lity of PAG neurons (change in excitation was found by Chang et al., 1985; Samson et al., 1980, no effect was found by Ogawa et al., this volume), overall there is little questioning of a prominent role for PAG-LHRH in the function of the lordotic reflex. In the hypothalamus LHRH fibers were particularly abundant in a ventral periventricular area. A dorsal fiber contingent passed caudally through the hypothalamus in a subependymal position to the PAG.

Throughout the PAG, LHRH positive fibers were mainly found in a ven- tral position. In the rostral parts of the PAG (Bregma -4.8 to -5.8), many fibers were running in the frontal plane of the sections, as opposed to the middle (Bregma -5.8 to -7.8) and caudal (Bregma -7.8 to -8.8) parts of the PAG, where the fibers were orientated in a longitudinal direction (Fig. 8A,B).

Irrespective of their location in the PAG, LHRH fibers were found in close association with the ependyma (Fig. 8). In the middle part of the PAG, many of the LHRH fibers were preferentially located in two or three ventro-medial or ventro-lateral, rostro-caudally orientated thickenings of the ependymal lining (Fig. 8A-E). More caudally, these thickenings were not present, and the number of LHRH fibers diminished. In the pontine grey substance LHRH positive fibers could no longer be found.

In semithin and electron microscopical sections, many profiles of LHRH fibers were found in their subependymal position (Fig. 8C-H). Particularly in the ependymal thickenings, LHRH fibers were also found in an intraependymal position (Fig. 8C-E). Electronmicroscopically, PAP-complexes, indicating the pre- sence of LHRH-immunoreactivity, were only found in axonal varicosities (Fig. 8F-H). Most of the reaction product was located in dense-core secretory granules (diameter ca. 110 nm). All varicosities contained many of such granules, and smaller numbers of clear vesicles (Fig. 8H). Secretory granules and clear vesicles were apparently dispersed in the varicosities without clustering in particular areas. Serial sections never showed ultrastructural characteristics of synapses, like electron dense appositions against the cytoplasmic side of each synaptic membrane, and a widened synaptic cleft filled with electron dense material.

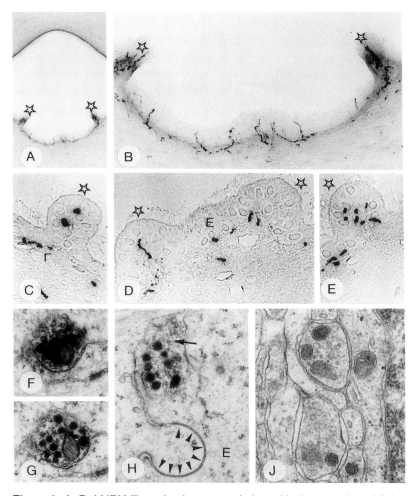

Figure 8. A, B. LHRH fibers in close association with the ependymal lining of the aqueduct of the middle PAG (bregma -6.8) in the male rat. Note preferential location of fibers in ependymal thickenings (asterisks). A x 75. B Enlargement of A. x 280. C-J. LHRH fibers in unstained semithin (C-E) and electronmicroscopical (F-J) sections of the PAG of the male (C, D, H) and female (E, F-G, J) rat. C-E. Note sub- and intraependymal location of LHRH fibers. Asterisks indicate ependymal thickenings. x 600. F-H. Immunoreactive LHRH fibers in middle PAG. Note electron dense PAP-reaction product located over secretory granules. Note location of clear vesicles (arrow). Arrow heads indicate location of gap junction between ependymal cells (E). F-H x 40000. J Routine electron micrograph of fibers in ependymal thickening. x 50000. Reprinted with permission from Buma (1989a) Neuroendocrinology.

Profiles of dendrites were never found in the direct vicinity of the varicosities. The absence of synaptic contacts was confirmed by routine electron microscopic examination of the ependymal thickenings. Fibers, resembling the LHRH fibers, never made synapses with dendrites of PAG neurons (Fig. 8J). Non-immuno-reactive synaptic contacts were visible in the neuropil of the PAG. Furthermore, other types of cell to cell contacts, e.g. tight- and gap junctions could be easily found in the same tissue sections between ependymal cells (Fig. 8H), and in the neuropil of the PAG, indicating that the fixation was sufficient.

No apparent differences in distribution and number of LHRH fibers were found between males and female rats. Thus, these results are in favor of a non-synaptic neuromodulatory action of LHRH on PAG neurons. From these results it is not clear how far the LHRH can reach in the PAG. However, until now it is not established whether the receptors for LHRH are located on the cell bodies or on the dendrites of the PAG cells. If LHRH influences PAG cells via the dendrites, the cell bodies may be located throughout the PAG, since many PAG cells have dendrites that strongly project to the aqueduct (unpublished results).

ACTH-Immunoreactivity

ACTH is derived from a common precursor molecule, pro-opiomelanocortine (POMC). In the rat, detailed light microscopical anatomical mapping studies of the ACTH fiber distribution showed that ACTH cells project extensively throughout the hypothalamus, to the limbic system (Joseph, 1980), and to the brain stem (Buma et al., 1989, for references). It is believed that these projections play a role in the regulation of various neuroendocrine, behavioral and autonomic functions (Piekut, 1985). Immuno-electron microscopical details of the POMC cell bodies have been described (Lamberts and Goldsmith, 1985; Léranth et al., 1980). Ultrastructural studies of the nature of the synaptic contacts of the neuronal projections of the POMC cells are confined to the preoptic area (Léranth et al., 1988). ACTH-immunoreactive fibers are very prominent in the PAG (for detailed light microscopical descriptions see Joseph, 1980; Romagnano and Joseph, 1983), a key brain stem structure in various behavioral responses induced by the POMC derived peptides: e.g., the induction of the excessive grooming response by ACTH peptides (De Wied, 1980), the mediation of analgesia and the inhibition of sexual behavior by endorphins (Buma et al., 1989, for references).

The light microscopy of ACTH fibers in the PAG has been described elsewhere in detail (see Buma et al., 1989: Fig. 9A,B). Irrespective of the area investigated in the PAG, two types of axonal varicosity could be distinguished on the basis of the vesicular content at the ultrastructural level. Type 1 closely resembled the Met-enkephalin, dynorphin and CRF synaptic contacts. They always contained numerous electron-lucent synaptic vesicles (Fig. 9C,E).

All varicosities of this type made synaptic contacts with dendrites of PAG neurons (Fig. 9C,E). Most of the electron lucent synaptic vesicles were located in the direct vicinity of the active zone. The dense-core granules, if present, were located in parts of the terminals that were relatively heavily stained. In none of the serial sectioned ACTH-positive synapses of this type were dense-core secre-

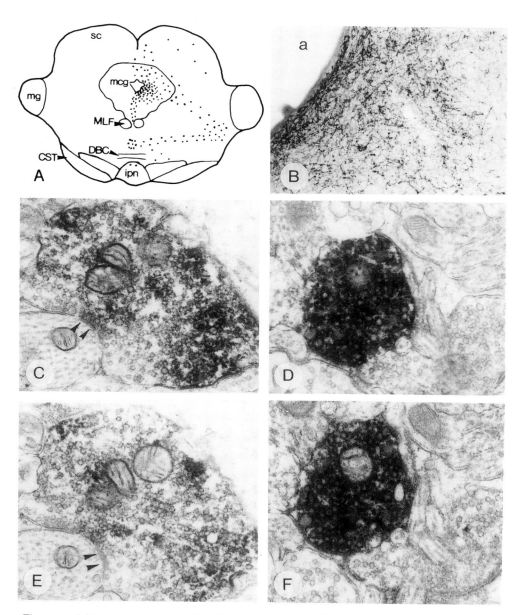

Figure 9. A Line drawing illustrating the distribution of ACTH-immunoreactive fibers in the mesencephalon of the rat at Bregma -6.8. B. Light micrograph taken from immuno-stained brain section, at bregma -6.8. a, Cerebral aqueduct. C-F. Electronmicrographs of serial sections of varicosities with (C,E) and without synaptic specializations (D, F). C, E x 55000. Arrowheads indicate synaptic specializations. D, F x 38000. Reprinted with permission from Buma et al. (1989) European Journal of Neuroscience.

tory granules or high levels of immunoreactivity observed in the direct vicinity of the synaptic contact zone (Fig. 9C,E). All postsynaptic dendrites are varicose. No spines were present. The dendrites did not show postsynaptic membrane appositions ("Taxi-bodies"). In many cases non-immunoreactive axonal boutons contacted the same dendrites.

The type 2 ACTH varicosity closely resembled the LHRH varicosities, i.e. they always contained many large, round dense-core secretory granules; electron lucent synaptic vesicles were scarce (Fig. 9D,F). The mean diameter of the secretory granules was approximately 115 nm. All varicosities were completely filled with an intense PAP-reaction product. Again, the label was predominantly located around the dense-core secretory granules (Fig. 9D,F). Occasionally, a type 1 varicosity was found to be connected by a thin axonal shaft with a profile of type 2 varicosity, indicating that the different types of varicosity may occur alternatively along the same axon.

Synaptic Versus Non-Synaptic Communication

Thus, in the present study three types of fiber were found in the PAG (Table I), viz 1) varicose fibers that made synaptic contacts with PAG neurons on every varicosity; 2) fibers with two types of varicosity: synapse bearing varicosities and varicosities without synaptic specializations; and 3) varicose fibers without any synaptic specializations.

Table I. Summarizing table with respect to the ultrastructural characteristics of immuno-stained fibres in the PAG. SS, synaptic specialization.

Peptide	Varicosities with SS	Varicosities without SS
MET	+	
DYN	+	
CRF	+	
LHRH		+
ACTH	+	+

The synapse bearing varicosities are characterized by a presynaptically located cluster of synaptic vesicles containing a transmitter, electron-dense appositions to pre- and postsynaptic membranes, and a widened synaptic cleft (for references see Buma, 1989b). It is generally believed that the contents of the synaptic vesicles are released by exocytosis (e.g., Heuzer et al., 1979), after which the released transmitter substance diffuses to receptors located in the postsynaptic membrane. Recently, the ultrastructural localization of these receptor molecules became possible by means of immunocytochemistry with monoclonal antibodies raised to the isolated receptor complex (for references see Buma, 1989b). However, until now only receptor molecules for small transmitter molecules as acetylcholine, glycine and GABA could be detected in this way. The postsynaptic localization of the receptor molecules for these small classical neurotransmitter molecules, strongly suggest that a synaptic way of information transfer predominates with these non-peptidergic transmitters.

In the present study we showed that peptidergic neurotransmitters are localized in dense-core secretory granules in synapse-bearing or non-synaptic varicosities. In the presence of a synaptic contact the peptidergic secretory granules are almost exclusively located at a certain distance from the active site. Varicosities without membrane specializations have also been described in the trigeminal nucleus (Zhu et al., 1986), and in the spinal cord (substance-P- and serotonin-immunoreactive fibers, e.g. Ulfhake et al., 1987). It has been suggested that transmitters released from nonspecialized parts of the axolemma may travel for some distance through the extracellular space of the CNS, and act on particular distant targets that possess the appropriate receptors (Beaudet and Descarries, 1978; Nieuwenhuys, 1985; Buma and Roubos, 1986). Although in the PAG the CRF, Met-enkephalin and dynorphin varicosities displayed synaptic specializations, a non-synaptic action of these transmitter seems more plausible than a synaptic action on the following arguments. If there is exocytosis of dense-core secretory granules, it occurs at non-specialized parts of the axolemma and may involve release of peptidergic contents (Buma, 1989b). Peptides are generally located in dense-cored secretory granules. Peptides have been previously found to be involved in non-synaptic transmission in lower animals (see Buma and Roubos, 1986). Finally the location of peptidergic secretory granules in synapse-bearing varicosities does not exclude the synaptic action of a different transmitter from the same peptidergic varicosity. It is well known that many neuropeptides are colocalized with classical neurotransmitters such as acetylcholine (ACh), γ-aminobutyric acid (GABA), catecholamines (see for references Hökfelt et al., 1987). Some classical neurotransmitters are localized in small clear vesicles (see for instance: ACh: Ceccarelli and Hurlbot, 1980; Matteoli et al., 1988; Glycine: Altschuler et al., 1986; GABA: Van Den Pol, 1985; Somogyi and Hodgson, 1985). The location of the peptidergic neurotransmitter in synaptic varicosities containing many of such clear electron lucent synaptic vesicles strongly suggest that these terminals also contain, in addition to the peptides, a classical neurotransmitter. This raises the possibility that one synaptic terminal releases two or more neuroactive substances.

However, particular peptides would be involved in non-synaptic communication (for references see Buma and Roubos, 1985). In this light it is very interesting that Nieuwenhuys (1985) considered the PAG as part of the core of the neuraxis, a number of brain regions, fiber tracts and nuclei, that are extremely rich in (peptidergic) neuromediators. This core coincides largely with the limbic system, and with the oestrogen- and androgen-concentrating neuronal groups. In the light of the high number of peptidergic neurotransmitters, it is very tempting to speculate that particularly in this (paracrine?) core of the brain non-synaptic communication could play an important (predominant?) role (Nieuwenhuys, 1985; Nieuwenhuys et al., 1989). The present observations on the ultrastructure of LHRH and ACTH fibers in the PAG support the concept of paracrine communication in this particular brain area. It may be expected, however, that the same ultrastructural characteristics of peptidergic fibers are found not only in other peptidergic fibers located in the PAG, but also are present in other nuclei and fiber tracts of the neuraxis. If such fibers influence neurons in such a non-synaptic way, it would constitute a highly effective mode of information transfer in which relatively few peptidergic fibers could influence many neurons. This type of information transfer offers also the unique possibility that one messenger has inhibitory and/or stimulatory influences on more than one fiber system, depending on the localization of the specific receptor molecules, and thus can have more than one (behavioral) effect. That this is not just a theoretical assumption is illustrated by the effect of LHRH infusion in the PAG. Besides the well known influence on the lordotic reflex, it was recently found (Gargiulo et al., 1989) that LHRH infusion also induces a selective modification of distinct motor activities involved in grooming behavior.

Abbreviations

CGCe: central zone; CGD:dorsomedial part; CGDL: dorsolateral part; CGL: lateral part; CGPv: periventricular zone; CGVL: ventrolateral part; Dk:nucleus Darkschewitsch; DR: nucleus raphe dorsalis; EW: nucleus Edinger-Westphal; flm: fasciculus longitudinalis medialis; MA3: medial accessory oculomotor nucleus; PAG: periaqueductal (central) gray; PHA-L: Phaseolus-L; PDR: nucleus raphe paradorsalis; Su3: supraoculomotor part; 3: nucleus oculomotorius; 3PC: nucleus oculomotorius pars parvocellularis

References

Altschuler, R.A., Betz, H., Parakkal, M.H., Reeks, K.A. and Wenthold, R.J., Identification of glycinergic synapses in the cochlear nucleus through immunocytochemical localization of the postsynaptic receptor, Brain Res., 369 (1986) 316-320.

Bandler, R., Brain mechanisms of aggression as revealed by electrical and chemical stimulation: suggestions of a central role for the midbrain periaqueductal grey region, Prog. Psychobiol. Physiol. Psychol., 13 (1988) 67-154.

Bandler, R. and McCulloch, T., Afferents to a midbrain periaqueductal grey region involved in the 'defense reaction' in the cat as revealed by horseradish peroxidase. II. The diencephalon, Behav. Brain Res., 13 (1984) 279-285.

Barrett, J.A., Edinger, H. and Siegel, A., Intrahypothalamic injections of norepinephrine facilitate feline affective aggression via α2-adrenoceptors, Brain Res., 525 (1990) 285-293.

Beaudet, A. and Descarries, L., The monoamine innervation of the cerebral cortex: synaptic and nonsynaptic axon terminals, Neuroscience, 3 (1978) 851-860.

Behbehani, M.M., Park, M.R. and Clement, C.E., Interactions between the lateral hypothalamus and the periaqueductal gray, J. Neurosci., 8 (1988) 2780-2787.

Beitz, A.J., The organization of afferent projections to the midbrain periaqueductal gray of the rat, Neuroscience, 7 (1982) 133-159.

Beitz, A.J., The midbrain periaqueductal gray in the rat. I. Nuclear volume, cell number, density, orientation, and regional subdivisions, J. Comp. Neurol., 237 (1985) 445-459.

Beitz, A.J., Possible origin of glutamatergic projections to the midbrain periaqueductal gray and deep layer of the superior colliculus, Brain Res. Bull., 23 (1989) 25-35.

Beitz, A.J. and Shepard, D., The midbrain periaqueductal gray in the rat. II. A Golgi analysis, J. Comp. Neurol., 237 (1985) 460-475.

Berk, M.L. and Finkelstein, J.A., Efferent connections of the lateral hypothalamic area of the rat: An autoradiographic investigation, Brain Res. Bull., 8 (1982) 511-526.

Buma, P. and Roubos, E.W., Ultrastructural demonstration of nonsynaptic release sites in the central nervous system of the snail Lymnaea stagnalis, the insect Periplaneta americana, and the rat, Neuroscience, 17 (1986) 867-879.

Buma, P., Characterization of luteinizing hormone releasing hormone (LHRH) fibers in the mesencephalic central grey substance of the rat, Neuroendocrinology, 49 (1989a) 623-630.

Buma, P., Synaptic and Non-synaptic release of neuromodulators in the CNS, Acta Morph. Neerl.-Scand., 26 (1989b) 81-113.

Buma, P., Veening, J. and Nieuwenhuys, R., Ultrastructural characterization of adrenocorticotrope hormone (ACTH) immunoreactive fibers in the mesencephalic central gray substance of the rat, Eur. J. Neurosci., 1 (1989) 659-672.

Ceccarelli, B. and Hurlbot, W.P., Vesicle hypothesis of the release of quanta acetylcholine, Physiol. Rev., 60 (1980) 369-441.

Chang, A., Dudley, C.A. and Moss, R.L., Hormonal modulation of the responsiveness of midbrain central grey neurons to LH-RH, Neuroendocrinology, 41 (1985) 163-168.

Clements, J.R., Beitz, A.J., Fletcher, T.F. and Mullett, M.A., Immunocytochemical localization of serotonin in the rat periaqueductal gray: a quantitative light and electron microscopical study, J. Comp. Neurol., 236 (1985) 60-70.

Clements, J.R., Madl, J.E., Johnson, R.L., Larson, A.A. and Beitz, A.J., Localization of glutamate, glutaminase, aspartate and aspartate aminotransferase in the rat midbrain periaqueductal grey, Exp. Brain Res., 67 (1987) 594-602.

Conrad, L.C.A. and Pfaff, D.W., Efferents from medial basal forebrain and hypothalamus in the rat. I. An autoradiographic study of the medial preoptic area, J. Comp. Neurol., 169 (1976a) 185-220.

Conrad, L.C.A. and Pfaff, D.W., Efferents from medial basal forebrain and hypothalamus in the rat. II. An autoradiographic study of the anterior hypothalamus, J. Comp. Neurol., 169 (1976b) 221-262.

Conti, F., Barbaresi, P. and Fabri, M., Cytochrome oxidase histochemistry reveals regional subdivisions in the rat periaqueductal gray matter, Neuroscience, 24 (1988) 629-633.

Endröczi, E. and Hilliard, J., Luteinizing hormone releasing activity in different parts of rabbit and dog brain, Endocrinology, 77 (1965) 667-673.

Gargiulo, P.A. and Donoso, A.O., Luteinizing Hormone Releasing Hormone (LHRH) in the periaqueductal gray substance increases some subcategories of grooming behavior in male rats, Pharmacol. Biochem. Behav., 32 (1989) 853-856.

Geeraedts, L.M.G., Nieuwenhuys, R. and Veening, J.G., Medial forebrain bundle of the rat. III. Cytoarchitecture of the rostral (telencephalic) part of the medial forebrain bundle bed nucleus, J. Comp. Neurol., 294 (1990a) 507-536.

Geeraedts, L.M.G., Nieuwenhuys, R. and Veening, J.G., Medial forebrain bundle of the rat. IV. Cytoarchitecture of the caudal (lateral hypothalamic) part of the medial forebrain bundle bed nucleus, J. Comp. Neurol., 294 (1990b) 537-568.

Gelsema, A.J., Roe, M.J. and Calaresu, F.R., Neurally mediated cardiovascular responses to stimulation of cell bodies in the hypothalamus of the rat, Brain Res., 482 (1989) 67-77.

Gerfen, C.R. and Sawchenko, P.E., An anterograde neuroanatomical method that shows the detailed morphology of neurons, their axons and terminals: immunohistochemical localization of an axonally transported plant lectin, Phaseolus vulgaris Leucoagglutinin (PHA-L), Brain Res., 290 (1984) 219-238.

Gioia, M., Bianchi, R. and Tredici, G., Cytoarchitecture of the periaqueductal gray matter in the cat: A quantitative Nissl study, Acta Anat., 119 (1984) 113-177.

Gioia, M., Tredici, G. and Bianchi, R., The ultrastructure of the periaqueductal gray matter of the cat, J. Submicrosc. Cytol., 15 (1983) 1013-1026.

Gray, T.S. and Magnuson, D.J., Galanin-like immunoreactivity within amygdaloid and hypothalamic neurons that project to the midbrain central grey in rat, Neurosci. Lett., 83 (1987) 264-268.

Hamilton, B.L., Cytoarchitectural subdivisions of the periaqueductal gray matter in the cat, J. Comp. Neurol., 149 (1973) 1-28.

Heuzer, J.E., Reeze, T.S., Dennis, M.J., Jan, Y., Jan, L. and Evans, L., Synaptic vesicle exocytosis captured with quick freezing and correlated with quantal transmitter release, J. Cell. Biol., 81 (1979) 275-300.

Hilton, S.M. and Redfern, W.S., A search for brainstem cell groups integrating the defensive reaction in the rat, J. Physiol., 378 (1986) 213-228.

Hinrichsen, C.F.L., Projections of the midlateral posterior hypothalamic area influencing cardiorespiratory function in rats, Brain Behav. Evol., 32 (1988) 108-118.

Hökfelt, T., Millhorn, D., Seroogy, K., Tsuruo, S., Ceccatelli, S., Lindh, B., Meister, B., Melander, T., Schalling, M., Bartfai, T. and Terenius, L., Coexistence of peptides with classical neurotransmitters, Experientia, 43 (1987) 768-780.

Holstege, G., Some anatomical observations on the projections from the hypothalamus to brainstem and spinal cord: an HRP and autoradiographic tracing study in the cat, J. Comp. Neurol., 260 (1987) 98-126.

Joseph, S.A., Immunoreactive adrenocorticotropin in rat brain: a neuroanatomical study using antiserum generated against synthetic ACTH[1-39], Am. J. Anat., 158 (1980) 533-548.

König, F.R. and Klippel, R.A., The Rat Brain, The Williams and Wilkins Co., Baltimore, 1963.

Krieger, M.S., Conrad, L.C.A. and Pfaff, D.W., An autoradiographic study of the efferent connections of the ventromedial nucleus of the hypothalamus, J. Comp. Neurol., 183 (1979) 785-816.

Kruk, M.R., Van Der Poel, A.M., Meelis, W., Hermans, J., Mostert, P.G., Mos, J. and Lohman, A.H.M., Discriminant analysis of the localization of aggression-inducing electrode placements in the hypothalamus of male rats, Brain Res., 260 (1983) 61-80.

Lamberts, R. and Goldsmith, P.C., Pre-embedding colloidal gold immunostaining of hypothalamic neurons: light and electron microscopic localization of β-endorphin-immunoreactive pericarya, J. Histochem. Cytochem., 33 (1985) 499-507.

Lammers, J.H.C.M., Meelis, W., Kruk, M.R. and Van der Poel, A.M., Hypothalamic substrates for brain stimulation-induced grooming, digging and circling in the rat, Brain Res., 418 (1987) 1-19.

Lammers, J.H.C.M., Kruk, M.R., Meelis, W. and Van der Poel, A.M., Hypothalamic substrates for brain stimulation-induced attack, teeth-chattering and social grooming in the rat, Brain Res., 449 (1988) 311-327.

Léranth, C., Williams, T.H., Chrétien, M. and Palkovits, M., Ultrastructural investigation of ACTH-immunoreactivity in arcuate and supraoptic nuclei of the rat, Cell Tiss. Res., 210 (1980) 11-19.

Léranth, C., MacLusky, N.J., Shanabrough, M. and Naftolin, F., Immunohistochemical evidence for synaptic connections between proopiomelanocortin-immunoreactive axons and LH-RH neurons in the preoptic area of the rat, Brain Res., 449 (1988) 167-176.

Liposits, Zs. and Sétáló, G., Descending luteinizing hormone-releasing hormone (LH-RH) nerve fibers to the midbrain of the rat, Neurosci. Lett., 20 (1980) 1-4.

Liu, R.P.C. and Hamilton, B.L., Neurons of the periaqueductal gray matter as revealed by Golgi study, J. Comp. Neurol., 189 (1980) 403-418.

Luiten, P.G.M., Ter Horst, G.J., Karts, H. and Steffens, A.B., The course of paraventricular hypothalamic efferents to autonomic structures in medulla and spinal cord, Brain Res., 329 (1985) 374-378.

Mantyh, P.W., The midbrain periaqueductal grey in the rat, cat and monkey: A Nissl, Weil and Golgi analysis, J. Comp. Neurol., 204 (1982a) 349-363.

Mantyh, P.W., Forebrain projections to the periaqueductal gray in the monkey, with observations in the cat and rat, J. Comp. Neurol., 206 (1982b) 146-158.

Marchand, J.E. and Hagino, N., Afferents to the periaqueductal gray in the rat. A horseradish peroxidase study, Neuroscience, 9 (1983) 95-106.

Matteoli, M., Haimann, C., Torri-Tarelli, F., Polak, J.M. and Ceccarelli, B., Differential effect of γ-latrotoxin on exocytosis from small synaptic vesicles and from large dense-core vesicles containing calcitonin gene-related peptide at the frog neuromuscular junction, Proc. Natl. Acad. Sci. U.S.A., 85 (1988) 7366-7370.

Meller, S.T. and Dennis, B.J., Afferent projections to the periaqueductal gray in the rabbit, Neuroscience, 19 (1986) 927-964.

Moss, M.S. and Basbaum, A.I., The fine structure of the caudal periaqueductal gray of the cat: morphology and synaptic organization of normal and immunoreactive enkephalin-labeled profiles, Brain Res., 289 (1983) 27-43.

Nieuwenhuys, R., Chemoarchitecture of the Brain, Springer Verlag, Berlin, 1985.

Nieuwenhuys, R., Veening, J.G. and van Domburg, P., Core and paracores; some new chemoarchitectural entities in the mammalian neuraxis, Acta Morph. Neerl.-Scand., 26 (1989) 131-163.

Paxinos, G. and Watson, C., The Rat Brain in Stereotaxic Coordinates, Academic Press, New York, 1986.

Pfaff, D.W., Lewis, C., Diakow, C. and Keiner, M., Neurophysiological analysis of mating behaviour responses as hormonal-sensitive reflexes, Prog. Physiol. Psychol., 5 (1972) 253-297.

Piekut, D.T., Relationship of ACTH[1-39]-immunostained fibers and magnocellular neurons in the paraventricular nucleus of the rat hypothalamus, Peptides, 6 (1985) 883-890.

Ramón y Cajal, S., Histologie du système nerveux de l'Homme et des Vertébrés, Vol.2., Maloine, Paris, 1911.

Reichling, D.B., Kwiat, G.C. and Basbaum, A.I., Anatomy, physiology and pharmacology of the periaqueductal gray contribution to antinociceptive controls, Prog. Brain Res., 77 (1988) 31-46.

Riskind, P. and Moss, R.L., Effects of lesions of putative LHRH-containing pathways and midbrain nuclei on lordotic behaviour and luteinizing hormone release in ovariectomized rats, Brain Res. Bull., 11 (1979) 493-500.

Roberts, W.W., [14C] Deoxyglucose mapping of first-order projections activated by stimulation of lateral hypothalamic sites eliciting gnawing, eating, and drinking in rats, J. Comp. Neurol., 194 (1980) 617-638.

Roeling, T.A.P., Veening, J.G., Kruk, M.R. and Nieuwenhuys, R., Grooming behaviour elicited by kainic acid evoked cell body stimulation in the hypothalamus of the rat, Neurosci. Res. Comm., 6 (1990) 111-118.

Roeling, T.A.P., Van Erp, A.M.M., Meelis, W., Kruk, M.R. and Veening, J.G., Behavioural effects of NMDA injected into the hypothalamic paraventricular nucleus of the rat, Brain Res., 550 (1991) 220-224.

Romagnano, M.A. and Joseph, S.A., Immunocytochemical localization of ACTH[1-39] in the brainstem of the rat, Brain Res., 276 (1983) 1-16.

Sakuma, Y. and Pfaff, D.W., Facilitation of female reproductive behaviour from mesencephalic central grey in the rat, Am. J. Physiol., R237 (1979a) 278-284.

Sakuma, Y. and Pfaff, D.W., Mesencephalic mechanisms for integration of female reproductive behaviour in the rat, Am. J. Physiol., R237 (1979b) 285-290.

Sakuma, Y. and Pfaff, D.W., LH-RH in the mesencephalic central grey can potentiate lordosis reflex of female rats, Nature, 283 (1980) 566-567.

Samson, W.K., McCann, S.M., Chud, L., Dudley, C.A. and Moss, R.L., Intra- and extrahypothalamic luteinizing hormone-releasing hormone (LHRH) distribution in the rat with special reference to mesencephalic sites which contain both LHRH and single neurons responsive to LHRH, Neuroendocrinology, 31 (1980) 66-72.

Saper, C.B., Loewy, A.D., Swanson, L.W. and Cowan, W.M., Direct hypothalamo-autonomic connections, Brain Res., 117 (1976a) 305-312.

Saper, C.B., Swanson, L.W. and Cowan, W.M., The efferent connections of the ventromedial nucleus of the hypothalamus of the rat, J. Comp. Neurol., 169 (1976b) 409-442.

Saper, C.B., Swanson, L.W. and Cowan, W.M., An autoradiographic study of the efferent connections of the lateral hypothalamic area in the rat, J. Comp. Neurol., 183 (1979) 689-706.

Shipley, M.T., McLean, J.H. and Behbehani, M.M., Heterogeneous distribution of neurotensin-like immunoreactive neurons and fibers in the midbrain periaqueductal gray of the rat, J. Neurosci., 7 (1987) 2025-2034.

Shivers, B.D., Harlan, R.E., Morrel, J.I. and Pfaff, D.W., Immunocytochemical localization of luteinizing hormone-releasing hormone in the male and female rat brains. Quantitative studies on the effects of gonadal steroids, Neuroendocrinology, 36 (1983) 1-12.

Siegel, A. and Pott, C.B., Neural substrates of aggression and flight in the cat, Prog. Neurobiol., 31 (1988) 261-283.

Simerly, R.B. and Swanson, L.W., Projections of the medial preoptic nucleus: A Phaseolus vulgaris leucoagglutinin anterograde tract-tracing study in the rat, J. Comp. Neurol., 270 (1988) 209-242.

Sirinathsinghji, D.J.S., Modulation of lordosis behaviour in the female rat by corticotropin releasing factor, ß-endorphin and gonadotropin releasing hormone in the mesencephalic central grey, Brain Res., 336 (1985) 45-55.

Sirinathsinghji, D.J.S., Rees, L.H., Rivier, J. and Vale, W., Corticotropin-releasing factor is a potent inhibitor of sexual receptivity in the female rat, Nature, 305 (1983) 232-235.

Somogyi, P. and Hodgson, A.J., Antisera to γ-aminobutyric acid. III. Demonstration of GABA in Golgi-impregnated neurons and in conventional electron microscopic sections of the cat striate cortex, J. Histochem. Cytochem., 33 (1985) 249-257.

Spencer, S.E., Sawyer, W.B. and Loewy, A.D., Cardiovascular effects produced by L-glutamate stimulation of the lateral hypothalamic area, Am. J. Physiol., 257 (1989) H540-H552.

Spencer, S.E., Sawyer, W.B. and Loewy, A.D., L-Glutamate mapping of cardioreactive areas in the rat posterior hypothalamus, Brain Res., 511 (1990) 149-157.

Swanson, L.W., An autoradiographic study of the efferent connections of the preoptic region in the rat, J. Comp. Neurol., 167 (1976) 227-256.

Taber, E., The cytoarchitecture of the brainstem of the cat. I. Brainstem nuclei of cat, J. Comp. Neurol., 116 (1961) 27-69.

Ter Horst, G.J., Luiten, P.G.M. and Kuipers, F., Descending pathways from hypothalamus to dorsal motor vagus and ambiguus nuclei in the rat, J. Auton. Nerv. Syst., 11 (1984) 59-75.

Ter Horst, G.J. and Luiten, P.G.M., The projections of the dorsomedial hypothalamic nucleus in the rat, Brain Res. Bull., 16 (1986) 231-248.

Ulfhake, B., Arvidsson, U., Cullheim, S., Hökfelt, T., Brodion, E., Verhofstad, A. and Visser, T., An ultrastructural study of 5-hydroxytryptamine-, thyrotropin-releasing hormone- and substance P-immunoreactive axonal boutons in the motor nucleus of spinal cord segments L7-S1 in the adult cat, Neuroscience, 23 (1987) 917-929.

Van Den Pol, A.N., Dual ultrastructural localization of two neurotransmitter-related antigens: colloidal gold-labeled neurophysin-immunoreactive supraoptic neurons receive peroxidase-labeled glutamate decarboxylase- or gold-labeled GABA-immunoreactive synapses, J. Neurosci., 5 (1985) 2940-2954.

Van Erp, A.M.M., Kruk, M.R., Willekens-Bramer, D.C., Bressers, W.M.A., Roeling, T.A.P., Veening, J.G. and Spruyt, B.M., Grooming induced by intrahypothalamic injection of ACTH in the rat: comparison with grooming induced by intrahypothalamic electrical stimulation and i.c.v. injection of ACTH, Brain Res., 538 (1991) 203-210.

Veening, J.G., The Lie, S., Posthuma, P., Geeraedts, L.M.G. and Nieuwenhuys, R., A topographical analysis of the origin of some efferent projections from the lateral hypothalamic area in the rat, Neuroscience, 22 (1987) 537-551.

Veening, J.G., Stroeken, P.C., Posthuma, P. and Steinbusch, H.W.M., A comparison of the descending hypothalamic and some other diencephalic projections to the nucleus raphe magnus and the mesencephalic periaqueductal gray in the rat, Neurosci. Res. Comm., 7 (1990) 123-132.

Waldrop, T.G., Bauer, R.M. and Iwamoto, G.A., Microinjection of GABA antagonists into the posterior hypothalamus elicits locomotor activity and a cardiorespiratory activation, Brain Res., 444 (1988) 84-94.

Watanabe, K. and Kawana, E., The cells of origin of the incertofugal projections to the tectum, thalamus, tegmentum and spinal cord in the rat: a study using the autoradiographic and horseradish peroxidase methods, Neuroscience, 7 (1982) 2389-2406.

Wied, D. de., Behavioral effects of neuropeptides related to ACTH and ß-LPH, In: Neuropeptides and Neurotransmission, Agmone C. and Traczyk W.Z. (Eds.), Raven Press, New York, 1980, pp. 217-226.

Yardley, C.P. and Hilton, S.M., The hypothalamic and brainstem areas from which the cardiovascular and behavioural components of the defense reaction are elicited in the rat, J. Auton. Nerv. Syst., 15 (1986) 227-244.

Yoshida, M. and Taniguchi, Y., Projection of proopiomelanocortin neurons from the rat arcuate nucleus to the midbrain central gray as demonstrated by double staining with retrograde labeling and immunohistochemistry, Arch. Histol. Cytol., 51 (1988) 175-183.

Zhu, P.C., Thureson-Klein, Å. and Klein, R.L., Exocytosis from large and dense-core vesicles outside the active synaptic zones of terminals within the trigeminal subnucleus caudalis: a possible mechanism for neuropeptide release, Neuroscience, 19 (1986) 43-54.

Topographical Specificity of Forebrain Inputs to the Midbrain Periaqueductal Gray: Evidence for Discrete Longitudinally Organized Input Columns

Michael T. Shipley, Matthew Ennis,
Tilat A. Rizvi and Michael M. Behbehani*

Department of Anatomy and Cell Biology
and
*Department of Physiology and Biophysics
University of Cincinnati College of Medicine
Cincinnati, U.S.A.

Introduction

Over two decades ago it was discovered that electrical stimulation of the periaqueductal gray (PAG) caused profound analgesia (Reynolds, 1969). It was subsequently found that "PAG-analgesia" is, at least in part, mediated by opiate and neurotensin systems acting via PAG projections to the rostral medulla (Basbaum and Fields, 1984; Behbehani, 1981; Behbehani and Fields, 1979; Behbehani and Pert, 1984; Behbehani et al., 1987; Lakos and Basbaum, 1988; Reichling et al., 1988; Shipley et al., 1987). As a result of the observation that a discrete CNS structure exerted such a potent regulation of pain, much subsequent research has focused on the role of PAG in antinociception. At the same time there has been growing evidence that PAG plays a key role in the "defense reaction" (Bandler and Carrive, 1988; Bandler and Depaulis, this volume; Bandler et al., 1985a, 1991; Depaulis and Vergnes, 1986; Depaulis et al., 1989; Zhang et al., 1990), vocalization (Jürgens, 1976; Jürgens and Richter, 1986; Larson, 1985; Larson and Kistler, 1984; 1986), and in certain sexual behaviors (Ogawa et al., this volume; Sakuma and Pfaff; 1979a,b).

Although widely viewed as an important integrative element for these functions, scattered, often anecdotal, observations have suggested that PAG influences autonomic responses, such as control of blood pressure, heart rate and respiration. However, most functionally oriented studies of PAG have not appreciated that autonomic adjustments are integral to the analgesic, defensive, and sexual responses elicited from PAG. As a result, PAG's role in autonomic regulation has, until recently, received surprisingly little attention. New results show that PAG exerts potent, bidirectional influences on blood pressure. Specifically, electrical microstimulation or microinjection of excitatory amino acids into discrete regions of PAG causes rapid elevation or depression of blood pressure (Bandler et al., 1991; Carrive, this volume; Carrive and Bandler, in press; Carrive et al., 1987; 1988; 1989a,b; Lovick, this volume). Tract tracing studies (Bandler et al., 1991; Carrive, this volume) suggest that PAG sites eliciting selective changes in blood pressure project to the rostral ventrolateral medulla (RVM). Thus, there is growing appreciation of the possibility that PAG is a major nodal point in CNS circuits regulating autonomic adjustments integral to the successful expression of behavioral and antinociceptive responses centering around the animals reaction to threatening external events.

Forebrain Inputs to PAG

Because the last two decades of PAG research have been dominated by studies of pain and the defense reaction, anatomical investigations of PAG circuitry have focused primarily on descending circuits from PAG to the medulla, including projections to the nucleus raphe magnus and the adjacent RVM. As a result, the connections of PAG with the forebrain have been largely ignored. Beitz (1982) published the first comprehensive retrograde tracing study of afferent inputs to PAG using horseradish peroxidase (HRP). This report demonstrated that inputs to PAG arise in an almost bewildering array of forebrain structures, but there has been scant appreciation of these observations in subsequent research on PAG. As a result, very little is known about the anatomical organization of forebrain inputs to PAG. There is, for example, almost no information about the magnitude or the topography of inputs from forebrain areas along the extensive rostrocaudal axis of PAG. This type of information is critical to understanding how forebrain areas articulate with and influence the activity of PAG neurons. For example, it is not known whether forebrain inputs project diffusely throughout PAG to raise or lower the activity of a broad population of PAG neurons or, alternatively, whether different forebrain areas project to distinct subregions of PAG to possibly trigger or modulate the activity of subsets of PAG neurons. We have been addressing these issues for the last two years (Ennis and Shipley; in preparation; Ennis et al., 1990; Rizvi et al., 1990; 1991). One major result that has emerged from our work is the finding that inputs to PAG from the cerebral cortex are far more extensive that previously reported. Earlier tract tracing studies revealed modest projections to PAG from medial prefrontal and insular cortex (Bandler et al., 1985b; Beitz, 1982; Hardy, 1986; Mantyh, 1982a; Marchand and Hagino, 1983; Meller and Dennis, 1986; Neafsey et al., 1986; Sesack et al., 1989; Terreberry and Neafsey, 1987; Wyss and Sripanidkulchai, 1984). Our retrograde tracing studies using more sensitive techniques, have confirmed these observations, but we have found that cortical inputs to PAG are far more substantial, and arise from a greater number of cortical fields than previously reported. More importantly, however, our results show that projections from distinct architectonic divisions of these cortical regions differentially terminate with a remarkable degree of topographic specificity along the rostrocaudal axis of PAG. Some cortical fields preferentially terminate in dorsal and lateral parts of PAG while other cortical fields terminate predominantly in lateral and ventrolateral PAG. Moreover, different cortical fields terminate with characteristic rostrocaudal gradients in PAG. These surprisingly specific terminal projections are consistent with the hypothesis that cortical inputs may preferentially target discrete subpopulations of PAG neurons and thereby selectively trigger or modulate the antinociceptive, defensive and sexual functions associated with PAG and the autonomic adjustments integral to the successful expressions of these functions. In this regard, it is noteworthy that most of the cortical fields projecting to PAG have been implicated in many of the same functions classically associated with PAG, such as autonomic regulation and antinociception (Burns and Wyss, 1985; Cechetto and Chen, 1990; Hardy, 1985; Hardy and Holmes, 1988; Ruggiero et al., 1987; Yasui et al., 1991). With this working hypothesis in mind, we next review some features of the organization and specificity of forebrain connections with PAG resulting from our recent anatomical investigations.

Results

Cortical fields projecting to PAG: retrograde tracing studies

As noted earlier, Beitz (1982) originally described a vast array of forebrain cortical and subcortical sites containing retrogradely labeled neurons following injection of HRP into the PAG. Since Beitz's study, there has been dramatic improvement of retrograde tracing methods, notably, the use of the lectin-HRP conjugate wheat germ agglutinin (WGA)-HRP which markedly increased the sensitivity of retrograde tracing. This methodological advance led to even greater improvement in the sensitivity of anterograde labeling. Thus, we decided to reinvestigate the forebrain connections of PAG in order to determine if the improved sensitivity of retrograde tracing would provide a more comprehensive picture of forebrain neurons projections to PAG, but even more importantly, to use the improved anterograde sensitivity of WGA-HRP-tetramethylbenzidine (TMB) to characterize the distribution and organization of projections from specific forebrain sites to PAG.

The PAG surrounds the cerebral aqueduct and parts of the IIIrd and IVth ventricles; in the rat, it extends for a rostrocaudal distance of approximately 5 mm. In order to provide a comprehensive picture of forebrain neurons projecting to PAG, we first performed a set of experiments in which WGA-HRP was iontophoretically injected from micropipettes (tip diam. approximately 10 μm) at a series of 3 equally spaced rostrocaudal sites in PAG (4.5 to 0.2 mm anterior to interaural line). At each site two injections were made, one in the dorsal and one in the ventral half of PAG. The purpose of these "multiple injection" cases was to optimize the likelihood of comprehensively labeling forebrain neurons projecting to PAG. The goal of this strategy, thus, was to put a frame around all the

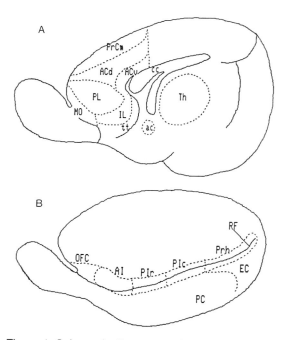

Figure 1. Schematic diagrams (adapted from Krettek and Price, 1977) of medial and lateral cortical fields projecting to PAG. A: Medial aspect of the rat brain showing the topography of fields in medial prefrontal cortex. Abbreviations: ACd, anterior cingulate dorsal; ACv, anterior cingulate ventral; ac, anterior commissure; cc, corpus callosum; IL, infralimbic; MO medial orbital; PL, prelimbic; PrCm, precentral medial; Th, thalamus; tt, tenia tecta. B: Lateral aspect of the rat brain showing the topography of cytoarchitectonic cortical fields located adjacent to the rhinal fissure (RF). Abbreviations: AI, anterior insular; EC, entorhinal cortex; OFC, orbital frontal cortex; PC, piriform cortex; PIr, posterior insular - rostral component; PIc, posterior insular - caudal component; Prh, perirhinal.

neurons that target PAG and then, in subsequent experiments, to use anterograde labeling from sites containing retrogradely labeled neurons to determine the organization of their terminal patterns in PAG.

The result of these multiple injection studies was the demonstration that the forebrain inputs to PAG are massive; inputs arise from the cerebral cortex, the amygdala, the basal telencephalon and hypothalamus. Although the sites containing labeled neurons were in general agreement with Beitz's (1982) earlier study, our use of more sensitive methods and multiple injections labeled neurons in all the areas reported by Beitz in addition to others that will be reported later. The most salient observation from these cases, however, is the sheer magnitude of neurons and the diversity of sites projecting to PAG. Characterization of these diverse, massive forebrain inputs to PAG will require great time and effort.

We have begun this task by analyzing cortical inputs to PAG. Several studies (Bandler et al., 1985b; Hardy, 1986; Mantyh, 1982a; Marchand and Hagino, 1983; Neafsey et al., 1986; Terreberry and Neafsey, 1987; Wyss and Sripanidkulchai, 1984), including Beitz's (1982), had noted retrograde labeling in medial frontal cortical fields. Our multiple injection retrograde tracing cases (Figs. 2-4) confirmed this but indicated that these projections were much heavier than previously reported. Most surprising, however, was the extent of retrograde labeling in the lateral parts of the cortex. Neurons were labeled in a continuous band extending from the frontal pole to nearly the caudal pole of the hemisphere - a stretch of nearly 10 mm of cortex (Figs. 3 and 4). At all rostrocaudal levels the labeled neurons were present in the cortical fields at and directly adjacent to the rhinal sulcus. Both the medial and lateral cortical neurons are distributed throughout a number of defined cortical architectonic fields (Fig. 1). In the medial frontal cortex neurons were present in the infralimbic cortex, prelimbic cortex, anterior cingulate cortex and in the precentral medial cortex (Figs. 1 and 2). In the lateral cortex, neurons were labeled in the anterior insular cortex, the posterior insular cortex and perirhinal cortex (Figs. 1, 3 and 4). In all of these cortical fields the neurons were primarily confined to layer V and appeared to be pyramidal cells. In most of these fields the magnitude of neurons labeled appeared comparable to the labeling of other parts of the neocortex following large pontine or medullary injections. An interesting question for future research is whether these medial and lateral cortical areas also project to other brainstem sites as heavily as they do to PAG. Based on the results of our anterograde studies to be described below, it appears that PAG is one of the major brainstem targets of these medial and lateral cortical fields.

The magnitude of the cortical retrograde labeling in these cases suggested that medial frontal and lateral cortical inputs to PAG were quite substantial. However, these studies left a number of questions unanswered. For example, despite the number of retrogradely labeled cortical neurons, could the projection to PAG represent minor collateral inputs to this structure? Do cortical inputs target all parts of PAG or do different fields selectively target, for example, different rostrocaudal sites of PAG, or different dorsoventral or mediolateral zones in PAG? Do medial cortical fields target different parts of PAG than lateral cortical fields?

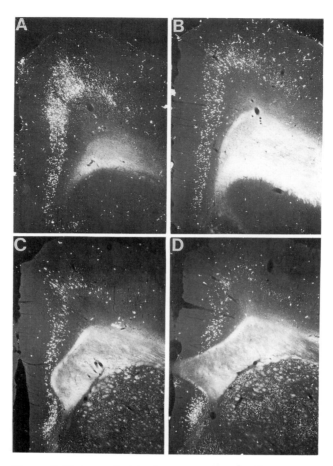

Figure 2. Retrograde labeling of medial prefrontal cortex neurons projecting to PAG. Photomicrographs show retrogradely labeled neurons from rostral (A) through caudal (D) levels of medial prefrontal cortex following a series of 6 injections of WGA-HRP into PAG in a single animal. Retrogradely labeled neurons form a continuous band spanning almost the entire medial wall and the dorsal convexity of frontal cortex. Labeled neurons in medial cortex are present in nearly all of the medial prefrontal cortical fields present at these levels, including infralimbic, prelimbic, anterior cingulate dorsal and precentral medial.

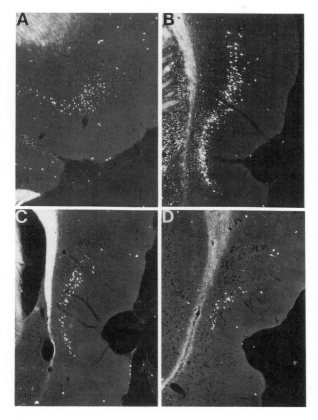

Figure 3. Retrograde labeling of lateral cortex neurons projecting to PAG. Photomicrographs show
retrograde labeling from rostral (A) through caudal
(D) levels of lateral cortex in the same case illustrated in Figure 2. Rostrally (A), labeled cells are located above the rhinal fissure, while at more caudal
levels (B-D), labeled cells are clustered along and
dorsal to the rhinal fissure

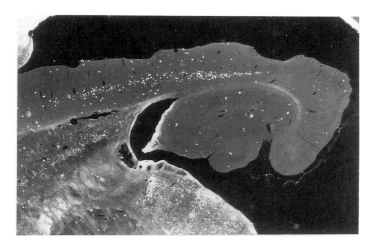

Figure 4. Photomicrograph of a horizontal section showing retrograde labeling of neurons in lateral cortex following multiple injections of WGA-HRP into PAG in a single animal. In this plane of section it can be seen that labeled neurons form a continuous band that comprises nearly the entire rostrocaudal extent of the lateral cortical mantle. Orientation: lateral is at the top, rostral is to the left.

Organization of cortical inputs to PAG: anterograde tracing studies

To begin to answer such questions we again utilized a "multiple injection" strategy. In one series of experiments, one to two iontophoretic injections of WGA-HRP were made into each of the medial cortical fields projecting to PAG in a single animal; in a second series of experiments, injections were made at equal intervals along the rostrocaudal extent of the lateral cortex. Thus, animals in each of these groups had WGA-HRP injections made either in all of the medial or all of the lateral fields projecting to PAG. The resulting anterograde labeling from the "multiple-medial" vs "multiple-lateral" cortical fields was compared. As shown in Figure 5, multiple injections encompassing infralimbic, prelimbic, anterior cingulate and precentral cortical sites produced dense and essentially uniform anterograde labeling throughout the entire rostrocaudal extent of PAG. The labeling was present from the rostral periventricular component of PAG to the caudal, pontine tegmental component of PAG. The labeling was bilateral although it was somewhat more extensive in the half of PAG ipsilateral to the cortical injections. At all rostrocaudal levels of PAG the anterograde labeling was essentially uniformly distributed with little suggestion of dorsal to ventral or central to peripheral gradients in the density of labeling. A very similar result was obtained from the cases with multiple lateral cortical injections encompassing the anterior and posterior insular cortical areas and the perirhinal cortex (Fig. 6). Anterograde labeling was densely and uniformly distributed in all parts of PAG at all rostrocaudal levels. The primary difference in labeling

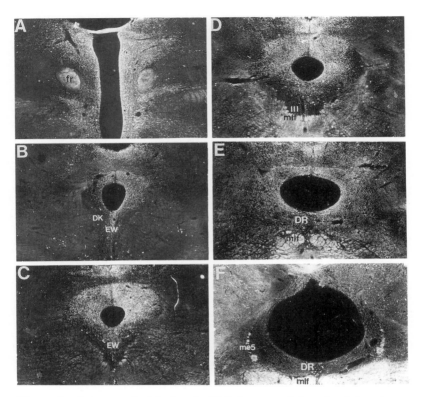

Figure 5. Anterograde labeling in PAG from medial prefrontal cortex: Aggregate projection. A series of photomicrographs of six representative rostrocaudal levels of PAG (A, rostral through F, caudal) from an experiment in which WGA-HRP was injected into each of the medial frontal cortical fields that project to PAG in a single animal. The aggregate projection from medial cortex to PAG is strongly bilateral, however, the ipsilateral projection (right side) is heavier. Note that the projection from medial cortex densely innervates PAG along most of its rostrocaudal axis although the projection is somewhat heavier to the rostral than the caudal half of PAG. At all rostrocaudal levels, the distribution of anterograde labeling is fairly uniform in all of the major subregions of PAG (i.e., dorsomedial, lateral, ventrolateral). DK, nucleus Darkschewitsch; DR, dorsal raphe; EW, Edinger-Westphal nucleus; fr, fasciculus retroflexus; III, oculomotor nucleus; mlf, medial longitudinal fasciculus; me5, mesencephalic trigeminal tract.

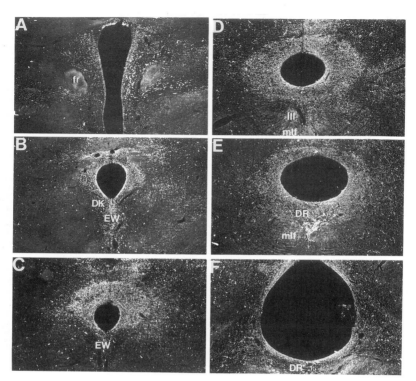

Figure 6. Anterograde labeling in PAG from lateral cortex: Aggregate projection. A series of photomicrographs of 6 rostrocaudal levels of PAG (A,
rostral through F, caudal) from an experiment in which WGA-HRP was
injected at intervals along the rostrocaudal extent of the lateral cortex to
include all the lateral fields projecting to PAG. As for the projections arising from the sum of the medial cortical fields (Fig. 5), the aggregate projection from lateral cortex is heaviest to the half of PAG ipsilateral to the
injection sites (right side), but there is also a strong projection to contralateral PAG. Like the aggregate medial projection, inputs from lateral cortex densely and uniformly innervate PAG along its entire rostrocaudal
extent. However, in contrast to the medial projection, the projections
from lateral cortex are heavier to the caudal than the rostral half of PAG.
DK, nucleus Darkschewitsch; DR, dorsal raphe; EW, Edinger-Westphal
nucleus; fr, fasciculus retroflexus; III, oculomotor nucleus; mlf, medial longitudinal fasciculus; me5, mesencephalic trigeminal tract.

from the "multiple medial" vs "multiple lateral" cases was a tendency for the anterograde labeling from medial cortical cases to be somewhat denser in the rostral half and the labeling from lateral cortex to be somewhat denser in the caudal half of PAG. This medial to rostral, lateral to caudal gradient was also observed in subsequent studies, to be described presently, of the projections from individual medial or lateral cortical fields to PAG. It should be emphasized, however, that this difference in medial vs lateral cortical projections is a moderate gradient and not a dichotomous difference in the organization of cortical projections to PAG. It should also be reemphasized that these aggregate projections were bilateral but with heavier labeling ipsilateral to the injection sites. A final comment concerns the density of the labeling. Although multiple injections were made, this should not be construed to mean that the injections involved all or even a majority of the cortical neurons projecting to PAG. The multiple injections insured that the contribution from each of the medial or lateral cortical fields was labeled, but the individual injections were usually confined to only a limited part of each specific field. Thus, the aggregate projections from medial and lateral cortex are doubtless underrepresented in these multiple injection cases. Notwithstanding this important proviso, it is nonetheless clear that the cortical inputs to PAG are quite substantial. It is equally clear from this material that there is no part of PAG - dorsal, dorsolateral, lateral, ventrolateral, central vs peripheral - nor any part along the entire rostrocaudal extent of PAG that lacks cortical inputs.

These multiple injection cases, thus, demonstrated that all parts of PAG receive cortical inputs. However, these results provided little insight as to organization of inputs from individual medial and lateral fields. One possibility, consistent with the dense uniform labeling resulting from the multiple injection cases, was that each cortical field projects diffusely to PAG with the result that all cortical fields terminate in all parts of PAG. Were this true, one might expect that the functional role of cortical inputs would be of a global nature; the number of active cortical inputs might somehow raise or lower the general level of activity in PAG neurons. At the other extreme, the dense but diffuse labeling resulting from multiple injections could arise if each of the cortical fields projected to discrete but complimentary subregions of PAG; each input would target different, focal parts of PAG such that the collective input from the medial or lateral cortical fields would target all parts of PAG. In this case, the selective activation of one cortical field might preferentially influence specific subregions in PAG. The issue then is whether inputs from discrete cortical fields terminated diffusely or focally in PAG. It was also conceivable that some cortical fields terminated diffusely and others focally, or that each field has both focal and diffuse components. In each of these scenarios, the potential functional significance of cortical inputs to PAG would be different.

To address these issues, a series of experiments was performed in which a single, discrete iontophoretic or pressure injection of WGA-HRP was made in only one of the component medial and lateral cortical fields projecting to PAG.

MEDIAL FRONTAL CORTEX. In the rat medial prefrontal cortex, the architectonic boundaries between component cortical fields are not sharp. To facilitate parcellation of medial cortical fields that contain neurons that project to PAG, we used a combined technique for acetylcholinesterase and Nissl staining (Shipley et al., 1989). Our parcellation of medial prefrontal cortex (Fig. 1) is essentially similar to that of Krettek and Price (1977), thus we have used their nomenclature. As shown in Figure 2, neurons projecting to PAG from medial frontal cortex comprise a continuous band of cells running along the medial wall and dorsal convexity of the precallosal wall of the hemisphere. More caudally the emergence of the callosum splits the band of neurons into a subcallosal and supracallosal group. The subcallosal group comprises the caudal extension of the infralimbic and prelimbic cortices; these fields extend only a short distance below the rostrum of the callosum at the emergence of the anterior hippocampal continuation (AHC). The AHC is devoid of labeled neurons. The supracallosal labeling extends somewhat further caudally than the subcallosal cells but the density of supracallosal labeling decreases as the anterior cingulate cortex gives way to the

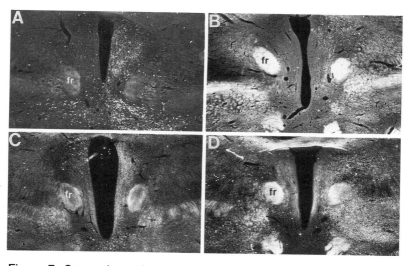

Figure 7. Comparison of the complimentary innervation patterns from each of the medial cortical fields at a single periventricular level of PAG. A: The projection from infralimbic cortex terminates in a broad zone along the lateral walls of the aqueduct. B: Inputs from anterior cingulate cortex, in contrast, preferentially target the most dorsal and dorsolateral parts of PAG, terminating in the area ignored by the infralimbic projection. C: The projection from prelimbic cortex selectively innervates a narrow zone alongside the wall of the cerebral aqueduct and does not extend appreciably into the lateral parts of PAG. D: The projection from precentral medial cortex is restricted to a discrete zone in the lateral most parts of PAG, and is complimentary to the projection from prelimbic cortex (C). DK, nucleus Darkschewitsch; EW, Edinger-Westphal nucleus; fr, fasciculus retroflexus.

posterior cingulate cortex. At precallosal levels, retrogradely labeled neurons extend beyond the dorsal limits of the anterior cingulate cortex around the dorsal convexity of the hemisphere. Thus, labeled cells are present throughout most of the field recognized as precentral medial cortex. This continuous expanse of retrogradely labeled neurons extends smoothly through several architechtonically defined cortical fields, including the infralimbic cortex, prelimbic cortex, anterior cingulate cortex and precentral medial cortex.

A single discrete iontophoretic injection of WGA-HRP was targeted by stereotaxic coordinates into one of each of these cortical areas. The difficult part of this experiment was the successful restriction of injection sites to each of the specific fields. Thus a number of cases was generated and only those with successfully targeted injections as judged by cytoarchitecture and patterns of cortical and subcortical labeling were used to analyze anterograde labeling patterns in PAG.

As illustrated in Figures 7 and 8, the most salient finding in this study was that each of the medial frontal cortical fields projected densely and focally to restricted parts of PAG. Thus, while the "aggregate" projection from medial frontal cortex terminates uniformly throughout PAG, the individual fields giving rise to the projection terminate with highly circumscribed patterns. These patterns are partially overlapping but are mainly complimentary such that the input from each cortical field selectively innervates its own unique zone in PAG (Figs. 7 and 8). Each cortical field projects along most, if not all, of the rostrocaudal axis of PAG.

In general, the projections from the medial frontal cortical fields project more heavily to the rostral two-thirds of PAG than to the caudal third. Thus, the individual fields reflect the rostral to caudal gradient seen in the multiple injection cases. This gradient was more prominent in the projections of some of the individual fields than in the aggregate projection.

The projections from individual medial frontal fields exhibited a remarkable specificity and selectivity. For one thing, the cortical projections began to terminate abruptly at the rostralmost limits of the periventricular parts of PAG but each projection had a unique pattern of termination (Fig. 7). For example, the projection from infralimbic cortex terminated in a broad zone along the lateral walls of the ventricle; the labeling extended throughout most of the centrolateral parts of the periventricular component of PAG but did not extend to the lateralmost limits of PAG, nor did the labeling extend appreciably into the dorsal parts of the periventricular component of PAG. The projection from the anterior cingulate cortex was essentially the converse of this pattern: this projection terminated in the dorsal most parts of the periventricular component of PAG occupying the territory ignored by the infralimbic projection. The inputs from prelimbic and precentral medial cortex were equally complementary. As illustrated in Figures 7C and 8, the prelimbic labeling was largely restricted to the dorsal part of the circumventricular PAG and did not extend into the lateral parts of dorsal PAG. By contrast the projection from precentral medial cortex was restricted to the lateralmost parts of the rostral periventricular PAG. Thus, the projections from

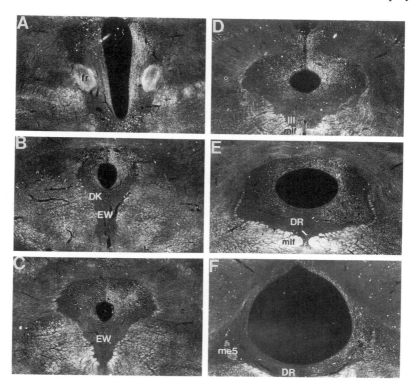

Figure 8. Projections to PAG from prelimbic cortex. Photomicrographs of 6 rostrocaudal levels of PAG (A, rostral through F, caudal) from an experiment in which a single, restricted WGA-HRP injection was made into prelimbic cortex. Characteristic of individual cortical fields projecting to PAG, the projection from prelimbic cortex is bilateral, with weaker contralateral homotypic labeling. The projection is heavier to the rostral than the caudal half of PAG. The projection from prelimbic cortex densely innervates the dorsomedial and dorsolateral parts of PAG through the rostral one-third of PAG (A, B). At mid to caudal levels (C-F), two separate zones in dorsomedial and lateral or ventrolateral PAG are selectively targeted by inputs from prelimbic cortex. DK, nucleus Darkschewitsch; DR, dorsal raphe, EW, Edinger-Westphal; fr, fasciculus retroflexus; III, oculomotor nucleus; mlf, medial longitudinal fasciculus; me5, mesencephalic trigeminal tract.

infralimbic, prelimbic, anterior cingulate and precentral medial cortex to the rostral periventricular PAG terminate as four distinctive, partially overlapping, but largely complimentary terminal fields. This organizational principle was characteristic for the terminal projections from these medial cortical fields throughout most of the rostrocaudal extent of PAG.

Proceeding caudally through PAG a second principle emerged. Several of the medial frontal cortical projections began to exhibit a bipartite terminal field.

The projections from a given cortical area often gave rise to one terminal field in the dorsal or dorsomedial part of PAG and a second terminal field in the lateral or ventrolateral part of PAG. For some of the cortical areas this bipartite terminal field was present over a distance of 1-2 mm in the middle parts of PAG then usually one of the fields gradually dissipated. Thus, the projections of each cortical area terminates as a discrete longitudinal column running through a specific zone in rostral PAG; then, starting at about the rostral pole of the oculomotor complex, this column bifurcates or is joined by a second longitudinal column. These two columns extend throughout the middle, and for some cortical fields, through the caudal parts of PAG. The location of the longitudinal columns from each medial cortical field is characteristic for that cortical field and the columns from different cortical fields partially overlap but are largely complementary to those of other medial cortical fields. Thus, the "aggregate" projection as seen in the multiple injection cases appears to be composed of discrete complementary projections from each of the contributing cortical fields. Two other organizational features should be noted. For each cortical area, the location of the second, ventral or ventrolateral terminal column often showed a gradual lateral to ventrolateral angular shift at progressively caudal levels of PAG; by contrast, the dorsal or dorsomedial terminal field showed less angular displacement along the rostrocaudal axis. As a result of this, as one moves along the rostrocaudal axis of PAG the projection from any given medial frontal cortical field resembled the two hands of a clock in which the lateral/ventrolateral "hand" moves from 1 to 4 o'clock or from 2 to 5 o'clock, etc, at progressively caudal sites in PAG. The other tendency is that the projections from some cortical fields are more dense in the central circumaqueductal part of PAG while others were more dense in the more peripheral parts of PAG. Thus, there is considerable angular and radial specificity in these cortical projections.

This anterograde tracing analysis of the projection from medial frontal cortical fields to PAG answered several of the questions posed by examination of the aggregate projections seen in the multiple injection cases. Thus, the projections from individual fields do not terminate diffusely in PAG. To the contrary, each field projects with a high degree of axial and radial specificity along the rostrocaudal axis of PAG. Based on these highly specific patterns of terminal input, therefore, it is not unreasonable to suggest that different medial frontal cortical fields could selectively or preferentially influence specific populations of PAG neurons along the rostrocaudal axis of PAG. Thus, it is possible that each medial cortical field could selectively influence specific PAG functions.

LATERAL CORTEX. As shown in Figures 3 and 4, neurons projecting to PAG from lateral cortex extend as a continuous band from the ventrolateral orbital and anterior insular cortex through the rostral and caudal components of the posterior insular cortex and throughout the extent of the perirhinal cortex. Lateral cortical projections to PAG thus extend from the rostral to the caudal pole of the hemisphere over a distance of more than 10 mm. While these neurons constitute a continuous band, the rostrocaudal distribution of neurons projecting to PAG is not entirely uniform. At rostral levels, corresponding to ventrolateral orbital and anterior insular cortical areas, retrogradely labeled neurons are clustered above the rhinal fissure (Fig. 3A). Caudally, as the invagination of the rhinal fissure decreases, the number of labeled cells increases and they begin to occupy a position directly adjacent to the rhinal fissure (Fig. 3B). Thus, from the genu of the corpus callosum to the genu of the anterior commissure, retrogradely labeled neurons are located directly alongside the rhinal fissure. Caudal to the genu of the anterior commissure (corresponding to posterior insular cortex and in the caudal parts of the perirhinal cortex), labeled neurons are primarily located in a similar juxtarhinal position but additional neurons begin to extend more dorsal to the rhinal fissure; the number of labeled neurons per section is maximal at these levels (Fig. 3B and C). In the caudalmost perirhinal cortex, labeled neurons are fewer in number and are primarily located alongside the dorsal half of the rhinal fissure and just dorsal to the rhinal fissure (Fig. 3D). Neurons in the piriform and entorhinal cortices were never retrogradely labeled from PAG.

Based on the "multiple cases" in which injections were made into each of these lateral cortical fields in the same brain, the "aggregate lateral" projection to PAG, like the aggregate projection from medial frontal cortex, appeared to terminate densely and uniformly in all parts of PAG throughout its rostrocaudal axis (Fig. 6). As noted earlier, however, the aggregate lateral projections tended to be slightly heavier to the caudal half of PAG in contrast the medial frontal aggregate projections which tended to be somewhat heavier in the rostral half of PAG.

To explore the terminal organization of projections from specific individual lateral cortical fields to PAG, single injections of WGA-HRP were made into each of the lateral cortical areas in different animals. Because labeled neurons at any given level of lateral cortex are not as densely packed as labeled neurons in the medial cortical fields, pressure injections were used to obtain slightly more spread of the tracer than obtained by iontophoresis. Once again, numerous cases were generated in order to obtain a population of cases in which the injections were successfully restricted to a single cortical field as judged by cytoarchitecture and patterns of retrograde and anterograde labeling with other cortical and subcortical structures.

The organizational principles that characterized the projections from medial frontal cortex to PAG also typified the projections from lateral cortical areas to PAG. Thus, the projection from each lateral cortical field terminated with a characteristic radial and axial specificity throughout the rostrocaudal axis of PAG. The projections from anterior insular cortex and the rostral component of posterior insular cortex (Fig. 9) were much lighter to the rostral periventricular parts of PAG than those from the caudal component of posterior insular and perirhinal cortex or from any of the medial cortical fields. The projections from all the lateral cortical fields was denser to the caudal third of PAG than those from any of the medial frontal cortical areas except the prelimbic cortex whose projections to the caudal parts of PAG was comparable to those of any of the lateral cortical areas with the possible exception of the rostral component of the posterior insular cortex which, of all cortical fields examined, had the heaviest

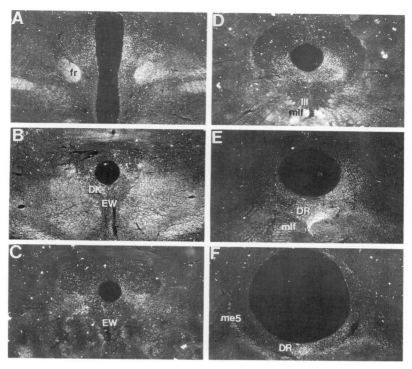

Figure 9. Projections to PAG from the rostral component of posterior insular cortex. Photomicrographs of 6 rostrocaudal levels of PAG (A, rostral through F, caudal) from an experiment in which a single, restricted WGA-HRP injection was made into the rostral component of posterior insular cortex. The projection from this cortical field is heaviest to the lateral and ventrolateral parts of PAG and, as is characteristic of most of the lateral cortical fields projecting to PAG, the projection is heaviest to the caudal half of PAG. DK, nucleus Darkschewitsch; DR, dorsal raphe; EW, Edinger-Westphal nucleus; fr, fasciculus retroflexus; III, oculomotor nucleus; mlf, medial longitudinal fasciculus; me5, mesencephalic trigeminal tract.

terminal projections to the caudal third of PAG. Thus, the tendency, as seen in the multiple injection cases, for the "aggregate" lateral cortical projection to be somewhat heavier to the caudal parts of PAG than the aggregate projection from medial frontal cortex, was also reflected by the projections from the individual lateral fields. Once again, however, this tendency was not absolute as the more caudal lateral cortical areas (caudal components of posterior insular cortex and perirhinal cortex) projected to the rostral periventricular parts of PAG with nearly the same density as medial frontal inputs. The caudal lateral cortical fields (caudal component of posterior insular cortex and perirhinal cortex), like the medial cortical projections, exhibited bipartite terminal fields at mid-caudal levels of PAG but the more rostral lateral cortex areas had only one terminal field in the lateral or ventrolateral part of PAG throughout most of its rostrocaudal extent.

The organization of the terminal projections from the caudal lateral cortical fields, thus, was very similar to those from the medial cortical fields while the more rostral lateral cortical areas tended to have somewhat "simpler" terminal fields. Notwithstanding, the somewhat simpler terminal organization of the rostral lateral cortical areas, all the lateral and medial cortical areas shared the common principle of projecting selectively to discrete, longitudinally continuous zones running along the rostrocaudal axis of PAG. This, then, is the principle feature that characterizes the projections from all medial and lateral cortical areas to PAG: each cortical area terminates as one or two discrete, continuous, longitudinally organized input columns extending throughout much or all of the rostrocaudal extent of PAG. Taken together, these cortical projections to PAG constitute a set of partially overlapping but largely complementary, rostrocaudally oriented columns of cortical inputs spanning the longitudinal axis of PAG. Although additional analysis is needed to determine the degree to which inputs arising from different medial and lateral cortical areas overlap along the rostrocaudal axis of PAG, the results of the present observations support the hypothesis that these highly focal, longitudinally organized cortical input columns comprise a potential anatomical substrate for selectively influencing discrete columns of neurons along the rostrocaudal axis of PAG. That such longitudinal columns of PAG neurons might exist and represent discrete output channels for at least some of PAG's functional manifestations will be discussed shortly.

Forebrain subcortical projections to PAG

The existence of robust and highly organized cortical inputs to PAG represents a new and perhaps somewhat surprising aspect of PAG circuitry as it suggests the possibility that there are discretely organized channels by which integrated cortical activity may directly influence the neural operations of PAG, a phylogenetically primitive structure which has historically been associated with relatively stereotyped functions presumed to be somewhat remote from cortical or cognitive influence. However, notwithstanding the novelty of these surprisingly extensive and organized cortical inputs, it would be a mistake to ignore the massive contributions of other subcortical forebrain projections to PAG because, based on our material and previous reports of distribution of retrograde labeling in the forebrain, many of these subcortical projections may terminate more densely in PAG than do cortical inputs. Analysis of the organization of these inputs along lines similar to those just described for the cortical projections, will require much time and effort and the task will be more difficult than for the cortical projections because of technical reasons. Our study of cortical inputs was somewhat complicated by the fact that the definitions of many of the cortical areas is difficult using cytoarchitectonic criteria alone. For example, we have been unable to distinguish the rostral and caudal components of the posterior insular cortex using Nissl staining alone, yet the cortical and subcortical projection patterns of these two adjacent cortical fields differ significantly and thus provides a basis for distinguishing them and their terminal fields in PAG. In the case of the medial frontal cortex the architectonic boundaries between the component cortical fields are not sharp at all rostrocaudal cortical levels containing these fields but the use of histochemical stains for the enzyme acetylcholinesterase combined with analysis of Nissl architecture (Shipley et al., 1989) facilitated the determination of medial frontal cortical boundaries. On the positive side, however, one great advantage in the cortical studies was that the anterograde and retrograde labeling patterns resulting from cortical injection sites could be fairly safely interpreted because there is very little risk that labeling patterns were confounded by fibers passing through these cortical fields. As we shift our attention to the study of forebrain subcortical projections to PAG we are confronted not only by an increase in the complexity of defining nuclei and subnuclei in basal telencephalic and hypothalamic areas and by the relatively small size and complex geometries of subcortical nuclei, but also by the increased potential that tracer injections may label fibers-of-passage running through the nuclei of interest en route to PAG. Thus, it will be necessary to use methods less prone (but not immune) to uptake by fibers of passage. However, most of these methods pose their own problems of interpretation. For example isotopically labeled amino acid tracing results depend critically on post-injection survival and autoradiographic exposure times. Thus, the actual size of the injection site and the density of the ensuing anterograde labeling patterns are difficult to assess. There are also problems with newer methods like PHA-L as it is not yet clear why some neurons at the site of injection take up and transport this lectin and others do not (Gerfen and Sawchenko, 1985); in addition, the relatively sparse but all-or-none anterograde fiber labeling with PHA-L makes it difficult to assess the significance of different patterns of terminal labeling or to distinguish axonal from terminal arbors. Thus, multiple tracing methods and much interpretive care will be required to analyze the organization of forebrain subcortical inputs to PAG. We have initiated such studies and here report the results for two prominent telencephalic inputs to PAG, the central nucleus of the amygdala and the medial preoptic area.

CENTRAL NUCLEUS OF THE AMYGDALA. The central nucleus of the amygdala (CNA) is a nodal point for intraamygdalar circuits (Price et al, 1987) and receives extensive afferents from several structures including the lateral cortex, specifically the anterior insular cortex and the rostral component of posterior insular cortex (Price et al, 1987; Shipley and Ennis, unpublished observations). CNA has been implicated in several functions including aggressive and defensive behaviors and antinociception (Frenk et al., 1978; Kalivas et al., 1982), and has recently been shown to be an important component in the circuits that mediate the autonomic reactions accompanying conditioned fear responses (Hitchcock et al., 1986; Kaada et al., 1972; Kapp et al., 1982). Thus, many of the functions of CNA are similar to the responses elicited from PAG.

It is known that CNA projects to PAG (Beitz, 1982; Hopkins and Holstege, 1978; Hopkins et al., 1981; Post and Mai, 1980), but the organization of CNA's

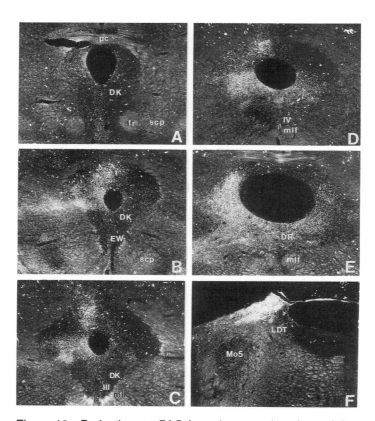

Figure 10. Projections to PAG from the central nucleus of the
amygdala (CNA). Darkfield micrographs of anterograde labeling along the rostrocaudal axis (A, rostral through F, caudal) of
PAG following a WGA-HRP injection centered in CNA. At the
rostral most level of PAG (A), a narrow band of labeled fibers
are aggregated alongside the lateral wall of the aqueduct. At
more caudal levels (B-D), inputs from CNA focally and densely
target dorsomedial and lateral/ventrolateral PAG; these two
zones are sharply separated by dorsolateral PAG which
contains only weak labeling. Caudally (E), at the level of the
dorsal raphe (DR), there is heavy labeling throughout all of
PAG except for the dorsomedial region. DK, nucleus
Darkschewitsch; EW, Edinger-Westphal nucleus; fr, fasciculus
retroflexus; lateral dorsal tegmental nucleus (LDT); III, oculomotor nucleus; IV, trochlear nucleus; mlf, medial longitudinal
fasciculus; pc, posterior commissure; Mo5, motor trigeminal
nucleus; scp, superior cerebellar peduncle. Reprinted with permission from Rizvi et al. (1991) Journal of Comparative
Neurology.

connections with PAG have not been characterized in detail. We recently completed a systematic analysis of the CNA to PAG pathway (Rizvi et al., 1991), and additionally, disclosed that PAG sends a substantial reciprocal projection back to CNA. After multiple injections of WGA-HRP throughout PAG the vast majority of retrogradely labeled amygdalar cells were in the medial subdivision of CNA. These same cases contained anterograde labeling in CNA. Discrete iontophoretic injections of WGA-HRP into the medial subdivision of CNA produced robust anterograde labeling throughout the rostrocaudal extent of PAG. As for the cortico-PAG projections, the CNA projection exhibited a marked degree of topographical specificity in PAG. As shown in Figure 10, the anterograde labeling was significantly heavier in the caudal than the rostral half of PAG. At rostral levels the projection was moderate and exhibited a bipartite terminal field with the inputs aggregated in the dorsomedial and lateral parts of PAG. Beginning at the mid-rostrocaudal level of the oculomotor complex, the projection increased markedly in density; this heavier labeling density persisted through all subsequent caudal levels. The bipartite terminal field organization was most pronounced at mid-levels of the oculomotor complex. Beginning at the level of the dorsal raphe, the dorsomedial terminal field dropped out and the projection densely terminated throughout most of the lateral and ventrolateral parts of PAG. Like the cortical inputs, the projection from CNA was heaviest in the half of PAG ipsilateral to the injection site but there was also moderate contralateral labeling at most rostrocaudal levels of PAG.

WGA-HRP injections in CNA also produced retrograde labeling of neurons in PAG, particularly in its rostral half. Many of these neurons were in the lateral and dorsal parts of PAG and thus were not dorsal raphe neurons which had been previously reported to project to CNA and were also labeled in our material. Thus, there is a direct, projection from PAG to CNA arising from non-raphe neurons; these non-raphe neurons give rise to moderate to a heavy terminal field in CNA.

Two points should be made about the potential significance of this PAG to CNA projection. (1) Electrical stimulation has been used by some investigators to study the functions of CNA and role of CNA in modulating PAG functions. Caution should be exercised in the interpretation of such experiments as electrical stimulation of CNA could cause antidromic activation of PAG to CNA projection neurons. These antidromic effects may be difficult to dissociate from the effects produced by orthodromic activation of the CNA to PAG projection. The use of excitatory amino acid stimulation of CNA may be required to selectively evoke orthodromic effects. (2) The existence of projections from PAG to CNA suggests the possibility that activity in PAG directly influences CNA neurons including, possibly, the same neurons that project to back to PAG and/or those that project to other subcortical sites. Thus, the PAG to CNA projection provides a potential anatomical substrate for PAG to influence CNA neurons that project to other important autonomic regulatory sites. As PAG is known to trigger defensive responses and modulate pain thresholds and as CNA is also implicated in these same functions, the PAG to CNA projections may provide a direct feedforward linkage whereby PAG could initiate CNA activation during such functional states.

MEDIAL PREOPTIC AREA. The medial preoptic area (MPO) is a complex region consisting of several distinct nuclei lateral to the walls of the third ventricle at the base of the telencephalon (Simerly and Swanson, 1986). MPO has been strongly implicated in neuroendocrine (gonadal steroid) regulation, thermal regulation and sexual behavior (McGinty and Szymusiak, 1990). Recent stu-

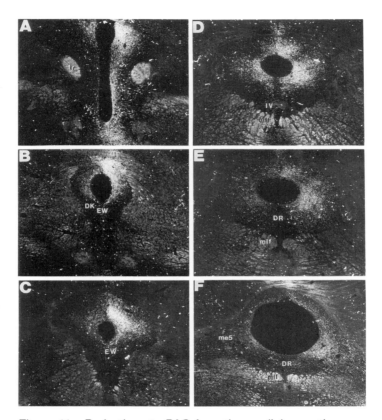

Figure 11. Projections to PAG from the medial preoptic area (MPO): WGA-HRP method. A-F: Darkfield photomicrographs (A, rostral to F, caudal) of PAG following a WGA-HRP injection centered in MPO. Projections from MPO to PAG are heaviest along the walls of the aqueduct and are less dense in the peripheral parts of PAG. At rostral levels (A-B) inputs from MPA selectively and heavily terminate in the dorsal and dorsolateral PAG. At the middle levels of PAG (C-E), inputs from MPA focally target two discrete longitudinal bands in dorsomedial and lateral PAG. Caudally (F), dense labeling is observed in ventrolateral PAG. DK, nucleus Darkschewitsch; DR, dorsal raphe; EW, Edinger-Westphal nucleus; fr-fasciculus retroflexus; IV, trochlear nucleus; me5, mesencephalic trigeminal tract; mlf-medial longitudinal fasciculus.

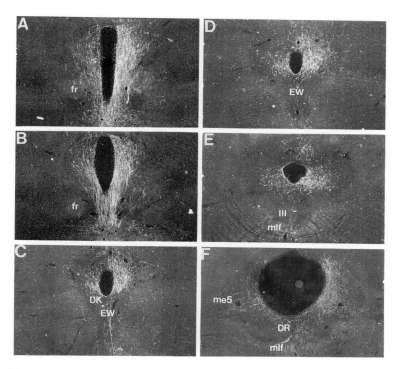

Figure 12. Projections to PAG from the medial preoptic area (MPO): PHA-L method. A-F: Darkfield photomicrographs (A, rostral to F, caudal) of PAG following PHA-L injections centered in MPO. At rostral levels of PAG (A-B), fibers and terminals from MPO are particularly dense in dorsal PAG; note longitudinally-oriented fibers coursing from ventral to dorsal PAG. At middle levels of PAG (C-F), inputs from MPO, in the form of both fibers and terminals, discretely target dorsomedial and lateral/ventrolateral PAG. While the overall distribution of anterograde labeling in PAG with WGA-HRP (Fig. 11) and PHA-L is very similar, more fiber labeling and less terminal-like labeling is produced with PHA-L. For example, in rostral PAG, PHA-L injections label many fibers that course dorsoventrally through PAG (A,B), while at this same level (Fig. 11A) WGA-HRP labeling is relatively sparse and predominantly punctuate. At higher magnification (not shown), the PHA-L labeled fibers in rostral PAG are not beaded and appear to contain few boutons and varicosities. This contrasting result with PHA-L and WGA-HRP indicates that care must be taken in equating PHA-L labeled fibers with innervation. DK, nucleus Darkschewitsch; DR, dorsal raphe; EW, Edinger-Westphal nucleus; fr, fasciculus retroflexus; III, oculomotor nucleus; mlf, medial longitudinal fasciculus; me5, mesencephalic trigeminal tract.

dies, for example, have shown that MPO is one of the most strikingly sexual dimorphic structures in the brain (Gorski et al., 1980; Simerly et al., 1984). There are also reports suggesting that MPO plays a role in cardiovascular regulation (Wang and Ranson, 1941) and antinociception (Carstens et al., 1982). Thus, like CNA, MPO is implicated in functions also associated with PAG.

Following multiple WGA-HRP injections in PAG, retrogradely labeled neurons are present in all of the nuclear groups comprising MPO expect for the medial subdivision of the medial preoptic nucleus. The majority of these neurons exhibited intense retrograde labeling suggesting that they give rise to a rich terminal projection to PAG. This was confirmed in a series of experiments in which discrete iontophoretic injections of WGA-HRP (Fig. 11) or the lectin PHA-L (Fig. 12) were deposited in MPO. The projections from MPO to PAG exhibited a striking degree of topographical specificity throughout the entire rostrocaudal extent of PAG. The projection was somewhat denser in the caudal than the rostral half of PAG but this is a fairly moderate gradient. As illustrated in Figure 11, at the most rostral levels of PAG beginning at the periventricular component of PAG, the projection, as seen in WGA-HRP material, is focally restricted to the dorsomedial part of PAG. When the projection at these same rostral levels is visualized by PHA-L labeling (Fig. 12), it appears to be more diffuse with a pronounced labeling of fibers running parallel to the walls of the ventricle. However, upon closer inspection it is seen that most of this PHA-L labeling is present in smooth fibers; only when these fibers reach the dorsomedial parts of PAG are varicosities and bouton-like terminals seen. This suggests that WGA-HRP labeling more accurately reflects the terminal field in this projection and further suggests that care should be taken in the analysis and schematic charting of PHA-L labeled material because this tracer's remarkable ability to label axons as well as terminals can lead to a misleading representation of terminal field localization and organization. Notwithstanding this proviso, the fact that the terminal labeling with PHA-L corresponded so remarkably to the labeling patterns seen with WGA-HRP suggests that the labeling patterns were not contaminated to an appreciable extent by uptake and transport by fibers passing through MPO.

The MPO terminal field remains confined to the dorsomedial part of PAG throughout most of the rostral third of PAG but then begins to expand dorsolaterally at the rostralmost level of the oculomotor complex. At mid-occulomotor levels, the projection becomes distinctly bipartite with one dense terminal field in the dorsomedial and a second in the lateral PAG. This bipartite distribution continues caudally to about the level of the cuneiform nucleus where the projection becomes very dense; the heaviest labeling at these levels is in the ventrolateral PAG. A constant feature of the projection at all rostrocaudal levels is that it is densest in the central parts of PAG. The labeling thus shows a central to peripheral (radial) gradient being heaviest nearer the ventricle or aqueduct and falling off to nearly no labeling in the peripheral half of PAG. This preferentially central labeling is a marked contrast to the inputs from CNA but was also seen in the projection from some of the medial and lateral cortical fields.

As was the case for CNA, the MPO to PAG projection is reciprocal.

Following WGA-HRP injections into MPO, retrogradely labeled neurons were present at all rostrocaudal levels of PAG. More neurons were labeled in the rostral than the caudal half of PAG but this difference was graded. The neurons showed little topographic specificity at any rostrocaudal level of PAG; the majority of the PAG neurons projecting to MPO were non-raphe cells.

Discussion

The major result of the studies reported here is that cortical and subcortical inputs to PAG are extensive and surprisingly highly organized. Inputs from the cortex arise from no fewer than eight distinct cortical areas in medial frontal cortex (infralimbic, prelimbic, anterior cingulate and precentral medial cortices) and lateral cortex (anterior insular, posterior insular - rostral and caudal components and perirhinal cortices). It should also be noted that sparser inputs arise from all parts of orbital (lateral and ventral) cortex. These orbital cortical neurons labeled by PAG injections are much fewer in number than those in the medial and lateral areas, but they appear to form a kind of bridge between the medial and lateral cortical areas. Thus, PAG is innervated by neurons extending from the precentral medial cortex on the dorsal convexity of the frontal pole, along the entire medial wall of the frontal cortex, extending (sparsely) through the orbital areas and then extending along the entire lateral margin of the cortical mantle.

Projections from each of the cortical fields terminates with a complex but characteristic focal pattern in PAG. Different cortical fields terminate with partially overlapping but largely complementary patterns to other cortical fields. Similarly, the two subcortical forebrain projections that we have characterized thus far from CNA and MPO, also terminate with characteristic focal patterns organized as longitudinal columns throughout much or all of the rostrocaudal axis of PAG.

Taken together, these observations suggest that forebrain inputs to PAG may preferentially or selectively influence discrete, longitudinally organized columns of PAG tissue spanning the entire rostrocaudal extent of this structure. This implies that the midbrain PAG may have a previously unsuspected degree of anatomical and/or functional specificity that is organized as an as yet to be determined number of longitudinally organized cylindrical "modules". These hypothetical longitudinal modules may be selectively addressed by afferent input columns arising from multiple forebrain areas; forebrain inputs may differentially trigger and/or modulate the activity of discrete subsets of PAG neurons to generate functionally integrated patterns of responses ranging from defense reactions, antinociception, lordosis or vocalization, depending upon the weighted patterns of activation or modulation in different forebrain input columns.

To what features of intrinsic PAG organization might such longitudinally organized modules correspond? It is difficult, given our present state of knowledge of the intrinsic anatomy of PAG, to give more than suggestive answers to this question. Our understanding of PAG organization has gained little from the study of its architecture using classical Nissl and Golgi methods. Until the last

decade there was little agreement on whether PAG could be consistently subdivided on the basis of Nissl architectonics. Hamilton's (1973) studies in the cat
and Beitz's (1985) analysis in the rat yielded an architectonic parcellation scheme
that was fairly similar for these two species yet anyone who has looked at Nissl
stained PAG sections realizes that these parcellations are often tenuous.

The development of immunocytochemical methods led to the discovery of
several classes of transmitter/peptide containing neurons and terminal fields in
PAG (reviewed in Shipley et al., 1987). The distributions of these classes of neurons appears to be characteristic but they often bear little systematic relation to
the Nissl defined subdivisions. More recently, several workers have identified
the locations of PAG neurons giving rise to descending projections to the medulla (Beitz et al., 1983; Carlton et al., 1983; Carrive, this volume; Carrive et al., 1988;
Fardin et al., 1984; Van Bockstaele et al., 1991). The distributions of these descending projection neurons are also characteristic but, to date, any systematic relationship between the distributions of medullary projection neurons, transmit-
ter/peptides specific neurons and Nissl defined architectonic subdivisions of
PAG is not compelling. Adding to the complexity of the situation is the fact that
the few Golgi studies of PAG indicate that while many of the neurons in PAG
have relatively restricted dendritic arbors (Beitz and Shepard, 1985; Gioia et al.,
1985; Mantyh, 1982b), many others have extensive dendritic fields in the coronal
plane (Beitz and Shepard, 1985). Thus, at the risk of overgeneralizing, it may be
fair to say that our current picture of the intrinsic organization of PAG does not
provide a compelling framework for postulating any obvious dimensions of
intrinsic organizational specificity that corresponds to the surprisingly discrete
longitudinal specificity exhibited by the terminal fields of the forebrain inputs
reported here.

On the other hand, there is new evidence from recent functional studies
that at least some behavioral and autonomic responses show a remarkable
degree of site specificity within PAG. These studies further suggest that sites of
similar functional specificity show a marked degree of longitudinal organization
along the rostrocaudal axis of PAG. Using focal applications of excitatory amino
acids, work from Bandler's laboratory (Bandler et al., 1991; Carrive, this volume;
Carrive et al., 1987; 1988; 1989a,b) has shown that sites yielding pressor or
depressor responses are discretely localized in parallel but distinct, longitudinal
zones stretching several millimeters along the rostrocaudal axis of PAG in the
cat. More recently these investigators have provided anatomical evidence suggesting that these functionally mapped, longitudinally organized pressor-
depressor zones give rise to remarkably selective projections to medullary cardiovascular regulatory sites (Carrive, this volume; Carrive and Bandler, 1991;
Carrive et al., 1988; 1989b). They have shown further, in the cat and in the rat,
that these same longitudinally organized functional units trigger distinct, site
specific behavioral responses related to defense reactions (Bandler and Depaulis,
this volume; Bandler et al., 1991). These new lines of evidence are consistent with
the idea that there is a degree of functional specificity in PAG and that functio-

nally specific units may be discretely organized as longitudinal columns or modules. At the present time it is difficult to more than speculate what might be the relationship between these longitudinally organized functional modules mapped in the cat and in the rat, to the longitudinally organized forebrain input columns demonstrated in our anatomical studies in the rat. It is reasonable, however, to at least suggest that inputs from some forebrain areas might preferentially or selectively target longitudinally organized functional modules. The experimental methods to begin to assess this possibility are available. For example, functionally mapped sites in PAG could be marked with a persisting dye or latex microspheres and compared to the distribution of terminals from cortical or subcortical sites projecting to PAG. For those forebrain input columns that co-distribute with physiologically defined longitudinal functional columns, subsequent experiments could employ focal stimulation in the forebrain site of origin to determine if responses similar to those elicited from PAG are triggered or modulated; reversible lesions could be employed to determine if PAG neurons are necessary to the expression of forebrain initiated responses. Experiments of this kind are currently in progress in this laboratory.

The use of combined functional and anatomical methods, thus, may provide important new insights about the degree to which the longitudinal organization suggested by our forebrain anterograde tracing studies are functionally related to the longitudinal organization suggested by the physiological micromapping studies of Bandler, Carrive and Depaulis. Beyond this, however, there is an obvious need to reevaluate our traditional views of the anatomical organization of PAG and at the same time, the conceptual frameworks we have used to try to piece together the seemingly diverse functions associated with this midbrain structure.

On the anatomical side, there is a need to seek specificity in the anatomical organization of PAG and its connectional affiliations. So long as this enigmatic midbrain structure is viewed as a relatively primitive, homogeneous reticular element, then dimensions of specificity in its intrinsic and extrinsic organization may escape our attention. It must also be remembered that PAG has an appreciable rostrocaudal extent. Observations of the distributions of terminal inputs, transmitter-specific neurons, receptor binding sites or output neurons charted at only a few rostrocaudal levels of PAG may give little indication of schemes of anatomical specificity that are organized along the entire longitudinal dimension of this structure. Thus, future anatomical mapping studies might strive to use regularly spaced levels along the rostrocaudal axis of PAG without regard to preconceived functional or anatomical biases that "only the middle parts are the true PAG". We should also search for intrinsic PAG connections that span the longitudinal axis of PAG as these may provide critical linkages among longitudinally distributed anatomical and functional modules. For example, preliminary studies in our laboratory suggests that there may be a remarkable degree of polarity of intrinsic PAG connections with more rostral levels preferentially cascading onto more caudal levels. Similarly, we should search for patterns of local axial connections that might link longitudinal modules. Combined anatomical and physiological approaches should be embraced (e.g., simultaneous mapping of afferent inputs in relation to pressor-depressor zones, retrogradely labeled output neurons or immunocytochemically identified populations of intrinsic neurons) as these types of studies may yield organizational principles difficult to infer from comparisons of distributions mapped in different brains. And, as noted earlier, we should be prepared to search for anatomical maps that address functionally defined units as the intrinsic manifestations of anatomical specificity in PAG may be subtle or complex.

Finally, our conceptual frameworks for PAG may need to become more integrative. For too long, research on PAG has been fractionated among workers who view PAG through conceptual filters set to detect antinociception, the defense reaction, fear, lordosis, vocalization or autonomic functions. PAG is doubtless central to all of these important functions, but how and why? Our own conceptual orientation is also based on a filter but it is a somewhat different filter shaped by our still growing appreciation of the massive projections of the forebrain to PAG. Our research is guided by the working hypothesis that PAG is a structure that plays a central role in the generation of certain stereotypical behaviors (defense reactions, lordosis, vocalization) critical to the animal's survival. These behaviors require rapid, profound autonomic adjustments and, at the same time, significant alterations of pain thresholds. From this perspective the analgesic and autonomic functions of PAG are integral components of stereotypical, organismic responses to potentially threatening stimuli. From this vantage point, even lordosis can be seen as one alternative response to threatening stimuli. In this regard it is noteworthy that Bandler and collaborators (Bandler and Carrive, 1988; Bandler and Depaulis, this volume; Bandler et al., 1985a; 1991; Carrive, this volume; Carrive and Bandler, 1991; Carrive et al., 1987; 1988; 1989a,b; Depaulis et al., 1989; Zhang et al., 1990) recently showed that DLH-induced pressor responses evoked from PAG are invariably associated with "flight-fight" behavioral reactions, while depressor responses are accompanied by cessation of locomotion and general movements. Thus, there appears to be a coordinate expression of both skeletal and autonomic adjustments appropriate to the animals response strategy.

The PAG appears to be a cross-roads for a multitude of circuits that integrate the animals' response to potentially threatening stimuli. Central to these responses are rapid, profound, coordinated autonomic adjustments. These adjustments are critical to the successful expression of behavioral responses necessary to the animals' survival. From this perspective it is reasonable to consider that more recently elaborated forebrain structures articulate with PAG to coordinate and, possibly, trigger or even inhibit behavioral and autonomic responses in concert with the increased role of the forebrain in sensory processing and cognitive and emotional responses. The organization of the anatomical circuits that mediate forebrain influences on PAG function, thus, may provide clues to the mechanism by which this important midbrain structure expresses its vital functions.

Acknowledgements

This work was supported by PHS Grants HL08097, NS120643, NS24698. We thank Philip Pfalzgraf for technical and photographic assistance. We thank Clifford Saper for helpful discussions and sharing prepublication data on cortical projections.

References

Bandler, R. and Carrive, P., Integrated defence reaction elicited by excitatory amino acid microinjection in the midbrain periaqueductal grey region of the unrestrained cat, Brain Res., 439 (1988) 95-106.

Bandler, R., Carrive, P. and Zhang, S.P., Integration of somatic and autonomic reactions within the midbrain periaqueductal grey: Viscerotopic, somatotopic and functional organization, Prog. Brain. Res., 87 (1991) 269-305.

Bandler, R., Depaulis, A. and Vergnes, M., Identification of midbrain neurons mediating defensive behavior in the rat by microinjections of excitatory amino acids, Behav. Brain Res., 15 (1985a) 107-119.

Bandler, R., McCulloch, T. and Dreher, B., Afferents to a midbrain periaqueductal grey region involved in the "defense reaction" in the cat as revealed by horseradish peroxidase. I. The telencephalon, Brain Res., 330 (1985b) 109-119.

Basbaum, A.I. and Fields, H.L., Endogenous pain control systems: brainstem spinal pathways and endorphin circuitry, Annu. Rev. Neurosci., 7 (1984) 309-338.

Behbehani, M.M., Effect of chronic morphine treatment on the interaction between the periaqueductal grey and the nucleus raphe magnus of the rat, Neuropharmacol., 20 (1981) 581-586.

Behbehani, M.M. and Fields, H.L., Evidence that an excitatory connection between the periaqueductal gray and nucleus raphe magnus mediates stimulation produced analgesia, Brain Res., 170 (1979) 85-93.

Behbehani, M.M. and Pert, A., A mechanism for the analgesic effect of neurotensin as revealed by behavioral and electrophysiological techniques, Brain Res., 324 (1984) 35-42.

Behbehani, M.M., Shipley, M.T. and McLean, J.H., Effect of neurotensin on neurons in the periaqueductal gray: An in vitro study, J. Neurosci., 7 (1987) 2035-2040.

Beitz, A.J, The midbrain periaqueductal gray in the rat. I. Nuclear volume, cell number, density, orientation, and regional subdivisions, J. Comp. Neurol., 237 (1985) 444-469.

Beitz, A.J, The organization of afferent projection to the midbrain periaqueductal gray of the rat, Neurosci., 7 (1982) 133-159.

Beitz, A.J., Mullett, M.A. and Weiner, L.L., The periaqueductal gray projections to the rat spinal trigeminal, raphe magnus, gigantocellular pars alpha and paragigantocellular nuclei arise from separate neurons, Brain Res., 288 (1983) 307-314.

Beitz, A.J. and Shepard, R.D., The midbrain periaqueductal gray in the rat. II. A golgi analysis, J. Comp. Neurol., 237 (1985) 460-475.

Burns, S.M. and Wyss, M.J, The involvement of the anterior cingulate cortex in blood pressure control, Brain Res., 340 (1985) 71-77.

Carlton, S.M., Leichnetz, G.R., Young, E.G. and Mayer, D.J., Supramedullary afferents of the nucleus raphe magnus in the rat: a study using the transcannula HRP gel and autoradiographic techniques, J. Comp. Neurol., 214 (1983) 43-58.

Carrive, P. and Bandler, R., Viscerotopic organization of neurons subserving hypotensive reactions within the midbrain periaqueductal grey: A correlative functional and anatomical study, Brain Res., 541 (1991) 206-215.

Carrive, P., Bandler, R. and Dampney, R.A, Anatomical evidence that hypertension associated with the defence reaction in the cat is mediated by a direct projection from a restricted portion of the midbrain periaqueductal grey to the subretrofacial nucleus of the medulla, Brain Res., 460 (1988) 339-345.

Carrive, P., Bandler, R. and Dampney, R.A., Somatic and autonomic integration in the midbrain of the unanesthetized decerebrate cat: a distinctive pattern evoked by excitation of neurones in the subtentorial portion of the midbrain periaqueductal grey, Brain Res., 483 (1989a) 251-258.

Carrive, P., Bandler, R. and Dampney, R.A, Viscerotopic control of regional vascular beds by discrete groups of neurons within the midbrain periaqueductal gray, Brain Res., 493 (1989b) 385-390.

Carrive, P., Dampney, R.A.L. and Bandler, R., Excitation of neurones in a restricted portion of the midbrain periaqueductal grey elicits both behavioral and cardiovascular components of the defence reaction in the unanaesthetised decerebrate cat, Neurosci. Lett., 81 (1987) 273-278.

Carstens, C., MacKinnon, J.D. and Guinan, M.J., Inhibition of spinal dorsal horn neuronal responses to noxious skin heating by medial preoptic and septal stimulation in the cat, J. Neurophysiol., 46 (1982) 981-991.

Cechetto, D.F. and Chen, S.J, Subcortical sites mediating sympathetic responses from insular cortex in rats, Am. J. Physiol., 258 (1990) R245-R255.

Clemens, J.A., Smalstig, E.B. and Sawyer, C.H., Studies on the role of the preoptic area in the control of reproductive function in the rat, Endocrinol., 99 (1976) 728-735.

Depaulis, A., Bandler, R. and Vergnes, M., Characterization of pretentorial periaqueductal gray matter neurons mediating intraspecific defensive behaviors in the rat by microinjections of kainic acid, Brain Res., 486 (1989) 121-132.

Depaulis, A. and Vergnes, M., Elicitation of intraspecific defensive behaviors in the rat by microinjection of picrotoxin, a gamma-aminobutyric acid antagonist, into the midbrain periaqueductal gray matter, Brain Res., 367 (1986) 87-95.

Ennis, M. and Shipley, M.T, Cortical inputs to the midbrain periaqueductal gray in the rat. I. Distribution and architectonics, in preparation.

Ennis, M. and Shipley, M.T., Organization of cortical inputs to the midbrain periaqueductal gray in the rat. II. Topography of cortical innervation, in preparation.

Ennis, M. Rizvi, T., Shipley, M. and Behbehani, M., Organization of some forebrain inputs to the midbrain periaqueductal gray, Soc. Neurosci. Abstr., 16 (1990) 635.

Fardin, V., Oliveras, J.L. and Besson, J.M., Projections from the periaqueductal gray matter to the B3 cellular area (nucleus raphe magnus and nucleus reticularis paragigantocellularis) as revealed by the retrograde transport of horseradish peroxidase in the rat, J. Comp. Neurol., 223 (1984) 483-500.

Frenk, H., McCarty, B.L. and Liebeskind, J.C., Different brain areas mediate the analgesic and epileptic properties of enkephalin, Science, 200 (1978) 335-337.

Gerfen, C.R. and Sawchenko, P.E., An anterograde neuroanatomical tracing method that shows the detailed morphology of neurons, their axons and terminals: Immunohistochemical localization of an axonally transported plant lectin, Phaseolus vulgaris-leucoagglutinin (PHA-L), Brain Res., 290 (1985) 219-238.

Gioia, M., Tredici, G. and Bianchi, R., A Golgi study of the periaqueductal gray matter in the cat. Neuronal types and their distribution, Exp. Brain Res., 58 (1985) 318-332.

Gorski, R.A., Harlan, R.E., Jacobson, C.D., Shryne, J.E. and Southam, A.M., Evidence for the existence of a sexually dimorphic nucleus in the preoptic area of the rat, J. Comp. Neurol., 193 (1980) 529-539.

Hamilton, B.L., Cytoarchitectural subdivisions of the periaqueductal gray matter in the cat, J. Comp. Neurol.,149 (1973) 1-28.

Hardy, S.G.P., Analgesia elicited by prefrontal stimulation, Brain Res., 339 (1985) 281-284.

Hardy, S.G.P., Projections to the midbrain from the medial versus the lateral prefrontal cortices of the rat, Neurosci. Lett., 63 (1986) 159-164.

Hardy, S.G.P. and Holmes, D.E., Prefrontal stimulus-produced hypotension in rat, Exp. Brain Res., 73 (1988) 249-255.

Hitchcock, J. and Davis, M., Lesions of the amygdala, but not of the cerebellum or red nucleus, block conditioned fear as measured with a potentiated startle paradigm, Behav. Neurosci., 100 (1986) 11-22.

Hopkins, D.A. and Holstege, G., Amygdaloid projections to the mesencephalon, pons and medulla oblongata in the cat, Exp. Brain Res., 32 (1978) 529-547.

Hopkins, D.A., McLean, J.H. and Takeuchi, Y., Amygdalotegmental projections: Light and electron microscopic studies utilizing anterograde degeneration and anterograde and retrograde transport of HRP, In: The amygdaloid complex, Ben Ari Y. (Ed.), INSERM Symposium No. 20., Elsevier- Biomedical press, North Holland, 1981, pp. 133-147.

Jürgens, U., Reinforcing concomitants of electrically elicited vocalizations, Exp. Brain Res., 26 (1976) 203-214.

Jürgens, U. and Richter, K., Glutamate-induced vocalization in the squirrel monkey, Brain Res., 373 (1986) 349-358.

Kaada, B., Stimulation and regional ablation of the amygdaloid complex with reference to functional representations, In: The Neurobiology of the Amygdala, Eleftheriou B.E. (Ed.), Plenum press, New York, 1972, pp. 145-204.

Kalivas, P.W., Gau, B.A., Nemeroff, C.B. and Prange, A.J.Jr., Antinociception after microinjection of neurotensin into the central nucleus of the amygdala, Brain Res., 243 (1982) 279-286.

Kapp, B.S., Gallaghen, M., Underwood, M.D., McNall, C.L., and Whitehorn, D., Cardiovascular responses elicited by electrical stimulation of the amygdalar central nucleus in rabbit, Brain Res., 234 (1982) 251-262.

Krettek, J.E. and Price, J.L., The cortical projections of the mediodorsal nucleus and adjacent thalamic nuclei in the rat, J. Comp. Neurol., 171 (1977) 157-192.

Lakos, S. and Basbaum, A.I., An ultrastructural study of the projections from the midbrain periaqueductal gray to spinally projecting, serotonin-immunoreactive neurons of the medullary nucleus raphe magnus in the rat, Brain Res., 443 (1988) 383-388.

Larson, C.R., The midbrain periaqueductal gray: A brainstem structure involved in vocalization, J. Speech. Hear. Res., 28 (1985) 241-249.

Larson, C.R. and Kistler, M.K., Periaqueductal gray neuronal activity associated with laryngeal EMG and vocalization in the awake monkey, Neurosci. Lett., 46 (1984) 261-266.

Larson, C.R. and Kistler, M.K., The relationship of periaqueductal gray neurons to vocalization and laryngeal EMG in the behaving monkey, Exp. Brain Res., 63 (1986) 596-606.

Mantyh, P.W., Forebrain projections to the periaqueductal gray in the monkey, with observations in the cat and rat, J. Comp. Neurol., 206 (1982a) 146-158.

Mantyh, P.W., The midbrain periaqueductal gray in the rat, cat, and monkey: a nissl, weil, and golgi analysis, J. Comp. Neurol., 204 (1982b) 349-363.

Marchand, J.E. and Hagino, N., Afferents to the periaqueductal gray in the rat. A horseradish peroxidase study, Neurosci., 9 (1983) 95-106.

Marlsbury, C.W., Facilitation of male copulatory behavior by electrical stimulation of the medial preoptic area, Physiol. Behav., 7 (1971) 797-805.

McGinty, D. and Szymusiak, R., Keeping cool: a hypothesis about the mechanisms and functions of slow-wave sleep, Trends Neurosci., 13 (1990) 480-486.

Meller, S.T. and Dennis, B.J., Afferent projection to the periaqueductal gray in the rabbit, Neurosci., 19 (1986) 927-964.

Nance, D.M., Christensen, L.W., Shryne, J.E. and Gorski, R.A., Modifications in gonadotropin control and reproductive behavior in the female rat by hypothalamic and preoptic lesions, Brain Res. Bull., 2 (1969) 307-312.

Neafsey, E.J., Hurley-Guis, K.M. and Arvanitis, D., The topographical organization of neurons in the rat medial frontal, insular and olfactory cortex projecting to the solitary nucleus, olfactory bulb, periaqueductal gray and superior colliculus, Brain Res., 377 (1986) 261-270.

Post, S., and Mai, K., Contribution to the amygdaloid projection fields in the rat: A quantitative autoradiographic study, J. Hirnforsch., 21 (1980) 199-225.

Price, J.L., Russchen, F.T. and Amaral, D.G., The limbic region. II. The amygdaloid complex, In: Handbook of Chemical Neuroanatomy. Vol.5: Integrated Systems in the CNS, Part I. Hypothalamus, Hippocampus, Amygdala, and Retina, Björklund A., Hökfelt T. and Swanson L.W. (Eds.), Elsevier, Amsterdam, 1987, pp. 279-388.

Reichling, D.B., Kwiat, G.C. and Basbaum, A.I., Anatomy, physiology and pharmacology of the periaqueductal gray contribution to antinociceptive controls, Prog. Brain Res., 77 (1988) 31-46.

Reynolds, D., Surgery in the cat during electrical analgesia induced by focal brain stimulation, Science, 164 (1969) 444-445.

Rizvi, T., Ennis, M., Behbehani, M. and Shipley, M., Reciprocal projections between the medial preoptic area and midbrain periaqueductal gray, Soc. Neurosci. Abstr., 16 (1990) 562.

Rizvi, T. A., Ennis, M., Shipley, M.T. and Behbehani, M., Connections between the central nucleus of the amygdala and the midbrain periaqueductal gray: topography and reciprocity, J. Comp. Neurol., 303 (1991) 121-131.

Ruggiero, D.A., Mraovitch, S., Granata, A.R., Anwar, M. and Reis, D.J., A role of insular cortex in cardiovascular function, J. Comp. Neurol., 257 (1987) 189-207.

Sakuma, Y. and Pfaff, D.W., Facilitation of female reproductive behavior from mesencephalic central grey in the rat, Am. J. Physiol., 236 (1979a) R278-R284.

Sakuma, Y. and Pfaff, D.W., Mesencephalic mechanisms for integration of female reproductive behavior in the rat, Am. J. Physiol., 237 (1979b) R285-R290.

Sesack, S.R., Deutch, A.Y., Roth, R.H. and Bunney, B.S., Topographical organization of the efferent projections of the medial prefrontal cortex in the rat: An anterograde tract-tracing study with Phaseolus vulgaris leucoagglutinin, J. Comp. Neurol., 290 (1989) 213-242.

Shipley, M.T., Ennis, M. and Behbehani, M., Acetylcholinesterase and Nissl staining in the same histological section, Brain Res., 504 (1989) 347-353.

Shipley, M.T., McLean, J.H. and Behbehani, M.M., Heterogeneous distribution of neurotensin-like immunoreactive fibers in the midbrain periaqueductal grey of the rat, J. Neurosci., 7 (1987) 2025-2034.

Simerly, R.B. and Swanson, L.W., The organization of neural inputs to the medial preoptic nucleus of the rat, J. Comp. Neurol., 246 (1986) 312-341.

Simerly, R.B., Swanson, L.W. and Gorski, R.A., Demonstration of a sexual dimorphism in the distribution of serotonin-immunoreactive fibers in the medial preoptic nucleus of the rat, J. Comp. Neurol., 225 (1984) 151-166.

Terreberry, R.R. and Neafsey, E.J., The rat medial frontal cortex projects directly to autonomic regions of the brainstem, Brain Res. Bull., 19 (1987) 639-649.

Van Bockstaele, E., Aston-Jones, G., Pieribone, V.A., Ennis, M. and Shipley, M.T., Periaqueductal gray and perioculomotor neurons innervate specific subdivisions of the rostral ventrolateral medulla in the rat: Anterograde studies using phaseolus vulgaris - leucoagglutinin, J. Comp. Neurol., 1991, in press.

Wang, S.Q. and Ranson, S.W., The role of the hypothalamus and preoptic region in the regulation of heart rate, Am. J. Physiol., 132 (1941) 5-8.

Wyss, J.M. and Sripanidkulchai, K., The topography of the mesencephalic and pontine projections from the cingulate cortex of the rat, Brain Res., 293 (1984) 1-15.

Yasui, Y., Breder, C.D., Saper, C.B. and Cechetto, D.F., Autonomic responses and efferent pathways from the insular cortex in the rat, J. Comp. Neurol., 303 (1991) 355-374.

Zhang, S.P., Bandler, R. and Carrive, P., Flight and immobility evoked by excitatory amino acid microinjection within distinct parts of the subtentorial midbrain periaqueductal gray of the cat, Brain Res., 520 (1990) 73-82.

Regional Subdivisions in the Midbrain Periaqueductal Gray of the Cat Revealed by *In Vitro* Receptor Autoradiography

Andrew L. Gundlach

The University of Melbourne
Clinical Pharmacology and Therapeutics Unit
Department of Medicine, Austin Hospital
Heidelberg, Victoria, Australia

Introduction

The periaqueductal gray (PAG) is a midline structure that encircles the mesencephalic aqueduct. While appearing homogeneous by some neuroanatomical criteria (Mantyh, 1982), recent studies in the rat of cell clustering and cytochrome oxidase activity have suggested the existence of four subdivisions in the PAG with different neuronal diameters and neuronal packing densities (Beitz, 1985; Conti et al., 1988). [Earlier studies described somewhat different subdivisions of the PAG in the cat (Hamilton, 1973)]. These regions also have differential connectivities with descending (hypothalamic, cortical and limbic) and ascending (medullary and spinal) afferents and reciprocal efferent connections, as well as populations of interneurons (e.g., Blomqvist and Craig; Veening et al.; Shipley et al.; Holstege, this volume). Consequently the PAG has a large number of established and putative transmitters present in its perikarya and afferent nerve terminals including the amino acids glutamate, γ-aminobutyric acid (GABA) and glycine (e.g., Belin et al., 1979; Clements et al., 1985), the amines acetylcholine, noradrenaline and 5-hydroxytryptamine (e.g., Clements et al., 1986) and various neuropeptides (e.g., Moss and Basbaum, 1983; Moss et al., 1983; Reichling et al., 1988; 1991).

Despite this diversity and the central involvement of the midbrain PAG in regulating various sensory and autonomic functions, relatively little is known about many aspects of its neurochemical organization. Even though considerable information is available about the localization of neurotransmitters, neuropeptides and related enzymes in many parts of the brain including the PAG, our knowledge of transmitter receptor distribution is less comprehensive. The localization of receptors in particular nuclei or loci of the central nervous system delineates potential sites of innervation by neurons containing specific neurotransmitters or sites of action of neuromodulatory peptides and hormones. Despite the large number of autoradiographic studies of various neurotransmitter and neuropeptide binding sites in rodent brain, there are few, if any, complete reports of the distribution of receptors throughout the extent of the midbrain PAG. This makes it difficult to predict or evaluate the likely location of the many transmitter interactions that must occur at the postsynaptic level in an area with such a major integrative function. Thus, in an attempt to further our understanding of the neurochemical organization of the midbrain PAG, *in vitro* autoradiographic localization of several [³H]radioligands was used to determine whether any significant regional variations occur in the distribution of a number of ubiquitous transmitter receptors within the PAG. Based on the criteria of probable presence of the receptor in the PAG and the availability in the laboratory of an appropriate high affinity and/or selective ligand, the studies reported here detail the distribution of muscarinic cholinergic, excitatory kainate, $GABA_A$/benzodiazepine, inhibitory glycine and adenosine A_1 receptor binding sites in the PAG of the cat.

Methodology

Cats of either sex (2.5-4.5 kg, n = 6) were used in these studies. Under deep halothane anesthesia, the animal was perfused transcardially with 2 l of 0.1M phosphate buffered saline, with the descending aorta clamped. The cortex was then dissected away and the midbrain removed and frozen in embedding mixture using plastic embedding moulds in a dry ice-ethanol bath. Tissue blocks were stored at -70 °C. Frozen tissue sections (10 µm; Berman (1968) co-ordinates A 6.4 to P 1.5) were cut at -15 °C on a cryostat and thaw-mounted onto chrome alum-gelatin subbed slides. Sections were left to dry at room temperature and then stored at -20 °C until processing for receptor binding studies. Serial coronal sections of the entire rostro-caudal extent of the PAG were divided into equal-sized groups to assist the comparative analysis of the localization of different receptors at all levels of the midbrain PAG.

The binding characteristics of [³H]quinuclindinyl benzilate (QNB), [³H]kainic acid (KA), [³H]flunitrazepam (FNZ), [³H]strychnine (STR), [³H]phenylisopropyladenosine (PIA), and [³H]nitrobenzylthioinosine (NBI) in tissue sections *in vitro* have been described in previous reports (see refs in Table I). In the present study comparable binding conditions have been used and ligands were generally used at a concentration near or equal to their dissociation constant (K_d), to maximize selectivity. Tissue sections were warmed to room temperature and preincubated in buffer to remove any endogenous ligand present in the tissue. Slides were then placed in incubation buffer containing the appropriate [³H]ligand (Table I). Non-specific binding was measured by incubating sections and [³H]ligand in the presence of a high concentration of a related unlabeled displacing drug, which displaced the [³H]ligand from its specific receptor sites. Following incubation, slides were washed in buffer to remove unbound ligand,

Table I: Conditions for receptor autoradiography

Receptor	Muscarinic	Kainate	Benzodiazepine	Glycine	Adenosine A_1	Adenosine Uptake
Ligand (conc)	[³H]QNB (1 nM)	[³H]KA (15 nM)	[³H]FNZ (1 nM)	[³H]STR (4 nM)	[³H]PIA (10 nM)	[³H]NBI (1 nM)
S. A. (Ci/mmol)	44	60	90	30	40	23
Displacer (conc)	Atropine (1 µM)	Glutamate (0.1 mM)	Diazepam (1 µM)	Glycine (10 mM)	CHA (10 µM)[e]	NBI (1µM)
Preincubation	None	20 min, RT Tris-citrate[b]	20 min, RT Tris-HCl[d]	20 min, RT PBS	30 min, RT +ADA[f]	20 min, RT Tris-HCl
Incubation	120 min, RT PBS[a]	120 min, 0°C Tris-acetate[c]	40min, 0°C Tris-HCl	30 min, 0°C PBS	90 min, RT Tris-HCl	30 min, RT Tris-HCl
Wash	2 x 5 min, 0°C PBS	3 min, 0°C Tris-acetate	2 min, 0°C Tris-HCl	4 min, 0°C PBS	5 min, 0°C Tris-HCl	2 x 3 min, 0°C Tris-HCl
Reference	Wamsley et al. (1981)	Unnerstall & Wamsley (1983)	Unnerstall et al. (1981)	Zarbin et al. (1981)	Hosli & Hosli (1988)	Bisserbe et al. (1985)

a Phosphate buffered saline, pH 7.4. b 50 mM Tris-citrate, pH 7.1. c 50 mM Tris-acetate, pH 7.1. d 50 mM Tris-HCl, pH 7.4. e Cyclohexyladenosine. f Adenosine deaminase, 1 unit/ml.

dipped in ice-cold distilled water and dried under a stream of air. The amount of total and non-specific binding was quantitated by wiping sections from some of the slides with filter papers and counting the filter-bound radioactivity by scintillation spectrometry. Labeled tissue sections were apposed to Hyperfilm tritium-sensitive film for the appropriate time, which ranged from 6-12 weeks for the different ligands used. Following exposure films were developed, fixed and washed. The resulting autoradiograms were printed directly on photographic paper; thus labeled areas appear white on a black background. Autoradiograms were analysed with the aid of the atlas of Berman (1968). [³H]Quinuclindinyl benzilate, [³H]vinylidene kainic acid and [³H]flunitrazepam were obtained from NEN-Dupont, MA, U.S.A., [³H]phenylisopropyladenosine and [³H]strychnine were from Amersham, U.K. and [³H]nitrobenzylthioinosine was from Moravek Biochemicals, CA, U.S.A..

Receptor Localization

Technical considerations
All radioligands exhibited a high degree of specific binding with generally a very low, undetectable signal of non-specific binding on tritium sensitive film with the exposure times used. Binding was generally confined to gray matter areas with no appreciable binding in the white matter. Specific binding of all radioligands tested was present throughout the PAG, but the concentration of binding was usually heterogenous. It should be noted that differences in binding conditions (e.g., buffers, incubation conditions, ligand affinity) can dramatically alter the amount of binding, irrespective of the absolute number of receptors. Therefore the autoradiograms reliably illustrate the differences in the distribution throughout the PAG, but absolute differences in the number of receptors are not represented. It is important to remember that this technique favours the detection of high affinity, slowly-dissociating binding and less readily detects low affinity binding sites (see Herkenham, 1987). For those receptors studied, maximal differences in relative receptor density (determined by measurements of film image optical density) are only of the order of 2-fold. This is perhaps not surprising in light of the general distribution in brain of the particular receptor types examined and the central integrative role of the midbrain PAG. Another potential technical problem in experiments of this type is the differential quenching of [³H]-emissions in areas containing different densities of myelin (white matter) with consequent artificial differences in the detected density of receptors. The midbrain PAG is, however, reported to be very lightly myelinated (e.g., Beitz, 1985). The overall results obtained here with several ligands suggests that this is not a significant problem in this study.

[³H]QNB binding
Specific [³H]QNB binding was present throughout the PAG (Fig. 1). Rostrally, the density of [³H]QNB binding was highest in the lateral and dorsolateral portions (Fig. 1A-C). This is similar to the limited descriptions of muscari-

nic receptor distribution in the rat PAG (Wamsley et al., 1981; Beitz et al., 1982). At the level of the superior colliculus a bilateral circular area in the ventrolateral region of the PAG contained a higher concentration of muscarinic receptors than the surrounding areas (Fig. 1B-C). A moderate density of receptors was present over the Edinger-Westphal nucleus while low levels of binding were present over the oculomotor nucleus (Fig. 1A-C). More caudally, at the level of the inferior colliculus, [³H]QNB binding was more evenly distributed within the PAG with lower levels easily discernible along the dorsal midline region (Fig. 1D-E). At this rostrocaudal level the density of receptors in the dorsal tegmental nucleus and nucleus coeruleus was higher than the adjacent medial dorsal raphe

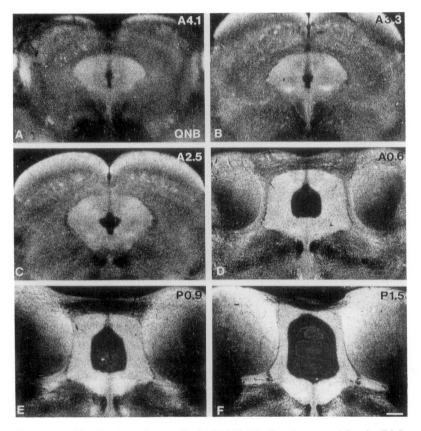

Figure 1: Distribution of specific [³H]QNB binding in cat midbrain PAG. White areas represent muscarinic binding sites. Non-specific binding of [³H]QNB (in the presence of 1 µM atropine) produced images not above film background (e.g. see upper edges of each plate). Note the higher density of binding in the lateral (A), ventrolateral (B), and dorsolateral (C) regions. Approximate co-ordinates according to Berman (1968) are given in the upper right corner of each plate. Bar = 1 mm.

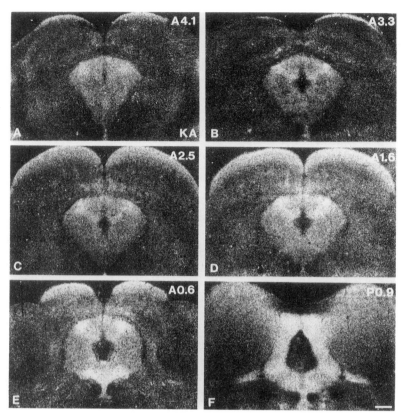

Figure 2: Distribution of specific [³H]KA binding in cat midbrain PAG. Binding is highest in the dorsolateral area, rostrally (A-D) and in the dorsolateral and ventrolateral areas and dorsal raphe nucleus more caudally (E-F). Non-specific binding of [³H]KA (in the presence of 100 μM glutamate) was not above film background (data not shown). Bar = 1 mm.

nucleus and all regions of the PAG (Fig. 1E-F). Moderate levels of binding were seen in the adjacent cuneiform nucleus, ventral to the inferior colliculus and in the ventral tegmental nucleus, ventrolateral to the medial longitudinal bundle.

[³H]KA binding

Specific [³H]KA binding was also present throughout the PAG (Fig. 2). Rostrally, the density of [³H]KA binding was higher in the dorsolateral area (Fig. 2A-C). This is similar to the distribution of [³H]KA binding observed in the PAG of the rat (Depaulis, personal communication; Unnerstall and Wamsley, 1983; Albin et al., 1990). More caudally, in addition to higher levels in the dorsolateral area, binding was high in the ventrolateral area and dorsal raphe nucleus (Fig. 2E). The juxta-aqueductal area was discernible by its lower density of

[³H]KA binding. The superficial layers of the superior colliculus and the cuneiform nucleus had similar or higher levels of [³H]KA binding to those in the PAG,
while other adjacent areas had a low density of binding.

[³H]FNZ binding

In the rostral PAG at the level of the posterior commissure, both high and
low densities of benzodiazepine receptors were found in different regions of the
ventral PAG. High concentrations were seen around the ventrolateral border of
the PAG in the area of the nucleus of Darkschewitsch, with low levels in the area
below the aqueduct (Fig. 3A). More caudally at the more rostral levels of the
superior colliculus, the relative concentration of specific [³H]FNZ binding in the
lateral regions running the complete dorsoventral extent of the aqueduct were
somewhat similar to the distribution of muscarinic receptors (Fig. 3B). More caudally a region of high density binding was present in the dorsolateral area (Fig.
3C). An area of quite low density was located laterally, directly adjacent to the
aqueduct just dorsal to the location of an area of high muscarinic receptor binding (see Fig. 1B-C). The dorsal midline extension of the aqueduct had a slightly
lower density of binding and a thin layer of higher density binding was discernible around the aqueduct, particularly the dorsal part (Fig. 3D). At the level of
the inferior colliculus, slightly higher densities of binding were again seen in the
dorsal and dorsolateral regions and in the medial dorsal raphe nucleus. In similar studies, high affinity $GABA_A$ receptors in the PAG have been labeled by
[³H]muscimol (data not shown). The distribution was very similar to that of benzodiazepine receptors labelled by [³H]FNZ with highest densities in the ventrolateral region in the most rostral PAG, and in the dorsolateral and dorsal regions of
the rostral and caudal PAG respectively. The distribution in the surrounding
mesencephalon was also similar.

[³H]STR binding

Low concentrations of specific [³H]STR binding were present throughout
the PAG relative to the high concentrations in pontine and medullary nuclei
(data not shown). Within the PAG (at the level of the superior colliculus), higher
concentrations were seen in the lateral region especially in the more ventral parts
of this area. Areas of lower than average density were seen in an area corresponding to the area of low density of [³H]FNZ binding in the ventral PAG (see Fig.
3C) and in the region immediately surrounding the aqueduct. At more caudal
levels higher concentrations persisted in the lateral areas and were also associated with the dorsal tegmental nucleus. While [³H]STR binding levels were also
lower throughout the mesencephalon than in the adjacent pons and medulla,
moderate binding was seen over the paramedian interpeduncular nucleus and
the superficial layers of the superior colliculus. These results are in agreement
with the descriptions of high levels of [³H]STR binding "in lateral and dorsal
parts" rostrally and in peripheral aspects of the dorsolateral PAG caudally in
human midbrain (Probst et al., 1986), and with the limited mapping shown
recently in cat PAG (Glendenning and Baker, 1988).

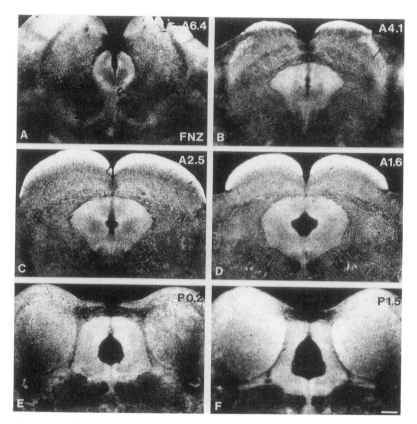

Figure 3: Distribution of specific [³H]FNZ binding in cat midbrain PAG. A higher density of binding is seen in the ventrolateral (A), lateral (B), and dorsolateral (C) regions, while a low density is seen in a circular area adjacent to the aqueduct (C). Non-specific binding of [³H]FNZ (in the presence of 1 µM diazepam) was not above film background. Bar = 1 mm.

[³H]PIA and [³H]NBI binding

Specific [³H]PIA binding to adenosine A1 receptors was also found in the entire rostrocaudal extent of the PAG, albeit at levels well below those present in cortex and other forebrain areas (Fig. 4). The density of binding was fairly even throughout apart from a higher than average concentration surrounding the aqueduct, particularly on the lateral and ventral walls. Moderate levels of binding are also seen in the nucleus of Darkschewitsch, the oculomotor, the trochlear nuclei and the raphe nuclei. [³H]PIA binding in the surrounding mesencephalon was relatively low apart from moderate levels in the superficial layer of the superior colliculus, the dorsomedial inferior colliculus, substantia nigra pars compacta, cuneiform nucleus and pontine gray (see Fig. 4).

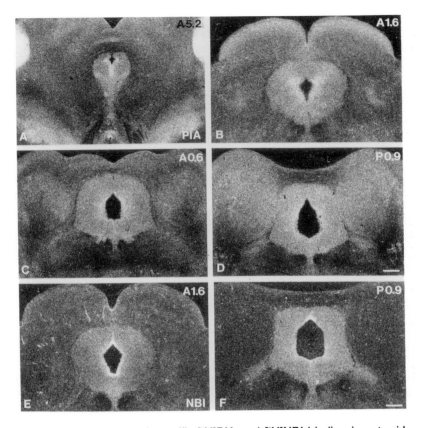

Figure 4: Distribution of specific [³H]PIA and [³H]NBI binding in cat midbrain PAG. [³H]PIA binding was evenly distributed throughout all levels of the PAG (and the surrounding midbrain), apart from a slightly higher density in the area around the aqueduct (A-D). The distribution of [³H]NBI binding in the PAG was somewhat similar to that of [³H]PIA (E-F). [³H]NBI binding was relatively low in the surrounding areas, especially the inferior colliculus. Non-specific binding of both ligands was not above film background. Bar = 1 mm.

Similar results were illustrated for the cat midbrain by Fastbom et al. (1987) using [³H]cyclohexyladenosine to label A_1 receptors.

The availability of [³H]NBI, a ligand which labels a population of adenosine uptake carrier molecules (Bisserbe et al., 1985) allowed the comparison of the distribution of adenosine uptake sites with that of A_1 receptors. Interestingly, the distribution of the two binding sites was somewhat similar within the PAG, but differences were apparent in other regions of the mesencephalon. Thus, [³H]NBI binding was concentrated around the aqueduct and evenly distributed throughout remaining areas of the PAG and nuclei contained therein (Fig. 4E-F). However, outside the PAG, in areas such as the superior and inferior colliculi, quite low densities of [³H]NBI were observed, in contrast to the higher relative abundance of [³H]PIA binding. The interpeduncular nucleus, an area where adenosine receptors were also relatively abundant, also contained a high concentration of [³H]NBI binding (the highest in the mesencephalon).

Discussion

It is clear, even from the relatively few "binding entities" examined in these studies, that valuable information about the localization of all types of membrane receptors, uptake carriers and even cellular enzymes and other proteins in the PAG can be obtained using *in vitro* autoradiographic methods. As mentioned in the introduction, based on cell cluster and enzymatic studies (in the rat), the PAG has previously been described as containing four subdivisions (Beitz, 1985; Conti et al., 1988). Other studies in rat and cat have further delineated the PAG on anatomical (connectivity) and functional grounds (e.g., Beitz, 1985; Holstege, 1988; Beart et al., 1990; Bandler et al., 1991). In fact, recent attempts have again been made, based on functional and anatomical data, to agree on a somewhat different description of the PAG's four main subdivisions as dorsomedial, dorsolateral, lateral and ventrolateral, with the latter excluding the area ventral to the aqueduct containing the oculomotor complex and dorsal raphe nucleus and not including the region immediately surrounding the aqueduct which contains very few cells (see Bandler and Depaulis, introduction of this volume).

Thus it is desirable to consider receptor distributions in relation to these subdivisions at different rostrocaudal levels. Unfortunately very few functional or anatomical studies have focussed on the rostral third of the PAG, but in the middle and caudal thirds a clear correlation exists between the definition or delineation of the dorsolateral area from the adjacent dorsomedial and lateral/ventrolateral areas on the basis of receptor concentration (e.g., GABA$_A$/benzodiazepine and kainate receptors) and the extent of these areas based on anatomical differences such as the distribution of PAG-caudal brainstem projection neurons (e.g., Holstege, 1988). The dorsolateral area is largely devoid of neurons projecting to pontine and medullary sites (Holstege, 1988), while possessing a high density of GABA immunoreactive cells (Reichling et al., 1991) which are presumably GABAergic interneurons. This region is also delineated by aspects of its afferent input. It receives a large number of projections from the superior colliculus, cuneiform nucleus, prepositus hypoglossal nucleus and from some regions of the cortex (Cowie, personal communication; Shipley et al., this volume). In the lateral and ventrolateral areas where large numbers of descending projection neurons are located (e.g., Carrive; Holstege, this volume), the density of particular receptors is lower (e.g., benzodiazepine and glycine receptors). In the caudal third, where histologically* and in anatomical studies the dorsolateral area is

---

* Along with regions of the ventromedial and lateral PAG, the dorsolateral subdivision is clearly visible as a distinct area under darkfield or transverse illumination in wet cryostat sections.

revealed to be very small and located only in the extreme dorsolateral edge, this is clearly evident in particular receptor mapping studies (e.g., kainate binding sites). In contrast, the distributions of adenosine receptor and uptake sites are quite uniform throughout the divisions of the PAG apart from a higher density in the region immediately adjacent to the aqueduct. This may be in line with adenosine's suggested role as a neuromodulator with binding sites strategically located close to the circulating cerebrospinal fluid, which under some conditions may contain high concentrations of adenosine.

Earlier studies (Beitz, 1985) in the rat have also described variations in receptor density. For instance, opiate and neurotensin receptor binding sites are dense in the dorsolateral subdivision and low in the adjacent dorsal (dorsome-dial) area (Beitz et al., 1982; Herkenham and Pert, 1982; Young and Kuhar, 1979), while muscarinic cholinergic receptors are found in highest density in the dorso-lateral subdivision with lower levels in the ventrolateral and dorsal areas (Beitz et al., 1982; Wamsley et al., 1981). A high density of all subtypes of excitatory amino acid receptors has been demonstrated in the dorsolateral area of the rat PAG (Beitz et al., 1982; Greenamyre et al., 1984; Albin et al., 1990). Clearly then, on the basis of the results presented in this study, there are similarities between the PAG of the cat and the rat, where the dorsolateral region is delineated by higher concentrations of muscarinic, kainate and benzodiazepine binding sites. Opiate receptors (mu and delta) are also found in high density in the dorsolate-ral PAG of the cat (Walker et al., 1988).

Functional Implications

The PAG is an integrative and relay centre of the brain important in control of an animal's reaction to threatening and stressful stimuli (Bandler et al., 1991). It is involved in the mediation of pain-processing mechanisms (e.g., Basbaum and Fields, 1984) and integrates defensive behaviour patterns like "fight-or-flight" and aversive motivational states (e.g., Audi and Graeff, 1984; Bandler and Depaulis, this volume). There is good evidence that the binding sites mapped in these studies represent functional receptor sites for the respecti-ve transmitter (neuromodulatory) substances, as many pharmacological, physio-logical and behavioural studies have documented the effects of acetylcholine (e.g., Richmond and Clemens, 1986), GABA (and benzodiazepine) (e.g., Brandão et al., 1982; Audi and Graeff 1984; 1987), and glutamate-like agents (e.g., Bandler, 1982; Bandler et al., 1991) on PAG function. Furthermore most of these transmit-ters and their synthetic enzymes have been detected in the PAG (e.g., Barbaresi and Manfrini, 1986; Clements et al., 1987; Reichling et al., 1990; Beitz and Williams, this volume).

However several authors have previously discussed the apparent mis-match that can exist between the distribution of drug binding sites, so-called receptors, and the distribution of their transmitters determined by immunocyto-chemical or enzymatic methods (see e.g., Kuhar, 1985; Herkenham, 1987). So, while there is evidence for the neuronal localization of glutamate/aspartate and

GABA and their synthetic enzymes in the PAG, the lack of abundance of choline acetyltransferase-immunoreactive cells and nerve terminals in the cat PAG (Vincent and Reiner, 1987) does not appear to match with the presence of a high density of muscarinic receptors. More generally, however, legitimate reasons for such mismatches probably exist, including an incomplete knowledge of all transmitters and receptors within a related group, the expression of transmitter and receptors throughout the extent of different neuron populations, and technical limitations of receptor and transmitter/enzyme detection (see Herkenham, 1987 for review). Another specific example emphasizes that it is not always possible to equate binding sites with functional receptors. Recent evidence from studies in frog brain reveal that pharmacologically defined kainate binding sites are confined to axons and that the kainate receptor is the same entity as the quisqualate receptor and has a different distribution (Wenthold et al., 1990). This point also raises the possible existence of high affinity, non-synaptic receptors which may be present in high densities in different brain areas, including the PAG, and function in the presence of levels of transmitter far lower than those present in synaptic areas. [For further discussions of receptor and transmitter mismatch and the concept of parasynaptic transmission see Herkenham, 1987; Veening et al., this volume)].

Various recent investigations have begun to integrate the molecular, cellular and functional organization of the mammalian PAG and already several groups have described functional/topographical subdivisions within the borders of the PAG (e.g., Carrive; Lovick; Bandler and Depaulis, this volume; Bandler et al., 1991). Results of autoradiographic receptor studies can potentially provide a better neuroanatomical framework for future biochemical, pharmacological and behavioural investigations of the role of the intrinsic and afferent PAG neuronal systems.

Future directions

The current studies report the distribution of receptor subtypes for a number of the major mammalian CNS "classical neurotransmitters" in the cat PAG. As the amino acid transmitters, GABA, glycine and glutamate (aspartate) are recognized as the predominate inhibitory and excitatory transmitters in virtually all areas of the CNS, it is predicted that receptors for these transmitters should be broadly distributed in all areas, particularly in integrative areas such as the midbrain PAG. However while the distribution of the major neurotransmitter receptors are perhaps predictably quite uniformly widespread in the PAG, more restricted distribution patterns may be observed within different subclasses of receptors. For instance, recent molecular and pharmacological studies have identified multiple, potential receptor subtypes of muscarinic (Buckley et al., 1988), $GABA_A$/benzodiazepine (Wisden, et al., 1988) and glutamate (Keinänen et al., 1990) receptors. Thus as more selective ligands are developed and novel molecular techniques such as immunohistochemistry and *in situ* hybridization histochemistry are applied to receptor localization studies, the overall discrimination of variations in chemoarchitecture of the PAG by "receptor mapping" may rival that produced by more traditional techniques. In addition, given the far more restricted distribution of various peptide-containing neurons in the PAG (e.g., substance P, enkephalin, Moss and Basbaum, 1983; Moss et al., 1983; calcitonin gene-related peptide, Conti and Sterniur, 1989; and somatostatin, Finley et al., 1981), one would predict that future detailed analysis of the distribution of receptors for these putative neurotransmitters/neuromodulators would reveal unique patterns of overlap with each other and the more widely distributed receptors (e.g., substance P receptors; Lui and Swenberg, 1988).

In summary, *in vitro* autoradiographic studies revealed that muscarinic, kainate, GABA$_A$/benzodiazepine, inhibitory glycine and adenosine-A$_1$ receptors were present throughout the rostrocaudal extent of the PAG with some subtle variations in density in different regions of the PAG. The dorsolateral area of the PAG contained higher densities of [^{3}H]QNB, [^{3}H]KA and [^{3}H]FNZ binding than were seen in the adjacent dorsomedial and lateral areas. [^{3}H]QNB binding was also relatively high in the lateral and parts of the ventrolateral regions rostrally. [^{3}H]FNZ binding was present in somewhat higher density in the medial region surrounding the aqueduct throughout the PAG, in the lateral and ventrolateral regions at the most rostral levels, as well as in the dorsolateral region, more caudally. Distinct areas of low density [^{3}H]FNZ binding were discernible ventrolaterally at the level of the superior colliculus. Highest relative concentrations of [^{3}H]STR binding were observed in the lateral PAG, especially at rostral levels, while low concentrations were found in the ventral area which also had a low density of [^{3}H]FNZ binding, and in the area immediately adjacent to the aqueduct. [^{3}H]PIA binding were evenly distributed throughout the PAG, apart from somewhat higher densities in the medial region surrounding the aqueduct. The topographical patterns generated by these apparent differences in the density of various transmitter receptors in subregions of the cat PAG coincide largely with previously and more recently described subdivisions of the PAG. These differences in receptor distribution most likely reflect differences in the distribution of intrinsic neuronal cell groups and afferent neuronal terminals within the PAG.

Acknowledgements

I would like to acknowledge the technical assistance of Corinne Grabara and the continuing support of Professor Graham Johnston. I would also like to thank Richard Bandler and Pascal Carrive for their assistance and encouragement. This research was supported by a Program Grant from the National Health and Medical Research Council of Australia while the author was in the Department of Pharmacology, The University of Sydney, NSW, Australia.

References

Albin, R.L., Makoweic, R.L., Hollingsworth, Z., Dure, L.S., Penney, J.B., and Young, A.B., Excitatory amino acid receptors in the periaqueductal gray of the rat, Neurosci. Lett., 118 (1990) 112-115.

Audi, E.A. and Graeff, F.G., Benzodiazepine receptors in the periaqueductal grey mediate anti-aversive drug action, Eur. J. Pharmacol., 103 (1984) 279-285.

Audi, E.A. and Graeff, F.G., GABA$_A$ receptors in the midbrain central grey mediate the antiaversive action of GABA, Eur. J. Pharmacol., 135 (1987) 225-229.

Bandler, R., Induction of rage following microinjections of glutamate into midbrain but not hypothalamus of cats, Neurosci. Lett., 30 (1982) 183-188.

Bandler, R., Carrive, P. and Zhang, S.P., Integration of somatic and autonomic reactions within the midbrain periaqueductal grey: viscerotopic, somatotopic and functional organization, Prog. Brain Res., 87 (1991) 269-305.

Barbaresi, P. and Manfrini, E., Glutamate decarboxylase-immunoreactive neurons and terminals in the periaqueductal gray of the rat, Neuroscience, 27 (1988) 183-191.

Basbaum, A.I. and Fields, H.L., Endogenous pain control systems: brainstem spinal pathways and endorphin circuitry, Ann. Rev. Neurosci., 7 (1984) 309-338.

Beart, P.M., Summers, R.J., Stephenson, J.A., Cook, C.J. and Christie, M.J., Excitatory amino acid projections to the periaqueductal gray in the rat: a retrograde transport study utilizing D[^{3}H]aspartate and [^{3}H]GABA, Neuroscience, 34 (1990) 163-176.

Beitz, A.J., The midbrain periaqueductal gray in the rat. I. Nuclear volume, cell number, density, orientation, and regional subdivisions, J. Comp. Neurol., 237 (1985) 445-459.

Beitz, A.J., Buggy, J., Terracio, L. and Wells, W.E., Autoradiographic localization of opiate, beta-adrenergic, cholinergic and GABA receptors in the midbrain periaqueductal gray, Soc. Neurosci. Abstr., 8 (1982) 265.

Belin, M.F., Aguera, M., Tappaz, M., McRae-Degueurce, A., Bobillier, P. and Pujol, J.F., GABA-accumulating neurons in the nucleus raphe dorsalis and periaqueductal gray in the rat: a biochemical and radioautographic study, Brain Res., 170 (1979) 279-297.

Berman, A.L., The Brainstem of the Cat. A Cytoarchitectonic Atlas with Stereotaxic Coordinates, The University of Wisconsin Press, Madison, 1968.

Bisserbe, J.C., Patel, J. and Marangos, P.J., Autoradiographic localization of adenosine uptake sites in rat brain using [^{3}H]nitrobenzylthioinosine, J. Neurosci., 5 (1985) 544-550.

Brandão, M.L., De Aguiar, J.C. and Graeff, F.G., GABA mediation of the anti-aversive action of minor tranquillizers, Pharmacol. Biochem. Behav., 16 (1982) 397-402.

Buckley, N.J., Bonner, T.I. and Brann, M.R., Localization of a family of muscarinic receptor mRNAs in rat brain, J. Neurosci., 8 (1988) 4646-4652.

Clements, J.R., Beitz, A.J., Fletcher, T.F., Mullett, M.A., Immunocytochemical localization of serotonin in the rat midbrain periaqueductal gray: A quantitative light and electron microscopic study, J. Comp. Neurol., 236 (1985) 60-70.

Clements, J.R., Madl, J.E., Johnson, R.L., Larson, A.A. and Beitz, A.J., Localization of glutamate, glutaminase, aspartate and aspartate aminotransferase in the rat midbrain periaqueductal grey, Exp. Brain Res., 67 (1987) 594-602.

Conti, F., Barbaresi, P. and Fabri, M., Cytochrome oxidase histochemistry reveals regional subdivisions in the rat periaqueductal gray matter, Neuroscience, 24 (1988) 629-633.

Conti, F. and Sternini, C., Calcitonin gene-related peptide (CGRP)-positive neurons and fibers in the cat periaqueductal grey matter, Somatosens. Motor Res., 5 (1989) 497-511.

Fastbom, J., Pazos, A. and Palacios, J.M., The distribution of adenosine A$_1$ receptors and 5'-nucleotidase in the brain of some commonly used experimental animals, Neuroscience, 22 (1987) 813-826.

Finley, J.C.W., Madedrut, J.L., Roger, L.J. and Petrusz, P., The immunocytochemical localization of somatostatin-containing neurons in rat central nervous system, Neuroscience, 6 (1981) 2173-2192.

Glendenning, K.K. and Baker, B.N., Neuroanatomical distribution of receptors for three potential inhibitory neurotransmitters in the brainstem auditory nuclei of the cat, J. Comp. Neurol., 275 (1988) 288-308.

Greenamyre, J.T., Young, A.B. and Penney, J.B., Quantitative autoradiographic distribution of L-[^{3}H]glutamate binding sites in rat central nervous system, J. Neurosci., 4 (1984) 2133-2144.

Hamilton, B.L., Cytoarchitectural subdivisions of the periaqueductal gray matter in the cat, J. Comp. Neurol., 149 (1973) 1-28.

Herkenham, M., Mismatches between neurotransmitter and receptor localizations in brain: observations and implications, Neuroscience, 23 (1987) 1-38.

Herkenham, M. and Pert, C.B., Light microscopic localization of brain opiate receptors: A general autoradiographic method which preserves tissue quality, J. Neurosci., 2 (1982) 1129-1149.

Holstege, G., Direct and indirect pathways to lamina I in the medulla oblongata and spinal cord of the cat, Prog. Brain Res., 77 (1988) 47-94.

Hösli, E. and Hösli, L., Autoradiographic studies on the uptake of adenosine and on the binding of adenosine analogues in neurons and astrocytes of cultured cerebellum and spinal cord, Neuroscience, 24 (1988) 621-628.

Keinänen, K., Wisden, W., Sommer, B., Werner, P., Herb, A., Verdoorn, T.A., Sakmann, B. and Seeburg, P.H., A family of AMPA-selective glutamate receptors, Science, 249 (1990) 556-560.

Kuhar, M.J., The mismatch problem in receptor mapping studies, Trends Neurosci., 8 (1985) 190-191.

Lui, R.P.C. and Swenberg, M.-L., Autoradiographic localization of substance P ligand binding sites and distribution of immunoreactive neurons in the periaqueductal gray of the rat, Brain Res., 475 (1988) 73-79.

Mantyh, P.W., The midbrain periaqueductal gray in the rat, cat and monkey: a Nissl, Weil and Golgi analysis, J. Comp. Neurol., 204 (1982) 349-363.

Moss, M.S. and Basbaum, A.I., The peptidergic organization of the cat periaqueductal gray. II. The distribution of immunoreactive substance P and vasoactive intestinal peptide, J. Neurosci., 3 (1983) 1437-1449.

Moss, M.S., Glazer, E.J. and Basbaum, A.I., The peptidergic organization of the cat periaqueductal gray. I. The distribution of immunoreactive enkephalin-containing neurons and terminals, J. Neurosci., 3 (1983) 603-616.

Probst, A., Cortés, R. and Palacios, J.M., The distribution of glycine receptors in the human brain. A light microscopic autoradiographic study using [³H]strychnine, Neuroscience, 17 (1986) 11-35.

Reichling, D.B., Kawashima, Y. and Basbaum, A.I., β-Endorphin, enkephalin, dynorphin, and GABA immunoreactivity in the rat midbrain periaqueductal gray, Brain Res., 1991, in press.

Reichling, D.B., Kwiat, G.C., and Basbaum, A.I., Anatomy, physiology and pharmacology of the periaqueductal gray contribution to antinociceptive controls, Prog. Brain Res., 77 (1988) 31-46.

Richmond, G. and Clemens, L.G., Evidence for involvement of midbrain central gray in cholinergic mediation of female sexual receptivity in rats, Behav. Neurosci., 100 (1986) 376-380.

Unnerstall, J.R., Kuhar, M.J., Neihoff, D.L. and Palacios, J.M., Benzodiazepine receptors are coupled to a subpopulation of γ-aminobutyric acid (GABA) receptors: Evidence from a quantitative autoradiographic study, J. Pharmacol. Exp. Ther., 218 (1981) 797-804.

Unnerstall, J.R. and Wamsley, J.K., Autoradiographic localization of high affinity [³H]kainic acid binding sites in the rat forebrain, Eur. J. Pharmacol., 86 (1983) 361-371.

Vincent, S.R. and Reiner, P.B., The immunohistochemical localization of choline acetyltransferase in the cat brain, Brain Res. Bull., 18 (1987) 371-415.

Walker, J.M., Bowen, W.D., Thompson, L.A., Frascella, J., Lehmkuhle, S. and Hughes, H.C., Distribution of opiate receptors within visual structures of the cat brain, Exp. Brain Res., 73 (1988) 523-532.

Wamsley, J.K, Lewis, M.S., Young, W.S. and Kuhar, M.J., Autoradiographic localization of muscarinic cholinergic receptors in rat brainstem, J. Neurosci., 1 (1981) 176-191.

Wenthold, R.J., Hampson, D.R., Wada, K., Hunter, C., Oberdorfer, M.D. and Dechesne, C.J., Isolation, localization and cloning of a kainic acid binding protein from frog brain, J. Histochem. Cytochem., 38 (1990) 1717-1723.

A.L. Gundlach

Wisden, W., Morris, B.J., Darlison, M.G., Hunt, S.P. and Barnard, E.A., Distinct GABA$_A$ receptor alpha-subunit mRNAs show differential patterns of expression in bovine brain, Neuron, 1 (1988) 937-947.

Young III, W.S. and Kuhar, M.J., Neurotensin receptors: Autoradiographic localization in rat CNS, Eur. J. Pharmacol., 59 (1979) 161-163.

Zarbin, M.A., Wamsley, J.K., Kuhar, M.J., Glycine receptor: light microscopic autoradiographic localization with [^{3}H]strychnine, J. Neurosci., 1 (1981) 532-547.

PARTICIPANTS

R. Bandler
Dept. of Anatomy
University of Sydney
Sydney, NSW 2006
Australia

P. Barbaresi
Istituto di Fisiologia Umana
Facoltà di Medicina e Chirurgia
Via Ranieri-Monte d'Ago
Ancona 60131
Italy

S. R.G. Beckett
Medical School
Queens Medical Centre
Nottingham NG7 2UH
U.K.

A.S. Beitz
Dept. of Veterinary Biology
University of Minnesota
1988 Fitch Avenue
St. Paul, MN 55108
USA

J.M. Besson
INSERM, Unité 161
2, rue d'Alésia
75014 Paris
France

R. Bianchi
Istituto di Anatomia Umana Normale
Università di Milano
Via Mangiagalli 31
20133 Milano
Italy

A. Blomqvist
Institutionen för Cellbiologi
Universitetet I Linköping
Hälsouniversitetet
S-581 85 Linköping
Sweden

D. Bouhassira
INSERM U-161
2 rue d'Alésia
75014 Paris
France

M. Brandão
Laboratorio de Psicobiologia
FFCLRP
Universidade de Sao Paulo
14049, Ribeirao Preto, SP
Brasil

P. Carrive
Dept. of Anatomy
University of Sydney
Sydney, NSW 2006
Australia

R. J. Cowie
Dept. of Anatomy
College of Medicine
Howard University
520 W. St., N.W.
Washington, DC, 20059
USA

P. J. Davis
School of Communication Disorders
Cumberland College of Health Sciences
Sydney University
Sydney, NSW 2006
Australia

A. Depaulis
L.N.B.C.
Centre de Neurochimie du CNRS
5 rue Blaise Pascal
67084 Strasbourg Cedex
France

R. Depoortère
Department of Pharmacology
Texas College of Osteopathic medicine
3500 Camp Bowie Boulevard
Forth North, Texas 76107
U.S.A.

G. Di Scala
L.N.B.C.
Centre de Neurochimie du CNRS
5 rue Blaise-Pascal
67084 Strasbourg Cedex
France

A. W. Duggan
Dept. of Preclinical Veterinary Sciences
University of Edinburgh
Summerhall, Edinburgh EH9 1QH
U.K.

M. Ennis
Dept. Anatomy & Cell Biology
University of Cincinnati
Coll. Med.
231 Bethesda Ave.
Cincinnati, OH 45267-0521
USA

M. Fanselow
Dept. of Psychology
UCLA, 405 Hilgard Ave.
Los Angeles, CA 90024-1563
USA

M. Gioia
Istituto di Anatomia Umana Normale
Università di Milano
Via Mangiagalli 31
20133 Milano
Italy

A. Gundlach
University of Melbourne
Clinical Pharmacology and Therapeutics
Unit
Austin Hospital
Heidelberg, Victoria 3084
Australia

R. M. Harper
Dept. of Anatomy and Cell Biology
UCLA
Los Angeles, CA 90024-1763
USA

G. Holstege
Dept. of Anatomy
Rijksuniversiteit Groningen
9713 EZ Groningen
The Netherlands

F. Jenck
Pharmaceutical Research Dept.
Hoffmann-La Roche Ltd.
CH- 4002 Basel
Switzerland

U. Jürgens
Max Planck Institut für Psychiatrie
Kraeplinstrasse 2, Postfach 40 12 40
8000 München 40
Germany

C. Larson
Dept. of Communication Sciences and
Disorders
Northwestern University
2299 Sheridan Rd
Evanston, IL 60208-3570
USA

A. Lawrence
Department of Physiology and
Pharmacology
Queens Medical Centre
Nottingham NG7 2UH
U.K.

D. Lima
Institute of Histology & Embryology
Faculty of Medicine
4200 Porto
Portugal

T. Lovick
Dept. of Physiology
The Medical School
University of Birmingham
Birmingham B15 2TJ
U.K.

P. Mason
Dept. of Neurology
M-794, UCSF
San Francisco, CA 94143-0114
USA

M. M. Morgan
Dept. of Neurology
M-794, UCSF
San Francisco, CA 94143-0114
USA

V. Pieribone
Karolinska Institutet
Department of Histology and Neurobiology
104 01 Stockholm
Sweden

M. Portavella Garcia
Laboratorio de Psicobiología
Departamento de Fisiología y Biología
Animal
Universidad de Sevilla
c/ San Fco. Javier s/n
41005 Sevilla
Spain

P. Redgrave
Dept. of Psychology
The University of Sheffield
Sheffield, S10 2TN
U.K.

D. Reichling
Dept. Physiology
Columbia University
630 West 168 Street
New York, NY 10032
USA

J. Ribas
Dept. de Fisiología Médica y Biofísica
Facultad de Medicina
Avda Sanchez Pizjuan n° 4
41009 Sevilla
Spain

D. Sanchez Romero
Dept. de Fisiología Médica y Biofísica
Facultad de Medicina
Avda. Sanchez Pizjuan n° 4
41009 Sevilla
Spain

J. Sandkühler
II. Physiologisches Institut
Universität Heidelberg
Im Neuenheimer Feld 326
6900 Heidelberg
Germany

G. Sandner
L.N.B.C.
Centre de Neurochimie du CNRS
5 rue Blaise-Pascal
67084 Strasbourg Cedex
France

S. Schwartz-Giblin
The Rockefeller University
1230 York Avenue
New York, NY 10021-6399
USA

M. T. Shipley
Dept. of Anatomy & Cell Biology
Neurobiology division
University of Cincinnati
231 Bethesda Ave.
Cincinnati, OH 45267-0521
USA

A. Siegel
Dept. of Neuroscience
New Jersey Medical School
185 South Orange Avenue
Newark, NJ 07103-2757
USA

J. Somogyi
United Research Organization of
Hungarian Academy of Sciences
and Semmelveis University
Laboratory of Neurobiology
1094 Budapest Tüzolto U. 58
Hungary

E. Van Bockstaele
Hahnemann University
Dept. of Mental Health Sciences
Mail Stop 403, Broad & Vine Streets
Philadelphia, PA 19102
USA

J. G. Veening
Vakgroep Anatomie/Embryologie
Faculteit der Geneeskunde en
Tandheelkunde
Geert Grooteplein noord 21, Postbus
9101
6500 HB Nijmegen
The Netherlands

J. A. Wada
Division of Neurosciences
University Hospital
University of British Columbia
Vancouver, B.C., V6T 2A1
Canada

R. Weber
Neuroimmunology Unit
Laboratory of Medicinal Chemistry
National Institute of Diabetes
and Digestive and Kidney Diseases
NIH, Bethesda MD 20892
USA

F. G. Williams
Dept. of Veterinary Biology
University of Minnesota
1988 Fitch Avenue
St. Paul, MN 55108
USA

R.P. Yezierski
Dept. of Neurological Surgery
University of Miami
1600 N.W. 10th Ave, R-48
Miami, FL 33136
USA

From left to right:
1st row: Th. Lovick, S. Schwartz-Giblin, J. Somogyi, A. Gundlach, J. Veening, G. Holstege, A. Siegel
2nd row: M. Gioia, E. Van Bockstaele, J. Wada, A. Depaulis, R. Bandler, J.-M. Besson, R. Yezierski, R. Bianchi
3rd row: M. Shipley, P. Barbaresi, P. Carrive, D. Reichling, P. Davis, P. Mason, M. Morgan
4th row: M. Portavella-Garcia, M. Fanselow, J. Ribas, F. Williams, R. Cowie
5th row: G. Sandner, V. Pieribone, A. Blomqvist, A. Duggan, A. Lawrence, S. Beckett, J. Sandkühler, U. Jürgens, M. Brandão
6th row: M. Ennis, P. Redgrave, R. Harper, C. Larson

INDEX

Index (by chapter)

Acethylcholine, 11, 449
ACTH, 387
Adenosine, 449
Amygdala, 151, 305, 329, 365, 417
Analgesia, 101, 121, 139, 151, 287, 305, 365
Anterograde tracing, 67, 199, 345, 387, 417
Antinociception, 101, 121, 139, 151, 287, 305, 329, 365
Anxiety, 151
Aspartate, 305
Autonomic function, 41, 67, 101, 365, 417
Autoradiography, 449
Aversion, 121, 139
Avoidance, 175, 199
Axo-dendritic synapses, 345
Benzodiazepine, 139, 151, 449
Blood pressure, 67, 101, 239
Brain electrical stimulation, 11, 23, 101, 121, 139, 151, 199, 211, 287, 365
Cardiovascular function, 41, 67, 101, 365
Cat, 41, 57, 67, 121, 287, 345, 365, 449
Central nucleus of the amygdala, 417
CRH, 387
Cricothyroid, 23, 57
Cuneiform nucleus, 199, 239
Cytoarchitecture, 387, 417
Defense, 67, 101, 139, 151, 175, 199, 211, 417
Diaphragm, 23, 41, 57
D,L-homocysteic acid, 57, 67, 101, 175, 199, 305
Dopamine, 11
Dorsal column nuclei, 345
Dorsal raphe nucleus, 121, 139
Dorsomedial hypothalamic nucleus, 387
Dorsoventral interactions, 101, 139
Dynorphin, 387
Electromyographic, 23, 57
Electron microscopy, 305, 387
Electrophysiology, 23, 41, 101, 211, 305, 365
Emotion, 67, 101, 151, 175, 199
Escape, 175, 199
Excitatory amino acids, 11, 57, 67, 101, 139, 175, 199, 305, 449
Extracellular recording, 23, 41, 211, 365
Fear, 151
Freezing, 139, 151, 175, 211

GABA, 11, 199, 239, 329, 449
Glutamate, 11, 305, 449
Glycine, 11, 449
Histamine, 11
Hypothalamus, 211, 267, 329, 387
Immunocytochemistry, 305, 329, 387
Immuno-electronmicroscopy, 329, 387
Insular cortex, 417
Intercostal, 23, 57
Intermediolateral cell column, 239
Interneuron, 211, 329
Kainic acid, 139, 175, 449
Laryngeal, 23, 57
Lateral cervical nucleus, 345
Lateral hypothalamic area, 387
LHRH, 211, 387
Locus coeruleus, 267
Longitudinal columnar organization, 1, 67, 175, 239, 417
Medial prefrontal cortex, 417
Met-Enkephaline, 211, 387
Microdialysis, 305
Microinjections, 11, 57, 67, 101, 139, 175, 199, 211, 305
Monkey, 11, 23, 345
Morphine, 287, 305, 329
Motor reactions, 121
Naloxone, 151
Naltrexone, 151
Narcotic, 329
Nociception, 121, 139, 239, 287, 365
Nociceptive facilitation, 121, 287, 365
Non-synaptic communication, 211, 387
Nonopioid analgesia, 139, 151
Noradrenaline, 11
Nucleus raphe magnus, 121, 239, 267, 287, 305, 329
Nucleus raphe pallidus, 239
Off-cells, 287
On-cells, 287
Opioid analgesia, 139, 151
Opioids 11, 139, 211, 287, 329
Pain, 121, 139, 287, 365
Premotor neurons, 57, 67, 239
Pedunculopontine nucleus, 239
Perirhinal cortex, 417
Phaseolus-L, 387, 417
Posterior cricoarytenoid, 23, 57
Preoptic region, 417
Quiet sleep, 41

Rat, 101, 121, 139, 151, 175, 199, 211, 305, 387, 417
Receptive fields, 365
Receptors, 449
Rectus abdominus, 23, 57
Recuperative behavior, 101
REM sleep, 41
Respiration, 23, 41, 57
Response selection, 151
Retroambiguus nucleus, 57, 239
Retrograde tracing, 67, 199, 329, 345, 387, 417
Sensory function, 175, 199, 345, 365
Serotonin, 11, 101, 121, 239, 365
Sexual behavior, 211, 239, 417
Single unit recording, 23, 41, 365
Sleep, 41
Solitary tract nucleus, 67, 267
Somatosensory projections, 345, 365
Somatotopic organization, 345
Spinal cord, 239, 345, 365
Spinal trigeminal nucleus, 345

Spinomesencephalic tract, 175, 345, 365
Spino-PAG, 345, 365
Stimulation-produced analgesia, 101, 121, 139, 287, 305
Stress, 121, 151
Subretrofacial nucleus, 67, 239
Superficial layers of the dorsal horn, 345, 365
Superior colliculus, 175, 199, 267
Sympathoexcitation, 67, 101
Sympathoinhibition, 67, 101
Thalamus, 329, 365
Thyroarythnoid, 23
Ventrolateral medulla, 67, 101, 239, 287, 305
Ventromedial Hypothalamic Nucleus, 387
Ventromedial medulla, 67, 101, 139, 239, 287, 305, 329
Viscerotopic organization, 67
Vocalization, 11, 23, 57, 121, 175, 239
Zona incerta, 267, 387